Lecture Notes in Computer Science 2561

Edited by G. Goos, J. Hartmanis, and J. van Leeuwen

Springer-Verlag Berlin Heidelberg GmbH

Harrie C.M. de Swart (Ed.)

Relational Methods in Computer Science

6th International Conference, RelMiCS 2001
and 1st Workshop of COST Action 274 TARSKI
Oisterwijk, The Netherlands, October 16-21, 2001
Revised Papers

Springer

Volume Editor

Harrie C.M. de Swart
Tilburg University
P.O. Box 90153
5000 LE Tilburg
The Netherlands
E-mail: H.C.M.deSwart@uvt.nl

Cataloging-in-Publication Data applied for

A catalog record for this book is available from the Library of Congress.

Bibliographic information published by Die Deutsche Bibliothek
Die Deutsche Bibliothek lists this publication in the Deutsche Nationalbibliografie; detailed bibliographic data is available in the Internet at http://dnb.ddb.de

CR Subject Classification (1998): F.4, I.1, I.2.3, D.2.4

ISSN 0302-9743
ISBN 978-3-540-00315-1 ISBN 978-3-540-36280-7 (eBook)
DOI 10.1007/978-3-540-36280-7

http://www.springer.de

Originally published by Springer-Verlag Berlin Heidelberg New York in 2002.

Typesetting: Camera-ready by author, data conversion by PTP Berlin, Stefan Sossna e. K.
Printed on acid-free paper SPIN: 10871819 06/3142 5 4 3 2 1 0

Preface

This volume contains the papers presented at **RelMiCS 2001**, the 6th International Conference on Relational Methods in Computer Science, and the First Workshop of COST Action 274 **TARSKI**, Theory and Application of Relational Structures as Knowledge Instruments. The conference was held in conference centre Boschoord, Oisterwijk near Tilburg, The Netherlands, from October 16 till October 21, 2001. The conference attracted interest from many parts of the world with contributions from many countries.

This conference was a continuation of international conferences/workshops on Relational Methods in Computer Science held in: Schloss Dagstuhl, Germany, January 1994; Parati near Rio de Janeiro, September 1995; Hammamet, Tunisia, January 1997; the Stefan Banach Center, Warsaw, September 1998; and Quebec, Canada, January 2000.

The purpose of these conferences/workshops is to bring together researchers from various subdisciplines of Computer Science, Mathematics, and Philosophy, all of whom use relational methods as a conceptual and methodological tool in their work. Topics include, but are not limited to: relational, cylindric, fork, and Kleene algebras; relational proof theory and decidability issues; relational representation theorems; relational semantics; applications to programming, databases, and analysis of language; and computer systems for relational knowledge representation. With respect to applications one can think of: relational specifications and modeling; relational software design and development techniques; programming with relations; and implementing relational algebra.

The RelMiCS'6 conference had three *invited lectures*, one of which has been included in these Proceedings. In addition, three *tutorials* were presented: Ivo Düntsch and Günther Gediga, Rough' Sets: tools for non-invasive data analysis; Wolfram Kahl and Eric Offermann, Programming with and in relational categories; Gheorghe Ştefănescu, An introduction to network algebra.

After a thorough refereeing process the Program Committee selected 21 papers for inclusion in these Proceedings. The papers have been classified into the different Work Areas (WAs) of Cost Action TARSKI: WA 1, Algebraic and logical foundations of "real-world" relations; WA 2, Mechanization of relational reasoning; WA 3, Relational scaling and preferences.

September 2002 Harrie de Swart

Organization

Program Committee

Harrie de Swart (Chair):	Tilburg University (The Netherlands)
Ewa Orlowska:	Institute of Telecommunications (Warsaw)
Gunther Schmidt:	Universität der Bundeswehr (München)
Chris Brink:	University of Wollongong (Australia)
Ivo Düntsch:	Brock University (St. Catharines, Canada)

Invited Speakers

Michel Grabisch:	Université Pierre et Marie Curie (Paris)
Wendy MacCaull:	St. Francis Xavier University (Antigonish, Canada)
Wolfram Kahl:	Fakultät für Informatik (UniBw, München)

Referees

Philippe Balbiani
Wojchiech Buszkowski
Anthony Cohn
Stephane Demri
Thorsten Ehm
Melvin Fitting
Günther Gediga
Wolfram Kahl
Joachim Lambek
Wendy MacCaull
Larisa Maksimova
Bernhard Möller
Eric Offermann
Ewa Orlowska
Jochen Pfalzgraf
Anna Radzikowska
Anna Romanowska
Marc Roubens
Gunther Schmidt
Bruce Spencer
Harrie de Swart
Dimiter Vakarelov
Michael Winter
Andrzej Wronski

Sponsoring Institutions

Faculty of Philosophy of Tilburg University
COST Action 274: TARSKI
Royal Netherlands Academy of Arts and Sciences
Dutch Research School in Logic
TUE-UvT Research Group on Logic and Information Systems

Table of Contents

Invited Lecture

Contributed Papers

WA 1: Algebraic and Logical Foundations of 'Real World' Relations

A Relation-Algebraic Approach to Graph Structure Transformation

(Invited Talk — Extended Abstract)

Wolfram Kahl*

Department of Computing and Software, McMaster University

kahl@cas.mcmaster.ca

1 Introduction

Graph transformation is a rapidly expanding field of research, motivated by a wide range of applications.

Individual graph transformation steps can be specified at different levels of abstraction. On one end, there are "programmed" graph transformation systems with very fine control over manipulation of individual graph items. Different flavours of rule-based graph transformation systems match their patterns in more generic ways and therefore interact with the graph structure at slightly higher levels.

A rather dominant approach to graph transformation uses the abstractions of category theory to define matching and replacement almost on a black-box level, using easily understandable basic category-theoretic concepts, like pushouts and pullbacks. However, some details cannot be covered on this level, and most authors refrain from resorting to the much more advanced category-theoretic tools of topos theory that are available for graphs, too — topos theory is not felt to be an appropriate language for specifying graph transformation.

We show that the language of relations is better suited for this purpose and can be used very naturally to cover all the problems of the categoric approach to graph transformation. Although much of this follows from the well-known fact that every graph-structure category is a topos, very little of this power has been exploited before, and even surveys of the categoric approach to graph transformation state essential conditions outside the category-theoretic framework.

One achievement is therefore the capability to provide descriptions of all graph transformation effects on a suitable level of abstraction from the concrete choice of graph structures.

* Most of this work was completed during the author's appointment at Institute of Software Technology, Universität der Bundeswehr München.

H. de Swart (Ed.): RelMiCS 2001, LNCS 2561, pp. 1–14, 2002.

Another important result is the definition of a graph rewriting approach where relational matchings can match rule parameters to arbitrary subgraphs, which then can be copied or deleted by rewriting. At the same time, the rules are still as intuitive as in the double-pushout approach, and there is no need to use complicated encodings as in the pullback approaches.

In short: A natural way to achieve a double-pushout-like rewriting concept that incorporates some kind of "graph variable" matching and replication is to amalgamate pushouts and pullbacks, and the relation-algebraic approach offers appropriate abstractions that allow to formalise this in a fully component-free yet intuitively accessible manner.

This extended abstract is a short overview of the central results of [Kah01]. The next section contains a summary of the conventional approaches to describing graph transformation in the setting of categories, which are a generalisation of total functions between sets. Section 3 moves to the abstract setting of Dedekind categories, one particular abstraction from the setting of relations between sets. We first present natural *relational* characterisations of the central ingredients of the category theoretic approaches, and then use the flexibility of the relational approach to amalgamate the relational versions of pushouts and pullbacks into the original *pullout approach* to graph structure transformation.

2 Conventional Categoric Graph Transformation

The so-called "algebraic approach to graph transformation" is really a collection of approaches that essentially rely on category-theoretic abstractions.[1]

A *category* consists of a collection of objects and a collection of morphisms, where every morphism has a source and a target; we introduce a morphism F with source $\mathcal{A}$ and target $\mathcal{B}$ by writing $F : \mathcal{A} \to \mathcal{B}$. In addition, for every object $\mathcal{A}$ there is an identity morphism $\mathbb{I}_{\mathcal{A}} : \mathcal{A} \to \mathcal{A}$ usually just written $\mathbb{I}$, and the associative *composition* operator combines two morphisms $F : \mathcal{A} \to \mathcal{B}$ and $G : \mathcal{B} \to \mathcal{C}$ into their composition $F;G$, which is a morphism from $\mathcal{A}$ to $\mathcal{C}$.

2.1 Pushout Rewriting

Historically, Ehrig, Pfender, and Schneider developed the double-pushout approach as a way to generalise Chomsky grammars from strings to graphs, using pushouts as "gluing construction" to play the rôle of concatenation on strings [EPS73], see also [CMR+97]. The name "algebraic approach" derives from the fact that graphs can be considered as a special kind of algebras, and that the pushout in the appropriate category of algebras was perceived more as a concept from universal algebra than from category theory.

[1] An accessible introduction to category theory in computing science is [BW90]. However, in the context of the algebraic approach to graph transformation, usually only a very limited amount of category theory is employed, and most of the literature is accessible starting for example from the handbook chapters [CMR+97,EHK+97].

The *pushout* for a span (i.e., a pair of morphisms starting from a common source) $\mathcal{B} \xleftarrow{P} \mathcal{A} \xrightarrow{Q} \mathcal{C}$ in some category consists of a *pushout object* $\mathcal{D}$ and two morphisms R and S with $\mathcal{D}$ as target, such that the resulting square commutes, i.e., $P{;}R = Q{;}S$, and that for every other completion $\mathcal{B} \xrightarrow{R'} \mathcal{D}' \xleftarrow{S'} \mathcal{C}$ to a commuting square, i.e. with $P{;}R' = Q{;}S'$, there is a unique morphism $Y : \mathcal{D} \to \mathcal{D}'$ such that the candidate morphisms can be *factorised* via Y, that is, $R' = R{;}Y$ and $S' = S{;}Y$.

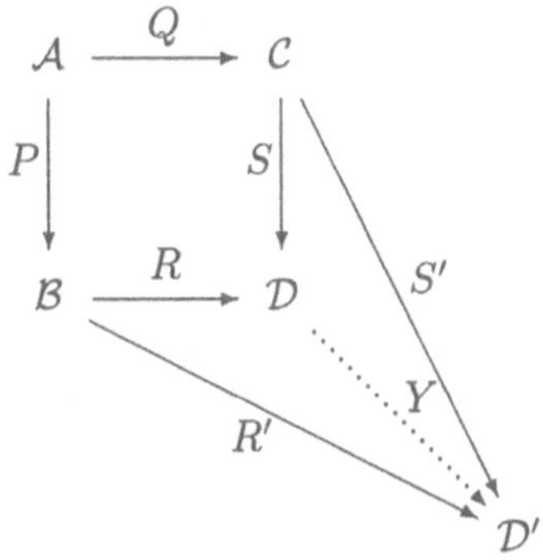

The following example in the category of unlabelled directed graphs shows a pushout that "glues" together two directed graphs along a common interface consisting of two nodes connected by a single directed edge (for anchoring the edge component of graph homomorphisms, we draw edges as small squares with incoming and outgoing source and target "tentacles"). The pushout object can be considered as containing "copies" of the two graphs $\mathcal{B}$ and $\mathcal{C}$, where the two "instances" of the "interface" $\mathcal{A}$ have been "identified".

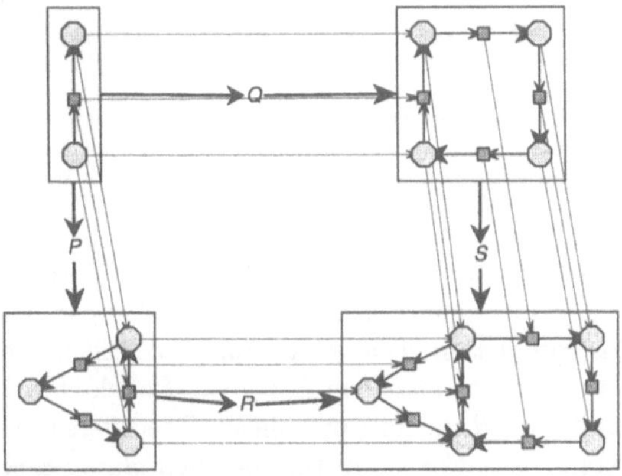

In the double-pushout approach, a rewriting rule is a *span* $\mathcal{L} \xleftarrow{\Phi_L} \mathcal{G} \xrightarrow{\Phi_R} \mathcal{R}$ of morphisms starting from the *gluing object* $\mathcal{G}$.

As an example consider the following graph rewriting rule, which deletes a path of length two between the two interface nodes, and inserts a four-node cycle with one edge connecting the interface nodes.

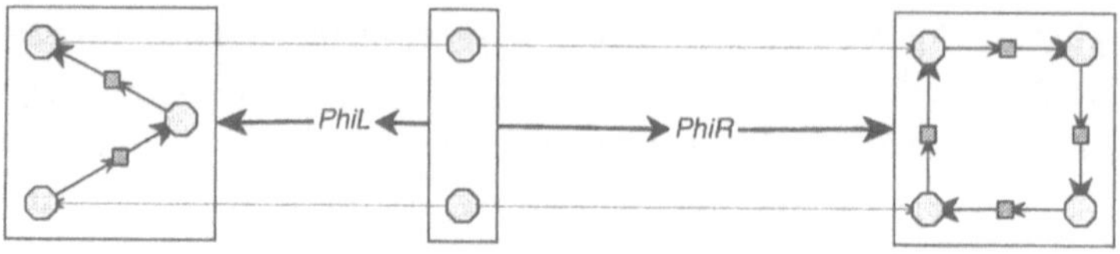

A *redex* for such a rule is a morphism $X_{\mathrm{L}} : \mathcal{L} \to \mathcal{A}$ from the rule's left-hand side $\mathcal{L}$ into some *application graph* $\mathcal{A}$. In the following example, the redex is injective, although, under certain circumstances, non-injective redex morphisms are allowed, too.

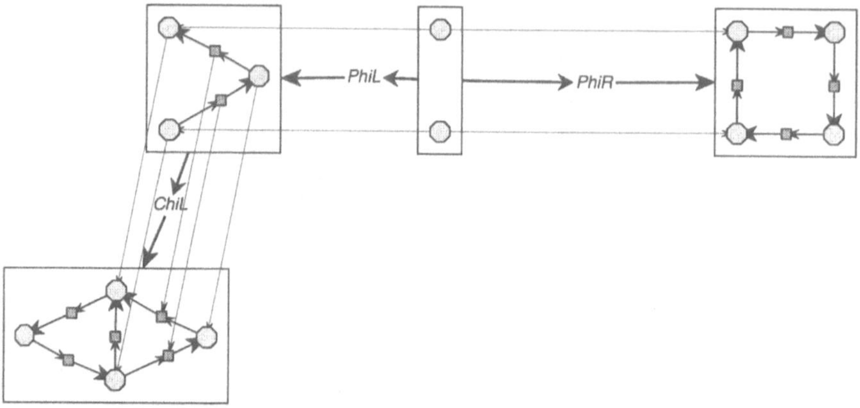

Application of the rule has to establish a *double-pushout diagram* of the following shape:

$$\begin{array}{ccccc} \mathcal{L} & \xleftarrow{\Phi_{\mathrm{L}}} & \mathcal{G} & \xrightarrow{\Phi_{\mathrm{R}}} & \mathcal{R} \\ {\scriptstyle X_{\mathrm{L}}}\downarrow & & \downarrow{\scriptstyle \Xi} & & \downarrow{\scriptstyle X_{\mathrm{R}}} \\ \mathcal{A} & \xleftarrow{\Psi_{\mathrm{L}}} & \mathcal{H} & \xrightarrow{\Psi_{\mathrm{R}}} & \mathcal{B} \end{array}$$

Note that for the left-hand side pushout, the "wrong" arrows are given, so the completion to a pushout square is not a universally characterised categorical construction. This completion is called a *pushout complement*, which consists of a *host object* $\mathcal{H}$ and a *host morphism* $\Xi : \mathcal{G} \to \mathcal{H}$ such that the pushout of the rule's left-hand side Φ_{L} and the host morphism Ξ recovers the application graph. Such a pushout complement exists iff the so-called *gluing condition* holds — for items from left-hand side outside the image of the interface, the redex must not perform any identification or mapping to nodes incident with "dangling edges". This gluing condition is usually stated using the concrete language of graphs with nodes and edges; the only abstract formalisation given so far is that of Kawahara [Kaw90], using the language of toposes (categories with "internal logic"), which may be the reason why this seems not to have caught on.

Once a host morphism is found, the right-hand-side pushout completes the rewriting step, which in our example then looks as follows:

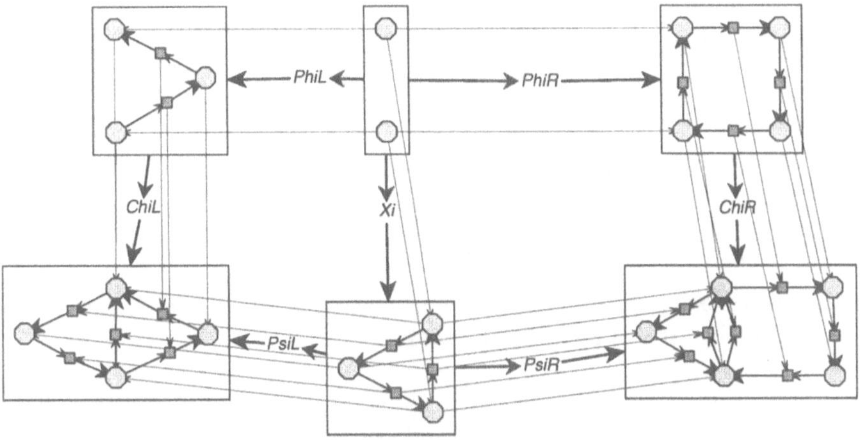

Note that in contrast with the first pushout example, there is no edge in the interface graph here; this produces a pushout object $\mathcal{B}$ that has two edges between the images of the interface nodes — only the interface nodes themselves are forced to be identified here.

As this example illustrates, application of double-pushout rules is quite easy to understand. This makes the double-pushout approach a useful tool also for *specifying* all kinds of transition and transformation systems, with applications ranging from applied software systems and systems modelling to aspects of parallelism, distributed systems and synchronisation mechanisms, and operational semantics of programming languages from a wide range of paradigms, including functional, logic, and object-oriented.

However, pushout rewriting does not easily lend itself to extensions that include some kind of *graph variable* that could be matched to whole subgraphs, such that rewriting might replicate or delete these subgraphs.

2.2 Pullback Rewriting

Bauderon proposed a different categorical setting for graph rewriting that overcomes the lack of replication capabilities in the pushout approaches. Starting from the fact that the most natural replication mechanism in category theory is the categorical product, and that, in graph-structure categories, pullbacks are subobjects of products, Bauderon and Jacquet introduced a setup that uses pullbacks in place of pushouts [Bau97,BJ01].

The pullback is the *dual* concept of the pushout, that is, defined with "all arrows turned around": Starting from two morphisms R and S with common target $\mathcal{D}$, a pullback consists of a *pullback object* $\mathcal{A}$ and two morphisms P and Q starting from $\mathcal{A}$ such that the resulting square commutes, and such that each commuting candidate square can be uniquely factorised via $\mathcal{A}$.

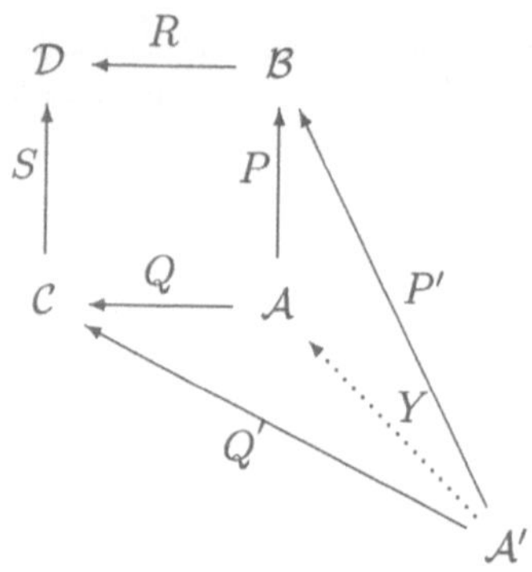

The following example shows how a simple *single-pullback rule* R can be used to specify duplication of application graphs.

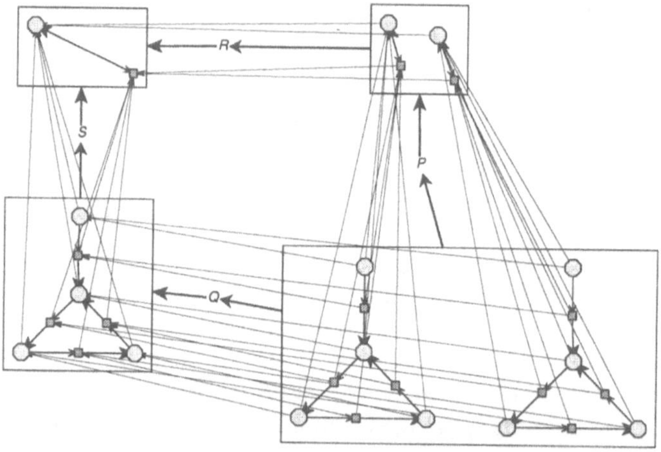

In general, however, pullback rules tend to be less intuitive.

The natural approach is to consider different parts of the target of the rule arrow R as standing for different "rôles" in the rewriting step. The rule morphism R decides the behaviour of each rôle, while the morphism S from the application graph into $\mathcal{D}$ assigns rôles to all parts of the application graph. For maximum flexibility in the treatment of these rôles, Bauderon and Jacquet therefore provide an *alphabet graph* that contains items that each represent a different treatment via rewriting, such as preservation, duplication, or deletion.

The following shows the translation of a simple NLC rule ("node-label-controlled graph rewriting", one variant of vertex replacement [ER97]), where $\mathcal{D}$ is the alphabet graph for three labels, which are represented by the horizontally flattened three-node clique in the middle. The top node is the "context", and the bottom node is the "redex". The rule morphism R rewrites the redex by splitting it into two nodes connected by a single edge, redirecting incoming edges from one of the three kinds of neighbours to the source of that new edge, and redirecting the source tentacles of outgoing edges directed at the same kind of neighbours to the target of the new edge.

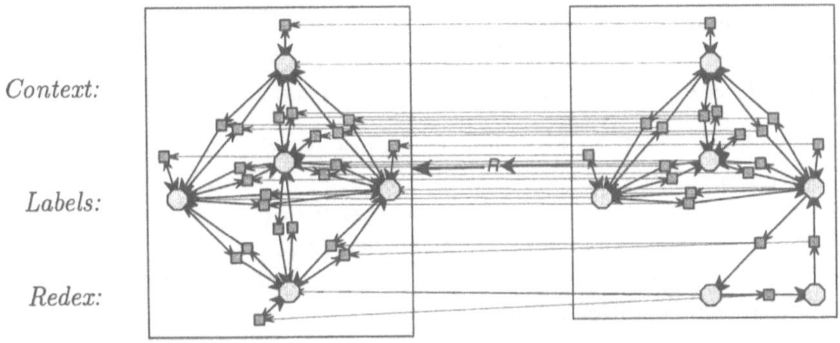

Such a rule is applied by establishing what Bauderon and Jacquet call a "label" on an application graph, that is, a graph morphism from the application graph to the alphabet graph such that exactly one node is mapped to the "redex" node, precisely its neighbours are mapped to label nodes in the alphabet graph (which are the neighbours of the redex node), and all other nodes are mapped to the "context" node.

Since just the rule above even sufficiently demonstrates that pullback rules are much less intuitive than pushout rules, we refrain from showing an application of this rule.

Even though the expressivity of pullback rewriting encompasses most well-known approaches to graph rewriting, including node replacement, (hyper-)edge replacement, and pushout rewriting, the lacking intuitive accessibility of its rewriting concept has severely limited interest in the pullback approach.

3 Amalgamating Pushouts and Pullbacks to Pullouts

We are going to present a formalism that allows the abstract specification of graph transformation with "graph variables" and replication of their images. It will superficially be modelled at the double-pushout approach, but works in the setting of relational graph morphisms, and it fully incorporates both the pushout and the pullback approaches. For details, see [Kah01].

The key to the relation-algebraic approach to graph transformation is the fact the graph structures with *relational graph-structure homomorphisms* give rise to *Dedekind categories.* Dedekind categories are relational categories that differ from heterogeneous relation algebras only in that the lattice of relations between two objects need not be Boolean. Therefore, most of the laws of relation algebras that do not involve complement still hold.

An even weaker, but in many circumstances still useful structure is an *allegory* [FS90], which is essentially a category with converse and intersection. Dedekind categories are also known as "locally complete distributive allegories".

We call morphisms of allegories and Dedekind categories *relations*, and we introduce a relation R from object $\mathcal{A}$ to object $\mathcal{B}$ by writing $R : \mathcal{A} \leftrightarrow \mathcal{B}$; and for relations $R, S : \mathcal{A} \leftrightarrow \mathcal{B}$ we use converse $R^{\smile}$ (which is a relation from

$\mathcal{B}$ to $\mathcal{A}$), union $R \sqcup S$, intersection $R \sqcap S$; the domain of definition of R is $\mathsf{dom}\, R = \mathbb{I}_{\mathcal{A}} \sqcap R{:}R^{\smile}$, and the range is $\mathsf{ran}\, R = \mathsf{dom}\,(R^{\smile})$. For a *homogeneous* relation $Q : \mathcal{A} \leftrightarrow \mathcal{A}$, we write Q^* for the reflexive and transitive closure of Q.

We consider a graph to be a unary algebra $(\mathcal{N}, \mathcal{E}, s, t)$ consisting of a node set $\mathcal{N}$, an edge set $\mathcal{E}$, and two mappings $s, t : \mathcal{E} \to \mathcal{N}$ which map edges to their source resp. target nodes. Given two graphs $\mathcal{G}_1 = (\mathcal{N}_1, \mathcal{E}_1, s_1, t_1)$ and $\mathcal{G}_2 = (\mathcal{N}_2, \mathcal{E}_2, s_2, t_2)$, a *relational graph homomorphism* $R : \mathcal{G}_1 \leftrightarrow \mathcal{G}_2$ is a pair $(R_{\mathcal{N}}, R_{\mathcal{E}})$ of relations $R_{\mathcal{N}} : \mathcal{N}_1 \leftrightarrow \mathcal{N}_2$ and $R_{\mathcal{E}} : \mathcal{E}_1 \leftrightarrow \mathcal{E}_2$ such that the following relational homomorphism condition holds:

$$R_{\mathcal{E}}{:}s_2 \sqsubseteq s_1{:}R_{\mathcal{N}} \qquad \text{and} \qquad R_{\mathcal{E}}{:}t_2 \sqsubseteq t_1{:}R_{\mathcal{N}}$$

This definition is easily generalised to arbitrary many-sorted algebras. In general, algebras with such relational homomorphisms give rise to allegories, while unary algebras, also called *graph structures*, give rise to Dedekind categories.

The following example of a relational graph homomorphism is neither total nor univalent:

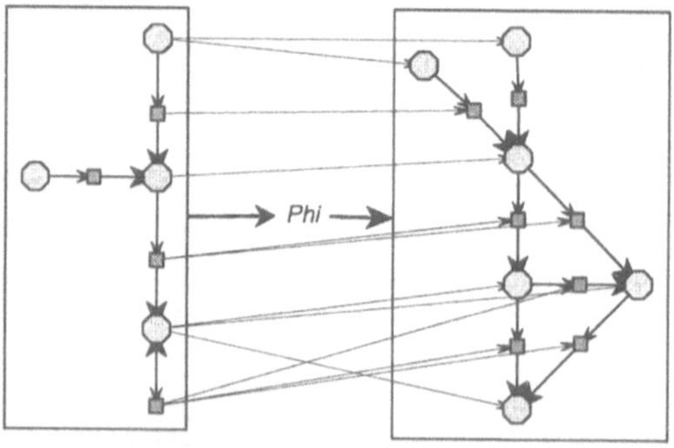

In a Dedekind category of *relational* graph structure homomorphisms, conventional graph structure homomorphisms are recovered as the *mappings* in these Dedekind categories, that is, as total and univalent relational graph structure homomorphisms.

Pushouts and pullbacks of mappings in Dedekind categories have nice relational characterisations; there are even useful generalisations that correspond to starting from the diagonals $U := P^{\smile}{:}Q$ resp. $V := R{:}S^{\smile}$.

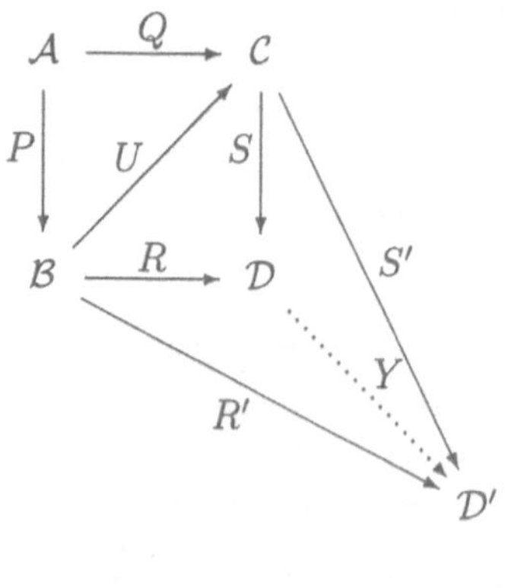

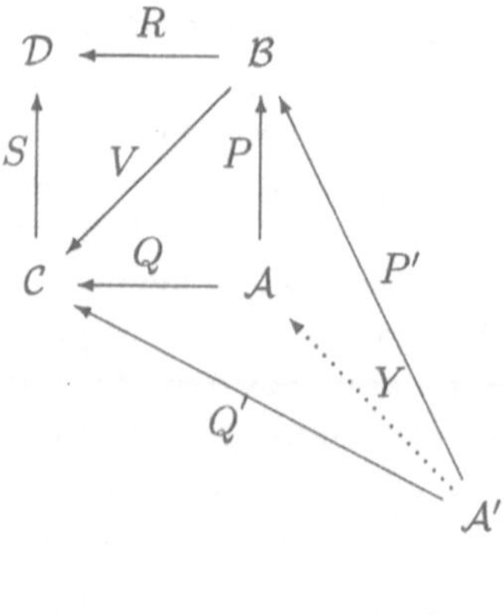

The generalisation of pullbacks is called *tabulation* by Freyd and Scedrov [FS90, 2.14] and is defined already in allegories; we slightly rearrange the conditions:

Definition 3.1 In an allegory **D**, let $V : \mathcal{B} \leftrightarrow \mathcal{C}$ be an arbitrary relation.

The span $\mathcal{B} \xleftarrow{P} \mathcal{A} \xrightarrow{Q} \mathcal{C}$ in the *allegory* **D** (i.e., P and Q are not yet specified as mappings) is a *direct tabulation* for V iff the following equations hold:

$$P^{\smile};Q = V \qquad \begin{array}{c} P^{\smile};P = \mathsf{dom}\, V \\ Q^{\smile};Q = \mathsf{ran}\, V \end{array} \qquad P;P^{\smile} \sqcap Q;Q^{\smile} = \mathbb{I} \qquad \square$$

For pushouts, the following conditions have first been stated by Kawahara [Kaw90], working directly with $P^{\smile};Q$ instead of with U:

Definition 3.2 In a Dedekind category **D**, let $U : \mathcal{B} \leftrightarrow \mathcal{C}$ be an arbitrary relation.

The cospan $\mathcal{B} \xrightarrow{R} \mathcal{D} \xleftarrow{S} \mathcal{C}$ in the Dedekind category **D** is a *direct gluing* for U iff the following equations hold:

$$R;S^{\smile} = U;(U^{\smile};U)^{*} \qquad \begin{array}{c} R;R^{\smile} = (U;U^{\smile})^{*} \\ S;S^{\smile} = (U^{\smile};U)^{*} \end{array} \qquad R^{\smile};R \sqcup S^{\smile};S = \mathbb{I} \qquad \square$$

A tabulation for a universal relation is a direct product, and a gluing for an empty relation is a direct sum. Although the usual relational characterisation of direct sums is dual to that of direct products, the gluing conditions cannot be obtained from the tabulation conditions via naïve relational dualisation. The reason for this is that for a cospan $\mathcal{B} \xrightarrow{R} \mathcal{D} \xleftarrow{S} \mathcal{C}$ of mappings, the relation $R;S^{\smile}$ is always *difunctional* (and this difunctionality is recovered via $U;(U^{\smile};U)^{*}$, which is the difunctional closure of U), while for a span $\mathcal{B} \xleftarrow{P} \mathcal{A} \xrightarrow{Q} \mathcal{C}$ the relation $P^{\smile};Q$ can essentially be arbitrary.

For our relational rewriting concept we are going to use a construction that can be understood as amalgamated from a pushout and a pullback.

This construction starts with a span $\mathcal{R} \xleftarrow{\Phi} \mathcal{G} \xrightarrow{\Xi} \mathcal{H}$ of two *relational* morphisms, and in addition a partial identity $u_0 \sqsubseteq \mathbb{I}_{\mathcal{G}}$ designating the *interface component* of the gluing object $\mathcal{G}$. From this interface component, one may calculate the *variable* (or *parameter*) *component* v_0 as the smallest partial identity such that $u_0 \sqcup v_0 = \mathbb{I}_{\mathcal{G}}$; we call this a *semi-complement.*

An intuitive explanation of the amalgamated construction is to start with producing a pushout for the interface parts $u_0;\Phi$ and $u_0;\Xi$, and a pullback for the converses of the parameter parts, i.e., for $\Phi^{\smile};v_0$ and $\Xi^{\smile};u_0$. Then the two result objects have to be glued together along a gluing relation B induced by the restrictions of the pushout and pullback morphisms to the borders between interface and parameter parts.

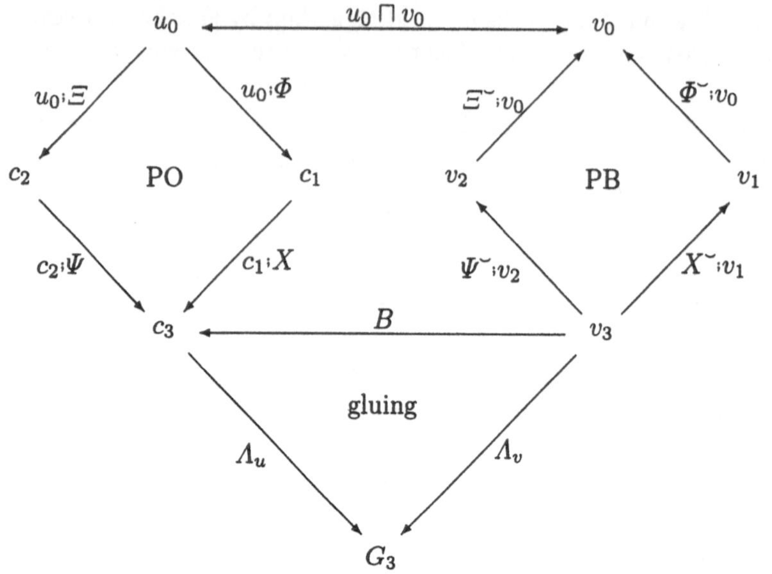

One may observe that with this construction, most of the details about Φ and Ξ are completely irrelevant, and we only need access to the resulting parameter and interface parts. In the spirit of the definitions of tabulations and gluings, we therefore can abstract away from Φ and Ξ, and substitute

$$U := \Phi^{\smile};u_0;\Xi \ , \qquad V := \Phi^{\smile};v_0;\Xi \ .$$

With a few *interface preservation* conditions that we leave out here, this setup allows to define that a cospan $G_1 \xrightarrow{X} G_3 \xleftarrow{\Psi} G_2$ is called a *glued tabulation for V along U on c_1 and c_2* (where c_1 and c_2 are partial identities comprising the respective interface parts and contexts) iff the following conditions hold:

$$\begin{array}{rcccc} X;\Psi^{\smile} & = & U;(U^{\smile};U)^* & \sqcup & V \\ X;X^{\smile} & = & c_1;(U;U^{\smile})^*;c_1 & \sqcup & \mathsf{dom}\, V \\ \Psi;\Psi^{\smile} & = & c_2;(U^{\smile};U)^*;c_2 & \sqcup & \mathsf{ran}\, V \\ \mathbb{I} & = & (X^{\smile};c_1;X \sqcup \Psi^{\smile};c_2;\Psi) & \sqcup & (X^{\smile};v_1;X \sqcap \Psi^{\smile};v_2;\Psi) \end{array} \qquad \square$$

Starting with a span $\mathcal{R} \xleftarrow{\Phi} \mathcal{G} \xrightarrow{\Xi} \mathcal{H}$ and a partial identity $u_0 \sqsubseteq \mathbb{I}_{\mathcal{G}}$ designating the interface component, a glued tabulation for V along U as defined above is called a *pullout for* $\mathcal{R} \xleftarrow{\Phi} (\mathcal{G}, u_0) \xrightarrow{\Xi} \mathcal{H}$.

Obviously, the pullout conditions are an amalgamation of the gluing and tabulation conditions. In addition, the construction sketched above satisfies these conditions.

The ease with which this amalgamation is possible is the essential advantage of the relational approach.

Once we have pullouts, we naturally obtain a *double-pullout approach* to rewriting.

A *double-pullout rule* consists of left- and right-hand side morphisms as usual, but these are now *relational* morphisms; and the rule has an additional component: a partial identity u_0 on the gluing graph that indicates the *interface*. The following is an example rule, where the interface consists only of the top-most node:

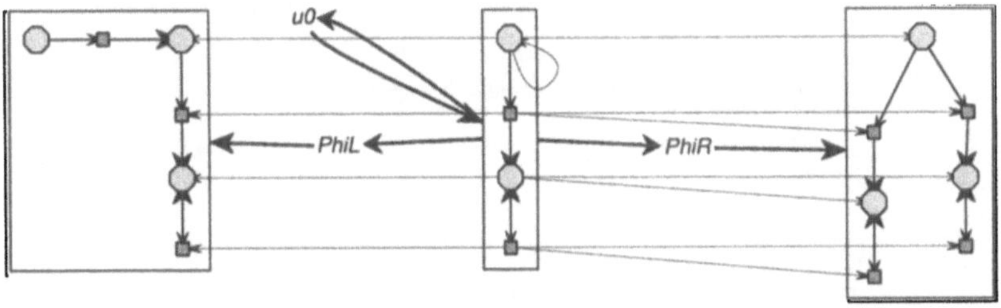

Everything besides the interface is considered as *parameter*, and we see that the right-hand side of the example rule duplicates the parameter part.

A redex is a relational graph morphism, too, and is restricted to be total and univalent besides the parameter part. (For simplicity, we ignore further more technical conditions.) It may map the parameter part to a larger subgraph, as in the following example, where the redex relation $X_{\mathcal{L}}$ is univalent only on the interface part:

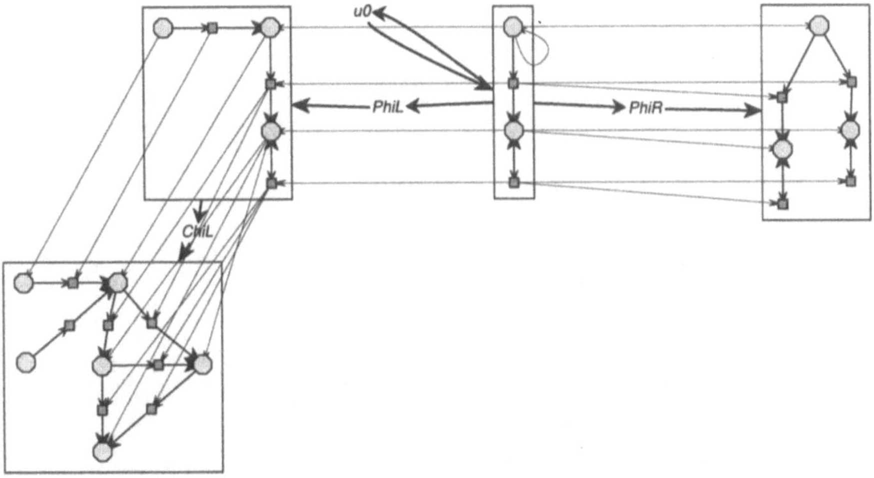

The host construction can easily be expressed using relation-algebraic constructions. Here, the rule's left-hand side $\Phi_{\mathcal{L}}$ is univalent on the parameter part, which allows us to use essentially the same host construction that was identified by Kawahara [Kaw90] for the double-pushout approach. (If the rule's left-hand side $\Phi_{\mathcal{L}}$ is not univalent on the parameter part, then the rule can be applied only if the redex documents appropriate replication of the respective parameter parts, which is equivalent with the existence of a certain partial equivalence relation Θ on the application object $\mathcal{A}$. In such cases a non-trivial *pullback complement* has to be constructed for the parameter part; this becomes easy once Θ has been identified.)

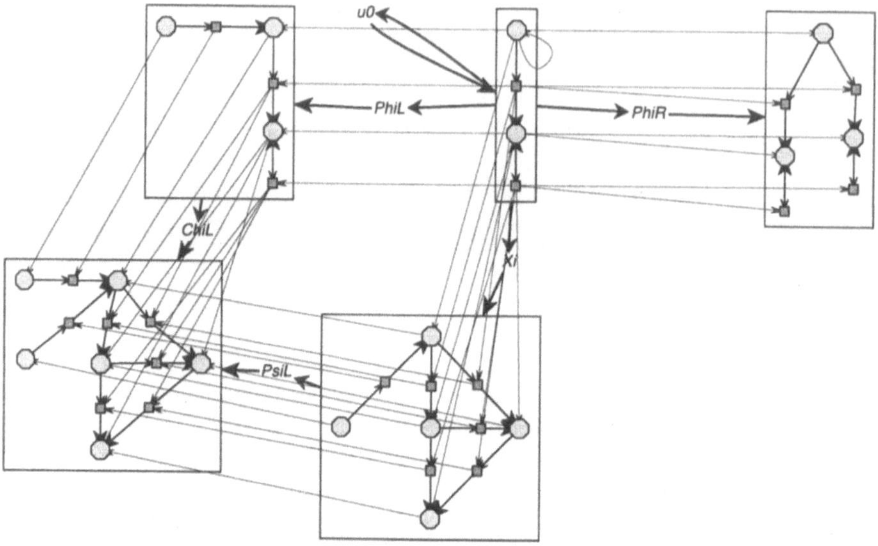

For the result, the parameter part is replicated according to the prescription of the rule's right-hand side:

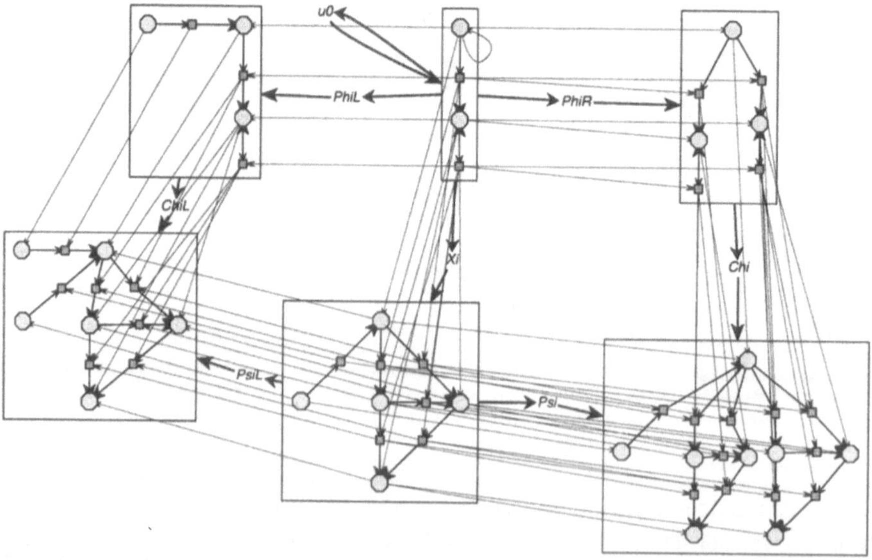

This concludes the application of the double-pullout rule. Although we have oriented our nomenclature and the direction of the morphisms at the double-pushout approach, it should be obvious that this choice is really arbitrary since the double-pullback approach is embedded in our approach to an equal extent; the pullback component made the replication of the parameter possible while the pushout component guaranteed the rigid matching of the interface component.

In the next picture we show an example how one may use a single rule to achieve several effects at once. This time, we use directed hypergraphs: besides nodes (big octagons) and edges (big squares), we now also have source- and target-tentacles (small upwards and downwards triangles) going from nodes to edges and vice versa.

Starting from the left-hand side of the rule, we see that the upper and lower "ridges" are preserved, but the vertical "bridge" in the middle of the left-hand side is deleted. Furthermore, the left and right parts of the left-hand side each contain a unit hypergraph and act as parameters; the left parameter is deleted.

In the right-hand side, we see that the two nodes the left parameter was incident with are identified. Furthermore, the right parameter is duplicated, and the "new copy" is connected with the "original copy" via two new edges.

In the example application of this rule, we only show the matching and result morphisms, and omit the host part for the sake of readability.

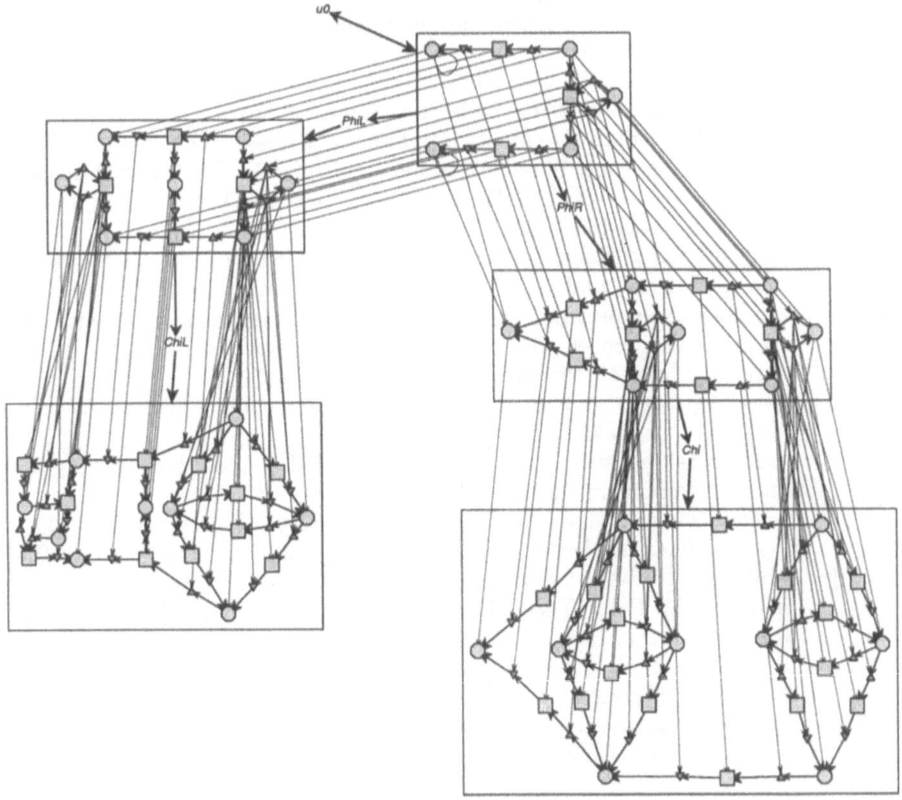

4 Conclusion

We have introduced an abstract approach to graph structure transformation that is based on the language of relations and leads to a rewriting concept where rewriting rules can contain parameters as well as "rigid" parts, and can specify not only addition and deletion of the rigidly mapped parts, but also deletion or replication of the parts matched to parameters, thus giving rise to a useful "graph variable" concept.

The important point of the presented work is that we are able to specify this kind of rewriting with fully abstract definitions that never need to mention edges or nodes. Nevertheless, we obtain a rewriting concept that unifies the intuitive elegance of the pushout approaches with the expressive and replicative power of the pullback approaches. This is made possible by working in a relation-algebraic setting, which, we claim, is the right level of abstraction for specifying graph-structure transformation.

References

[Bau97] Michel Bauderon. A uniform approach to graph rewriting: The pullback approach. In Manfred Nagl, editor, *Graph Theoretic Concepts in Computer Science, WG '96*, volume 1017 of *LNCS*, pages 101–115. Springer, 1997.

[BJ01] Michel Bauderon and Hélène Jacquet. Pullback as a generic graph rewriting mechanism. *Applied Categorical Structures*, 9(1):65–82, 2001.

[BW90] Michael Barr and Charles Wells. *Category Theory for Computing Science.* Prentice Hall International Series in Computer Science. Prentice Hall, 1990.

[CMR+97] Andrea Corradini, Ugo Montanari, Francesca Rossi, Hartmut Ehrig, Reiko Heckel, and Michael Löwe. Algebraic approaches to graph transformation, part I: Basic concepts and double pushout approach. In Rozenberg [Roz97], chapter 3, pages 163–245.

[EHK+97] Hartmut Ehrig, Reiko Heckel, Martin Korff, Michael Löwe, Leila Ribeiro, Annika Wagner, and Andrea Corradini. Algebraic approaches to graph transformation, part II: Single pushout approach and comparison with double pushout approach. In Rozenberg [Roz97], chapter 4, pages 247–312.

[EPS73] Hartmut Ehrig, M. Pfender, and H. J. Schneider. Graph grammars: An algebraic approach. In *Proc. IEEE Conf. on Automata and Switching Theory, SWAT '73*, pages 167–180, 1973.

[ER97] Joost Engelfriet and Grzegorz Rozenberg. Node replacement graph grammars. In Rozenberg [Roz97], chapter 1, pages 1–94.

[FS90] Peter J. Freyd and Andre Scedrov. *Categories, Allegories*, volume 39 of *North-Holland Mathematical Library.* North-Holland, Amsterdam, 1990.

[Kah01] Wolfram Kahl. A relation-algebraic approach to graph structure transformation, 2001. Habil. Thesis, Fakultät für Informatik, Univ. der Bundeswehr München, Techn. Bericht 2002-03.

[Kaw90] Yasuo Kawahara. Pushout-complements and basic concepts of grammars in toposes. *Theoretical Computer Science*, 77:267–289, 1990.

[Roz97] Grzegorz Rozenberg, editor. *Handbook of Graph Grammars and Computing by Graph Transformation, Vol. 1: Foundations.* World Scientific, Singapore, 1997.

Emptiness Relations in Property Systems

Philippe Balbiani

Institut de recherche en informatique de Toulouse
118 route de Narbonne, 31062 Toulouse Cedex 4, France

Abstract. In this paper, we carry out the modal analysis of emptiness through a modal logic which modalities correspond to emptiness relations in property systems. We mainly address the questions of axiomatization/completeness and decidability/complexity of our modal logic.

1 Introduction

The concept of informational relation has been introduced by Pawlak [10] within the context of attribute systems — knowledge-based systems which describe objects in terms of attributes — and Vakarelov [16] within the context of property systems — knowledge-based systems which describe objects in terms of properties. It has been furthered by Demri [2], Demri, Orłowska and Vakarelov [3], Orłowska [6,7], Orłowska and Pawlak [9] and Vakarelov [13,14,15, 16] who have considered the questions of axiomatization/completeness and decidability/complexity of modal logics which modalities correspond to informational relations. The aim of these logics is to provide a formal method for reasoning about uncertain knowledge discovered from attribute systems or property systems.

All the informational relations in attribute systems and property systems are either indistinguishability relations or distinguishability relations. Indistinguishability relations indicate the way objects share attributes or properties whereas distinguishability relations indicate the way attributes and properties differentiate objects. This paper is devoted to the modal analysis of emptiness relations in property systems. We give in section 2 the formal definition of emptiness relations in property systems. We provide a mathematical definition of the syntax and semantics of our modal logic in section 3 where we mainly address the questions of completeness and decidability. In section 4 we show that the satisfiability problem for any given formula of our modal logic requires polynomial space whereas in section 5 we show that the satisfiability problem for any given formula of our modal logic is decidable in polynomial space.

2 Emptiness Relations

Adapted from Vakarelov [16], a property system will be any algebraic structure (Obj, Pro, f) where:

H. de Swart (Ed.): RelMiCS 2001, LNCS 2561, pp. 15–34, 2002.

- Obj is a nonempty set of objects;
- Pro is a nonempty set of properties;
- f is a function with domain Obj and range the power set of Pro.

The property system (Obj, Pro, f) will be defined to be non-trivial if for all $x \in Obj$, $f(x) \neq \emptyset$ and $f(x) \neq Pro$. We should consider, for example, the property system $S = (Obj, Pro, f)$ defined as follows. Define:

- Obj is $\{Ann, Bob, Cindy, Daniel, Emma\}$;
- Pro is $\{Arabic, Bulgarian, Castilian, Dutch, English\}$;
- f is the function defined by table 1.

Table 1. Example of a property system.

Ann	Bob	Cindy	Daniel	Emma
$\{Arabic,$ $Bulgarian,$ $Castilian\}$	$\{Arabic,$ $Bulgarian,$ $Castilian,$ $Dutch\}$	$\{Arabic,$ $Bulgarian\}$	$\{English\}$	$\{Castilian,$ $Dutch,$ $English\}$

Property systems constitute the starting point for the formal examination of sentences of the form "object x is indistinguishable from object y" or sentences of the form "object x is distinguishable from object y". In this respect, the emptiness relations might play an important role. Given any property system $S = (Obj, Pro, f)$, the emptiness relations over S are defined by the following universal conditions:

$x \preceq_S y$ iff $f(x) \cap \overline{f(y)} = \emptyset$;
$x \succeq_S y$ iff $\overline{f(x)} \cap f(y) = \emptyset$;
$x \Delta_S y$ iff $f(x) \cap f(y) = \emptyset$;
$x \nabla_S y$ iff $\overline{f(x)} \cap \overline{f(y)} = \emptyset$.

Within the context of attribute systems, Orłowska [8] has named "forward inclusion", "backward inclusion", "right orthogonality", "left orthogonality" our emptiness relations $\preceq_S$, $\succeq_S$, Δ_S, ∇_S. The property system of table 1 is such that $Ann \preceq_S Bob$, $Ann \succeq_S Cindy$, $Ann \Delta_S Daniel$ and $Ann \nabla_S Emma$. We leave it to the reader to prove the following lemma.

Lemma 1. *Let $S = (Obj, Pro, f)$ be a non-trivial property system. The following universal conditions are satisfied.*

$x \preceq_S x$;
If $x \preceq_S y$ *then* $y \succeq_S x$;
If $x \preceq_S y$ *and* $y \preceq_S z$ *then* $x \preceq_S z$;
If $x \preceq_S y$ *and* $y \Delta_S z$ *then* $x \Delta_S z$;
$\overline{x \Delta_S x}$;
If $x \Delta_S y$ *then* $y \Delta_S x$;
If $x \Delta_S y$ *and* $y \succeq_S z$ *then* $x \Delta_S z$;
If $x \Delta_S y$ *and* $y \nabla_S z$ *then* $x \preceq_S z$;
$x \succeq_S x$;
If $x \succeq_S y$ *then* $y \preceq_S x$;
If $x \succeq_S y$ *and* $y \succeq_S z$ *then* $x \succeq_S z$;
If $x \succeq_S y$ *and* $y \nabla_S z$ *then* $x \nabla_S z$;
$\overline{x \nabla_S x}$;
If $x \nabla_S y$ *then* $y \nabla_S x$;
If $x \nabla_S y$ *and* $y \preceq_S z$ *then* $x \nabla_S z$;
If $x \nabla_S y$ *and* $y \Delta_S z$ *then* $x \succeq_S z$.

Lemma 1 motivates the following definition. An emptiness structure will be any relational structure $(W, \preceq, \succeq, \triangle, \bigtriangledown)$ where:

- W is a nonempty set of possible worlds;
- $\preceq$, $\succeq$, $\triangle$, $\bigtriangledown$ are binary relations on W subject to the universal conditions of lemma 1.

The following theorem explains the connection between emptiness structures and non-trivial property systems.

Theorem 1. (Characterization theorem for emptiness structures) *Let $(W, \preceq, \succeq, \triangle, \bigtriangledown)$ be an emptiness structure. There is a non-trivial property system $S = (Obj, Pro, f)$ such that $Obj = W$ and the following universal conditions are satisfied.*

$x \preceq_S y$ iff $x \preceq y$; *$x \succeq_S y$ iff $x \succeq y$;*
$x \triangle_S y$ iff $x \triangle y$; *$x \bigtriangledown_S y$ iff $x \bigtriangledown y$.*

Proof. It follows immediately from the definition that the following universal conditions are satisfied:

$x \preceq x$;
If $x \preceq y$ then $y \succeq x$;
If $x \preceq y$ and $y \preceq z$ then $x \preceq z$;
If $x\overline{\triangle}y$ then $x\overline{\triangle}x$;
$x\overline{\triangle}x$ or $x \preceq y$;
If $x\overline{\triangle}y$ then $y\overline{\triangle}x$;
If $x\overline{\triangle}y$ and $y \preceq z$ then $x\overline{\triangle}z$;
$x\overline{\triangle}y$ or $y\overline{\bigtriangledown}z$ or $x \preceq z$;

$x \succeq x$;
If $x \succeq y$ then $y \preceq x$;
If $x \succeq y$ and $y \succeq z$ then $x \succeq z$;
If $x\overline{\bigtriangledown}y$ then $x\overline{\bigtriangledown}x$;
$x\overline{\bigtriangledown}x$ or $x \succeq y$;
If $x\overline{\bigtriangledown}y$ then $y\overline{\bigtriangledown}x$;
If $x\overline{\bigtriangledown}y$ and $y \succeq z$ then $x\overline{\bigtriangledown}z$;
$x\overline{\bigtriangledown}y$ or $y\overline{\triangle}z$ or $x \succeq z$.

By Vakarelov [16], there is a property system $S = (Obj, Pro, f)$ such that $Obj = W$ and the following universal conditions are satisfied:

$x \preceq_S y$ iff $x \preceq y$; $x \succeq_S y$ iff $x \succeq y$;
$x \triangle_S y$ iff $x \triangle y$; $x \bigtriangledown_S y$ iff $x \bigtriangledown y$.

In view of the fact that for all $x \in W$, $x\overline{\triangle}x$ and $x\overline{\bigtriangledown}x$, for all $x \in Obj$, $f(x) \neq \emptyset$ and $f(x) \neq Pro$. Consequently S is non-trivial.

3 Modal Analysis

Let us be clear that our emptiness relations are nothing but the relations of informational inclusion and the complementary relations of the relations of similarity introduced by Vakarelov [16]. In this section, their modal analysis is presented for the very first time. An abstract structure will be any structure $(W, \preceq, \succeq, \triangle, \bigtriangledown, \equiv, R)$ where:

- W is a nonempty set of possible worlds;
- $\preceq$, $\succeq$, $\triangle$, $\bigtriangledown$ are binary relations on W subject to the universal conditions of lemma 1;

- $\equiv$, R are binary relations on W subject to the following universal conditions:
  - $x \equiv y$ iff $x \preceq y$ and $x \succeq y$;
  - xRy iff $x \bigtriangleup y$ and $x \bigtriangledown y$.

The proof of the following lemma is left as an exercise for the reader.

Lemma 2. *Let $(W, \preceq, \succeq, \bigtriangleup, \bigtriangledown, \equiv, R)$ be an abstract structure. The following universal conditions are satisfied.*

If $x \equiv y$ then $x \preceq y$;	*If xRy then $x \bigtriangleup y$;*
If $x \equiv y$ then $x \succeq y$;	*If xRy then $x \bigtriangledown y$;*
$x \equiv x$;	
If $x \equiv y$ then $y \equiv x$;	*If xRy then yRx;*
If $x \equiv y$ and $y \equiv z$ then $x \equiv z$;	*If xRy and $y \equiv z$ then xRz;*
If $x \equiv y$ and yRz then xRz;	*If xRy and yRz then $x \equiv z$.*

Lemma 2 motivates the following definition. A nonstandard abstract structure will be any structure $(W, \preceq, \succeq, \bigtriangleup, \bigtriangledown, \equiv, R)$ where:

- W is a nonempty set of possible worlds;
- $\preceq$, $\succeq$, $\bigtriangleup$, $\bigtriangledown$ are binary relations on W subject to the universal conditions of lemma 1 but the universal conditions "$x\overline{\bigtriangleup}x$", "$x\overline{\bigtriangledown}x$";
- $\equiv$, R are binary relations on W subject to the universal conditions of lemma 2.

The linguistic basis of our modal logic is the propositional calculus enlarged with the modalities $[\preceq]$, $[\succeq]$, $[\bigtriangleup]$, $[\bigtriangledown]$, $[\equiv]$, $[R]$. We define the set of all formulas as follows:

- $A ::= p \mid \neg A \mid (A \vee B) \mid [\preceq]A \mid [\succeq]A \mid [\bigtriangleup]A \mid [\bigtriangledown]A \mid [\equiv]A \mid [R]A$;

where p ranges over a countably infinite set of propositional variables. The other standard connectives are defined by the usual abbreviations. In particular, $\langle\preceq\rangle A$ is $\neg[\preceq]\neg A$, $\langle\succeq\rangle A$ is $\neg[\succeq]\neg A$, $\langle\bigtriangleup\rangle A$ is $\neg[\bigtriangleup]\neg A$, $\langle\bigtriangledown\rangle A$ is $\neg[\bigtriangledown]\neg A$, $\langle\equiv\rangle A$ is $\neg[\equiv]\neg A$, $\langle R\rangle A$ is $\neg[R]\neg A$. We follow the standard rules for omission of the parentheses. The number of occurrences of symbols in formula A is denoted $length(A)$. A model (respectively: a nonstandard model) will be any structure $(W, \preceq, \succeq, \bigtriangleup, \bigtriangledown, \equiv, R, V)$ where:

- $(W, \preceq, \succeq, \bigtriangleup, \bigtriangledown, \equiv, R)$ is an abstract structure (respectively: a nonstandard abstract structure);
- V is a function with domain the set of all propositional variables and range the power set of W.

Let $M = (W, \preceq, \succeq, \bigtriangleup, \bigtriangledown, \equiv, R, V)$ be either a model or a nonstandard model. We define the relation "formula A is true at possible world x in M", denoted $M, x \models A$, as follows:

- $M, x \models p$ iff $x \in V(p)$;
- $M, x \models \neg A$ iff $M, x \not\models A$;
- $M, x \models A \vee B$ iff $M, x \models A$ or $M, x \models B$;
- $M, x \models [\preceq]A$ iff for all $y \in W$, if $x \preceq y$ then $M, y \models A$;

- $M, x \models [\succeq]A$ iff for all $y \in W$, if $x \succeq y$ then $M, y \models A$;
- $M, x \models [\triangle]A$ iff for all $y \in W$, if $x \triangle y$ then $M, y \models A$;
- $M, x \models [\triangledown]A$ iff for all $y \in W$, if $x \triangledown y$ then $M, y \models A$;
- $M, x \models [\equiv]A$ iff for all $y \in W$, if $x \equiv y$ then $M, y \models A$;
- $M, x \models [R]A$ iff for all $y \in W$, if xRy then $M, y \models A$.

Let LER — logic of emptiness relations — be the smallest normal modal logic that contains the following axioms:

$[\preceq]A \to A$;
$A \to [\preceq]\langle\succeq\rangle A$;
$[\preceq]A \to [\preceq][\preceq]A$;
$[\triangle]A \to [\preceq][\triangle]A$;
$A \to [\triangle]\langle\triangle\rangle A$;
$[\triangle]A \to [\triangle][\succeq]A$;
$[\preceq]A \to [\triangle][\triangledown]A$;
$[\preceq]A \to [\equiv]A$;
$[\succeq]A \to [\equiv]A$;
$[\equiv]A \to A$;
$A \to [\equiv]\langle\equiv\rangle A$;
$[\equiv]A \to [\equiv][\equiv]A$;
$[R]A \to [\equiv][R]A$;

$[\succeq]A \to A$;
$A \to [\succeq]\langle\preceq\rangle A$;
$[\succeq]A \to [\succeq][\succeq]A$;
$[\triangledown]A \to [\succeq][\triangledown]A$;
$A \to [\triangledown]\langle\triangledown\rangle A$;
$[\triangledown]A \to [\triangledown][\preceq]A$;
$[\succeq]A \to [\triangledown][\triangle]A$;
$[\triangle]A \to [R]A$;
$[\triangledown]A \to [R]A$;

$A \to [R]\langle R\rangle A$;
$[R]A \to [R][\equiv]A$;
$[\equiv]A \to [R][R]A$.

A typical result is the following.

Theorem 2. (Completeness theorem for LER) *LER is complete with respect to the class of all models, the class of all finite nonstandard models and the class of all nonstandard models, i.e. the following conditions are equivalent.*

1. *A is true at all possible worlds in all models;*
2. *A is true at all possible worlds in all finite nonstandard models;*
3. *A is true at all possible worlds in all nonstandard models;*
4. *A is a theorem of LER.*

Proof. (1 implies 2): Let $M = (W, \preceq, \succeq, \triangle, \triangledown, \equiv, R, V)$ be a finite nonstandard model and $M' = (W', \preceq', \succeq', \triangle', \triangledown', \equiv', R', V')$ be the structure defined as follows. Define:

– W' is $W \times \mathbb{Z}^\star \times \mathbb{Z}^\star$;
– For all $x, y \in W$, for all $i_1, i_2 \in \mathbb{Z}^\star$ and for all $j_1, j_2 \in \mathbb{Z}^\star$, $(x, i_1, i_2) \preceq' (y, j_1, j_2)$ iff $x \preceq y$, $i_1 = j_1$, $i_2 \preceq j_2$ and if $i_2 \succeq j_2$ then $x \equiv y$;
– For all $x, y \in W$, for all $i_1, i_2 \in \mathbb{Z}^\star$ and for all $j_1, j_2 \in \mathbb{Z}^\star$, $(x, i_1, i_2) \succeq' (y, j_1, j_2)$ iff $x \succeq y$, $i_1 = j_1$, $i_2 \succeq j_2$ and if $i_2 \preceq j_2$ then $x \equiv y$;
– For all $x, y \in W$, for all $i_1, i_2 \in \mathbb{Z}^\star$ and for all $j_1, j_2 \in \mathbb{Z}^\star$, $(x, i_1, i_2) \triangle' (y, j_1, j_2)$ iff $x \triangle y$, $i_1 = -j_1$, $i_2 \preceq -j_2$ and if $i_2 \succeq -j_2$ then xRy;
– For all $x, y \in W$, for all $i_1, i_2 \in \mathbb{Z}^\star$ and for all $j_1, j_2 \in \mathbb{Z}^\star$, $(x, i_1, i_2) \triangledown' (y, j_1, j_2)$ iff $x \triangledown y$, $i_1 = -j_1$, $i_2 \succeq -j_2$ and if $i_2 \preceq -j_2$ then xRy;
– For all $x, y \in W$, for all $i_1, i_2 \in \mathbb{Z}^\star$ and for all $j_1, j_2 \in \mathbb{Z}^\star$, $(x, i_1, i_2) \equiv' (y, j_1, j_2)$ iff $x \equiv y$, $i_1 = j_1$ and $i_2 = j_2$;

- For all $x, y \in W$, for all $i_1, i_2 \in \mathbb{Z}^\star$ and for all $j_1, j_2 \in \mathbb{Z}^\star$, $(x, i_1, i_2)R'(y, j_1, j_2)$ iff xRy, $i_1 = -j_1$ and $i_2 = -j_2$;
- For all propositional variables p, $V'(p)$ is $V(p) \times \mathbb{Z}^\star \times \mathbb{Z}^\star$.

It is a simple matter to check that M' is a model. Moreover, M is a p-morphic image of M', i.e., the following conditions are satisfied:

- For all $x, y \in W$, for all $i_1, i_2 \in \mathbb{Z}^\star$ and for all $j_1, j_2 \in \mathbb{Z}^\star$, if $(x, i_1, i_2) \preceq' (y, j_1, j_2)$ then $x \preceq y$;
- For all $x, y \in W$, for all $i_1, i_2 \in \mathbb{Z}^\star$ and for all $j_1, j_2 \in \mathbb{Z}^\star$, if $(x, i_1, i_2) \succeq' (y, j_1, j_2)$ then $x \succeq y$;
- For all $x, y \in W$, for all $i_1, i_2 \in \mathbb{Z}^\star$ and for all $j_1, j_2 \in \mathbb{Z}^\star$, if $(x, i_1, i_2) \bigtriangleup' (y, j_1, j_2)$ then $x \bigtriangleup y$;
- For all $x, y \in W$, for all $i_1, i_2 \in \mathbb{Z}^\star$ and for all $j_1, j_2 \in \mathbb{Z}^\star$, if $(x, i_1, i_2) \bigtriangledown' (y, j_1, j_2)$ then $x \bigtriangledown y$;
- For all $x, y \in W$, for all $i_1, i_2 \in \mathbb{Z}^\star$ and for all $j_1, j_2 \in \mathbb{Z}^\star$, if $(x, i_1, i_2) \equiv' (y, j_1, j_2)$ then $x \equiv y$;
- For all $x, y \in W$, for all $i_1, i_2 \in \mathbb{Z}^\star$ and for all $j_1, j_2 \in \mathbb{Z}^\star$, if $(x, i_1, i_2)R'(y, j_1, j_2)$ then xRy;

and the following conditions are satisfied:

- For all $x, y \in W$ and for all $i_1, i_2 \in \mathbb{Z}^\star$, there is $j_1, j_2 \in \mathbb{Z}^\star$ such that if $x \preceq y$ then $(x, i_1, i_2) \preceq' (y, j_1, j_2)$;
- For all $x, y \in W$ and for all $i_1, i_2 \in \mathbb{Z}^\star$, there is $j_1, j_2 \in \mathbb{Z}^\star$ such that if $x \succeq y$ then $(x, i_1, i_2) \succeq' (y, j_1, j_2)$;
- For all $x, y \in W$ and for all $i_1, i_2 \in \mathbb{Z}^\star$, there is $j_1, j_2 \in \mathbb{Z}^\star$ such that if $x \bigtriangleup y$ then $(x, i_1, i_2) \bigtriangleup' (y, j_1, j_2)$;
- For all $x, y \in W$ and for all $i_1, i_2 \in \mathbb{Z}^\star$, there is $j_1, j_2 \in \mathbb{Z}^\star$ such that if $x \bigtriangledown y$ then $(x, i_1, i_2) \bigtriangledown' (y, j_1, j_2)$;
- For all $x, y \in W$ and for all $i_1, i_2 \in \mathbb{Z}^\star$, there is $j_1, j_2 \in \mathbb{Z}^\star$ such that if $x \equiv y$ then $(x, i_1, i_2) \equiv' (y, j_1, j_2)$;
- For all $x, y \in W$ and for all $i_1, i_2 \in \mathbb{Z}^\star$, there is $j_1, j_2 \in \mathbb{Z}^\star$ such that if xRy then $(x, i_1, i_2)R'(y, j_1, j_2)$.

As a consequence, see Hughes and Cresswell [4], if A is true at some possible world in M then A is true at some possible world in M'.
(2 implies 3): Let $M = (W, \preceq, \succeq, \bigtriangleup, \bigtriangledown, \equiv, R, V)$ be a nonstandard model and $M' = (W', \preceq', \succeq', \bigtriangleup', \bigtriangledown', \equiv', R', V')$ be the structure defined as follows. We define the smallest set Γ_A of formulas such that:

- $A \in \Gamma_A$;
- Γ_A is closed under subformulas.

It should be remarked that $Card(\Gamma_A) \leq length(A)$. Let $=_A$ be the equivalence relation on W defined as follows. For all $x, y \in W$, define:

- $x =_A y$ iff for all formulas B, if $B \in \Gamma_A$ then $M, x \models B$ iff $M, y \models B$.

For all $x \in W$, the equivalence class of x modulo $=_A$ is denoted $x_{|=_A}$. The quotient set of W modulo $=_A$ is denoted $W_{|=_A}$: For all propositional variables p, the quotient set of $V(p)$ modulo $=_A$ is denoted by $V(p)_{|=_A}$. Define:

- W' is $W_{|=_A}$;
- For all $x, y \in W$, $x_{|=_A} \preceq' y_{|=_A}$ iff for all formulas B, if $B \in \Gamma_A$ then:
  - If $M, x \models [\preceq]B$ then $M, y \models [\preceq]B$;
  - If $M, x \models [\triangle]B$ then $M, y \models [\triangle]B$;
  - If $M, y \models [\succeq]B$ then $M, x \models [\succeq]B$;
  - If $M, y \models [\triangledown]B$ then $M, x \models [\triangledown]B$;
- For all $x, y \in W$, $x_{|=_A} \succeq' y_{|=_A}$ iff for all formulas B, if $B \in \Gamma_A$ then:
  - If $M, x \models [\succeq]B$ then $M, y \models [\succeq]B$;
  - If $M, x \models [\triangledown]B$ then $M, y \models [\triangledown]B$;
  - If $M, y \models [\preceq]B$ then $M, x \models [\preceq]B$;
  - If $M, y \models [\triangle]B$ then $M, x \models [\triangle]B$;
- For all $x, y \in W$, $x_{|=_A} \triangle' y_{|=_A}$ iff for all formulas B, if $B \in \Gamma_A$ then:
  - If $M, x \models [\preceq]B$ then $M, y \models [\triangledown]B$;
  - If $M, x \models [\triangle]B$ then $M, y \models [\succeq]B$;
  - If $M, y \models [\preceq]B$ then $M, x \models [\triangledown]B$;
  - If $M, y \models [\triangle]B$ then $M, x \models [\succeq]B$;
- For all $x, y \in W$, $x_{|=_A} \triangledown' y_{|=_A}$ iff for all formulas B, if $B \in \Gamma_A$ then:
  - If $M, x \models [\succeq]B$ then $M, y \models [\triangle]B$;
  - If $M, x \models [\triangledown]B$ then $M, y \models [\preceq]B$;
  - If $M, y \models [\succeq]B$ then $M, x \models [\triangle]B$;
  - If $M, y \models [\triangledown]B$ then $M, x \models [\preceq]B$;
- For all $x, y \in W$, $x_{|=_A} \equiv' y_{|=_A}$ iff for all formulas B, if $B \in \Gamma_A$ then:
  - If $M, x \models [\preceq]B$ then $M, y \models [\preceq]B$;
  - If $M, x \models [\succeq]B$ then $M, y \models [\succeq]B$;
  - If $M, x \models [\triangle]B$ then $M, y \models [\triangle]B$;
  - If $M, x \models [\triangledown]B$ then $M, y \models [\triangledown]B$;
  - If $M, x \models [\equiv]B$ then $M, y \models [\equiv]B$;
  - If $M, x \models [R]B$ then $M, y \models [R]B$;
  - If $M, y \models [\preceq]B$ then $M, x \models [\preceq]B$;
  - If $M, y \models [\succeq]B$ then $M, x \models [\succeq]B$;
  - If $M, y \models [\triangle]B$ then $M, x \models [\triangle]B$;
  - If $M, y \models [\triangledown]B$ then $M, x \models [\triangledown]B$;
  - If $M, y \models [\equiv]B$ then $M, x \models [\equiv]B$;
  - If $M, y \models [R]B$ then $M, x \models [R]B$;
- For all $x, y \in W$, $x_{|=_A} R' y_{|=_A}$ iff for all formulas B, if $B \in \Gamma_A$ then:
  - If $M, x \models [\preceq]B$ then $M, y \models [\triangledown]B$;
  - If $M, x \models [\succeq]B$ then $M, y \models [\triangle]B$;
  - If $M, x \models [\triangle]B$ then $M, y \models [\succeq]B$;
  - If $M, x \models [\triangledown]B$ then $M, y \models [\preceq]B$;
  - If $M, x \models [\equiv]B$ then $M, y \models [R]B$;
  - If $M, x \models [R]B$ then $M, y \models [\equiv]B$;
  - If $M, y \models [\preceq]B$ then $M, x \models [\triangledown]B$;
  - If $M, y \models [\succeq]B$ then $M, x \models [\triangle]B$;
  - If $M, y \models [\triangle]B$ then $M, x \models [\succeq]B$;
  - If $M, y \models [\triangledown]B$ then $M, x \models [\preceq]B$;
  - If $M, y \models [\equiv]B$ then $M, x \models [R]B$;
  - If $M, y \models [R]B$ then $M, x \models [\equiv]B$;
- For all propositional variables p, $V'(p)$ is $V(p)_{|=_A}$.

It is a rather remarkable fact that M' is a finite nonstandard model. The interesting result is that M' is a filtration of M, i.e., the following conditions are satisfied:

- For all $x, y \in W$, if $x \preceq y$ then $x_{|=_A} \preceq' y_{|=_A}$;
- For all $x, y \in W$, if $x \succeq y$ then $x_{|=_A} \succeq' y_{|=_A}$;
- For all $x, y \in W$, if $x \bigtriangleup y$ then $x_{|=_A} \bigtriangleup' y_{|=_A}$;
- For all $x, y \in W$, if $x \bigtriangledown y$ then $x_{|=_A} \bigtriangledown' y_{|=_A}$;
- For all $x, y \in W$, if $x \equiv y$ then $x_{|=_A} \equiv' y_{|=_A}$;
- For all $x, y \in W$, if xRy then $x_{|=_A} R' y_{|=_A}$;

and the following conditions are satisfied:

- For all $x, y \in W$ and for all formulas B, if $x_{|=_A} \preceq' y_{|=_A}$, $B \in \Gamma_A$ and $M, x \models [\preceq]B$ then $M, y \models B$;
- For all $x, y \in W$ and for all formulas B, if $x_{|=_A} \succeq' y_{|=_A}$, $B \in \Gamma_A$ and $M, x \models [\succeq]B$ then $M, y \models B$;
- For all $x, y \in W$ and for all formulas B, if $x_{|=_A} \bigtriangleup' y_{|=_A}$, $B \in \Gamma_A$ and $M, x \models [\bigtriangleup]B$ then $M, y \models B$;
- For all $x, y \in W$ and for all formulas B, if $x_{|=_A} \bigtriangledown' y_{|=_A}$, $B \in \Gamma_A$ and $M, x \models [\bigtriangledown]B$ then $M, y \models B$;
- For all $x, y \in W$ and for all formulas B, if $x_{|=_A} \equiv' y_{|=_A}$, $B \in \Gamma_A$ and $M, x \models [\equiv]B$ then $M, y \models B$;
- For all $x, y \in W$ and for all formulas B, if $x_{|=_A} R' y_{|=_A}$, $B \in \Gamma_A$ and $M, x \models [R]B$ then $M, y \models B$.

As a consequence, see Hughes and Cresswell [4], if A is true at some possible world in M then A is true at some possible world in M'.
(3 implies 4): The proof can be obtained by the canonical model construction.
(4 implies 1): The proof is trivial because models satisfy the conditions which are needed to verify the axioms of LER.

We now turn our attention to the decidability of the satisfiability problem for any given formula of LER.

Theorem 3. (Decidability theorem for LER**)** *The satisfiability problem for any given formula of* LER *is decidable.*

Proof. By theorem 2, LER is a finitely axiomatizable normal modal logic which has the finite model property. As a consequence, the satisfiability problem for any given formula of LER is decidable.

A less obvious result concerns the inherent difficulty of the satisfiability problem for any given formula of LER. What we have in mind is to prove that the satisfiability problem for any given $\{[\equiv], [R]\}$-free formula A of LER is $PSPACE$-complete in $length(A)$. We outline how this result can be proved, but leave the details to the rest of the paper. In section 4, we show how the satisfiability problem for any given formula of $S4$ can be linearly reduced to the satisfiability problem for any given $\{[\equiv], [R]\}$-free formula of LER. In section 5, we show how the satisfiability problem for any given $\{[\equiv], [R]\}$-free formula of LER can be solved by means of a polynomial-space bounded nondeterministic algorithm.

4 *PSPACE*-Hardness

We define the set of all formulas of $S4$ as follows:

- $A ::= p \mid \neg A \mid (A \vee B) \mid \Box A$;

where p ranges over a countably infinite set of propositional variables. We follow the standard rules for omission of the parentheses. The number of occurrences of symbols in formula A of $S4$ is denoted $length(A)$. A model of $S4$ is a structure $(W, \mid, V)$ where:

- W is a nonempty set of possible worlds;
- $\mid$ is a binary relation on W such that for all $x, y \in W$:
  - $x \mid x$;
  - If $x \mid y$ and $y \mid z$ then $x \mid z$;
- V is a function with domain the set of all propositional variables and range the power set of W.

Let $M = (W, \mid, V)$ be a model of $S4$. We define the relation "formula A of $S4$ is true at possible world x in M", denoted $M, x \models A$, as follows:

- $M, x \models p$ iff $x \in V(p)$;
- $M, x \models \neg A$ iff $M, x \not\models A$;
- $M, x \models A \vee B$ iff $M, x \models A$ or $M, x \models B$;
- $M, x \models \Box A$ iff for all $y \in W$, if $x \mid y$ then $M, y \models A$.

Let t be the function that assigns to each formula A of $S4$ the $\{[\equiv], [R]\}$-free formula $t(A)$ of LER as follows:

- $t(p) = p$;
- $t(\neg A) = \neg t(A)$;
- $t(A \vee B) = t(A) \vee t(B)$;
- $t(\Box A) = [\preceq]t(A)$.

The interesting result is the following.

Lemma 3. *For all formulas A of $S4$, the following conditions are equivalent.*

1. *A is true at some possible world in some model of $S4$;*
2. *$t(A)$ is true at some possible world in some model of LER.*

Proof. (1 implies 2): Let $M = (W, \mid, V)$ be a model of $S4$ and $M' = (W', \preceq', \succeq', \Delta', \nabla', V')$ be the structure defined as follows. Define:

- W' is W;
- For all $x, y \in W$, $x \preceq' y$ iff $x \mid y$;
- For all $x, y \in W$, $x \succeq' y$ iff $y \mid x$;
- For all $x, y \in W$, $x \overline{\Delta'} y$;
- For all $x, y \in W$, $x \overline{\nabla'} y$;
- For all propositional variables p, $V'(p)$ is $V(p)$.

It is a simple matter to check that M' is a model of LER. The reader may easily verify by induction on the *length* of formula A of $S4$ that if A is true at some possible world in M then $t(A)$ is true at some possible world in M'.
(2 implies 1): Let $M = (W, \preceq, \succeq, \triangle, \triangledown, V)$ be a model of LER and $M' = (W', |', V')$ be the structure defined as follows. Define:

- W' is W;
- For all $x, y \in W$, $x \mid' y$ iff $x \preceq y$;
- For all propositional variables p, $V'(p)$ is $V(p)$.

It is a rather remarkable fact that M' is a model of $S4$. The reader may easily verify by induction on the *length* of formula A of $S4$ that if $t(A)$ is true at some possible world in M then A is true at some possible world in M'.

Seeing that the satisfiability problem for any given formula A of $S4$ is $PSPACE$-hard in *length*(A), see Ladner [5], we infer immediately the following.

Theorem 4. **(*PSPACE*-hardness theorem for *LER*)** *The satisfiability problem for any given* $\{[\equiv], [R]\}$*-free formula A of LER is PSPACE-hard in length*(A).

5 *PSPACE*-Completeness

Following the line of reasoning suggested by Spaan [12], let Cl be the function that assigns to each string $\omega \in \{\preceq, \succeq, \triangle, \triangledown\}^*$ and to each $\{[\equiv], [R]\}$-free formula A of LER the smallest set $Cl(\omega, A)$ of $\{[\equiv], [R]\}$-free formulas of LER as follows:

- $A \in Cl(\epsilon, A)$;
- $Cl(\omega, A)$ is closed under subformulas;
- For all $\{[\equiv], [R]\}$-free formulas B of LER:
  - If $[\preceq]B \in Cl(\omega, A)$ then $[\preceq]B \in Cl(\omega \preceq, A)$ and $[\triangledown]B \in Cl(\omega\triangle, A)$;
  - If $[\succeq]B \in Cl(\omega, A)$ then $[\succeq]B \in Cl(\omega \succeq, A)$ and $[\triangle]B \in Cl(\omega\triangledown, A)$;
  - If $[\triangle]B \in Cl(\omega, A)$ then $[\triangle]B \in Cl(\omega \preceq, A)$ and $[\succeq]B \in Cl(\omega\triangle, A)$;
  - If $[\triangledown]B \in Cl(\omega, A)$ then $[\triangledown]B \in Cl(\omega \succeq, A)$ and $[\preceq]B \in Cl(\omega\triangledown, A)$;

where ϵ denotes the empty string. The following lemma is basic.

Lemma 4. *Let A be a* $\{[\equiv], [R]\}$*-free formula of LER.*

- $Card(Cl(\epsilon, A)) \leq length(A)$;
- $Card(Cl(\preceq, A)) \leq length(A)$;
- $Card(Cl(\succeq, A)) \leq length(A)$;
- $Card(Cl(\triangle, A)) \leq length(A)$;
- $Card(Cl(\triangledown, A)) \leq length(A)$;
- *For all strings* $\omega \in \{\preceq, \succeq, \triangle, \triangledown\}^*$:
  - $Card(Cl(\omega \preceq\preceq, A)) = Card(Cl(\omega \preceq, A))$;
  - $Card(Cl(\omega \preceq\succeq, A)) < Card(Cl(\omega \preceq, A))$;
  - $Card(Cl(\omega \preceq \triangle, A)) = Card(Cl(\omega \preceq, A))$;
  - $Card(Cl(\omega \preceq \triangledown, A)) < Card(Cl(\omega \preceq, A))$;
  - $Card(Cl(\omega \succeq\preceq, A)) < Card(Cl(\omega \succeq, A))$;
  - $Card(Cl(\omega \succeq\succeq, A)) = Card(Cl(\omega \succeq, A))$;

- $Card(Cl(\omega \succeq \triangle, A)) < Card(Cl(\omega \succeq, A))$;
- $Card(Cl(\omega \succeq \bigtriangledown, A)) = Card(Cl(\omega \succeq, A))$;
- $Card(Cl(\omega\triangle \preceq, A)) < Card(Cl(\omega\triangle, A))$;
- $Card(Cl(\omega\triangle \succeq, A)) = Card(Cl(\omega\triangle, A))$;
- $Card(Cl(\omega \triangle \triangle, A)) < Card(Cl(\omega\triangle, A))$;
- $Card(Cl(\omega \triangle \bigtriangledown, A)) = Card(Cl(\omega\triangle, A))$;
- $Card(Cl(\omega\bigtriangledown \preceq, A)) = Card(Cl(\omega \bigtriangledown A))$;
- $Card(Cl(\omega\bigtriangledown \succeq, A)) < Card(Cl(\omega\bigtriangledown, A))$;
- $Card(Cl(\omega \bigtriangledown \triangle, A)) = Card(Cl(\omega\bigtriangledown, A))$;
- $Card(Cl(\omega \bigtriangledown \bigtriangledown, A)) < Card(Cl(\omega\bigtriangledown, A))$.

Proof. We first observe that $Cl(\epsilon, A)$ is the set of all subformulas of A. As a consequence $Card(Cl(\epsilon, A)) \leq length(A)$. It is a simple matter to check that $Cl(\preceq, A)$ does not contain more formulas than $Cl(\epsilon, A)$. Hence we have $Card(Cl(\preceq, A)) \leq length(A)$. Similarly $Card(Cl(\succeq, A)) \leq length(A)$, $Card(Cl(\triangle, A)) \leq length(A)$ and $Card(Cl(\bigtriangledown, A)) \leq length(A)$. It is a rather remarkable fact that $Cl(\omega \preceq\preceq, A)$ does not contain less formulas than $Cl(\omega \preceq, A)$. Hence we have $Card(Cl(\omega \preceq\preceq, A)) = Card(Cl(\omega \preceq, A))$. Similarly $Card(Cl(\omega \preceq \triangle, A)) = Card(Cl(\omega \preceq, A))$, $Card(Cl(\omega \succeq\succeq, A)) = Card(Cl(\omega \succeq, A))$, $Card(Cl(\omega \succeq \bigtriangledown, A)) = Card(Cl(\omega \succeq, A))$, $Card(Cl(\omega\triangle \succeq, A)) = Card(Cl(\omega\triangle, A))$, $Card(Cl(\omega \triangle \bigtriangledown, A)) = Card(Cl(\omega\triangle, A))$, $Card(Cl(\omega\bigtriangledown \preceq, A)) = Card(Cl(\omega \bigtriangledown A))$ and $Card(Cl(\omega \bigtriangledown \triangle, A)) = Card(Cl(\omega\bigtriangledown, A))$. The reader can readily check that $Cl(\omega \preceq\succeq, A)$ contains less formulas than $Cl(\omega \preceq, A)$. Hence we have $Card(Cl(\omega \preceq\succeq, A)) < Card(Cl(\omega \preceq, A))$. Similarly $Card(Cl(\omega \preceq \bigtriangledown, A)) < Card(Cl(\omega \preceq, A))$, $Card(Cl(\omega \succeq\preceq, A)) < Card(Cl(\omega \succeq, A))$, $Card(Cl(\omega \succeq \triangle, A)) < Card(Cl(\omega \succeq, A))$, $Card(Cl(\omega\triangle \preceq, A)) < Card(Cl(\omega\triangle, A))$, $Card(Cl(\omega \triangle \triangle, A)) < Card(Cl(\omega\triangle, A))$, $Card(Cl(\omega\bigtriangledown \succeq, A)) < Card(Cl(\omega\bigtriangledown, A))$ and $Card(Cl(\omega \bigtriangledown \bigtriangledown, A)) < Card(Cl(\omega\bigtriangledown, A))$.

To continue we introduce the function *depth* that assigns to each string $\omega \in \{\preceq, \succeq, \triangle, \bigtriangledown\}^\star$ the positive integer $depth(\omega)$ as follows:

- $depth(\epsilon) = 0$;
- $depth(\preceq) = 0$;
- $depth(\succeq) = 0$;
- $depth(\triangle) = 0$;
- $depth(\bigtriangledown) = 0$;
- For all strings $\omega \in \{\preceq, \succeq, \triangle, \bigtriangledown\}^\star$:
  - $depth(\omega \preceq\preceq) = depth(\omega \preceq)$;
  - $depth(\omega \preceq\succeq) = depth(\omega \preceq) + 1$;
  - $depth(\omega \preceq \triangle) = depth(\omega \preceq)$;
  - $depth(\omega \preceq \bigtriangledown) = depth(\omega \preceq) + 1$;
  - $depth(\omega \succeq\preceq) = depth(\omega \succeq) + 1$;
  - $depth(\omega \succeq\succeq) = depth(\omega \succeq)$;
  - $depth(\omega \succeq \triangle) = depth(\omega \succeq) + 1$;
  - $depth(\omega \succeq \bigtriangledown) = depth(\omega \succeq)$;
  - $depth(\omega\triangle \preceq) = depth(\omega\triangle) + 1$;
  - $depth(\omega\triangle \succeq) = depth(\omega\triangle)$;

- $depth(\omega \vartriangle \vartriangle) = depth(\omega\vartriangle) + 1$;
- $depth(\omega \vartriangle \triangledown) = depth(\omega\vartriangle)$;
- $depth(\omega\triangledown \preceq) = depth(\omega\triangledown)$;
- $depth(\omega\triangledown \succeq) = depth(\omega\triangledown) + 1$;
- $depth(\omega \triangledown \vartriangle) = depth(\omega\triangledown)$;
- $depth(\omega \triangledown \triangledown) = depth(\omega\triangledown) + 1$.

For our purpose, the crucial property of the function *depth* is the following.

Lemma 5. *For all strings $\omega \in \{\preceq, \succeq, \vartriangle, \triangledown\}^\star$ and for all $\{[\equiv], [R]\}$-free formulas A of LER, if $depth(\omega) \geq length(A)$ then $Cl(\omega, A) = \emptyset$.*

Proof. The proof is conducted by induction on the *length* of $\{[\equiv], [R]\}$-free formula A of LER.

Basis: Let p be a propositional variable. For all strings $\omega \in \{\preceq, \succeq, \vartriangle, \triangledown\}^\star$, if $depth(\omega) \geq length(p)$ then $depth(\omega) \geq 1$. Hence $Cl(\omega, p) = \emptyset$.

Hypothesis: Let B, C be $\{[\equiv], [R]\}$-free formulas of LER such that for all strings $\omega \in \{\preceq, \succeq, \vartriangle, \triangledown\}^\star$, if $depth(\omega) \geq length(B)$ then $Cl(\omega, B) = \emptyset$ and if $depth(\omega) \geq length(C)$ then $Cl(\omega, C) = \emptyset$.

Step: For all strings $\omega \in \{\preceq, \succeq, \vartriangle, \triangledown\}^\star$, if $depth(\omega) \geq length(\neg B)$ then $depth(\omega) > length(B)$. It follows that $Cl(\omega, B) = \emptyset$ and $Cl(\omega, \neg B) \subseteq Cl(\omega, B)$. Therefore $Cl(\omega, \neg B) = \emptyset$.

For all strings $\omega \in \{\preceq, \succeq, \vartriangle, \triangledown\}^\star$, if $depth(\omega) \geq length(B \vee C)$ then $depth(\omega) > length(B)$ and $depth(\omega) > length(C)$. It follows that $Cl(\omega, B) = \emptyset$, $Cl(\omega, C) = \emptyset$ and $Cl(\omega, B \vee C) \subseteq Cl(\omega, B) \cup Cl(\omega, C)$. Therefore $Cl(\omega, B \vee C) = \emptyset$.

For all strings $\omega \in \{\preceq, \succeq, \vartriangle, \triangledown\}^\star$, if $depth(\omega) \geq length([\preceq]B)$ then $depth(\omega) > length(B)$. It follows that there is a string $\omega' \in \{\preceq, \succeq, \vartriangle, \triangledown\}^\star$ such that $\omega = \omega' \preceq\succeq$ or $\omega = \omega' \preceq \triangledown$ or $\omega = \omega' \succeq\preceq$ or $\omega = \omega' \succeq \vartriangle$ or $\omega = \omega'\vartriangle \preceq$ or $\omega = \omega' \vartriangle \vartriangle$ or $\omega = \omega'\triangledown \succeq$ or $\omega = \omega' \triangledown \triangledown$. Suppose that $\omega = \omega' \preceq\succeq$. Hence $depth(\omega' \preceq) \geq length(B)$. It follows that $Cl(\omega' \preceq, B) = \emptyset$ and $Cl(\omega, [\preceq]B) \subseteq Cl(\omega' \preceq, B)$. Therefore $Cl(\omega, [\preceq]B) = \emptyset$. The same line of reasoning applies if $\omega = \omega' \preceq \triangledown$ or $\omega = \omega' \succeq\preceq$ or $\omega = \omega' \succeq \vartriangle$ or $\omega = \omega'\vartriangle \preceq$ or $\omega = \omega' \vartriangle \vartriangle$ or $\omega = \omega'\triangledown \succeq$ or $\omega = \omega' \triangledown \triangledown$.

For all strings $\omega \in \{\preceq, \succeq, \vartriangle, \triangledown\}^\star$, the same line of reasoning applies if $depth(\omega) \geq length([\succeq]B)$ or $depth(\omega) \geq length([\vartriangle]B)$ or $depth(\omega) \geq length([\triangledown]B)$.

Let S be a set of $\{[\equiv], [R]\}$-free formulas of LER and A be a $\{[\equiv], [R]\}$-free formula of LER. A $(\preceq, A)$-witness of S will be any set T of $\{[\equiv], [R]\}$-free formulas of LER such that $A \notin T$ and for all $\{[\equiv], [R]\}$-free formulas B of LER:

- If $[\preceq]B \in S$ then $[\preceq]B \in T$;
- If $[\vartriangle]B \in S$ then $[\vartriangle]B \in T$;
- If $[\succeq]B \in T$ then $[\succeq]B \in S$;
- If $[\triangledown]B \in T$ then $[\triangledown]B \in S$.

A $(\succeq, A)$-witness of S will be any set T of $\{[\equiv], [R]\}$-free formulas of LER such that $A \notin T$ and for all $\{[\equiv], [R]\}$-free formulas B of LER:

- If $[\succeq]B \in S$ then $[\succeq]B \in T$;
- If $[\triangledown]B \in S$ then $[\triangledown]B \in T$;
- If $[\preceq]B \in T$ then $[\preceq]B \in S$;
- If $[\triangle]B \in T$ then $[\triangle]B \in S$.

A $(\triangle, A)$-witness of S will be any set T of $\{[\equiv], [R]\}$-free formulas of LER such that $A \notin T$ and for all $\{[\equiv], [R]\}$-free formulas B of LER:

- If $[\triangle]B \in S$ then $[\succeq]B \in T$;
- If $[\preceq]B \in S$ then $[\triangledown]B \in T$;
- If $[\triangle]B \in T$ then $[\succeq]B \in S$;
- If $[\preceq]B \in T$ then $[\triangledown]B \in S$.

A $(\triangledown, A)$-witness of S will be any set T of $\{[\equiv], [R]\}$-free formulas of LER such that $A \notin T$ and for all $\{[\equiv], [R]\}$-free formulas B of LER:

- If $[\triangledown]B \in S$ then $[\preceq]B \in T$;
- If $[\succeq]B \in S$ then $[\triangle]B \in T$;
- If $[\triangledown]B \in T$ then $[\preceq]B \in S$;
- If $[\succeq]B \in T$ then $[\triangle]B \in S$.

To test the satisfiability problem for any given $\{[\equiv], [R]\}$-free formula A of LER, let $SAT(A)$ be the nondeterministic algorithm defined as follows. Define:

```
function SAT(A) returns Boolean
begin
for all subsets S of Cl(ε, A) such that A ∈ S do
    WORLD(S, ε, A)
if all these calls return false then return false
return true
end
```

where $WORLD(\mathcal{L}, \omega, A)$ is the nondeterministic algorithm defined as follows. Our nondeterministic algorithm $WORLD$ is similar to Spaan's nondeterministic algorithm $S4_t - WORLD$ — a procedure developed by Spaan [12] to investigate the inherent difficulty of the satisfiability problem for any given formula of $S4_t$ — in the sense that we keep track of a list $\mathcal{L}$ of sets of $\{[\equiv], [R]\}$-free formulas and a string $\omega \in \{\preceq, \succeq, \triangle, \triangledown\}^*$. Define:

```
function WORLD(L, ω, A) returns Boolean
begin
let S be last(L)
for all {[≡],[R]}-free formulas B of LER such that ¬B ∈ Cl(ω, A) do
    if (¬B ∈ S and B ∈ S) or (¬B ∉ S and B ∉ S) then return false
for all {[≡],[R]}-free formulas B, C of LER such that B ∨ C ∈ Cl(ω, A) do
    if (B ∨ C ∈ S and B ∉ S and C ∉ S) or (B ∨ C ∉ S and B ∈ S) or (B ∨ C ∉ S
    and C ∈ S) then return false
for all {[≡],[R]}-free formulas B of LER such that [⪯]B ∈ Cl(ω, A) do
    begin
    if [⪯]B ∈ S and B ∉ S then return false
    if [⪯]B ∉ S and there is no T in L such that T is a (⪯, B)-witness of S then
```

```
        begin
        for all subsets T of Cl(ω ⪯, A) such that T is a (⪯, B)-witness of S do
            if ω = ε or ω is ended by ⪰ or △ then WORLD(T, ω ⪯, A) else
            WORLD(ℒ ∘ T, ω ⪯, A)
        if all these calls return false then return false
        end
    end
for all {[≡], [R]}-free formulas B of LER such that [⪰]B ∈ Cl(ω, A) do
    begin
    if [⪰]B ∈ S and B ∉ S then return false
    if [⪰]B ∉ S and there is no T in ℒ such that T is a (⪰, B)-witness of S then
        begin
        for all subsets T of Cl(ω ⪰, A) such that T is a (⪰, B)-witness of S do
            if ω = ε or ω is ended by ⪯ or ▽ then WORLD(T, ω ⪰, A) else
            WORLD(ℒ ∘ T, ω ⪰, A)
        if all these calls return false then return false
        end
    end
for all {[≡], [R]}-free formulas B of LER such that [△]B ∈ Cl(ω, A) do
    if [△]B ∉ S and there is no T in ℒ such that T is a (△, B)-witness of S then
        begin
        for all subsets T of Cl(ω△, A) such that T is a (△, B)-witness of S do
            if ω = ε or ω is ended by ⪰ or △ then WORLD(T, ω△, A) else
            WORLD(ℒ ∘ T, ω△, A)
        if all these calls return false then return false
        end
for all {[≡], [R]}-free formulas B of LER such that [▽]B ∈ Cl(ω, A) do
    if [▽]B ∉ S and there is no T in ℒ such that T is a (▽, B)-witness of S then
        begin
        for all subsets T of Cl(ω▽, A) such that T is a (▽, B)-witness of S do
            if ω = ε or ω is ended by ⪯ or ▽ then WORLD(T, ω▽, A) else
            WORLD(ℒ ∘ T, ω▽, A)
        if all these calls return false then return false
        end
return true
end
```

where $last(\mathcal{L})$ denotes the last element of $\mathcal{L}$ and $\mathcal{L} \circ T$ denotes the list obtained by extending $\mathcal{L}$ with T. The following lemma is basic.

Lemma 6. *For all $\{[\equiv], [R]\}$-free formulas A of LER and for all subsets S of $Cl(\epsilon, A)$, the following conditions are equivalent.*

1. *$WORLD(S, \epsilon, A)$ succeeds;*
2. *There is a model $M = (W, \preceq, \succeq, \triangle, \triangledown, V)$ of LER and there is a possible world $x \in W$ such that for all $\{[\equiv], [R]\}$-free formulas B of LER, if $B \in Cl(\epsilon, A)$ then $M, x \models B$ iff $B \in S$.*

Proof. (1 implies 2): Suppose $WORLD(S, \epsilon, A)$ succeeds and let $M = (W, \preceq, \succeq, \triangle, \triangledown, V)$ be the structure defined as follows. Define:

- W is the set of all pairs (T,ω) for which there is a finite sequence $(\mathcal{L}_0,\omega_0),\ldots,(\mathcal{L}_K,\omega_K)$ such that:
  - $\mathcal{L}_0 = S$ and $\omega_0 = \epsilon$;
  - $last(\mathcal{L}_K) = T$ and $\omega_K = \omega$;
  - For all positive integers k, if $k \leq K$ then $WORLD(\mathcal{L}_k,\omega_k,A)$ is successfully called in $WORLD(S,\epsilon,A)$ at some depth of the recursion;
  - For all positive integers k, if $k < K$ then $WORLD(\mathcal{L}_{k+1},\omega_{k+1},A)$ is called by $WORLD(\mathcal{L}_k,\omega_k,A)$;
- For all $(T,\omega),(T',\omega') \in W$, $(T,\omega) \ll (T',\omega')$ iff there is a finite sequence $(\mathcal{L}_0,\omega_0),\ldots,(\mathcal{L}_K,\omega_K)$ such that:
  - $\mathcal{L}_0 = S$ and $\omega_0 = \epsilon$;
  - $last(\mathcal{L}_K) = T$ and $\omega_K = \omega$;
  - For all positive integers k, if $k \leq K$ then $WORLD(\mathcal{L}_k,\omega_k,A)$ is successfully called in $WORLD(S,\epsilon,A)$ at some depth of the recursion;
  - For all positive integers k, if $k < K$ then $WORLD(\mathcal{L}_{k+1},\omega_{k+1},A)$ is called by $WORLD(\mathcal{L}_k,\omega_k,A)$;

  and there is a finite sequence $(\mathcal{L}'_0,\omega'_0),\ldots,(\mathcal{L}'_{K'},\omega'_{K'})$ such that:
  - $\mathcal{L}'_0 = S$ and $\omega'_0 = \epsilon$;
  - $last(\mathcal{L}'_{K'}) = T'$ and $\omega'_{K'} = \omega'$;
  - For all positive integers k, if $k \leq K'$ then $WORLD(\mathcal{L}'_k,\omega'_k,A)$ is successfully called in $WORLD(S,\epsilon,A)$ at some depth of the recursion;
  - For all positive integers k, if $k < K'$ then $WORLD(\mathcal{L}'_{k+1},\omega'_{k+1},A)$ is called by $WORLD(\mathcal{L}'_k,\omega'_k,A)$;

  and either $(\mathcal{L}_K,\omega_K) = (\mathcal{L}'_{K'},\omega'_{K'})$ or $WORLD(\mathcal{L}'_{K'},\omega'_{K'},A)$ is called by $WORLD(\mathcal{L}_K,\omega_K,A)$ in the $[\preceq]$-segment or the finite sequence $(\mathcal{L}'_0,\omega'_0),\ldots,(\mathcal{L}'_{K'},\omega'_{K'})$ is a prefix of the finite sequence $(\mathcal{L}_0,\omega_0),\ldots,(\mathcal{L}_K,\omega_K)$, the call $WORLD(\mathcal{L}_K,\omega_K,A)$ enters in the $[\preceq]$-segment and there is a $\{[\equiv],[R]\}$-free formula B of LER such that $[\preceq]B \in Cl(\omega_K,A)$, $[\preceq]B \notin T$ and T' is a $(\preceq,B)$-witness of T;
- For all $(T,\omega),(T',\omega') \in W$, $(T,\omega) \gg (T',\omega')$ iff there is a finite sequence $(\mathcal{L}_0,\omega_0),\ldots,(\mathcal{L}_K,\omega_K)$ such that:
  - $\mathcal{L}_0 = S$ and $\omega_0 = \epsilon$;
  - $last(\mathcal{L}_K) = T$ and $\omega_K = \omega$;
  - For all positive integers k, if $k \leq K$ then $WORLD(\mathcal{L}_k,\omega_k,A)$ is successfully called in $WORLD(S,\epsilon,A)$ at some depth of the recursion;
  - For all positive integers k, if $k < K$ then $WORLD(\mathcal{L}_{k+1},\omega_{k+1},A)$ is called by $WORLD(\mathcal{L}_k,\omega_k,A)$;

  and there is a finite sequence $(\mathcal{L}'_0,\omega'_0),\ldots,(\mathcal{L}'_{K'},\omega'_{K'})$ such that:
  - $\mathcal{L}'_0 = S$ and $\omega'_0 = \epsilon$;
  - $last(\mathcal{L}'_{K'}) = T'$ and $\omega'_{K'} = \omega'$;
  - For all positive integers k, if $k \leq K'$ then $WORLD(\mathcal{L}'_k,\omega'_k,A)$ is successfully called in $WORLD(S,\epsilon,A)$ at some depth of the recursion;
  - For all positive integers k, if $k < K'$ then $WORLD(\mathcal{L}'_{k+1},\omega'_{k+1},A)$ is called by $WORLD(\mathcal{L}'_k,\omega'_k,A)$;

  and either $(\mathcal{L}_K,\omega_K) = (\mathcal{L}'_{K'},\omega'_{K'})$ or $WORLD(\mathcal{L}'_{K'},\omega'_{K'},A)$ is called by $WORLD(\mathcal{L}_K,\omega_K,A)$ in the $[\succeq]$-segment or the finite sequence $(\mathcal{L}'_0,\omega'_0),\ldots,(\mathcal{L}'_{K'},\omega'_{K'})$ is a prefix of the finite sequence $(\mathcal{L}_0,\omega_0),\ldots,(\mathcal{L}_K,\omega_K)$, the call $WORLD(\mathcal{L}_K,\omega_K,A)$ enters in the $[\succeq]$-segment and there is a $\{[\equiv],[R]\}$-free formula B of LER such that $[\succeq]B \in Cl(\omega_K, A)$, $[\succeq]B \notin T$ and T' is a $(\succeq, B)$-witness of T;

- For all $(T,\omega), (T',\omega') \in W$, $(T,\omega) \smile (T',\omega')$ iff there is a finite sequence $(\mathcal{L}_0,\omega_0), \ldots, (\mathcal{L}_K,\omega_K)$ such that:
  - $\mathcal{L}_0 = S$ and $\omega_0 = \epsilon$;
  - $last(\mathcal{L}_K) = T$ and $\omega_K = \omega$;
  - For all positive integers k, if $k \leq K$ then $WORLD(\mathcal{L}_k, \omega_k, A)$ is successfully called in $WORLD(S, \epsilon, A)$ at some depth of the recursion;
  - For all positive integers k, if $k < K$ then $WORLD(\mathcal{L}_{k+1}, \omega_{k+1}, A)$ is called by $WORLD(\mathcal{L}_k, \omega_k, A)$;

  and there is a finite sequence $(\mathcal{L}'_0,\omega'_0), \ldots, (\mathcal{L}'_{K'},\omega'_{K'})$ such that:
  - $\mathcal{L}'_0 = S$ and $\omega'_0 = \epsilon$;
  - $last(\mathcal{L}'_{K'}) = T'$ and $\omega'_{K'} = \omega'$;
  - For all positive integers k, if $k \leq K'$ then $WORLD(\mathcal{L}'_k, \omega'_k, A)$ is successfully called in $WORLD(S, \epsilon, A)$ at some depth of the recursion;
  - For all positive integers k, if $k < K'$ then $WORLD(\mathcal{L}'_{k+1}, \omega'_{k+1}, A)$ is called by $WORLD(\mathcal{L}'_k, \omega'_k, A)$;

  and either $WORLD(\mathcal{L}'_{K'}, \omega'_{K'}, A)$ is called by $WORLD(\mathcal{L}_K, \omega_K, A)$ in the $[\triangle]$-segment or the finite sequence $(\mathcal{L}'_0,\omega'_0), \ldots, (\mathcal{L}'_{K'},\omega'_{K'})$ is a prefix of the finite sequence $(\mathcal{L}_0,\omega_0), \ldots, (\mathcal{L}_K,\omega_K)$, the call $WORLD(\mathcal{L}_K, \omega_K, A)$ enters in the $[\triangle]$-segment and there is a $\{[\equiv], [R]\}$-free formula B of LER such that $[\triangle]B \in Cl(\omega_K, A)$, $[\triangle]B \notin T$ and T' is a $(\triangle, B)$-witness of T;
- For all $(T,\omega), (T',\omega') \in W$, $(T,\omega) \frown (T',\omega')$ iff there is a finite sequence $(\mathcal{L}_0,\omega_0), \ldots, (\mathcal{L}_K,\omega_K)$ such that:
  - $\mathcal{L}_0 = S$ and $\omega_0 = \epsilon$;
  - $last(\mathcal{L}_K) = T$ and $\omega_K = \omega$;
  - For all positive integers k, if $k \leq K$ then $WORLD(\mathcal{L}_k, \omega_k, A)$ is successfully called in $WORLD(S, \epsilon, A)$ at some depth of the recursion;
  - For all positive integers k, if $k < K$ then $WORLD(\mathcal{L}_{k+1}, \omega_{k+1}, A)$ is called by $WORLD(\mathcal{L}_k, \omega_k, A)$;

  and there is a finite sequence $(\mathcal{L}'_0,\omega'_0), \ldots, (\mathcal{L}'_{K'},\omega'_{K'})$ such that:
  - $\mathcal{L}'_0 = S$ and $\omega'_0 = \epsilon$;
  - $last(\mathcal{L}'_{K'}) = T'$ and $\omega'_{K'} = \omega'$;
  - For all positive integers k, if $k \leq K'$ then $WORLD(\mathcal{L}'_k, \omega'_k, A)$ is successfully called in $WORLD(S, \epsilon, A)$ at some depth of the recursion;
  - For all positive integers k, if $k < K'$ then $WORLD(\mathcal{L}'_{k+1}, \omega'_{k+1}, A)$ is called by $WORLD(\mathcal{L}'_k, \omega'_k, A)$;

  and either $WORLD(\mathcal{L}'_{K'}, \omega'_{K'}, A)$ is called in $WORLD(\mathcal{L}_K, \omega_K, A)$ in the $[\nabla]$-segment or the finite sequence $(\mathcal{L}'_0,\omega'_0), \ldots, (\mathcal{L}'_{K'},\omega'_{K'})$ is a prefix of the finite sequence $(\mathcal{L}_0,\omega_0), \ldots, (\mathcal{L}_K,\omega_K)$, the call $WORLD(\mathcal{L}_K, \omega_K, A)$ enters in the $[\nabla]$-segment and there is a $\{[\equiv], [R]\}$-free formula B of LER such that $[\nabla]B \in Cl(\omega_K, A)$, $[\nabla]B \notin T$ and T' is a (∇, B)-witness of T;
- For all propositional variables p, $V(p)$ is the set of all $(T,\omega) \in W$ such that $p \in T$.

Now the definition of M can be completed. Firstly define:

- $\preceq_0$ is $\ll \cup \gg^{-1}$;
- $\succeq_0$ is $\gg \cup \ll^{-1}$;
- Δ_0 is $\smile \cup \smile^{-1}$;
- ∇_0 is $\frown \cup \frown^{-1}$.

Secondly define for all positive integers k:

- $\preceq_{k+1}$ is $(\preceq_k \circ \preceq_k) \cup (\Delta_k \circ \nabla_k)$;
- $\succeq_{k+1}$ is $(\succeq_k \circ \succeq_k) \cup (\nabla_k \circ \Delta_k)$;
- Δ_{k+1} is $(\preceq_k \circ \Delta_k) \cup (\Delta_k \circ \succeq_k)$;
- ∇_{k+1} is $(\succeq_k \circ \nabla_k) \cup (\nabla_k \circ \preceq_k)$.

Thirdly define:

- $\preceq$ is $\bigcup\{\preceq_k$: k is a positive integer$\}$;
- $\succeq$ is $\bigcup\{\succeq_k$: k is a positive integer$\}$;
- Δ is $\bigcup\{\Delta_k$: k is a positive integer$\}$;
- ∇ is $\bigcup\{\nabla_k$: k is a positive integer$\}$.

It is a simple matter to check that M is a model of LER. What is more one can establish by induction on the positive integer k the remarkable result that for all $\{[\equiv], [R]\}$-free formulas B of LER and for all $(T, \omega), (T', \omega') \in W$:

- If $[\preceq]B \in T$ and $(T, \omega) \preceq_k (T', \omega')$ then $[\preceq]B \in T'$;
- If $[\succeq]B \in T$ and $(T, \omega) \succeq_k (T', \omega')$ then $[\succeq]B \in T'$;
- If $[\Delta]B \in T$ and $(T, \omega) \preceq_k (T', \omega')$ then $[\Delta]B \in T'$;
- If $[\nabla]B \in T$ and $(T, \omega) \succeq_k (T', \omega')$ then $[\nabla]B \in T'$;
- If $[\Delta]B \in T$ and $(T, \omega) \Delta_k (T', \omega')$ then $[\succeq]B \in T'$;
- If $[\nabla]B \in T$ and $(T, \omega) \nabla_k (T', \omega')$ then $[\preceq]B \in T'$;
- If $[\preceq]B \in T$ and $(T, \omega) \Delta_k (T', \omega')$ then $[\nabla]B \in T'$;
- If $[\succeq]B \in T$ and $(T, \omega) \nabla_k (T', \omega')$ then $[\Delta]B \in T'$.

It follows immediately by induction on the *length* of $\{[\equiv], [R]\}$-free formula B of LER that for all $(T, \omega) \in W$, if $B \in Cl(\omega, A)$ then:

- $M, (T, \omega) \models B$ iff $B \in T$.

Hence we have that for all $\{[\equiv], [R]\}$-free formulas B of LER, if $B \in Cl(\epsilon, A)$ then:

- $M, (S, \epsilon) \models B$ iff $B \in S$.

(2 implies 1): The reader may easily prove by induction on the *depth* of string $\omega \in \{\preceq, \succeq, \Delta, \nabla\}^*$ that for all sets T of $\{[\equiv], [R]\}$-free formulas of LER, for all models $M = (W, \preceq, \succeq, \Delta, \nabla, V)$ of LER and for all possible worlds $x \in W$, if $T \subseteq Cl(\omega, A)$ and:

- For all $\{[\equiv], [R]\}$-free formulas B of LER, if $B \in Cl(\omega, A)$ then $M, x \models B$ iff $B \in T$;

then $WORLD(T, \omega, A)$ succeeds. These considerations prove that if there is a model $M = (W, \preceq, \succeq, \Delta, \nabla, V)$ of LER and there is a possible world $x \in W$ such that:

- For all $\{[\equiv],[R]\}$-free formulas B of LER, if $B \in Cl(\epsilon, A)$ then $M, x \models B$ iff $B \in S$;

then $WORLD(S, \epsilon, A)$ succeeds.

It follows immediately from lemma 6 that the nondeterministic algorithm $SAT(A)$ works correctly, i.e., $SAT(A)$ succeeds iff A is true at some possible world in some model of LER. Much more difficult than lemma 6 are the following important results.

Lemma 7. *For all $\{[\equiv],[R]\}$-free formulas A of LER, for all lists $\mathcal{L}$ of sets of $\{[\equiv],[R]\}$-free formulas and for all strings $\omega \in \{\preceq, \succeq, \triangle, \triangledown\}^\star$, if $WORLD(\mathcal{L}, \omega, A)$ is called in $WORLD(S, \epsilon, A)$ at some depth of the recursion then $length(\mathcal{L}) = \mathcal{O}(length(A)^2)$.*

Proof. Similar to the line of reasoning suggested by Demri [2] at pages 225–226.

Lemma 8. *For all $\{[\equiv],[R]\}$-free formulas A of LER, for all lists $\mathcal{L}$ of sets of $\{[\equiv],[R]\}$-free formulas and for all strings $\omega \in \{\preceq, \succeq, \triangle, \triangledown\}^\star$, if $WORLD(\mathcal{L}, \omega, A)$ is called in $WORLD(S, \epsilon, A)$ at some depth of the recursion then $depth(\omega) \leq length(A)$.*

Proof. Suppose that $depth(\omega) > length(A)$. It follows that there is a string $\omega' \in \{\preceq, \succeq, \triangle, \triangledown\}^\star$ such that $\omega = \omega' \preceq\succeq$ or $\omega = \omega' \preceq \triangledown$ or $\omega = \omega' \succeq\preceq$ or $\omega = \omega' \succeq \triangle$ or $\omega = \omega'\triangle \preceq$ or $\omega = \omega' \triangle \triangle$ or $\omega = \omega'\triangledown \succeq$ or $\omega = \omega' \triangledown \triangledown$. Suppose that $\omega = \omega' \preceq\succeq$. Hence $depth(\omega' \preceq) \geq length(A)$. It follows that $Cl(\omega' \preceq, A) = \emptyset$: a contradiction. Therefore $depth(\omega) \leq length(A)$. The same line of reasoning applies if $\omega = \omega' \preceq \triangledown$ or $\omega = \omega' \succeq\preceq$ or $\omega = \omega' \succeq \triangle$ or $\omega = \omega'\triangle \preceq$ or $\omega = \omega' \triangle \triangle$ or $\omega = \omega'\triangledown \succeq$ or $\omega = \omega' \triangledown \triangledown$.

By applying lemma 7 and lemma 8, we infer immediately that the nondeterministic algorithm $SAT(A)$ is polynomial-space bounded in $length(A)$. We can summarize the results proved above as follows for all $\{[\equiv],[R]\}$-free formulas A of LER:

- $SAT(A)$ succeeds iff A is true at some possible world in some model of LER;
- The total space to run $SAT(A)$ is polynomial in $length(A)$.

We get that the satisfiability problem for any given $\{[\equiv],[R]\}$-free formula A of LER is in $NPSPACE$ of $length(A)$. Seeing that $NPSPACE = PSPACE$, see Savitch [11], we therefore conclude the following.

Theorem 5. **($PSPACE$-completeness theorem for LER)** *The satisfiability problem for any given $\{[\equiv],[R]\}$-free formula A of LER is PSPACE-complete in $length(A)$.*

6 Conclusion

The key feature of this paper is the modal analysis of emptiness relations in property systems. The proof that the satisfiability problem for any given $\{[\equiv],[R]\}$-free formula A of LER is $PSPACE$-complete in $length(A)$ goes through a line of reasoning similar to that applied by Spaan [12] to investigate the inherent difficulty of the satisfiability problem for any given formula of $S4_t$. Let us note that our proof does not consider the full language of our modal logic, so it might not be suitable for studying the inherent difficulty of the satisfiability problem for any given formula of LER, a question that remains unsolved.

Previous modal analyses of informational relations in property systems have been given by Vakarelov [16] who has mainly considered the relations of informational inclusion — our emptiness relations $\preceq$ and $\succeq$ — and the relations of similarity — the complementary relations of our emptiness relations $\triangle$ and $\triangledown$. On that occasion, Vakarelov [16] gave a completeness theorem and a decidability theorem for a modal logic which modalities correspond to the relations of informational inclusion and the relations of similarity in property systems. Modal analyses of our emptiness relations in property systems together with other informational relations like similarity relations are not known.

References

1. Balbiani, P., Vakarelov, D.: First-order characterization and modal analysis of indiscernibility and complementarity in information systems. In Benferhat, S., Besnard, P. (Editors): Symbolic and Quantitative Approaches to Reasoning with Uncertainty. Springer-Verlag, Lecture Notes in Artificial Intelligence **2143** (2001) 772–781.
2. Demri, S.: The nondeterministic information logic NIL is PSPACE-complete. Fundamenta Informaticæ**42** (2000) 211–234.
3. Demri, S., Orłowska, E., Vakarelov, D.: Indiscernibility and complementarity relations in information systems. In Gerbrandy, J., Marx, M., de Rijke, M., Venema, Y. (Editors): JFAK: Essays Dedicated to Johan van Benthem on the Occasion of his 50th Birthday. Amsterdam University Press (1999) http://turing.wins.uva.nl/~j50/cdrom/contribs/demri/index.html.
4. Hughes, G., Cresswell, M.: A Companion to Modal Logic. Methuen (1984).
5. Ladner, R.: The computational complexity of provability in systems of modal propositional logic. SIAM Journal on Computing **6** (1977) 467–480.
6. Orłowska, E.: Logic of nondeterministic information. Studia Logica **44** (1985) 91–100.
7. Orłowska, E.: Kripke semantics for knowledge representation logics. Studia Logica **49** (1990) 255–272.
8. Orłowska, E. (Editor): Incomplete Information: Rough Set Analysis. Physica-Verlag, Studies in Fuzziness and Soft Computing **13** (1998).
9. Orłowska, E., Pawlak, Z.: Representation of nondeterministic information. Theoretical Computer Science **29** (1984) 27–39.
10. Pawlak, Z.: Information systems — theoretical foundations. Information Systems **6** (1981) 205–218.
11. Savitch, W.: Relationships between nondeterministic and deterministic tape complexities. Journal of Computer and System Sciences **4** (1970) 177–192.
12. Spaan, E.: The complexity of propositional tense logics. In de Rijke, M. (Editor): Diamonds and Defaults: Studies in Pure and Applied Intensional Logic. Kluwer, Synthese Library **229** (1993) 287–307.

13. Vakarelov, D.: A modal logic for similarity relations in Pawlak knowledge representation systems. Fundamenta Informaticæ**15** (1991) 61–79.
14. Vakarelov, D.: Modal logics for knowledge representation systems. Theoretical Computer Science **90** (1991) 433–456.
15. Vakarelov, D.: A duality between Pawlak's knowledge representation systems and BI-consequence systems. Studia Logica **55** (1995) 205–228.
16. Vakarelov, D.: Information systems, similarity relations and modal logics. In Orłowska, E. (Editor): Incomplete Information: Rough Set Analysis. Physica-Verlag, Studies in Fuzziness and Soft Computing **13** (1998) 492–550.

Pregroups: Models and Grammars

Wojciech Buszkowski

Faculty of Mathematics and Computer Science, Adam Mickiewicz University, Poznań, Poland
buszko@amu.edu.pl

Abstract. Pregroups, introduced in Lambek [12], are a generalization of partially ordered groups. In [5], we have proven several theorems on pregroups and grammars based on the calculus of free pregroups, in particular, the weak equivalence of these grammars and context-free grammars. In the present paper, we obtain further results of that kind. We consider left and right pregroups, study concrete left and right pregroups consisting of monotone functions on a poset and of monotone relations on a poset, and adjust the equivalence theorem to grammars based on left (right) pregroups.

1 Pregroups

A structure $(G, \leq, \cdot, 1)$ is called *a partially ordered monoid* (p.o. monoid), if $(G, \cdot, 1)$ is a monoid (i.e. a semigroup with unit 1), $(G, \leq)$ is a poset (i.e. a partially ordered set), and the following monotonicity condition holds:

$$\text{(MON) if } a \leq b \text{ then } ca \leq cb \text{ and } ac \leq bc,$$

for all $a, b, c \in G$.

A pregroup is defined as a structure $(G, \leq, \cdot, l, r, 1)$ such that $(G, \leq, \cdot, 1)$ is a p.o. monoid, and l, r are unary operations on G, satisfying the following inequalities:

$$a^l a \leq 1 \leq a a^l \text{ and } a a^r \leq 1 \leq a^r a \ , \tag{1}$$

for all $a \in G$. The element a^l (resp. a^r) is called *the left* (resp. *right*) *adjoint* of a.

Standard examples of pregroups are partially ordered groups (p.o. groups), i.e. structures $(G, \leq, \cdot, ^{\cup}, 1)$ such that $(G, \leq, \cdot, 1)$ is a p.o. monoid, and $^{\cup}$ is a unary operation on G, satisfying the equalities: $a^{\cup} a = 1 = a a^{\cup}$, for all $a \in G$. An extensive exposition of the theory of p.o. groups is given in Fuchs [8]. Clearly, in a p.o. group one can define $a^l = a^r = a^{\cup}$.

A pregroup G is said to be *commutative*, if $ab = ba$, for all $a, b \in G$. If G is commutative, then $a^l = a^r$, for all $a \in G$, and G is, actually, a p.o. group. Accordingly, the notion of a pregroup is a generalization of that of a p.o. group which can be innovatory for noncommutative structures only. A pregroup is said to be *proper*, if it is not a p.o. group (that means, $a^l \neq a^r$, for some element a).

H. de Swart (Ed.): RelMiCS 2001, LNCS 2561, pp. 35–49, 2002.

Pregroups are related to residuated monoids, abstract models for the Lambek calculus with possibly empty antecedents [11,4]. *A residuated monoid* is a structure $(G, \leq, \cdot, \backslash, /, 1)$ such that $(G, \leq, \cdot, 1)$ is a p.o. monoid, and $\backslash, /$ are binary operations on G, fulfilling the equivalences:

$$ab \leq c \text{ iff } b \leq a\backslash c \text{ iff } a \leq c/b \ , \tag{2}$$

for all $a, b, c \in G$. (Actually, (MON) follows from (2).) For a pregroup G, one defines $a\backslash b = a^r b$ and $a/b = ab^l$, which yields a residuated monoid. One can also define $a^l = 1/a$ and $a^r = a\backslash 1$. The latter definitions of adjoints, applied to an arbitrary residuated monoid, lead to inequalities $a^l a \leq 1$ and $aa^r \leq 1$, but the inequalities $1 \leq aa^l$ and $1 \leq a^r a$ need not hold.

The Lambek calculus with possibly empty antecedents is a formal system whose theorems are the inequalities valid in residuated monoids. Consequently, all theorems of the Lambek calculus are also valid in pregroups under the above interpretation of residuals $a\backslash b$ and a/b. The converse is not true even for product-free types [5]. Grammars based on the Lambek calculus are a kind of categorial grammars, extensively studied nowadays as a framework for deductive parsing (see [14,15,3]). Lambek [12] proposes alternative parsing strategies, based on the calculus of free pregroups (also see [2,7]). We discuss these matters in section 3.

Using (1), one easily proves the following equalities hold in every pregroup:

$$1^l = 1^r = 1,\ a^{lr} = a^{rl} = a,\ (ab)^l = b^l a^l,\ (ab)^r = b^r a^r \ . \tag{3}$$

By (1), we also derive:

$$aa^l a = a,\ aa^r a = a \ , \tag{4}$$

and (MON) with (1) yield:

$$\text{(MON')}\ a \leq b \text{ iff } b^l \leq a^l \text{ iff } b^r \leq a^r.$$

By $\mathbf{Z}$ we denote the set of integers. Following [12], for any pregroup G, $a \in G$, and $n \in \mathbf{Z}$, we define the element $a^{(n)} \in G$: $a^{(0)} = a$, $a^{(n+1)} = (a^{(n)})^r$, for $n \geq 0$, and $a^{(n-1)} = (a^{(n)})^l$, for $n \leq 0$. Using (3), one shows the latter equalities hold for all $n \in \mathbf{Z}$. Also:

$$a^{(n)}a^{(n+1)} \leq 1 \leq a^{(n+1)}a^{(n)},\ \text{for all } n \in \mathbf{Z} \ , \tag{5}$$

$$(ab)^{(n)} = a^{(n)}b^{(n)},\ \text{and if } a \leq b \text{ then } a^{(n)} \leq b^{(n)}\ (n \text{ is even}) \ , \tag{6}$$

$$(ab)^{(n)} = b^{(n)}a^{(n)},\ \text{and if } a \leq b \text{ then } b^{(n)} \leq a^{(n)}\ (n \text{ is odd}) \ . \tag{7}$$

Proposition 1. *Let G be a p.o. monoid. Then, for any $a \in G$, there exist at most one $b \in G$ such that $ba \leq 1 \leq ab$ and at most one $c \in G$ such that $ac \leq 1 \leq ca$.*

Proof. Assume $ba \leq 1 \leq ab$ and $b'a \leq 1 \leq ab'$. Then $b \leq bab' \leq b'$ and $b' \leq b'ab \leq b$, hence $b = b'$. □

As a consequence, a^l is the only element b such that $ba \leq 1 \leq ab$, and a^r is the only element c such that $ac \leq 1 \leq ca$ (this yields an easy proof for (3)). Further, $()^l, ()^r, ()^{(n)}$ can be regarded as partial operations on arbitrary p.o. monoids, and they satisfy (1), (MON'), (3), (4), (5), (6), (7), if the relevant elements exist; for instance, if $a^{(n)}, b^{(n)}$ exist, then $(ab)^{(n)}$ exists and the equation from (6) holds. In a pregroup, right and left adjoints are uniquely determined by the underlying p.o. monoid. Accordingly, we can also define pregroups as p.o. monoids in which every element admits the two adjoints (then, the signature of pregroups equals that of p.o. monoids, and we may say, for example: a given pregroup is a substructure of a given p.o. monoid).

Proposition 2. *For every p.o. monoid G, there exists a (unique) largest pregroup which is a substructure of G.*

Proof. Let H consist of all $a \in G$ such that $a^{(n)}$ exists, for all $n \in \mathbf{Z}$. By (6), (7), H is a submonoid of G. By the definition, H admits adjoints. Every pregroup H' which is a substructure of G must be contained in H. □

The pregroup H, defined above, will be denoted by $\mathrm{Pr}(G)$.

Let $(X, \leq)$ be a poset. A function $f : X \mapsto X$ is said to be *monotone*, if $x \leq y$ entails $f(x) \leq f(y)$. By $F(X, \leq)$ we denote the p.o. monoid of all monotone functions $f : X \mapsto X$ with the identity function I being the unit, the function composition being the operation $\cdot$, and the ordering being defined by: $f \leq g$ iff $f(x) \leq g(x)$, for all $x \in X$.

Proposition 3. *Every pregroup $(G, \leq, \cdot, l, r, 1)$ is embeddable into $Pr(F(G, \leq))$.*

Proof. For $a \in G$, the function $f_a : G \mapsto G$ is defined as follows: $f_a(x) = ax$. By (MON), f_a is monotone. We have: $f_{ab} = f_a \circ f_b$, $f_1 = I$, and $a \leq b$ iff $f_a \leq f_b$. Consequently, the mapping $h(a) = f_a$ is a monomorphism of p.o. monoids. Since $f_{a^l} \circ f_a \leq I \leq f_a \circ f_{a^l}$ and $f_a \circ f_{a^r} \leq I \leq f_{a^r} \circ f_a$, then $f_{a^l} = f_a^l$ and $f_{a^r} = f_a^r$, hence h is a monomorphism of pregroups. □

Consequently, every pregroup is isomorphic to a pregroup of monotone functions on a poset.

If $(X, \leq)$ is a poset and $Y \subseteq X$, then $\max Y$ and $\min Y$ will denote the greatest and the least, respectively, element of Y (thus, $\max Y$ is the l.u.b. of Y belongimg to Y, and $\min Y$ is the g.l.b. of Y belonging to Y).

Proposition 4. *For every function $f \in Pr(F(X, \leq))$, there hold the equalities:*
$f^l(x) = min\{y \in X : x \leq f(y)\}$, for all $x \in X$,
$f^r(x) = max\{y \in X : f(y) \leq x\}$, for all $x \in X$.

Further, if M is a submonoid of $F(X, \leq)$ and, for every $f \in M$, the functions f^l, f^r defined by these equations exist and belong to M, then M is a pregroup.

Proof. The first part is easy (see [5]). We prove the second part. Let M satisfy the condition. We prove (1). Since $f(x) \leq f(x)$, then $f^l(f(x)) \leq x$ (so, $f^l \circ f \leq I$). Also $x \leq f(f^l(x))$, since $f^l(x)$ is a y such that $x \leq f(y)$. □

These equalities can equivalently be expressed by the following Galois connections:

$$f^l(x) \leq y \text{ iff } x \leq f(y) \ , \tag{8}$$

$$y \leq f^r(x) \text{ iff } f(y) \leq x \ , \tag{9}$$

for all $x, y \in X$.

As a consequence, all functions in $\mathrm{Pr}(F(X,\leq))$ must be upward and downward unbounded, that means:

$$\forall x \exists y (x \leq f(y)),\ \forall x \exists y (f(y) \leq x) \ , \tag{10}$$

for all $x, y \in X$.

2 Left and Right Pregroups

A left (resp. *right*) *pregroup* is a p.o. monoid in which every element admits a left (resp. right) adjoint, that means, for every element a there exists an element a^l (resp. a^r) such that the left (resp. right) conjunct of (1) holds. Right pregroups are dual to left pregroups: $(G, \leq, \cdot, 1)$ is a left pregroup iff $(G, \geq, \cdot, 1)$ is a right pregroup. A pregroup is a p.o. monoid which is both a left pregroup and a right pregroup.

Everything which has been said above on pregroups can also be said on left (right) pregroups provided one confine herself to left (right) adjoints. In particular, every p.o. monoid G contains a largest left (resp. right) pregroup to be denoted by $\mathrm{LPr}(G)$ (resp. $\mathrm{RPr}(G)$); its elements are all $a \in G$ such that $a^{(n)}$ exists, for all $n \leq 0$ (resp. $n \geq 0$). Clearly, $\mathrm{Pr}(G)$ equals the meet of $\mathrm{LPr}(G)$ and $\mathrm{RPr}(G)$.

If we prove later on that no left (right) pregroup having a given property exists, then it follows that no pregroup having this property exists.

Of course, some arguments sound for pregroups need not be sound for left (right) pregroups. For example, to prove (MON') for pregroups one can prove: $a \leq b$ entails $b^l \leq a^l$ and $b^r \leq a^r$, and then, substitute b^l for a and a^l for b, and use the second equalities of (3) to obtain: $b^l \leq a^l$ entails $a \leq b$. For left pregoups, the latter entailment can be shown in another way. Assume $b^l \leq a^l$. Then:

$$a \leq bb^l a \leq ba^l a \leq b.$$

By an analogue of proposition 4, all functions in $\mathrm{LPr}(F(X,\leq))$ must be upward unbounded, and all functions in $\mathrm{RPr}(F(X,\leq))$ must be downward unbounded. These claims are dual to each other, since $\mathrm{LPr}(F(X,\leq))$ equals $\mathrm{RPr}(F(X,\geq))$.

Proposition 5. *Let a be the least (resp. greatest) element of poset $(X, \leq)$. Then, for every function $f \in RPr(F(X,\leq))$ (resp. $LPr(F(X,\leq))$) there holds: $f(x) = a$ iff $x = a$, for all $x \in X$.*

Proof. Let $f \in \mathrm{RPr}(F(X,\leq))$. Then, $f(a) = a$, since f is downward unbounded. Assume $f(x) = a$. Then $a = f^r(a) \geq x$, by proposition 4, hence $x = a$. □

In [5], it has been shown that, for any totally ordered set $(X, \leq)$ containing more than two elements, $\mathrm{Pr}(F(X, \leq))$ consists of all both upward and downward unbounded, monotone functions $f : X \mapsto X$ if and only if $(X, \leq)$ is isomorphic to the set of integers with the natural ordering. We prove a left (right) analogue of this theorem.

Theorem 1. *Let $(X, \leq)$ be a totally ordered set containing at least two elements. Then, $LPr(F(X, \leq))$ consists of all upward unbounded, monotone functions $f : X \mapsto X$ if and only if $(X, \leq)$ is a well ordered set without the greatest element. Dually, $RPr(F(X, \leq))$ consists of all downward unbounded, monotone functions $f : X \mapsto X$ if and only if $(X, \geq)$ is a well ordered set without the greatest element.*

Proof. Assume $(X, \leq)$ be a nonempty, well ordered set without the greatest element. Let M denote the set of all upward unbounded, monotone functions $f : X \mapsto X$. M is closed under function composition and $I \in M$, hence M is a submonoid of $F(X, \leq)$. Further, for any $f \in M$, the function f^l defined by the equation from proposition 4 exists and belongs to M (Caution: the nonexistence of the greatest element is needed to show that F^l is upward unbounded.) So, M is a left pregroup. But all functions in $\mathrm{LPr}(F(X, \leq))$ are upward unbounded, hence M equals the latter.

Now, let $(X, \leq)$ be a totally ordered set containing at least two elements. Assume $\mathrm{LPr}(F(X, \leq))$ consist of all upward unbounded, monotone functions $f : X \mapsto X$.

Suppose a to be the greatest element of X. Define $f(x) = a$, for all $x \in a$. Clearly, f is upward unbounded and monotone, hence $f \in \mathrm{LPr}(F(X, \leq))$, which contradicts proposition 5.

Suppose $(X, \leq)$ not to be well ordered. Then, there exists an infinite sequence $a_n \in X$, $n \geq 0$, such that $a_{n+1} < a_n$, for all n. Let Y denote the set of all $y \in X$ such that $a_{n+1} \leq y \leq a_n$, for some n. Define $f(x) = a_0$, for $x \in Y$, and $f(x) = x$, for $x \notin Y$. Clearly, f is upward unbounded and monotone. But $f^l(a_0) = \min(Y)$ does not exist. Contradiction. □

By ω we denote the set of nonnegative integers. It follows from theorem 1 that $\mathrm{LPr}(F(\omega, \leq))$, where $\leq$ is the natural ordering, contains all upward unbounded, monotone functions from ω into ω. It is easy to see that this left pregroup is proper (that means, it is not a p.o. group). Below we show how to do that.

An element a of a left (resp. right) pregroup is said to be *injective*, if $a^l a = 1$ (resp. $a^r a = 1$), and *surjective*, if $aa^l = 1$ (resp. $aa^r = 1$); it is said to be *bijective*, if it is both injective and surjective. In a left pregroup of functions, a function f is injective (resp. surjective) iff it is an injective (resp. a surjective) mapping. We prove this fact. Assume f to be injective. Then, $f^l \circ f = I$, so f is an injective mapping, since I is so. In a similar way, we show that, if f is surjective, then f is a surjective mapping. Assume f be an injective mapping. By (4), $f \circ f^l \circ f = f = f \circ I$, and consequently, $f^l \circ f = I$, so f is injective. In a similar way, we show that, if f is a surjective mapping, then f is surjective. The same holds true for right pregroups of functions.

Accordingly, in a left (right) pregroup of functions, a function f is bijective iff f is a bijective mapping. A left (right) pregroup is proper iff it contains at least one nonbijective element. Thus, $\mathrm{LPr}(F(\omega, \leq))$ is proper, since it contains nonbijective mappings, e.g. $f(x) = 2x$.

Proposition 6. *For any left pregroup G and all $a, b \in G$, the following conditions hold true:*

(i) a is injective (resp. surjective) iff a^l is surjective (resp. injective),

(ii) if ab is injective (resp. surjective), then b is injective (resp. a is surjective).

Proof. If $a^l a = 1$, then $a^l a^{ll} = 1$, and if $aa^l = 1$, then $a^{ll} a^l = 1$, by (3). The converse conditionals also hold, by (3) and (MON'). This proves (i). If $(ab)^l ab = 1$, then

$$1 = b^l a^l ab \leq b^l b \leq 1,$$

hence $b^l b = 1$, and if $ab(ab)^l = 1$, then

$$1 = abb^l a^l \geq aa^l \geq 1,$$

hence $aa^l = 1$, which proves (ii). □

An analogous fact is true for right pregroups.

Proposition 7. *If all elements of a left (right) pregroup are injective or surjective, then all elements are bijective.*

Proof. Assume that all elements of a left pregroup G be injective or surjective. Let $a \in G$. We prove that a is injective. Suppose not. Then, $a^l a$ is not injective, hence $a^l a$ is surjective, and consequently, a^l is surjective. So, a is injective. Contradiction. In a similar way, one proves that a is surjective. □

A left (right) pregroup is said to be *strictly proper*, if 1 is its only element which is injective or surjective. Every left (right) pregroup G contains a (unique) largest strictly proper left (right) pregroup G' which consists of all elements a of the former such that $a = 1$ or a is neither injective, nor surjective (by proposition 6, this set is closed under $\cdot$ and adjoints). If G is a pregroup, then G' is also a pregroup. By proposition 7, G is proper iff $G' \neq \{1\}$.

Theorem 2. *No totally ordered left (right) pregroup is proper.*

Proof. We show: $a^l a = 1$ iff $a \geq a^{ll}$. First, $a^l a = 1$ iff $a^l a \geq 1$, by (1). If $a^l a \geq 1$, then $a^{ll} \leq a^{ll} a^l a \leq a$.If $a \geq a^{ll}$, then $a^l a \geq a^l a^{ll} \geq 1$. In a similar way, one shows: $aa^l = 1$ iff $a \leq a^{ll}$.

In a totally ordered left pregroup, $a \leq a^{ll}$ or $a^{ll} \leq a$, for every element a. Consequently, all elements are injective or surjective, and we use proposition 7. By duality, the same holds for right pregroups. □

It is known [8] that there are no finite p.o. groups in which $a < b$ holds, for some elements a, b. For $a < b$ entails $1 < ba^{\cup}$, and $1 < c$ entails $1 < c < cc < ccc < \ldots$. This argument cannot be applied for pregroups. We nonetheless prove:

Theorem 3. *If G is a finite left (right) pregroup, then the ordering in G is the identity relation (hence, G is a group).*

Proof. Let $(G, \leq, \cdot, l, 1)$ be a finite left pregroup. The inequality $xa \leq b$ has a solution ($x = ba^l$), for all $a, b \in G$. For $a \in G$, define a mapping $f_a(x) = xa$. Let M denote the set of minimal elements of G. For any $b \in M$, there is $x \in G$ such that $f_a(x) = b$, hence there is $x \in M$ such that $f_a(x) = b$. Thus, f_a maps a subset of M onto M. Since M is finite, this subset equals M, and f_a restricted to M is a bijection of M onto itself. Consequently, $xa \in M$, for every $x \in M$.

The inequality $b \leq ax$ has a solution ($x = a^l b$), for all $a, b \in G$. For $a \in G$, define a mapping $g_a(x) = ax$. Let N denote the set of maximal elements of G. For any $b \in N$, there is $x \in G$ such that $g_a(x) = b$, hence there is $x \in N$ such that $g_a(x) = b$. As above, one shows that g_a restricted to N is a bijection of N onto itself. Consequently, $ax \in N$, for every $x \in N$.

Let $a \in N$. By the first paragraph, all minimal elements are of the form xa, for $x \in M$. By the second paragraph, these elements are maximal. We have shown that all minimal elements are maximal. Thus, the ordering is the identity relation. □

As a consequence, no finite left (right) pregroup can be proper. Below we obtain some results on the nonexistence of proper left or right pregroups of monotone functions on a poset.

Lemma 1. *Let $(X, \leq)$ be a poset, and $Y, Z \subseteq X$ be such that $X = Y \cup Z$, $Y \cap Z = \emptyset$. Let G be a left (right) pregroup contained in $F(X, \leq)$ such that, for all $f \in G$, $f[Y] \subseteq Y$ and $f[Z] \subseteq Z$. Let G_Y consist of all functions $f \in G$ restricted to Y, and G_Z consist of all $f \in G$ restricted to Z. Then, both G_Y and G_Z are left (right) pregroups contained in $F(Y, \leq)$ and $F(X, \leq)$, respectively (here, $\leq$ is the restricted ordering).*

Proof. Clearly, both G_Y and G_Z are closed under composition. Since $I \in G$, then $I_Y \in G_Y$ and $I_Z \in G_Z$ (here, I_Y is the identity function restricted to Y, and similarly for I_Z). If $f \in G$, then $f^l \in G$, and consequently, f^l restricted to Y belongs to G_Y. The condition $f^l(f(x)) \leq x \leq f(f^l(x)$, for all $x \in X$, holds true in G, hence it also holds true in G_Y and G_Z for f and f^l restricted to Y and Z, respectively. So, G_Y and G_Z admit adjoints. □

The direct product $G \times H$ of left (right) pregroups G, H is defined in the usual way. If G, H are left (right) pregroups, then $G \times H$ is a left (right) pregroup. In lemma 1, G is embeddable into $G_Y \times G_Z$. G is isomorphic to $G_Y \times G_Z$, if either $y \leq z$, for all $y \in Y$, $z \in Z$, or neither $y \leq z$, nor $z \leq y$. for all $y \in Y$, $z \in Z$.

A poset $(X, \leq)$ is called *a tree*, if $\{y \in X : y \leq x\}$ is well ordered, for all $x \in X$. It is said to be *well founded*, if every nonempty subset of X has a minimal element. Every tree is well founded.

Theorem 4. *If $(X, \leq)$ is well founded, then no right pregroup contained in $F(X, \leq)$ is proper.*

Proof. We fix a well founded poset $(X, \leq)$ and a right pregroup G contained in $F(X, \leq)$. It is sufficient to show that all functions in G are bijective.

We define a transfinite sequence of sets M_α. M_0 is the set of all minimal elements in $(X, \leq)$. For $\alpha > 0$, if $\bigcup_{\beta<\alpha} M_\beta$ is different from X, then M_α is the set of all minimal elements in $X - \bigcup_{\beta<\alpha} M_\beta$; otherwise our construction stops.

We show that, for all $f \in G$, f restricted to M_0 is a bijection of M_0 onto itself, and f restricted to $X - M_0$ maps $X - M_0$ into itself. First, we show $M_0 \subseteq f[M_0]$. Let $x \in M_0$. Since f is downward unbounded, then $f(y) = x$, for some $y \in X$. We take $z \in M_0$ such that $z \leq y$. Then, $f(z) \leq f(y)$, so $f(z) = x$. Second, we show $f[M_0] \subseteq M_0$. Let $x \in M_0$. Since f^r is downward unbounded, then we find $y \in M_0$ such that $f^r(y) = x$. Then, $f(x) = f(f^r(y)) \leq y$, so $f(x) \in M_0$. Third, we show $f[X - M_0] \subseteq X - M_0$. Assume $f(x) \in M_0$. Since $f^r[M_0] \subseteq M_0$ and $x \leq f^r(f(x))$, then $x \in M_0$. Finally, every $f \in G$ restricted to M_0 is one-to-one. Assume $f(x) = f(y)$, for $x, y \in M_0$. Denote $a = f^r(f(x))$. By the above, $a \in M_0$, and $x \leq a$, $y \leq a$, hence $x = y$.

Now, we can precisely formulate our inductive claim: for any α such that M_α is defined and for all $f \in G$, f restricted to M_α is a bijection of M_α onto itself, and f restricted to $X - \bigcup_{\beta \leq \alpha} M_\beta$ maps this set into itself. Assume this claim hold, for all $\gamma < \alpha$. Denote $Y = \bigcup_{\gamma < \alpha} M_\gamma$, $Z = X - Y$. Then, all functions $f \in G$ map bijectively Y onto Y and map Z into Z. By lemma 1, G_Z is a right pregroup contained in $F(Z, \leq)$. Since $(Z, \leq)$ is well founded, we can repeat the above arguments with G_Z in the place of G. Then, M_α takes the place of M_0. Consequently, all functions f in G_Z map bijectively M_α onto itself and restricted to $Z - M_\alpha$ map this set into itself. This proves our inductive claim.

Consequently, every $f \in G$ is a join of bijections on pairwise disjoint sets, hence it is a bijection. □

Actually, we can say more. Let f_α denote the restriction of f to M_α. For all $f, g \in G$, $f \leq g$ iff $f_\alpha \leq g_\alpha$, for all α, iff $f_\alpha = g_\alpha$, for all α, iff $f = g$. Consequently, every right pregroup of monotone functions on a well founded poset is a p.o. group whose ordering relation is the identity. Further, every well founded (right) pregroup is a p.o. group with the identity ordering (use theorem 4 and proposition 3), which yields a generalization of theorem 3.

RPr$(F(\omega, \leq))$ contains the only element I, since I is the only monotone bijection on this tree; evidently, the largest pregroup contained in $F(\omega, \leq)$ is the same structure. Thus, the largest (left, right) pregroups contained in a p.o. monoid need not include all injective or surjective elements of the p.o. monoid. For the full binary tree $\{0,1\}^*$, the largest right pregroup of monotone functions equals the group of all monotone bijections; one of its elements is the function: $f(\epsilon) = \epsilon$, $f(0x) = 1x$, $f(1x) = 0x$, for $x \in \{0,1\}^*$.

Let $(X, \leq)$ be a totally ordered set. A function $f : X \mapsto X$ is said to be *downward continuous* (resp. *upward continuous*), if, for all $x \in X$, $f(x) =$l.u.b.$\{f(y) : y < x\}$ (resp. $f(x) =$g.l.b.$\{f(y) : x < y\}$). The set $(X, \leq)$ is said to be *dense*, if, for all $x, y \in X$ such that $x < y$ there exists $z \in X$ such that $x < z < y$.

Lemma 2. *If $(X, \leq)$ is a dense, totally ordered set, then LPr$(F(X, \leq))$ (resp. RPr$(F(X, \leq))$) contains upward (resp. downward) continuous functions only.*

Proof. If $x < y$, then $f(x) \leq f(y)$, so $f(x)$ is a lower bound of $\{f(y) : x < y\}$. Assume $f(x)$ be not the g.l.b. of this set. Then, there exists z such that $f(x) < z$ and z is a lower bound of this set. The element $f^l(z)$ exists and equals $\min\{y : z \leq f(y)\}$. But $z \leq f(y)$ iff $x < y$, and the set $\{y : x < y\}$ has no least element. Contradiction. □

Theorem 5. *If $(X, \leq)$ is a dense, totally ordered set, then no left (right) pregroup contained in $F(X, \leq)$ is proper.*

Proof. Let G be a left pregroup contained in $F(X, \leq)$, and let $f \in G$. We show that f is injective. Assume $x < y$. Suppose $f(x) = f(y)$. Denote $a = f(y)$. By proposition 5, a is not the greatest element of X. We have $f^l(a) = f^l(f(x)) \leq x$, but, for all $z > a$, $f^l(z) = \min\{u : z \leq f(u)\} > y$. Thus, f^l is not upward continuous, which contradicts lemma 2. Accordingly, all functions in G are injective, and we use proposition 7. □

A poset $(X, \leq)$ is called *a lattice*, if, for all $a, b \in X$, there exist the g.l.b. and the l.u.b. of $\{a, b\}$; these elements will be denoted by $a \wedge b$ and $a \vee b$, respectively. A (left, right) pregroup is said to be *lattice ordered* (l.o.), if its underlying poset is a lattice. It is easy to show that in every l.o. pregroup the following equalities are valid:

$$(a \vee b)^l = a^l \wedge b^l, \ (a \wedge b)^l = a^l \vee b^l \ , \tag{11}$$

$$(a \vee b)^r = a^r \wedge b^r, \ (a \wedge b)^r = a^r \vee b^r \ , \tag{12}$$

$$(a \vee b)c = ac \vee bc, \ a(b \vee c) = ab \vee ac, \ (a \wedge b)c = ac \wedge bc, \ a(b \wedge c) = ab \wedge ac \ . \tag{13}$$

This is not the case for l.o. left (right) pregroups. On the other hand, if G is a p.o. monoid whose underlying poset is a lattice such that (13) are valid in G, then LPr(G) (resp. RPr(G)) is a l.o. left (resp. right) pregroup satisfying (11) (resp. (12)). We omit the proof. Examples can be found as follows. Take a totally ordered set $(X, \leq)$. Then, $F(X, \leq)$ is a lattice with $(f \vee g)(x) = f(x) \vee g(x)$, $(f \wedge g)(x) = f(x) \wedge g(x)$, and it satisfies (13). So, the largest left (resp. right) pregroup contained in $F(X, \leq)$ is a l.o. left (resp. right) pregroup satisfying (11) (resp. (12)). Accordingly, there exist proper l.o. (left, right) pregroups, e.g. the largest pregroup contained in $F(\mathbf{Z}, \leq)$ and the largest left pregroup contained in $F(\omega, \leq)$. We show that pregroups cannot be bounded.

Proposition 8. *If a left (right) pregroup contains $\bot$ or $\top$, then it is trivial.*

Proof. For pregroups the proof is a bit simpler than for left pregroups. Let G be a pregroup containing the least element $\bot$. For all $a \in G$, we have $\bot \leq a$, so $a^l \leq \bot^l$, and consequently, $a = (a^r)^l \leq \bot^l$. So, $\bot^l = \top$ is the greatest element of G. We have $\bot\bot \leq \bot$ and $\top \leq \top\top$, hence $\bot\bot = \bot$ and $\top\top = \top$. Also $\top^l = \bot$. Then, $\bot\top = \bot\bot\top = \bot\top^l\top \leq \bot$, so $\bot\top = \bot$, and $\bot\top = \bot\top\top = \bot\bot^l\top \geq \top$, so $\bot\top = \top$. Consequently, $\bot = \top$, hence G is trivial.

Now, let G be a left pregroup containing $\bot$. Again $a^l \leq \bot^l$, for all $a \in G$, but now we cannot infer $a \leq \bot^l$, for all $a \in G$. As above, $\bot\bot = \bot$, hence, by (3), $\bot^l\bot^l = \bot^l$. We have $\bot \leq \bot^{ll}$, so $\bot$ is surjective (see the proof of theorem 2) and $\bot\bot^l = 1$. Then, $\bot^l = \bot\bot^l\bot^l = \bot\bot^l = 1$, hence $\bot = \bot\bot^l = 1$. Since $a^l \leq \bot^l$, for all $a \in G$, then $a^l \leq \bot$, so $a^l = \bot$, for all $a \in G$. Consequently, $a^l = b^l$, which yields $a = b$, for all $a, b \in G$ (use (MON')).

In both cases, if G contains the greatest element $\top$, then the reasoning is dual. □

3 Pregroups of Relations

It has been proven in [1,6] that every residuated monoid is embeddable into the residuated monoid of binary relations on some set X. The latter is defined as follows. For $R, S \subseteq X^2$, one defines:

$$R \circ S = \{(x,y) : \exists z((x,z) \in R \text{ and } (z,y) \in S)\},$$

$$R\backslash S = \{(x,y) : \forall z((z,x) \in R \Rightarrow (z,y) \in S)\},$$

$$S/R = \{(x,y) : \forall z((y,z) \in R \Rightarrow (x,z) \in S)\}.$$

$P(W)$ denotes the powerset of set W. Then, $(P(X^2), \subseteq, \circ, \backslash, /, I)$ is a residuated monoid: the residuated monoid of binary relations on X.

Algebras of relations are important for many areas of logic in computer science (logics and semantics of programs, knowledge representation, logics in AI) [16]. In [10], residuated monoids of binary relations are used to model the notions of a weak prespecification and postspecification of programs (the underlying logic is the Lambek calculus with possibly empty antecedents, not explicitly quoted in [10]).

In this section we examine algebras of binary relations as models of pregroups. First, we observe that in order to model proper pregroups we must modify the basic monoid of binary relations. The p.o. monoid $(P(X^2), \subseteq, \circ, I)$ contains no proper left (right) pregroup. Consider the condition (1):

$$R^l \circ R \subseteq I \subseteq R \circ R^l.$$

By $R^{\cup}$ we denote the converse of R. By the second inclusion of (1), X equals both the domain of R and the codomain of R^l. We prove $R^{\cup} \subseteq R^l$. Assume $(y,x) \in R^{\cup}$. Then $(x,y) \in R$. So, there is z such that $(z,x) \in R^l$. By the first inclusion of (1), $z = y$, hence $(y,x) \in R^l$. We prove $R^l \subseteq R^{\cup}$. Assume $(x,y) \in R^l$. Then, there is z such that $(y,z) \in R$. By the first inclusion of (1), $x = z$, hence $(y,x) \in R$ and $(x,y) \in R^{\cup}$. We have shown $R^l = R^{\cup}$. Thus, in every left pregroup contained in the standard p.o. monoid the left adjoint of R equals the converse of R. Then, from $(x,y) \in R$ and $(x,y') \in R$ it follows that $y = y'$, again by the first inclusion of (1), so each relation R in this left pregroup is a function. But $R^{\cup}$ must also be a function, so R is a bijection from its domain (i.e. X) onto its codomain (again X, since X is the domain of $R^{\cup}$).

Let $(X, \leq)$ be a poset. A relation $R \subseteq X^2$ is said to be *monotone*, if it satisfies the following conditions: (i) xRy and $x' \leq x$ entail $x'Ry$, (ii) xRy and $y \leq y'$ entail xRy'. (We write xRy for $(x,y) \in R$.) By $R(X, \leq)$ we denote the set of all monotone relations $R \subseteq X^2$. This set is a p.o. monoid with the ordering $\leq$, the operation $\circ$ and the unit element $\leq$.

Proposition 9. *Every (left, right) pregroup $(G, \leq, \circ, 1)$ is embeddable into the largest (left, right) pregroup contained in $R(G, \leq)$.*

Proof. Fix a (left, right) pregroup $(G, \leq, \circ, 1)$. For $a \in G$, define $R_a \subseteq G^2$ by: xR_ay iff $x \leq ay$. Clearly R_a is monotone. If $xR_{ab}y$, then, for $z = by$, we have

xR_az and zR_by, so $R_{ab} \subseteq R_a \circ R_b$. Assume $xR_a \circ R_b y$. Then, there exists z such that $x \leq az$ and $z \leq by$, hence $x \leq aby$. We have shown $R_{ab} = R_a \circ R_b$. Clearly $R_1 =\leq$. Also $a \leq b$ iff $R_a \subseteq R_b$ (from the right to the left: if $R_a \subseteq R_b$, then aR_a1 implies aR_b1 which means $a \leq b$). Now, (1) hold in G, which yields:

$$R_{a^l} \circ R_a \subseteq \leq \subseteq R_a \circ R_{a^l} \text{ and } R_a \circ R_{a^r} \subseteq \leq \subseteq R_{a^r} \circ R_a,$$

for all $a \in G$. Consequently, R_{a^l} is the left adjoint of R_a, and R_{a^r} is the right adjoint of R_a. Then, the mapping $F(a) = R_a$ is the required embedding. □

The same construction can be performed for residuated monoids. Yet, in general, $R_{a\backslash b} \subseteq R_a \backslash R_b$ but the equality does not hold, and similarly for $/$.

Proposition 10. *Let G be a pregroup contained in $R(X, \leq)$. Then, for all $R \in G$ and all $x, y \in X$, the following equivalences hold true:*
$xR^ly \Leftrightarrow \forall z(yRz \Rightarrow x \leq z)$, $xR^ry \Leftrightarrow \forall z(zRx \Rightarrow z \leq y)$.

Proof. We prove the first equivalence. Since $R^l \circ R \subseteq \leq$, then:

$$xR^ly \Rightarrow \forall z(yRz \Rightarrow x \leq z).$$

Since $\leq \subseteq R \circ R^l$, then:

$$x \leq y \Rightarrow \exists z(xRz \text{ and } zR^ly.$$

Since $x \leq x$, then there exists z such that xRz and zR^lx, hence, for all u, if xRu then $z \leq u$. So, $z = \min\{u : xRu\}$. We have shown that, for all $x \in X$ there exists the least u such that xRu; this element u will be denoted by x_R. We have xRx_R and x_RR^lx. Assume $\forall z(yRz \Rightarrow x \leq z)$. Since yRy_R, then $x \leq y_R$, and since y_RR^ly, then xR^ly (R^l is monotone). We have proven the first equivalence.

The proof of the second equivalence is dual. Now, one shows that, for all $x \in X$, there exists the greatest element u such that uRx; this element u will be denoted by x^R. We have x^RRx and xR^rx^R. □

Since the left and the right part of this proof are independent of each other, then an analogous result holds for left (right) pregroups. The elements x_R and x^R defined in the proof satisfy the following equivalences:

$$xRy \text{ iff } x_R \leq y \text{ iff } x \leq y^R \ , \tag{14}$$

for all $x, y \in X$.

We have: $x \leq y$ entails $x_R \leq y_R$ and $x^R \leq y^R$. Also:

$$R \subseteq S \text{ iff } \forall x(x_S \leq x_R) \text{ iff } \forall x(x^R \leq x^S) \ , \tag{15}$$

$$(x_R)_S = x_{R \circ S},\ (x^R)^S = x^{S \circ R} \ . \tag{16}$$

We prove the second equality of (16). Assume $z \leq (x^R)^S$. Then, zSx^R, by (14), and x^RRx, so $zS \circ Rx$. which yields $z \leq x^{S \circ R}$, by (14). Assume $z \leq x^{S \circ R}$. By (14), $zS \circ Rx$, hence there is u such that zSu and uRx. By (14), $u \leq x^R$, so zSx^R, which yields $z \leq (x^R)^S$, by (14).

Denote G =Pr($R(X,\leq)$). To every relation $R \in G$ we assign a function $f^R : X \mapsto X$ by setting: $f^R(x) = x^R$. Clearly, f^R is monotone. By (14), $f^{\leq} = I$. By (16), $f^{S \circ R} = f^S \circ f^R$. By (15), $R \subseteq S$ iff $f^R \leq f^S$. Since (1) hold in G, then:

$$f^{R^l} \circ f^R \leq I \leq f^R \circ f^{R^l},\ f^R \circ f^{R^r} \leq I \leq f^{R^r} \circ f^R,$$

hence $f^{R^l} = (f^R)^l$ and $f^{R^r} = (f^R)^r$. Consequently, the mapping $F(R) = f^R$ is a monomorphism of G into Pr($F(X,\leq)$). Actually, F is an isomorphism. For the mapping $F'(f) = R_f$, where the relation $R_f \subseteq X^2$ is defined by: $xR_f y$ iff $x \leq f(y)$, is a monomorphism of Pr($F(X,\leq)$) into G converse to F. We omit an easy proof of this fact. Analogous arguments work for right pregroups. Accordingly, the following theorem is true.

Theorem 6. *For every poset $(X,\leq)$, the largest (resp. right) pregroup contained in $R(X,\leq)$ is isomorphic to the largest (resp. right) pregroup contained in $F(X,\leq)$.*

Interestingly, there is no full symmetry between the left and the right case. Denote G =LPr($R(X,\leq)$) and H =LPr($F(X,\leq)$). For $R \in G$, define: $f_R(x) = x_R$, for $x \in X$. Again, f_R is monotone. However, $R \subseteq S$ iff $f_S \leq f_R$, and $f_{S \circ R} = f_R \circ f_S$, by (15) and (16). One can show that the mapping $F(R) = f_R$ is a $\circ$ and $\leq$ inverting isomorphism of G onto H (it preserves left adjoints).

We show that, if $(X,\leq)$ is the poset $(\omega,\leq)$, then no isomorphism (i.e. $\leq$ and $\circ$ and l preserving bijection) from G onto H exists. For assume it exist. Then, by the composition of this morphism with the converse of F from the above paragraph, we obtain a $\circ$ and $\leq$ inverting and l preserving isomorphism h of H onto itself. But $g(f) = f^l$ defines a $\circ$ and $\leq$ inverting and l preserving monomorphism of H into itself (use (3) and (MON')). Denote $H' = g(H)$. Clearly, $g \circ h$ is an isomorphism of H onto H'. H' consists of all f^l, for $f \in H$. For every p.o. monoid, elements a^l are precisely those elements which admit right adjoints in this p.o. monoid. A function $f \in H$ admits a right adjoint iff f is downward unbounded, which is equivalent to the condition $f(0) = 0$. Now, all surjective elements of H satisfy this condition, hence they admit right adjoints in H. On the other hand, there are surjective elements in H' which do not admit right adjoints in H', e.g. $f(0) = 0$, $f(x) = x - 1$, for $x \neq 0$ (then, $f^r(0) = 1$, so $f^r \notin H'$). Consequently, H is not isomorphic to H', which falsifies our assumption. Observe that G is isomorphic to H'.

4 Grammars

In [5], it has been proven that grammars based on free pregroups are weakly equivalent to context-free grammars. In this section, we outline a proof of an analogous theorem for grammars based on free left (right) pregroups.

Let $(X,\leq)$ be a poset. *Terms* are expressions $a^{(n)}$, for $a \in X, n \in \mathbf{Z}$. *Types* are finite strings of terms; ϵ denotes the empty type.For types x, y, the concatenation of x and y is denoted by xy. The relation $\leq$ on the set of types is defined as the smallest reflexive and transitive relation, satisfying the following conditions:

(CON) $xa^{(n)}a^{(n+1)}y \leq xy$, (EXP) $xy \leq xa^{(n+1)}a^{(n)}y$, (IND) $xa^{(n)}y \leq xb^{(n)}y$, if either $a \leq b$ and n is even, or $b \leq a$ and n is odd, for all $a, b \in X$, types x, y and $n \in \mathbf{Z}$. We write $x \sim y$ if $x \leq y$ and $y \leq x$. Equivalence classes of $\sim$ are the members of *the free pregroup* generated by $(X, \leq)$. In this free pregroup, $1 = [\epsilon]$, $[x] \cdot [y] = [xy]$, $[x] \leq [y]$ iff $x \leq y$, and the adjoints of $[x] \neq 1$ are defined as follows:

$$[a_1^{(n_1)} \dots a_k^{(n_k)}]^l = [a_k^{(n_k-1)} \dots a_1^{(n_1-1)}],$$

$$[a_1^{(n_1)} \dots a_k^{(n_k)}]^r = [a_k^{(n_k+1)} \dots a_1^{(n_1+1)}].$$

This construction is due to Lambek [12]. (CON), (EXP) and (IND) can be treated as rewriting rules of a formal calculus: $x \leq y$ holds iff one can rewrite x into y by a finite number of applications of these rules. Lambek introduces more general rules: (GCON) $xa^{(n)}b^{(n+1)}y \leq xy$, (GEXP) $xy \leq xa^{(n+1)}b^{(n)}y$, both requiring that either n is even and $a \leq b$ in X, or n is odd and $b \leq a$ in X. Clearly, (GCON) is (IND) followed by (CON), and (GEXP) is (EXP) followed by (IND). A key result in [12] is the following Switching Lemma: if $x \leq y$, then there exist types u, v such that $x \leq u$ by (GCON) only, $u \leq v$ by (IND) only, and $v \leq y$ by (GEXP) only. Consequently, if $x \leq y$ and y is a term or $y = \epsilon$, then $x \leq y$ by (GCON) and (IND) only, that means, by (CON) and (IND) only.

As a consequence of the Switching Lemma, we show that free pregroups generated by posets are strictly proper. If $[x]$ is injective (resp. surjective), where $x = x_1 \dots x_k$, $x_1, \dots, x_k$ are terms, then $[x_k]$ is injective (resp. $[x_1]$ is surjective). Thus, it is sufficient to show that there exists no term x such that $[x]$ is injective or surjective. Take a term $a^{(n)}$. Assume $[a^{(n)}]$ be injective. Then $[a^{(n-1)}a^{(n)}] = 1$, so $\epsilon \leq a^{(n-1)}a^{(n)}$. By the Switching Lemma, the latter can be derived by (IND) and (GEXP) only, which is impossible. In a similar way, one shows that $[a^{(n)}]$ cannot be surjective.

By *a pregroup grammar* we mean a quadruple $(V, X, \leq, R)$ such that V is a nonempty, finite alphabet, $(X, \leq)$ is a finite poset, and R is a finite relation betwen elements of V and types on $(X, \leq)$. One says that the grammar assigns type x to string $v_1 \dots v_n$, $v_i \in V$, if there exists types x_i such that $v_i R x_i$, for $i = 1, \dots, n$, satisfying: $x_1 \dots x_n \leq x$ in the above sense. If G is a pregroup grammar and x is a type, then $L(G, x)$ denotes the set of all strings on V which are assigned type x by G. Using Switching Lemma, one proves that, for every pregroup grammar G and every term x, $L(G, x)$ is a context-free language (not containing ϵ). Conversely, for every context-free language L such that $\epsilon \notin L$, there exist a pregroup grammar G (whose poset $(X, \leq)$ is $(X, =)$) and a term $a \in X$ such that $L(G, a) = L$. Consequently, the languages generated by pregroup grammars are precisely the context-free languages not containing ϵ. Proofs of these results are given in [5].

Free left (resp. right) pregroups are defined as free pregroups except that one only admits terms $a^{(n)}$, for $n \leq 0$ (resp. $n \geq 0$) and drops right (resp. left) adjoints. It can easily be shown that free left (resp. right) pregroups are left (resp. right) pregroups. Further, if G is the free left (resp. right) pregroup generated by $(X, \leq)$ and H is an arbitrary left (resp. right) pregroup, then every $\leq$ preserving mapping h of X into H can uniquely be extended to a homomorphism of G into H, Consequently, $x \leq y$ (equivalently: $[x] \leq [y]$) holds G iff $x \leq y$ is true in every left (resp. right) pregroup H under every $\leq$ preserving assignment of members of H to elements of X.

If x is a type and y is a term, both in the sense of free left (resp. right) pregroups, then the following equivalence holds true: $x \leq y$ in the sense of free pregroups iff $x \leq y$ in the sense of free left (resp. right) pregroups. The conditional ($\Leftarrow$) is obvious, since the calculus of free pregroups is stronger than that of free left (resp. right) pregroups. To prove ($\Rightarrow$) assume $x \leq y$ in the sense of free pregroups. By Switching Lemma, $x \leq y$ by (CON) and (IND) only, and these steps can be performed in the calculus of free left (resp. right) pregroups, since they do not introduce any bad terms.

Grammars based on left (right) pregroups are defined as pregroup grammars except that one only admits types and rules in the sense of free left (right) pregroups. By the preceding paragraph and the results from [5], mentioned above, if G is a grammar based on left (right) pregroups and x is a term, then $L(G, x)$ is a context-free language. Conversely, if L is a context-free language such that $\epsilon \notin L$, then the pregroup grammar for L constructed in [5] involves left adjoints only (corresponding to types $p, p/q, (p/q)/r$), hence, by the preceding paragraph, it generates L as a grammar based on left pregroups. Also, one could use a grammar involving right adjoints only (corresponding to types $p, q\backslash p, r\backslash(q\backslash p)$) and construct a grammar based on right pregroups for L. Accordingly, grammars based on left (right) pregroups generate precisely the context-free languages not containing ϵ.

The same results can also be proven in a different, though essentially similar way. First, Switching Lemma can be proven for free left (right) pregroups by the argument from [12] (also see [5]). Further, the construction of a pregroup grammar for a given context-free language L, provided in [5], actually yields a grammar based on left (right) pregroups. Thus, one directly proves the weak equivalence of context-free grammars and grammars based on left (right) pregroups. Now, we have to prove the weak equivalence of grammars based on right pregroups and pregroup grammars. Switching Lemma for free pregroups implies that, for clauses $x \leq y$ such that x is a string of right terms and y is a right term, the calculus of free pregroups is conservative over the calculus of free right pregroups. Thus, every grammar based on right pregroups can be treated as a pregroup grammar. Conversely, given a pregroup grammar G, we define m to be the least $n \in \mathbf{Z}$ such that a term of the form $a^{(n)}$ appears in types involved in the relation R of G. If $m \geq 0$, then G can be treated as a grammar based on right pregroups. Consider the case $m < 0$. Define a mapping $f(a) = [a^{(-2m)}]$ from X into the free pregroup underlying G. By (MON'), f preserves $\leq$, so it can uniquely be extended to a homomorphism of this pregroup into itself. One easily proves that $f([x]) = [x]^{(-2m)}$, for all types x, hence f is an automorphism of this pregroup. Consequently, $[x] \leq [y]$ iff $f([x]) \leq f([y])$, for all types x, y. A grammar G' is constructed from G by replacing in R each $a^{(n)}$ by $a^{(n-2m)}$. By the above, for every term $a^{(n)}$, $L(G, a^{(n)}) = L(G', a^{(n-2m)})$. Yet, G' does not involve left terms, hence it can be treated as a grammar based on right pregroups. Therefore, grammars based on free right pregroups generate the same languages as pregroup grammars.

References

1. H. Andréka and S. Mikulaś, Lambek calculus and its relational semantics: completeness and incompleteness, *Journal of Logic, Language and Information* 3 (1994), 1-37.
2. D. Bargelli and J. Lambek, An algebraic approach to French sentence structure, in [9], 62-78.
3. W. Buszkowski, Mathematical Linguistics and Proof Theory, in [17], 683-736.
4. W. Buszkowski, Algebraic structures in categorial grammar, *Theoretical Computer Science* 199 (1998), 5-24.
5. W. Buszkowski, Lambek grammars based on pregroups, in [9], 95-109.
6. W. Buszkowski and M. Kołowska-Gawiejnowicz, Representation of residuated semigroups in some algebras of relations. (The method of canonical models.), *Fundamenta Informaticae* 31 (1997), 1-12.
7. C. Casadio and J. Lambek, An algebraic analysis of clitic pronouns in Italian, in [9], 110-124.
8. L. Fuchs, *Partially Ordered Algebraic Systems*, Pergamon Press, Oxford, 1963.
9. P. de Groote, G. Morrill and C. Retoré (eds.), *Logical Aspects of Computational Linguistics*, LNAI 2099, Springer, Berlin, 2001.
10. C. Hoare and H. Jifeng, The weakest prespecification, *Fundamenta Informaticae* 9 (1986), 51-84, 217-252.
11. J. Lambek, The mathematics of sentence structure, *American Mathematical Monthly* 65 (1958), 154-170.
12. J. Lambek, Type grammars revisited, in [13], 1-27.
13. A. Lecomte, F. Lamarche and G. Perrier (eds.), *Logical Aspects of Computational Linguistics*, LNAI 1582, Springer, Berlin, 1999.
14. M. Moortgat, *Categorial Investigations. Logical and Linguistic Aspects of the Lambek Calculus*, Foris, Dordrecht, 1988.
15. M. Moortgat, Categorial Type Logics, in [17], 93-177.
16. E. Orłowska and A. Szałas (eds.), *Relational Methods for Computer Science Applications*, Physica Verlag, Heidelberg, 2001.
17. J. van Benthem and A. ter Meulen (eds.), *Handbook of Logic and Language*, Elsevier, Amsterdam, The MIT Press, Cambridge Mass., 1997.

Algebraic Semantics of ER-Models in the Context of the Calculus of Relations. II: Dynamic View.

Ernst-Erich Doberkat[1] and Eugenio G. Omodeo[2]

[1] Department of Computer Science,
Universität Dortmund,
Dortmund, Germany
doberkat@acm.org
[2] Dipartimento di Informatica
Università degli Studi di L'Aquila
L'Aquila, Italy
omodeo@univaq.it

Abstract. We provide a detailed analysis of the insertion operations for an ER-model represented in terms of the map calculus. This continues our previous study of compiling an ER model into the abstract setting of what might be called *logic without variables*.

1 Introduction

Entity relationship modelling (*ER modelling* for short) is a widespread and powerful technique for data modelling. An ER model captures all the relationships between data using entities and relations together with attributes on them. A formal semantics of ER modelling, however, is not easy to come by: as usual, a popular technique is described more or less informally, and this is notoriously difficult to model formally. There are several approaches at describing the formal semantics of this modelling technique which are characterized in [4,9].

The present paper proposes formalizing ER modelling through the *algebra of (dyadic) relations* (which is akin, but different, from Codd's *relational algebra* so useful in data base programming languages!), a branch of Logic brought to flourish through the work of Ernst Schröder (see the historical introduction in [1]). Relation algebras have been used for decomposing relations in a database according to functional dependencies in [7]; these methods, however, have not yet been utilized for a systematic investigation of the dynamic behavior of a data base.

We separate the static structure (the topology) of the ER model from its dynamic counterpart, and we have shown already how to model the static view using the algebra of relations in a companion paper [9]. This is obviously not enough, because the dynamic nature of an ER model cannot be described using the static structure alone. Let us have a look at abstract data types for just conveying the flavor of our arguments.

H. de Swart (Ed.): RelMiCS 2001, LNCS 2561, pp. 50–65, 2002.

1.1 The ADT View

An abstract data type (ADT) encapsulates data and the operations (usually called *methods*) on it. This notion is fundamental because it supports data abstraction and program evolution by keeping data and their operation in a single well defined place. ADTs serve as templates, they are instantiated, and the instances of an ADT are the living capsules data and operations are kept in. *Design by Contract*, so forcefully advocated by Bertrand Meyer [8], goes one step beyond, associating with each ADT specific properties called *invariants*. Operations on an (instance of an) ADT have to respect these invariants in the sense that each operation that starts on an instance which satisfies the invariant leaves the instance in a state which also satisfies it. Each method m of an ADT is associated with a precondition pre_m and a postcondition post_m indicating a contract: entering m such that pre_m is satisfied guarantees leaving m with post_m satisfied. In Hoare's notation of predicate transforms,

$$\{\mathsf{inv} \wedge \mathsf{pre}_m\}\ m\ \{\mathsf{inv} \wedge \mathsf{post}_m\}.$$

Actually, Design by Contract entails more, but this need not concern us here.
An ER model $\mathcal{M}$ may be considered as such an ADT. The data to be stored in an instance are composed of the data stored in the entities, relations and attributes, and the invariant is provided by the conditions imposed on the model's validity (see Definition 6.5). We should look for these operations: initializing an instance of $\mathcal{M}$, inserting elements into entities and relations, and deleting elements from entities and relations.
The *invariant* to be maintained by these operations is the validity of the model; this means that the model before and after one of these families of operations has to conform to the model's declaration. The *post*conditions may in every case be set to true, because the operations are all geared towards maintaining the ADT's invariant. *Initialization* initializes every entity and every relation to the empty set, thus only the trivially valid precondition needs to be provided. The assumption is that we always start from an empty model, so we do not cater for this operation. The *insertion* of elements into an entity or a relation requires a set of conditions which will force the invariant to hold after the insertions took place. This will provide the precondition, see Proposition 6.2. Similarly, the *deletion* of elements requires a set of conditions which will help maintaining the invariant. The corresponding conditions are described in detail in [5].

1.2 Overview

What needs to be done then is to formulate the invariant and the precondition using the language we have chosen for our formalization. After we discuss the version of ER modelling we want to work with in Section 2, we introduce the algebra of relations (or *map algebra*, as we will call it usually) briefly in Section 3, where we will also provide some abbreviations that are helpful for the discussions to follow. Section 4 formulates essential pieces of an ER model in map algebra,

borrowing freely from [9]. Section 6 deals with a formulation of the preconditions for insertions; for reasons of reducing the complexity, this is split into the bare bones version of an ER model which does not entertain attributes. This leads to the notion of a *weakly valid* ER model, and it is shown under which conditions weak validity is maintained. Attributes are added to the discussions at that point, leading to the notion of a valid model, and strengthening the preconditions towards keeping validity invariant. A very similar procedure may be observed when discussing deletions, as outlined in the full report [6].
The present paper represents the second one in a series investigating the relationship between entity relationship modelling and map algebra [9,6]. Due to space limitations, it only provides the statements without providing a detailed rendering of the arguments substantiating the propositions. The proofs, however, can all be found in [6]. The authors have compiled the complete results of their investigations in the paper [5] which has been submitted for publication elsewhere. A prototype SWI-Prolog program which performs the translation of normalized Entity-Relationship models into the calculus of relation is available and is likely to evolve.

Acknowledgements. This research was in part supported through grants from the *Exchange Programme for Scientists between Italy and Germany* from the Italian Ministry of Foreign Affairs/Deutscher Akademischer Austauschdienst, from which both authors were funded at different times during Fall 2000 and Summer 2001. The first author's work was additionally partially funded through a grant from *Progetto speciale I.N.D.A.M./GNIM "Nuovi paradigmi di calcolo: Linguaggi e Modelli"* while visiting what is now the Dipartimento di Informatica at the University of L'Aquila. The second author was partly funded by the MURST/MIUR 40% project *Aggregate- and number-reasoning for computing: from decision algorithms to constraint programming with multisets, sets, and maps.* This work also benefited from the collaboration fostered by the European COST Action 274 (TARSKI, cf. http://www.tarski.org).

2 Entity Relationship Models

Entity Relationship modelling [3] is a popular and widespread technique for data modelling which we assume the reader to be familiar with. Many variants have been discussed [12]. We will restrict ourselves to a rather basic variant with the following properties: All relations are binary, and the only cardinality restriction that may be imposed on a relation is that it is left- or right- unique, inheritance is restricted to single inheritance, relations are assumed to be total (in fact, in the presence of inheritance non-total relations may be transformed into total ones by introducing additional entities for the domain, and for the range, respectively), and attributes are defined on entities only.
This is the version of ER modelling investigated in [9] and a bit more restrictive than the one investigated in [4]. These restrictions can be removed or refined at the cost of a more complicated technical development. We feel, however, that

the methods we develop here provide a way of modelling these more complicated situations.

We are given an instance $\mathcal{M}$ of an ER model which is *valid*, so all constraints formulated in the declaration of the model are satisfied. We want to investigate *change*, namely we want to investigate under which conditions insertions and deletions into $\mathcal{M}$ lead to a valid model again. We assume that we have complete information about the items to be inserted. Thus, if E is an entity, we know the items δ^+E to be inserted into E, yielding $E \cup \delta^+E$ as the new version of this entity. Similarly, we know for relations R the tuples δ^+R to be inserted, and we know for attributes α the changes in $\delta^+\alpha$. What we want to know is, under which conditions for $E, \delta^+E, R, \delta^+R$ and $\alpha, \delta^+\alpha$ the invariance of validity of the instance is maintained. The question arises *mutatis mutandis* for deletions.

Note that the assumption that the change sets δ^+ are given does not address the problem of constructing them. Postulating that complete information is available from the outset permits us to focus on the structural properties of these change sets.

3 Map Calculus

ER models will be formulated in terms of relational algebras. These algebras formalize axiomatically the usual operations on binary relations (like composition or inversion), so that binary relations appear as one of several models that are possible for these algebras. We will provide a very brief introduction to these algebras, and we will fix some notations for the reader's convenience.

A relational algebra (or *map algebra*) is defined as a Boolean algebra with additional properties that are imposed because a composition relation is available. The version of relational algebras we want to use is defined below; for variants and further developments the reader is encouraged to consult [11], [2, Ch. 2], or [1, Ch. 1].

Definition 1. $\langle \mathbb{L}, \mathbb{T}, \cap, \overline{\cdot}, \mathbb{I}, ;, \cdot^{\smile} \rangle$ *is called a* relational algebra *iff*

1. $\langle \mathbb{L}, \mathbb{T}, \cap, \overline{\cdot} \rangle$ *is a Boolean algebra with smallest element* $\mathbb{L}$*, largest element* $\mathbb{T}$*, intersection (meet)* $\cap$*, and complementation* $\overline{\cdot}$*; the associated order relation and union (join) are denoted by* $\subseteq$*, and* $\cup$*, resp.*
2. $;$ *is a binary associative operation on the Boolean algebra with* $\mathbb{I}$ *as the left- and right-neutral element,*
3. $\cdot^{\smile}$ *is a unary idempotent operation on the Boolean algebra,*
4. *the following properties hold:*
   a) $(P;Q)^{\smile} = Q^{\smile};P^{\smile}$,
   b) $(P \cap Q)^{\smile} = P^{\smile} \cap Q^{\smile}$,
   c) $P;(Q_1 \cap Q_2) \subseteq P;Q_1 \cap P;Q_2$ *($\cap$-subdistributivity),*
   d) $P;(Q_1 \cup Q_2) = P;Q_1 \cup P;Q_2$ *($\cup$-distributivity),*
5. $P \subseteq Q$ *implies* $P;R \subseteq Q;R$*,*
6. $(P;Q) \cap R = \mathbb{L}$ *implies* $(P^{\smile};R) \cap Q = \mathbb{L}$ *(Schröder's Rule).*

Map algebra consists of map equalities $P = Q$, where P and Q are map expressions:

Definition 2. Map expressions *are terms of the signature according to the table below, where we have added union* $\cup$ *as an associative operation, and the left-associative set difference* $\setminus$ *for convenience:*

Symbol	$\perp\!\!\!\perp$	$\top\!\!\!\top$	$\mathbb{I}$	r_i	$\cap$	$;$	$\cdot^{\smile}$	$\overline{\cdot}$	$\cup$	$\setminus$
Degree	0	0	0	0	2	2	1	1	2	2
Priority					5	6	7	2	2	2

Here r_i *is one of the countably many* map letters *which we assume to be available.*

Map letters are used to customizing map algebra by attaching additional properties through additional axioms for the relational algebra, as we will see in the sequel.

An *interpretation* $\mathcal{I}$ over a universe $\mathcal{U}$ maps each map expression to a subset of the Cartesian square $\mathcal{U}^2 \equiv_{\mathsf{Def}} \mathcal{U} \times \mathcal{U}$ such that e.g.

$$\begin{array}{r|c} \perp\!\!\!\perp^{\mathcal{I}} = \emptyset & (P \cap Q)^{\mathcal{I}} = P^{\mathcal{I}} \cap Q^{\mathcal{I}}, \\ \top\!\!\!\top^{\mathcal{I}} = \mathcal{U}^2 & (P;Q)^{\mathcal{I}} = P^{\mathcal{I}};Q^{\mathcal{I}} \\ \mathbb{I}^{\mathcal{I}} = \Delta & (Q^{\smile})^{\mathcal{I}} = (Q^{\mathcal{I}})^{\smile} \end{array}$$

Here Δ is the diagonal $\{\langle a, a\rangle \mid a \in \mathcal{U}\}$ of $\mathcal{U}$, and the operations on the right-hand side are the familiar ones manipulating relations over sets. Hence e.g. $\cup$-distributivity translates into the set equality $R;(S_1 \cup S_2) = R;S_1 \cup R;S_2$ that is familiar for the relations $R \subseteq A \times B$ and $S_1, S_2 \subseteq B \times C$ for sets A, B and C.

Adding new axioms through fixing properties of map letters has the effect of restricting interpretations: they have to satisfy the additional properties for the interpretation of the map letters, which in turn also have to be provided.

For convenience, we use some abbreviations which are listed in the table below.

Notation	Expression	Note
$\mathsf{Coll}(R)$	$R \subseteq \mathbb{I}$	$R^{\mathcal{I}}$ is a collection
$\mathsf{Total}(R)$	$R;\top\!\!\!\top = \top\!\!\!\top$	$R^{\mathcal{I}}$ is a total relation
$\mathsf{dom}(R)$	$R;\top\!\!\!\top \cap \mathbb{I}$	domain of R
$\mathsf{img}(R)$	$\top\!\!\!\top;R \cap \mathbb{I}$	range/image of R
$\mathsf{RUniq}(R)$	$\mathsf{Coll}\left(R^{\smile};R\right)$	$R^{\mathcal{I}}$ is a partial map
$\mathsf{LUniq}(R)$	$\mathsf{RUniq}\left(R^{\smile}\right)$	$\left(R^{\smile}\right)^{\mathcal{I}}$ is a partial map
$\mathsf{NonVoid}(R)$	$\mathsf{Total}(\top\!\!\!\top;R)$	$R^{\mathcal{I}} \neq \emptyset$
$\mathsf{Snglt}(R)$	$\mathsf{NonVoid}(R)$ & $\mathsf{RUniq}(\top\!\!\!\top;R)$ & $\mathsf{RUniq}\left(R^{\smile}\right)$	$R^{\mathcal{I}}$ is a singleton
$\mathsf{DomSub}(R, S)$	$R;\top\!\!\!\top \subseteq S;\top\!\!\!\top$	domain containment
$\mathsf{ImgSub}(R, S)$	$\top\!\!\!\top;R \subseteq \top\!\!\!\top;S$	range containment

For example, $\mathsf{Coll}(E)$ says that $E^{\mathcal{I}}$ is supposed to consist of pairs of the form $\langle a, a\rangle$, $\mathsf{Total}(E)$ indicates that $E^{\mathcal{I}}$ is (left-) total, hence that for each $a \in \mathcal{U}$ there is some $b \in \mathcal{U}$ with $\langle a, b\rangle \in E^{\mathcal{I}}$. The reader is invited to formulate these expressions in terms of set-theoretic relations.

4 Preparations

Now let an ER model $\mathcal{M}$ be given. All information concerning $\mathcal{M}$ can be found in a declaration which represents the static information about the model, and which permits stating the validity of an instantiation for $\mathcal{M}$. For the time being we concentrate on entities and relations. Attributes will be added later on.

4.1 The Basic Model

If E is the domain of relation R with F as its co-domain, then we will assume that E and F are tight, i.e., that for each entity e there exists f such that $\langle e, f\rangle$ is in R, similarly for F. This is indicated by $E \bullet\!\!- R -\!\!\bullet F$. In what follows, entities and relations will be considered as elements of a fixed (but anonymous) map algebra. An entity E is then represented through $\mathsf{Coll}(E)$, hence it consists of pairs the first and the second component of which agree, and $E \bullet\!\!- R -\!\!\bullet F$ translates to

$$\mathsf{DotDot}(E, R, F) \equiv_{\mathsf{Def}} \mathsf{Coll}(E) \;\&\; \mathsf{Coll}(F) \;\&\; \mathsf{DomSub}(E, R) \;\&\; \mathsf{ImgSub}(F, R).$$

Either relation $\bullet\!\!-$ or $-\!\!\bullet$ may be tightened to $\overset{1}{\bullet\!\!-}$ and $\overset{1}{-\!\!\bullet}$, resp., indicating uniqueness. Thus $E \overset{1}{\bullet\!\!-} R$ means in addition to $E \bullet\!\!- R$ that $\langle x, y\rangle \in R \wedge \langle x', y\rangle \in R \Rightarrow x = x'$ holds, which may be translated conveniently into $\mathsf{LUniq}(R)$. Similarly, $R \overset{1}{-\!\!\bullet} F$, which means $\langle x, y\rangle \in R \wedge \langle x, y'\rangle \in R \Rightarrow y = y'$ is translated into $\mathsf{RUniq}(R)$. Note that either of these conditions depends only on the relation, not on the domain or the codomain.

The different way a relation relates to its domain and its codomain may be captured through the suitable combination of macros which are comprehensively listed in the table below

Situation	Characterization
$E \overset{1}{\bullet\!\!-} R \overset{1}{-\!\!\bullet} F$	$\mathsf{DotDot}(E, R, F) \;\&\; \mathsf{LUniq}(R) \;\&\; \mathsf{RUniq}(R)$
$E \overset{1}{\bullet\!\!-} R -\!\!\bullet F$	$\mathsf{DotDot}(E, R, F) \;\&\; \mathsf{LUniq}(R)$
$E \bullet\!\!- R \overset{1}{-\!\!\bullet} F$	$\mathsf{DotDot}(E, R, F) \;\&\; \mathsf{RUniq}(R)$

4.2 Adding Place Holders

It may sometimes happen that information is incomplete: an element x is inserted into entity E, and $E \bullet\!\!- R -\!\!\bullet F$ holds, but there is no y in F such that $\langle x, y\rangle$ is to be inserted into R. This then would violate the condition $E; \mathbb{T} \subseteq R; \mathbb{T}$. There may even occur some unpleasant situations when place holders are not admitted; examples show that inserting an element may lead to a nonterminating loop of insertions, cf [5].

For enabling insertions also under somewhat problematic conditions, we postulate the existence of place holders which are collected in a relation P, so that in the situation considered $\langle x, *\rangle$ with $* \in P$ would be inserted into R. We assume that $\mathsf{Coll}(P)$ holds, and that the entities are free of place holders, thus

$E \cap P = \mathbb{L}$ is true for each entity E (note that this implies both $\mathbb{T}{;}E \cap \mathbb{T}{;}P = \mathbb{L}$ and $E{;}\mathbb{T} \cap P{;}\mathbb{T} = \mathbb{L}$). Let $\mathsf{Entity}(P, E) \equiv_{\mathsf{Def}} \mathsf{Coll}(E) \,\&\, E \cap P = \mathbb{L}$ denote that E is an entity.

Inheritance. Immediate inheritance between entities is given through the IsA-relation. Hence E IsA F translates into $\mathsf{Inherits}(P, E, F) \equiv_{\mathsf{Def}} \mathsf{Entity}(P, E)\,\&$ $\mathsf{Entity}(P, F) \,\&\, E \subseteq F$. There are some restrictions to be observed concerning the IsA-relation, mainly acyclicity and single inheritance, and the reader is referred to the companion paper [9] for details.

Constraints on place holders. In [9] some constraints on the use of place holders were formulated:

1. No placeholder occurs twice as the first or the second component of a pair in a relation R. Put $\mathsf{NoTwice}(P, R) \equiv_{\mathsf{Def}} P \cap (R \cap R{;}\overline{\mathbb{I}}){;}\mathbb{T} = \mathbb{L}$, then $\mathsf{NoTwice}(P, R)\,\&$ $\mathsf{NoTwice}(P, R^{\smile})$ should hold. (Literally, this constraint states that there are no placeholders in the domain of the *multivalent part* of either R or $R^{\smile}$, cf. [10, pp.60–61].)
2. No placeholder occurs in two distinct relations R, S as the first components of a pair, which is formulated as $\mathsf{NoBoth}(P, R, S) \equiv_{\mathsf{Def}} P \cap R{;}\mathbb{T} \cap S{;}\mathbb{T} = \mathbb{L}$. The analogous condition on second components holds.
3. No placeholder occurs both as the first component in relation R and as the second component in relation S, hence
$$\mathsf{NoFirstSecond}(P, R, S) \equiv_{\mathsf{Def}} \mathsf{NoBoth}\left(P, R, S^{\smile}\right).$$
4. No pair in a relation has place holders on both sides, thus
$$\mathsf{NoSamePair}(P, R) \equiv_{\mathsf{Def}} R \cap P{;}\mathbb{T} \cap \mathbb{T}{;}P = \mathbb{L}.$$
5. The situation $\langle *, y\rangle$ and $\langle x, y\rangle$ with $x \neq *$ (and, for symmetry, in the second component) does not occur; this is captured through
$$\mathsf{NoDoubleFirst}(P, R) \equiv_{\mathsf{Def}} R{;}R^{\smile} \cap P{;}\mathbb{T} \cap \mathbb{T}{;}\overline{P} = \mathbb{L}$$
and
$$\mathsf{NoDoubleSecond}(P, R) \equiv_{\mathsf{Def}} \mathsf{NoDoubleFirst}\left(P, R^{\smile}\right).$$

Summing up: If $\{R_1, \ldots, R_k\}$ are the identifiers for all the relations in play, the conjunction $\mathsf{PlaceHolder}(P, \{R_1, \ldots, R_k\})$ should hold, where

$$\begin{aligned}
\mathsf{PlaceHolder}(P, \{R_1, \ldots, R_k\}) \equiv_{\mathsf{Def}} \\
& \&_{i=1}^{k} \mathsf{NoTwice}(P, R_i) \,\&\, \mathsf{NoTwice}\left(P, R_i^{\smile}\right) \\
\& \; & \&_{i=1}^{k} \&_{j=i+1}^{k} \mathsf{NoBoth}(P, R_i, R_j) \\
\& \; & \&_{i=1}^{k} \&_{j=i+1}^{k} \mathsf{NoBoth}\left(P, R_i^{\smile}, R_j^{\smile}\right) \\
\& \; & \&_{i=1}^{k} \&_{j=1}^{k} \mathsf{NoFirstSecond}(P, R_i, R_j) \\
\& \; & \&_{i=1}^{k} \mathsf{NoSamePair}(P, R_i) \\
\& \; & \&_{i=1}^{k} \mathsf{NoDoubleFirst}(P, R_i) \\
\& \; & \&_{i=1}^{k} \mathsf{NoDoubleSecond}(P, R_i).
\end{aligned}$$

We reserve the map letter π for place holders.

5 Map Letters

We assume that we have countably many map letters $\mathsf{r}_1, \mathsf{r}_2, \ldots$ at our disposal, of which we reserve the first T initially for system purposes. We have reserved already π for place holders. Some additional reservations will have to be done.

5.1 Blocks

The map letters with indices beyond T will be used for the ER model under consideration in the following way. $\mathsf{r}_{T+1}, \ldots, \mathsf{r}_{T+S}$ will be reserved for entities, the next block of B map letters $\mathsf{r}_{T+S+1}, \ldots, \mathsf{r}_{T+S+B}$ will be reserved for relations, and finally we will reserve the next block of A map letters for attributes. In case of an insertion or a deletion, we reserve the next block of S map letters for the δ^+ resp. δ^--values for entities, the next block of size B for those values for relations, and finally the next block A map letters for attributes. We continue the sequence with the results, according to the following scheme (with $\Sigma := S + B + A$): if entity E corresponds to map letter r_{T+i} with δ^+E corresponding to $\mathsf{r}_{T+\Sigma+i}$, then $E \cup \delta^+E$ will be deposited at $\mathsf{r}_{T+2\cdot\Sigma+i}$. In the same linear way — proceeding in a block wise fashion — we deposit the changed values for relations and attributes. The arrangement of map letters is indicated in Fig. 1.

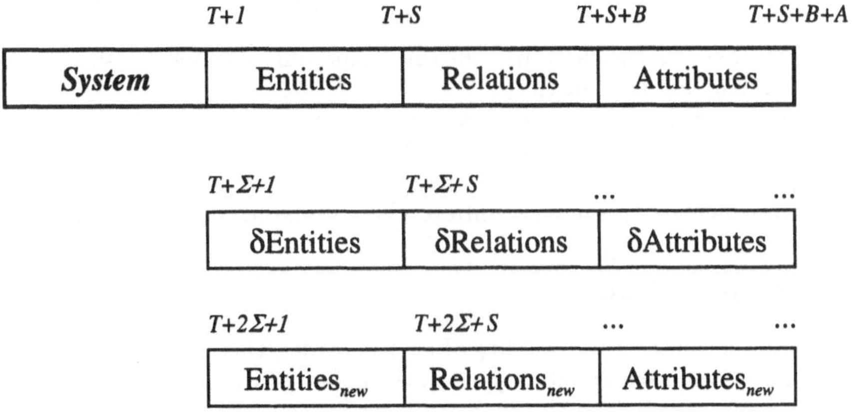

Fig. 1. Arrangement of Map Letters

5.2 Keeping Track

We keep a record the respective relations between entities and relations through a set-valued map $\mathsf{Track} : \{T+S+1, \ldots, T+S+B\} \to 2^{\{T+1,\ldots,T+S\}\times\{T+1,\ldots,T+S\}}$

upon setting $\langle i,j\rangle \in \mathsf{Track}(t) \Leftrightarrow \mathsf{r}_i \bullet\!\!-\, \mathsf{r}_t -\!\!\bullet \mathsf{r}_j$. Define for the relational index $t \in \{T+S+1,\ldots,T+S+B\}$ that $t \in \mathsf{LeftOne} \Leftrightarrow \exists i,j : \mathsf{r}_i \overset{1}{\bullet\!\!-}\, \mathsf{r}_t -\!\!\bullet \mathsf{r}_j$ and $t \in \mathsf{RightOne} \Leftrightarrow \exists i,j : \mathsf{r}_i \bullet\!\!-\, \mathsf{r}_t \overset{1}{-\!\!\bullet} \mathsf{r}_j$. Through these sets we get access to left- and right-unique relations. Note that $\mathsf{r}_i \overset{1}{\bullet\!\!-}\, \mathsf{r}_t -\!\!\bullet \mathsf{r}_j$ implies $\mathsf{r}_k \overset{1}{\bullet\!\!-}\, \mathsf{r}_t -\!\!\bullet \mathsf{r}_\ell$ for all r_k that form the domain of r_t.

Again, Track, LeftOne and RightOne can be shifted linearly along each Σ-block of indices.

The reflexive and transitive closure IsA^* of the inheritance relation is recorded through a reflexive and transitive relation Up on the set $\{T+1,\ldots,T+S\}$; note that this relation may be shifted linearly to the sets $\{T+k\cdot\Sigma+1,\ldots,T+k\cdot\Sigma+S\}$. The necessary properties of IsA^* are described in [9].

Attributes If entity E is represented by map letter r_i with $i \in \{T+1,\ldots,T+S\}$, then

$$\mathsf{Attributes}(i) \subseteq \{T+S+B+1,\ldots,T+S+B+A\}$$

is the set of map letters that are associated with E's attributes. Clearly,

$$\{\mathsf{Attributes}(i) \mid T+1 \leqslant i \leqslant T+S\}$$

forms a partition of the set $\{T+S+B+1,\ldots,T+S+B+A\}$. The set $\mathsf{Mandatory}(i) \subseteq \mathsf{Attributes}(i)$ contains the indices of all mandatory attributes (those attributes which are defined on all of E), and the set $\mathsf{Key}(i) \subseteq \mathsf{Mandatory}(i)$ contains all indices of the key attributes. We assume having only one set of key attributes per entity. It would be easy to work with a varying number of sets of keys for each entity, but this would only complicate the notation, without adding any new ideas.

When we execute an insertion or a deletion, we change the contents of the map letters by manipulating the extension of the corresponding data containers. Our block oriented scheme ensures that this process can be repeated without much ado by simply changing the base address where it all begins from T to $T+2\cdot\Sigma$.

6 Insertions: Validity

This section formulates the validity of an ER model; this is done first without taking attributes into account, leading to the notion of *weak validity*. Conditions are formulated under which the weak validity of an ER model is preserved. Then we add attributes to our discussion, and the notion of validity is formulated. Again, conditions are given under which the attributes of the model arising from insertions satisfy the constraints, this time leading to the instance of a valid ER model. The treatment of deletions along similar lines was carried out in [5], and it turned out to be simpler: we omit it for brevity here.

6.1 Weak Validity

An instance $\mathcal{M}$ of the ER model under consideration is *weakly valid* iff it satisfies all the constraints imposed on the entities and the relations laid down in the model's declaration. This can be described now formally:

Definition 3. *The instance* $\mathcal{M}$ *is called* weakly valid *iff*

$$\begin{aligned}
\&_{T+S+1\leqslant t\leqslant T+S+B} \&_{\langle i,j\rangle\in\mathsf{Track}(t)} \mathsf{DotDot}(\mathsf{r}_i,\mathsf{r}_t,\mathsf{r}_j) \;\; \& \\
\&_{t\in\mathsf{LeftOne}} \mathsf{LUniq}(\mathsf{r}_t) \;\; \& \\
\&_{t\in\mathsf{RightOne}} \mathsf{RUniq}(\mathsf{r}_t) \;\; \& \\
\&_{T+1\leqslant i\leqslant T+S} \; \mathsf{Entity}(\pi,\mathsf{r}_i) \;\; \& \\
\mathsf{PlaceHolder}(\pi,\{\mathsf{r}_{T+1},\ldots,\mathsf{r}_{T+S}\}) \;\; \& \\
\&_{\langle i,j\rangle\in\mathsf{Up}} \; \mathsf{r}_i \subseteq \mathsf{r}_j \;.
\end{aligned}$$

Note that weak validity is formulated using a fixed base address T, which, however, has not been incorporated into the notation that is already cluttered enough.

6.2 Maintaining Weak Validity

The insertions to be performed start from a weakly valid ER model and maintains weak validity as an invariant, cf. 1.1. We will need some preconditions, and elaborate on the insertions proper. If E is an entity, and δ^+E contains the insertions into E, then $E \cup \delta^+E$ will be formed, and this will be the new version of this entity. It is a bit more complicated with a relation R, since we cannot simply form $R \cup \delta^+R$ without running the risk of violating $\mathsf{NoDoubleFirst}(P, R \cup \delta^+R)$ or $\mathsf{NoDoubleSecond}(P, R \cup \delta^+R)$. Hence we have to clean up R by removing candidates for violations; they are easily seen to belong to

$$(\mathbb{T}{;}\pi \cap \delta^+R{;}\mathbb{T}) \cup (\pi{;}\mathbb{T} \cap \mathbb{T}{;}\delta^+R).$$

Thus we work with

$$[R, \delta^+R] \equiv_{\mathsf{Def}} R \setminus ((\mathbb{T}{;}\pi \cap \delta^+R{;}\mathbb{T}) \cup (\pi{;}\mathbb{T} \cap \mathbb{T}{;}\delta^+R))$$

instead of R and form $[R, \delta^+R] \cup \delta^+R$ as the new version of relation R. Occasionally we will replace the map letter π by the free variable P; the expression then will be denoted by $[R, \delta^+R]_P$.

For describing under which conditions weak validity is maintained, we need preparations.

Lemma 1. *Let* R *be a relation, and assume* $\mathsf{Entity}(P, E)$. *Then these implications hold:*

1.

$$\frac{\begin{array}{c}\mathsf{DomSub}(E, R) \;\; \mathsf{Entity}(P, E) \\ \mathsf{DomSub}(\delta^+E, \delta^+R) \;\; \mathsf{Entity}(P, \delta^+E)\end{array}}{\mathsf{DomSub}(E \cup \delta^+E, [R, \delta^+R]_P \cup \delta^+R)}$$

2.

$$\frac{\begin{array}{c}\mathsf{ImgSub}(F, R) \;\; \mathsf{Entity}(P, F) \\ \mathsf{ImgSub}(\delta^+F, \delta^+R) \;\; \mathsf{Entity}(P, \delta^+F)\end{array}}{\mathsf{ImgSub}(F \cup \delta^+F, [R, \delta^+R]_P \cup \delta^+R)}$$

In a similar way we can make sure that the new relation maintains its properties as a map, or as the inverse of a map:

Lemma 2. *Let R be a relation, then the following implications hold:*

1.

$$\frac{\mathsf{LUniq}(R)\ \ \mathsf{LUniq}(\delta^+R) \qquad [R,\delta^+R] \subseteq \overline{\overline{\mathbb{I}};\delta^+R}}{\mathsf{LUniq}([R,\delta^+R] \cup \delta^+R)}$$

2.

$$\frac{\mathsf{RUniq}(R)\ \ \mathsf{RUniq}(\delta^+R) \qquad [R,\delta^+R] \subseteq \overline{\delta^+R;\overline{\mathbb{I}}}}{\mathsf{RUniq}([R,\delta^+R] \cup \delta^+R)}$$

It may be noted that both implications above can be reversed.
Use in what follows as abbreviations

$$\begin{aligned}
\Gamma(A,B,C) &\equiv_{\mathsf{Def}} A \cap (B \cap (C;\overline{\mathbb{I}}));\top = \bot,\\
\Pi(A,B,C) &\equiv_{\mathsf{Def}} A \cap (B;\top \cap C;\top) = \bot\\
\Psi(A,B,C) &\equiv_{\mathsf{Def}} A;B^{\smile} \cup B;A^{\smile} \cup B;B^{\smile} \subseteq C
\end{aligned}$$

Lemma 3. *The following implications hold for any relation R:*

1.

$$\frac{\mathsf{NoTwice}(P,R)\ \ \mathsf{NoTwice}(P,\delta^+R,\delta^+R) \qquad \Gamma(P,[R,\delta^+R]_P,\delta^+R)\ \ \Gamma(P,\delta^+R,[R,\delta^+R]_P)}{\mathsf{NoTwice}(P,[R,\delta^+R]_P \cup \delta^+R)}$$

2.

$$\frac{\mathsf{NoBoth}(P,R,S)\ \ \mathsf{NoBoth}(P,\delta^+R,\delta^+S) \qquad \Pi(P,\delta^+R,[S,\delta^+S]_P)\ \ \Pi(P,[R,\delta^+R]_P,\delta^+S)}{\mathsf{NoBoth}(P,[R,\delta^+R]_P \cup \delta^+R,[S,\delta^+S]_P \cup \delta^+S)}$$

3.

$$\frac{\mathsf{NoSamePair}(P,R)\ \ \mathsf{NoSamePair}(P,\delta^+R)}{\mathsf{NoSamePair}(P,[R,\delta^+R]_P \cup \delta^+R)}$$

4.

$$\frac{\mathsf{NoDoubleFirst}(P,R)\ \ \mathsf{NoDoubleFirst}(P,\delta^+R) \qquad \delta^+R;[R,\delta^+R]_P^{\smile} \subseteq \overline{P;\top \cap \top;\overline{P}}}{\mathsf{NoDoubleFirst}(P,[R,\delta^+R]_P \cup \delta^+R)}$$

5.

$$\frac{\mathsf{NoDoubleSecond}(P,R)\ \ \mathsf{NoDoubleSecond}(P,\delta^+R) \qquad [R,\delta^+R]_P^{\smile};\delta^+R \subseteq \overline{\top;P \cap \overline{P};\top}}{\mathsf{NoDoubleSecond}(P,[R,\delta^+R]_P \cup \delta^+R)}$$

6.

$$\frac{\mathsf{Inherits}(P,E,F)\ \ \mathsf{Entity}(P,\delta^+E)\ \ \mathsf{Entity}(P,\delta^+F)\ \ \delta^+E \subseteq F \cup \delta^+F}{\mathsf{Inherits}(P,E \cup \delta^+E,F \cup \delta^+F)}$$

Define the set $\mathsf{Related}(t)$ as the smallest subset K of $\{T+1,\ldots,T+S+B\}$ sucht that $t \in K$, if $u \in K$ and $\langle u, v\rangle \in \mathsf{Up}$, then $v \in K$, and finally, if $\mathsf{r}_i \bullet\!\!- \mathsf{r}_j$ or $\mathsf{r}_j -\!\!\bullet \mathsf{r}_i$, then $i \in K$ iff $j \in K$.
Thus if we want to insert something into, say, entity E, and E corresponds to map letter r_i, then $\mathsf{Related}(i)$ contains the indices of exactly those entities and relations which are affected by this insertion.
Now let an entity or a relation correspond to map letter r_t.

Definition 4. *An insertion is called* local at t *iff*

$$s \in \{T+\Sigma,\ldots,T+2\cdot\Sigma+1\} \setminus \mathsf{Related}(t) \Rightarrow \mathsf{r}_s = \perp\!\!\!\perp.$$

Introducing this guard prevents the insertion or the deletion from violating the invariants for the model by letting properties creeping in that are not really controlled through our safety measures.
From the instance $\mathcal{M}$ a new instance $\mathcal{M}'$ is generated by performing the insertions. Put for each $j \in \{1,\ldots S\}$

$$\mathsf{r}_{T+2\cdot\Sigma+j} := \mathsf{r}_{T+j} \cup \mathsf{r}_{T+\Sigma+j}.$$

This accounts for insertions into entities. As far as relations are concerned, we set for each $j \in \{S+1,\ldots,B\}$

$$\mathsf{r}_{T+2\cdot\Sigma+j} := [\mathsf{r}_{T+j}, \mathsf{r}_{T+\Sigma+j}] \cup \mathsf{r}_{T+\Sigma+j},$$

accounting for the peculiar way we insert into a relation.
Upon shifting the base address from T to $T+2\cdot\Sigma$, the weak validity of $\mathcal{M}'$ can be investigated:

Proposition 1. *Let $\mathcal{M}$ be a weakly valid ER model, assume that an insertion is local at some index t, then the ER model arising from the insertions is weakly valid, provided that the following conditions are all satisfied:*

1. $\&_{s\in\{1,\ldots,S\}} \mathsf{Entity}(\pi, \mathsf{r}_{T+\Sigma+s})$,
2. $\&_{s\in\mathsf{Related}(t)\cap\{T+1,\ldots,T+B\}}$
   $\&_{\langle i,j\rangle\in\mathsf{Track}(s)} \mathsf{DomSub}(\mathsf{r}_{\Sigma+i}, \mathsf{r}_{\Sigma+s})$
   $\& \mathsf{Entity}(\mathsf{r}_{\Sigma+i}) \& \mathsf{ImgSub}(\mathsf{r}_{\Sigma+j}, \mathsf{r}_{\Sigma+s}) \& \mathsf{Entity}(\mathsf{r}_{\Sigma+j})$
3. $\&_{s\in\mathsf{LeftOne}\cap\mathsf{Related}(t)} \mathsf{r}_{\Sigma+s};\breve{\mathsf{r}}_{\Sigma+s} \subseteq \mathbb{I} \& [\mathsf{r}_s, \mathsf{r}_{\Sigma+s}] \subseteq \overline{\overline{\mathbb{I}};\mathsf{r}_{\Sigma+s}} \& \mathsf{r}_{\Sigma+s} \subseteq \overline{\overline{\mathbb{I}};[\mathsf{r}_s, \mathsf{r}_{\Sigma+s}]}$
4. $\&_{s\in\mathsf{RightOne}\cap\mathsf{Related}(t)} \breve{\mathsf{r}}_{\Sigma+s};\mathsf{r}_{\Sigma+s} \subseteq \mathbb{I} \& [\mathsf{r}_s, \mathsf{r}_{\Sigma+s}] \subseteq \overline{\mathsf{r}_{\Sigma+s};\overline{\mathbb{I}}} \& \mathsf{r}_{\Sigma+s} \subseteq \overline{[\mathsf{r}_s, \mathsf{r}_{\Sigma+s}];\overline{\mathbb{I}}}$
5. $\&_{s\in\mathsf{Related}(t)\cap\{T+S+1,\ldots,T+S+B\}}$
   $\pi \cap \left(\mathsf{r}_{\Sigma+s} \cap \mathsf{r}_{\Sigma+s};\overline{\mathbb{I}};\mathbb{T}\right) = \perp\!\!\!\perp \& \pi \cap \left(\breve{\mathsf{r}}_{\Sigma+s} \cap (\breve{\mathsf{r}}_{\Sigma+s};\overline{\mathbb{I}});\mathbb{T}\right) = \perp\!\!\!\perp$
   $\& \Gamma(\pi, [\mathsf{r}_s, \mathsf{r}_{\Sigma+s}], \mathsf{r}_{\Sigma+s}) \& \Gamma(\pi, [\breve{\mathsf{r}}_s, \breve{\mathsf{r}}_{\Sigma+s}], \breve{\mathsf{r}}_{\Sigma+s})$
   $\& \Gamma(\pi, \mathsf{r}_{\Sigma+s}, [\mathsf{r}_s, \mathsf{r}_{\Sigma+s}]) \& \Gamma(\pi, \breve{\mathsf{r}}_{\Sigma+s}, [\mathsf{r}_s, \breve{\ }\breve{\mathsf{r}}_{\Sigma+s}])$
6. $\&_{s\in\mathsf{Related}(t)\cap\{T+S+1,\ldots,T+S+B\}}$
   $\&_{v\in\mathsf{Related}(t)\cap\{s,\ldots,T+S+B\}}$
   $\& \pi \cap (\mathsf{r}_{\Sigma+s};\mathbb{T}) \cap (\mathsf{r}_{\Sigma+v};\mathbb{T}) = \perp\!\!\!\perp$
   $\Pi(\pi, \mathsf{r}_{\Sigma+s}, [\mathsf{r}_v, \mathsf{r}_{\Sigma+v}]) \& \Pi(\pi, [\mathsf{r}_s, \mathsf{r}_{\Sigma+s}], \mathsf{r}_v)$

7. $\&_{s \in \mathsf{Related}(t) \cap \{T+S+1,\ldots,T+S+B\}} \mathsf{r}_{\Sigma+s} \cap \pi ; \mathbb{T} \cap \mathbb{T} ; \pi = \perp\!\!\!\perp$
8. $\&_{s \in \mathsf{Related}(t) \cap \{T+S+1,\ldots,T+S+B\}}$
$\quad \mathsf{r}_{\Sigma+s} ; \breve{\mathsf{r}}_{\Sigma+s} \cap \pi ; \mathbb{T} \cap \mathbb{T} ; \overline{\pi} = \perp\!\!\!\perp$
$\quad\quad \& \; \mathsf{r}_{\Sigma+s} ; [\mathsf{r}_s, \breve{\mathsf{r}}_{\Sigma+s}] \subseteq \overline{\pi ; \mathbb{T} \cap \mathbb{T} ; \overline{\pi}}$
9. $\&_{s \in \mathsf{Related}(t) \cap \{T+S+1,\ldots,T+S+B\}}$
$\quad \breve{\mathsf{r}}_{\Sigma+s} ; \mathsf{r}_{\Sigma+s} \cap \mathbb{T} ; \pi \cap \overline{\pi} ; \mathbb{T} = \perp\!\!\!\perp$
$\quad\quad \& \; [\mathsf{r}_s, \breve{\mathsf{r}}_{\Sigma+s}] ; \mathsf{r}_{\Sigma+s} \subseteq \overline{\mathbb{T} ; \pi \cap \overline{\pi} ; \mathbb{T}}$
10. $\&_{\langle i,j \rangle \in \mathsf{Up} \cap \mathsf{Related}(T) \times \mathsf{Related}(T)} \mathsf{Entity}(\pi, \mathsf{r}_{\Sigma+i}) \; \& \; \mathsf{Entity}(\pi, \mathsf{r}_{\Sigma+j}) \; \& \; \mathsf{r}_{\Sigma+i} \subseteq \mathsf{r}_j \cup \mathsf{r}_{\Sigma+j}$

The conditions formulated above look certainly very technical, so let us interpret the second and the last of them. The former one states conditions under which $E \cup \delta^+ E$ and $F \cup \delta^+ F$ remain the tight domain and the tight codomain, resp., of $[R, \delta^+ R] \cup \delta^+ R$, provided E was the tight domain, and F was the tight codomain of R before the insertion. The conditions state that $\delta^+ E$ needs to be an entity such that $\mathsf{dom}(\delta^+ E) \subseteq \mathsf{dom}(\delta^+ R)$ is true, hence each element to be inserted into E should be the first component of a pair to be inserted into R. In the same way $\delta^+ F$ is required to be an entity such that $\mathsf{img}(\delta^+ F) \subseteq \mathsf{img}(\delta^+ R)$ holds. The last condition simply states that for $E \cup \delta^+ E$ to inherit from $R \cup \delta^+ R$ it is sufficient that E inherits from F, and that $\delta^+ E$ is a subset of $F \cup \delta^+ F$, and that the new sets are entities indeed. Similar interpretations are given for the other conditions; this is left to the reader.

6.3 Looking at Attributes

Attributes are defined on entities (this is one of our restrictions, cf. Sect. 2), they come in different flavors, as we will discuss now. An attribute α on entity E is a partial map, so $\mathsf{RUniq}(\alpha)$ should be satisfied, and its domain should be contained in (the domain of) E, thus $\mathsf{dom}(\alpha) \subseteq \mathsf{dom}(E)$ should hold. Moreover we assume attributes to have atomic values.

This requirement will be modelled as follows: We assume our universe $\mathcal{U}$ to be structured as $\mathcal{U} = \mathcal{A} \cup \mathcal{A}^*$, where $\mathcal{A} \neq \emptyset$ are the atomic values, and $\mathcal{A}^*$ denotes the set of all words over the alphabet $\mathcal{A}$, hence $\mathcal{A} \cap \mathcal{A}^* = \emptyset$ with ϵ as the empty word; as usual, we put $\mathcal{A}^+ := \mathcal{A}^* \setminus \{\epsilon\}$. We reserve a map letter $\varepsilon \in \{\mathsf{r}_1, \ldots, \mathsf{r}_T\}$ for representing ϵ (hence $\mathsf{Snglt}(\varepsilon)$ & $\mathsf{Coll}(\varepsilon)$) and permit only interpretations $\mathcal{I}$ that satisfy $\varepsilon^{\mathcal{I}} = \{\langle \epsilon, \epsilon \rangle\}$. The atomic entities in $\mathcal{A}$ are modelled through the map letter υ with $\mathsf{Coll}(\upsilon)$ & $\mathsf{NonVoid}(\upsilon)$ & $\upsilon \cap \varepsilon = \perp\!\!\!\perp$. In addition we postulate that $\pi \subseteq \upsilon$ holds.

Interpretations are restricted further by postulating that $\upsilon^{\mathcal{I}} = \{\langle a, b \rangle \in \mathcal{A}^2 | \, a = b\}$. Moreover we assume the existence of canonic projections CAR and CDR separating the head from the tail of a non-empty word, hence

$$\mathsf{CAR} : \mathcal{A}^+ \ni t_1 \ldots t_k \mapsto t_1 \in \mathcal{A},$$
$$\mathsf{CDR} : \mathcal{A}^+ \ni t_1 \ldots t_k \mapsto t_2 \ldots t_k \in \mathcal{A}^*.$$

These projections are represented through the map letters λ and ρ, corresponding to CAR and CDR, resp; their properties will not be discussed here, the reader is referred to [9, 3.1]. We abbreviate for later use the ith projection (hence the operation of extracting the ith component of a tuple) by

$$Z^{(i)} \equiv_{\mathsf{Def}} (i = 1\ ?\ \mathsf{CAR}:\ Z^{(i-1)};\mathsf{CDR})$$
$$\tau^{(i)} \equiv_{\mathsf{Def}} (i = 1\ ?\ \lambda:\ \tau^{(i-1)};\rho),$$

the latter abbreviation preparing for the use of map letters later on.
Returning to attributes: a mandatory attribute α on entity E is characterized through

$$\mathsf{dom}(\alpha) = \mathsf{dom}(E)\ \&\ \mathbb{T};\pi \cap \mathsf{img}(\alpha) = \mathbb{L}.$$

If $\{\alpha_0, \dots \alpha_w\}$ is a collection of key attributes on E, then [9] shows that this property means

$$\mathsf{RUniq}\left(\bigcap_{i=0}^{w} Z^{(i+1)};\alpha_i^{\smile}\right)$$

to hold.

Lemma 4. *The following properties hold:*

1.

$$\frac{\mathsf{RUniq}(\alpha)\Psi(\alpha, \delta^{+}\alpha, \mathbb{I})}{\mathsf{RUniq}(\alpha \cup \delta^{+}\alpha)}$$

2.

$$\frac{\begin{array}{c}\mathsf{dom}(\alpha) = \mathsf{dom}(E)\\ (\delta^{+}\alpha \setminus \alpha);\mathbb{T} = (\delta^{+}E;\mathbb{T}) \setminus (E;\mathbb{T})\end{array}}{\mathsf{dom}(\alpha \cup \delta^{+}\alpha) = \mathsf{dom}(E \cup \delta^{+}E)}$$

The conditions laid down in Lemma 4 permit stating conditions under which some attribute conditions persist under insertion. The exception is a condition which permits being a member of a family of key attributes stable under insertions. The criterion is formulated in Lemma 5. Abbreviate for the map expressions $A_0, \dots, A_k, B_0, \dots, B_k$ and for $J, K \subseteq \{0, \dots, k\}$

$$\Lambda(J, \langle A_0, \dots, A_k\rangle, \langle B_0, \dots, B_k\rangle) \equiv_{\mathsf{Def}} \bigcap_{j\in J} A_j; \left(Z^{(j+1)}\right)^{\smile} \cap$$
$$\bigcap_{j\notin J} (B_j \setminus A_j); \left(Z^{(j+1)}\right)^{\smile},$$
$$\Gamma(J, K, \langle A_0, \dots, A_k\rangle, \langle B_0, \dots, B_k\rangle) \equiv_{\mathsf{Def}} \Lambda(J, \langle A_0, \dots, A_k\rangle, \langle B_0, \dots, B_k\rangle)$$
$$;\Lambda(K, \langle A_0, \dots, A_k\rangle, \langle B_0, \dots, B_k\rangle)^{\smile}$$

With these notations we may formulate:

Lemma 5. *Invariance of a key under insertion is maintained by the following condition:*

$$\frac{\begin{array}{c}\&_{i=0}^{k}\mathsf{LUniq}(\alpha_i)\\ \&_{i=0}^{k}\mathsf{LUniq}(\delta^{+}\alpha_i)\\ \&_{J\subseteq\{0,\ldots,k\}}\&_{K\subseteq\{0,\ldots,k\}}\ \Gamma(J,K,\langle\alpha_0,\ldots,\alpha_k\rangle,\langle\delta^{+}\alpha_0,\ldots,\delta^{+}\alpha_k\rangle)\subseteq\mathbb{I}\end{array}}{\mathsf{LUniq}\left(\bigcap_{i=0}^{k} Z^{(i+1)};(\alpha_i\cup\delta^{+}\alpha_i)^{\smile}\right)}$$

It should be noted that the formulation above requires $\langle\alpha_0,\ldots,\alpha_k\rangle$ as well as the tuple $\langle\delta^{+}\alpha_0\setminus\alpha_0,\ldots,\delta^{+}\alpha_k\setminus\alpha_k\rangle$ to have the properties of key attributes.
The condition just formulated is exponential in the size of the key, consequently, it is not very convenient for practical purposes. On the other hand, it is exact, because a key can be extended if and only if the condition above is satisfied. It would be desirable to develop a more practical, if only sufficient condition for the invariance under insertions of the property being a key.
Now call an ER model $\mathcal{M}$ *valid* iff it is weakly valid, and if the conditions on attributes that have been laid down in the model's declaration are satisfied. Formally:

Definition 5. *The ER model $\mathcal{M}$ is called valid iff*

1. *$\mathcal{M}$ is weakly valid,*
2. *the attributes satisfy*
   $\&_{T+1\leqslant i\leqslant T+S}\&_{j\in\mathsf{Attributes}(i)}\ \mathsf{RUniq}(r_j)\ \&\ \mathsf{dom}(r_j)\subseteq\mathsf{dom}(r_i)\ \&\ \mathsf{img}(r_j)\subseteq\upsilon$
   $\&$
   $\&_{T+1\leqslant i\leqslant T+S}\&_{j\in\mathsf{Mandatory}(i)}\ \mathsf{dom}(r_j)=\mathsf{dom}(r_i)\ \&\ \mathbb{T};\pi\cap\mathsf{img}(r_j)=\mathbb{L}$
   $\&$
   $\&_{T+1\leqslant i\leqslant T+S}$ ***let*** $\{i_1,\ldots,i_j\}=\mathsf{Key}(i)$ ***in*** $\mathsf{RUniq}\left(\bigcap_{\ell=1}^{j} Z^{(\ell+1)};r_{i_\ell}^{\smile}\right)$

We will state now conditions under which the attributes of the changes ER model $\mathcal{M}'$ will cater for the model's validity after the construction process is extended to attributes in the obvious way. Investigating validity requires us to exploit properties of the change sets $\delta^{+}\cdot$ for attributes in the context of their relations to the change sets for entities (note that we do for the time being without attributes on the relations on $\mathcal{M}$).

Proposition 2. *Suppose that the ER model $\mathcal{M}$ is valid, and that in addition to the properties 1 – 10 from Proposition 1 the following properties are satisfied, when performing an insertion that is local at some index t:*

1. $\&_{i\in\mathsf{Related}(t)\cap\{T+1,\ldots,T+B\}}\&_{j\in\mathsf{Attributes}(i)}\ r_{\Sigma+j}^{\smile};r_{\Sigma+j}\subseteq\mathbb{I}\ \&\ \Psi(r_j,r_{\Sigma+j},\mathbb{I})\ \&$
   $\mathbb{T};r_{\Sigma+j}\subseteq\upsilon$
2. $\&_{i\in\mathsf{Related}(t)\cap\{T+1,\ldots,T+B\}}\&_{j\in\mathsf{Mandatory}(i)}\ (r_{\Sigma+j}\setminus r_j);\mathbb{T}=(r_{\Sigma+i};\mathbb{T})\setminus(r_i;\mathbb{T})$
3. $\&_{i\in\mathsf{Related}(t)\cap\{T+1,\ldots,T+B\}}\&_{j\in\mathsf{Mandatory}(i)}\ \mathbb{T};\pi\cap\mathbb{T};r_{\Sigma+j}=\mathbb{L}$
4. $\&_{i\in\mathsf{Related}(t)\cap\{T+1,\ldots,T+B\}}$
   let $\mathsf{Key}(i)=\{i_0,\ldots,i_k\}$ ***in***
   $\&_{J\subseteq\{0,\ldots,k\}}\&_{K\subseteq\{0,\ldots,k\}}\ \Gamma(J,K,\langle r_{i_0},\ldots,r_{i_k}\rangle,\langle r_{\Sigma+i_0},\ldots,r_{\Sigma+i_k}\rangle)\subseteq\mathbb{I}$

Then $\mathcal{M}'$ is a valid ER model.

7 Further Work

This work was performed under some simplifying assumptions: we did assume that we work only with attributes on entities, and that we have a rather scant selection of cardinality restrictions. Both assumptions are not essential for our approach, and we feel that they should be removed. Another technical issue addresses the fact that we work with binary relations only. The discussions concerning projections shows, however, that it should not be too difficult to extend our scenario for incorporating n-ary relations (although the notation then becomes slightly unbearable). From a modelling point of view, we work here in a somewhat untyped environment: we do not have sorts for different entities, but rather assume that one sort fits all. This is fairly problematic in applications, and not entirely practical. Introducing sorts is another step we feel should be undertaken (along with a detailed comparison of both approaches).
Finally, some temporal aspects may be considered: [5] shows briefly how techniques from model checking may be used for arguing about the properties of an ER model.

References

1. C. Brink, W. Kahl, and G. Schmidt, editors. *Relational Methods in Computer Science.* Advances in Computing. Springer-Verlag, Wien, New York, 1997.
2. D. Cantone, E. G. Omodeo, and A. Policriti. *Set Theory for Computing – From decision procedures to declarative programming with sets.* Texts and Monographs in Computer Science. Springer-Verlag, New York, Berlin, Heidelberg, 2001.
3. E. F. Codd. A relational model for large shared data banks. *Communications of the ACM*, 13(6):377–387, 1970.
4. E.-E. Doberkat. Generating an algebraic specification from an ER-model. *International Journal of Software Engineering and Knowledge Engineering*, 7(4):525–552, 1997.
5. E.-E. Doberkat and E. G. Omodeo. Algebraic semantics of ER-models from the standpoint of map calculus. Technical Report Forschungsbericht Nr.765, Fachbereich Informatik, University of Dortmund, October 2001.
6. E.-E. Doberkat and E. G. Omodeo. Algebraic semantics of ER-models in the context of the calculus of relations. Part II: Dynamic view. Technical Report Memo Nr.114, Chair for Software Technology, University of Dortmund, 2001.
7. A. Jaoua, N. Belkhiter, H. Ounalli, and T. Moukam. Databases. In *[1]*, pages 197 – 210. Springer-Verlag, 1997.
8. B. Meyer. *Object-oriented Software Construction.* Prentice-Hall, Englewood Cliffs, NJ, 5th edition, 1998.
9. E. G. Omodeo and E.-E. Doberkat. Algebraic semantics of ER-models in the context of the calculus of relations. Part I: Static view. *Electronic Notes in Theoretical Computer Science*, 44(3), 2001.
10. G. Schmidt and T. Ströhlein. *Relations and graphs.* Monographs on Theoretical Computer Science. Springer-Verlag, Berlin, 1993.
11. A. Tarski and S. Givant. *A formalization of set theory without variables*, volume 41 of *Colloquium Publications.* American Mathematical Society, 1987.
12. B. Thalheim. *Entity-Relationship Modeling: Foundations of Database Technology.* Springer-Verlag, 2000.

Interpretability of First–Order Dynamic Logic in a Relational Calculus

Marcelo F. Frias[1], Gabriel A. Baum[2], and Thomas S. E. Maibaum[3]

[1] Department of Computer Science, School of Sciences, Universidad de Buenos Aires, Pabellón I, Ciudad Universitaria, Buenos Aires, 1428, Argentina, and CONICET.[†]
mfrias@dc.uba.ar

[2] LIFIA, School of Informatics, Universidad Nacional de La Plata, C.C.11, Correo Central, 1900, La Plata, Provincia de Buenos Aires, Argentina, and CONICET.
gbaum@info.unlp.edu.ar

[3] Department of Computer Science, King's College, University of London, U.K.
tom@dcs.kcl.ac.uk

Abstract. Dynamic logic has become a very useful tool in Computer Science, with direct applications in system specification. Here we show how to interpret first-order dynamic logic in an extension of the relational calculus of fork algebras. That is, reasoning in first-order dynamic logic can be replaced by equational reasoning in the extended relational calculus. This allows to: (a) incorporate the features of dynamic logic in a relational framework, and, (b) provide an equational calculus for reasoning in first-order dynamic logic.

1 Introduction

The results reported in this article are part of a project on relational specification of computer systems. Dynamic logic is a framework suitable for the specification of dynamic properties of systems, and the interpretability of first-order dynamic logic in a relational calculus shows that this calculus is adequate for specifying those properties of systems captured by first-order dynamic logic. These results go a step further with respect to previous results on the interpretability of logics in relational frameworks. In [17], a proof is presented for the interpretability of classical first–order logic in a relational calculus. In [14], Orlowska uses relational proof systems, which are calculi following a Rasiowa–Sikorski style [15], in order to make deductions in propositional multimodal logics. Later, in [4], Frias and Orlowska present techniques similar to the ones used here in order to reason in multimodal logics, propositional dynamic logic and relevance logics.

We present an equational calculus (ω-CCFA^+) extending the relational calculus [16], and prove that semantic entailment in dynamic logic is fully captured by proofs in the calculus. To be precise, we prove the following theorem, in which $T_{\langle\rangle}$ maps sentences to relational expressions:

[†] Research Partially funded by Antorchas Foundation.

H. de Swart (Ed.): RelMiCS 2001, LNCS 2561, pp. 66–80, 2002.

Let Γ be a set of sentences in the language DL (of Dynamic Logic) and let φ be a sentence in the language DL. Then,

$$\Gamma \models_{DL} \varphi \iff \{ T_{\langle\rangle}(\gamma) = 1 : \gamma \in \Gamma \} \vdash_{\omega\text{-CCFA}^+} T_{\langle\rangle}(\varphi) = 1 .$$

Of course, a calculus whose language does not have a clear semantics runs a serious risk of not being used. This is why we introduce an intentional semantics for ω-CCFA$^+$ in terms of binary relations, and prove, as the second contribution in this paper, a representation theorem stating that models of the calculus are indeed isomorphic to the intended models.

Applying the results in the paper is very simple. For the reader who only wants to use the calculus the paper is self-contained, since the mapping from dynamic logic is given, the axioms of the calculus are available, and the inference rules are very easy to apply. The reader intending to fully grasp all the details present in proofs will require some knowledge on universal algebra, dynamic logic and relation algebras. As the source material in universal algebra, we suggest [1]. For results on dynamic logic, the book by Harel, Kozen and Tiuryn [6] contains all the required material. Finally, for results on relation algebras, the article [13] and other papers by Maddux contain all the background material required.

The paper is organized as follows. In Sect. 2 we present the syntax and semantics of dynamic logic. In Sect. 3 we present the calculus, its semantics and the representation theorem. In Sect. 4 we present the main results, including the interpretability theorem. In Sect. 5 we present our conclusions about this work. We also include an appendix, in which the proof of the representability theorem and a detailed sketch of the proof of interpretability are given.

2 Dynamic Logic

Dynamic logic is a formalism for reasoning about programs. From a set of primitive actions (also called programs), and using combinators, it is possible to build complex actions. The logic then allows to state properties of these actions, which may hold or not in a given structure. Actions can change (as usually programs do), the values of variables. We will assume that each action reads and/or modifies the value of finitely many state variables. When compared with classical first–order logic, the essential difference is the dynamic content of dynamic logic, which is clear in the notion of satisfiability. While satisfiability in classical first–order logic depends on the values of variables in one valuation, in dynamic logic it may be necessary to consider how actions modify the values of variables. This is done by considering two valuations: one valuation reflecting the values of variables *before* the action is performed, and another holding the values of variables *after* the action is performed.

Along the paper we will assume a fixed (but arbitrary) finite signature $\Sigma = \langle s, A, F, P \rangle$, where s is a sort, $A = \{ a_1, \ldots, a_n \}$ are the primitive action symbols, $F = \{ f_1, \ldots, f_k \}$ are the function symbols, and $P = \{ p_1, \ldots, p_m \}$ are the atomic predicate symbols. We will work in the monosorted case for simplicity, but the whole development extends strightforwardly to the manysorted case.

Definition 1. *The sets of* programs *and* formulas *on Σ are the smallest sets $Prg(\Sigma)$ and $For(\Sigma)$ satisfying:*

1. *$a \in Prg(\Sigma)$ for all $a \in A$.*
2. *If $r, s \in Prg(\Sigma)$, then $\{r^*, r \cup s, r;s\} \subseteq Prg(\Sigma)$.*
3. *If $\alpha \in For(\Sigma)$, then $\alpha? \in Prg(\Sigma)$, and is called a* test program.
4. *The set of classical first-order atomic formulas on the signature Σ is contained in $For(\Sigma)$.*
5. *If $\alpha, \beta \in For(\Sigma)$ and x is a variable, then $\{\neg\alpha, \alpha \vee \beta, (\exists x)\, \alpha\} \subseteq For(\Sigma)$.*
6. *If $\alpha \in For(\Sigma)$ and $p \in Prg(\Sigma)$, then $\langle p \rangle\, \alpha \in For(\Sigma)$.*

As is standard in dynamic logic, states are valuations of the (state) variables. The set of states will be denoted by S. For the rest of the paper we will assume a fixed (but arbitrary) structure $\mathcal{S} = \langle \mathbf{s}, m_{\mathcal{S}} \rangle$, where $\mathbf{s}$ is the *carrier* of the structure and $m_{\mathcal{S}}$ is the meaning function. The meaning function operates on functions and predicates as is standard in classical first-order logic. Given an action symbol $a \in A$, $m_{\mathcal{S}}(a) \subseteq S \times S$. Given a term t denoting an object from $\mathbf{s}$ and a state ν, $m_\nu(t)$ denotes the value of t in the state ν. When $\mathcal{S}$ is fixed, we will use just m instead of $m_{\mathcal{S}}$. The semantics of complex actions and formulas is given in the next definition. The notation $\mathcal{S}, \nu \models_{DL} \alpha$, is to be read "*the formula α is satisfied in the structure $\mathcal{S}$ by the state ν*". When $\mathcal{S}$ is clear from the context, we will use the notation $\nu \models_{DL} \alpha$ with the same meaning.

Definition 2. *The semantics of programs and formulas is given as follows.*

1. *If $a \in A$, then $m(a)$ is already defined.*
2. *If $a = b^*$, with $b \in Prg(\Sigma)$, then $m(a)$ is the reflexive-transitive closure of the binary relation $m(b)$.*
3. *If $a = b \cup c$, with $b, c \in Prg(\Sigma)$, then $m(a)$ is the union of the binary relations $m(b)$ and $m(c)$.*
4. *If $a = b;c$, with $b, c \in Prg(\Sigma)$, then $m(a)$ is the composition of the binary relations $m(b)$ and $m(c)$.*
5. *If $a = \alpha?$ with $\alpha \in For(\Sigma)$, then $m(a) = \{\langle \nu, \nu \rangle : \nu \models_{DL} \alpha\}$.*
6. *If $\varphi = p(t_1, \ldots, t_n)$ with $p \in P$, $\nu \models_{DL} \varphi$ if $\langle m_\nu(t_1), \ldots, m_\nu(t_n) \rangle \in m(p)$.*
7. *If $\varphi = \neg\alpha$, then $\nu \models_{DL} \varphi$ if $\nu \not\models_{DL} \alpha$.*
8. *If $\varphi = \alpha \vee \beta$, then $\nu \models_{DL} \varphi$ if $\nu \models_{DL} \alpha$ or $\nu \models_{DL} \beta$.*
9. *If $\varphi = (\exists x)\alpha$, then $\nu \models_{DL} \varphi$ if there exists $a \in \mathbf{s}$ such that $\nu_x^a \models_{DL} \alpha$ (ν_x^a denotes the valuation that agrees with ν in all variables but x, and satisfies $\nu_x^a(x) = a$).*
10. *If $\varphi = \langle p \rangle\, \alpha$, then $\nu \models_{DL} \varphi$ if there exists a state ν' such that $\langle \nu, \nu' \rangle \in m(p)$ and $\nu' \models_{DL} \alpha$.*

3 Omega Closure Fork Algebras

We begin this section presenting the *calculus for closure fork algebras* (CCFA), an extension of the *calculus of relations* (CR) [16] and of the *calculus of relations with fork* [2]. Because the theory of first-order dynamic logic is not recursively enumerable, the CCFA with its recursively enumerable theory cannot be the target of the interpretation. In order to overcome this restriction, in Def. 6 we will define the calculus ω-CCFA by extending the CCFA with an appropriate infinitary equational inference rule. In order to give a better understanding of the semantics of the calculi, in Def. 9 we define the class of *proper closure fork algebras*. In Thm. 1 we present a representation theorem showing that every model of ω-CCFA is isomorphic to some proper closure fork algebra.

Definition 3. *Given a set of relation symbols R, the set of* CCFA *terms on R is the smallest set T(R) satisfying:*

1. $R \cup \{0, 1, 1', 1'_s\} \subseteq T(R)$,
2. *If* $x, y \in T(R)$, *then* $\{\overline{x}, \breve{x}, x^*, x^\diamond, x+y, x \cdot y, x ; y, x \nabla y\} \subseteq T(R)$.

Definition 4. *Given a set of relation symbols R, the set of* CCFA *formulas on R is the set of identities* $t_1 = t_2$, *with* $t_1, t_2 \in T(R)$.

In order to define the calculus CCFA it only remains to provide the axioms and inference rules. Since the calculus of relations extends the Boolean calculus, we will denote by $\leq$ the ordering induced by the Boolean calculus in CCFA. As is usual, $x \leq y$ is a shorthand for $x+y = y$.

Definition 5. *The identities described in 1 to 5 below are the axioms of* CCFA.

1. *A set of identities axiomatizing the relational calulus [16].*
2. *The following three identities for the fork operator:*

$$x \nabla y = (x; (1' \nabla 1)) \cdot (y; (1 \nabla 1')), \qquad (Ax.\ 1)$$

$$(x \nabla y) ; (z \nabla w)^{\smile} = (x ; \breve{z}) \cdot (y ; \breve{w}), \qquad (Ax.\ 2)$$

$$(1' \nabla 1)^{\smile} \nabla (1 \nabla 1')^{\smile} \leq 1'. \qquad (Ax.\ 3)$$

3. *The following three axioms for the choice operator, taken from [11, p. 324]:*

$$x^\diamond ; 1 ; \breve{x}^\diamond \leq 1', \qquad (Ax.\ 4)$$

$$\breve{x}^\diamond ; 1 ; x^\diamond \leq 1', \qquad (Ax.\ 5)$$

$$1; (x \cdot x^\diamond) ; 1 = 1 ; x ; 1. \qquad (Ax.\ 6)$$

4. *The following two axioms for the Kleene star:*

$$x^* = 1' + x ; x^*, \qquad (Ax.\ 7)$$

$$x^* ; y \leq y + x^* ; (\overline{y} \cdot x ; y). \qquad (Ax.\ 8)$$

5. *Let us denote by* $1'_{\mathsf{U}}$ *the term* $\left(\left(\overline{1\nabla 1}\right)^{\smile};\overline{1\nabla 1}\right)\cdot 1'$. *Then, the following axiom is added*

$$1;1'_{\mathsf{U}};1 = 1\ . \qquad (Ax.\ 9)$$

The inference rules for the calculus CCFA are those of equational logic (see for instance [1, p. 94]). For the next definition, given $i > 0$, by x^i we denote the relation inductively defined as follows: $x^1 = x$, and $x^{i+1} = x;x^i$.

Definition 6. *We define the calculus* ω-CCFA *as the extension of the* CCFA *obtained by adding the following inference rule:*

$$\frac{\vdash 1' \leq y \qquad x^i \leq y \vdash x^{i+1} \leq y}{\vdash x^* \leq y} \qquad (i \in \mathbb{N})$$

Definition 7. *A model of the identities provable in* ω-CCFA *will be called an* omega closure fork algebras. *The class of omega closure fork algebras is denoted by* ω-CFA.

The intended (standard) models of the ω-CCFA are the *Proper Closure Fork Algebras* (PCFA for short). In order to define the class PCFA, we will first define the class of *Pre Proper Closure Fork Algebras*, denoted by $\bullet$PCFA.

Definition 8. *Let* E *be a binary relation on a set* U, *and let* R *be a set of binary relations satisfying:*

1. $\bigcup R \subseteq E$,
2. Id *(the identity relation on the set* A*),* $\emptyset$ *(the empty binary relation) and* E *belong to* R,
3. R *is closed under set union* ($\cup$)*, intersection* ($\cap$) *and complement relative to* E ($^-$)*,*
4. R *is closed under relational composition (denoted by* $\circ$*), converse (denoted by* $^\smile$*), and reflexive-transitive closure (denoted by* **).*
5. $\star : U \times U \to U$ *is one-to-one.*
6. R *is closed under the* fork *operator* (∇)*, defined by the condition*

$$S \nabla T = \{\langle x, y \star z\rangle : \langle x, y\rangle \in S \text{ and } \langle x, z\rangle \in T\}\ .$$

7. R *is closed under* $^\diamond$*, the set choice operator defined by the condition:*

$$x^\diamond \subseteq x \text{ and } |x^\diamond| = 1 \iff x \neq \emptyset\ .$$

Then, the structure $\langle R, U, \cup, \cap, ^-, \emptyset, E, \circ, Id, ^\smile, \nabla, ^\diamond, ^*, \star\rangle$ *is a* $\bullet$PCFA.

Notice that $x^\diamond$ denotes an arbitrary pair in x. This is why $x^\diamond$ is called a *choice* operator. We will call the set U in Def. 8 the *field* of the algebra, and will denote the field of an algebra $\mathfrak{A}$ by $U_{\mathfrak{A}}$.

Definition 9. *We define the class* PCFA *as* **Rd**$\bullet$PCFA *where* **Rd** *takes reducts to structures of the form* $\langle R, \cup, \cap, ^-, \emptyset, E, \circ, Id, ^\smile, \nabla, ^\diamond, ^*\rangle$.

Notice that given $\mathfrak{A} \in$ PCFA, the terms $(1' \nabla 1)^{\smile}$ and $(1 \nabla 1')^{\smile}$ denote respectively the binary relations $\{ \langle a \star b, a \rangle : a, b \in U_{\mathfrak{A}} \}$ and $\{ \langle a \star b, b \rangle : a, b \in U_{\mathfrak{A}} \}$. Thus, they behave as projections with respect to the injection $\star$. We will denote these terms by π and ρ, respectively. If we call *splitting* an object $a \in U_{\mathfrak{A}}$ for which there exist $b, c \in U_{\mathfrak{A}}$ such that $a = b \star c$, then $1'_U$ is a part of the identity relation whose domain contains all the non–splitting objects. Ax. 9 states that the set on non–splitting objects (that we will call *urelements*) is nonempty. We denote the set of urelements of a PCFA $\mathfrak{A}$ by $Urel_{\mathfrak{A}}$. From the fork operator we define the binary operator $\otimes$ (*cross*) by the condition $x \otimes y = (\pi ; x) \nabla (\rho ; y)$. When interpreted in an algebra $\mathfrak{B} \in$ PCFA, $\otimes$ behaves as a parallel product:

$$x \otimes y = \{ \langle a \star b, c \star d \rangle : \langle a, c \rangle \in x \ \wedge \ \langle b, d \rangle \in y \} .$$

Definition 10. *Given an algebra $\mathfrak{A}$ and a class of algebras K, $\mathfrak{A}$ is* representable *in K if there exists $\mathfrak{B} \in K$ such that $\mathfrak{A}$ is isomorphic to $\mathfrak{B}$. If h is such an isomorphism, then the pair $\langle \mathfrak{B}, h \rangle$ is a* representation of $\mathfrak{A}$ in K. *This notion generalizes as follows: a class of algebras K_1 is* representable in a class of algebras *K_2 if every member of K_1 is representable in K_2.*

We finally present the representation theorem for omega closure fork algebras, which is reproduced and proved as Thm 3 in the Appendix.

Theorem 1. *The class ω-CFA is representable in* PCFA.

4 Interpretability of *DL* in ω-CCFA

For the following definitions, σ and τ will be sequences of numbers increasingly sorted. Intuitively, these sequences will contain the indices of those variables that occur free in the formulas (or terms) being translated.

Notation 1.

- *$Pos(n, \sigma)$ denotes the position of the index n in the sequence σ.*
- *$\sigma \oplus n$ denotes the extension of the sequence σ with the index n.*
- *$\sigma(k)$ projects the element in the k-th position of σ.*
- *If τ is a subsequence of σ, $\sigma - \tau$ denotes the sequence containing those indices in σ but not in τ.*
- *Let τ and σ be disjoint and with $Length(\tau) = l_1$ and $Length(\sigma) = l_2$. Let $a = a_1 \star \cdots \star a_{l_1}$ and $b = b_1 \star \cdots \star b_{l_2}$ encode the values for the variables whose indices occur in τ and σ, respectively. $Merge_{\tau,\sigma}$, from the input $a \star b$, builds the encoding for the values of those variables whose indices occur either in τ or σ.*
  Example: Let $\tau = [3, 5]$ and $\sigma = [2, 4]$. Let $a = a_1 \star a_2$ and $b = b_1 \star b_2$. Then, $Merge_{\tau,\sigma}$ sends the object $a \star b$ to the object $b_1 \star a_1 \star b_2 \star a_2$. The relational expression for $Merge_{\tau,\sigma}$ (which of course depends on the specific τ and σ being considered) is $(\rho ; \delta_\sigma(v_2)) \nabla (\pi ; \delta_\tau(v_3)) \nabla (\rho ; \delta_\sigma(v_4)) \nabla (\pi ; \delta_\tau(v_5))$.
- *Finally, given a formula or term e, by σ_e we denote the sequence of all indices of variables with free occurrences in e.*

A relation x is right ideal if $x = x;1$. Intuitively, each element in the domain is related to all the elements in the universe.

Definition 11. *By* ω-CCFA$^+(\Sigma)$ *we denote the extension of* ω-CCFA *obtained by adding the following equations as axioms.*

1. $1'_s \leq 1'_U$ *(elements from s do not split).*
2. *In the first paragraph of Section 2 we assumed that actions modify finitely many state variables. If* $a \in A$ *reads or modifies k state variables, we add the equation*
$$\underbrace{(1'_s \otimes \cdots \otimes 1'_s)}_{k\text{-times}};a;\underbrace{(1'_s \otimes \cdots \otimes 1'_s)}_{k\text{-times}} = a\ .$$
3. *For each* $f : s^k \to s \in F$, *we add the equation*
$$\breve{f};\underbrace{(1'_s \otimes \cdots \otimes 1'_s)}_{k\text{-times}};f \leq 1'_s,$$
stating that f is a functional relation of the right arity.
4. *For each* $p \in P$ *of arity k, we add the equation*
$$\underbrace{(1'_s \otimes \cdots \otimes 1'_s)}_{k\text{-times}};p;1 = p,$$
stating that p is a binary right-ideal relation relating the right amount of inputs.

Notice that given finite A, F and P, only finitely many equations are introduced in items 1–4 above. As before, we will assume a fixed but arbitrary signature Σ, and use the notation ω-CCFA$^+$ as a shorthand for ω-CCFA$^+(\Sigma)$.

Definition 12. *A* model *for* ω-CCFA$^+$ *is a structure* $\mathcal{A} = \langle \mathfrak{A}, m_{\mathcal{A}} \rangle$ *where* $\mathfrak{A} \in$ ω-CFA *and* $m_{\mathcal{A}}$ *is the meaning function. When* $\mathcal{A}$ *is clear from context, we will note* $m_{\mathcal{A}}$ *simply by m. The meaning function operates on the language of* ω-CCFA$^+$ *as follows.*

1. $m(s)$ *is a partial identity belonging to* $\mathfrak{A}$.
2. *For each* $a \in A$ *reading or modifying k variables,* $m(a)$ *satisfies the typing condition in item 2 of Def. 11, and extends homomorphically to complex action terms.*
3. *For each* $f \in F$, $m(f) \in \mathfrak{A}$ *is a functional relation. If* $f : s^k \to s$, *then* $m(f)$ *satisfies the typing condition in item 3 of Def. 11.*
4. *For each* $p \in P$, $m(p)$ *is a right-ideal relation. If p has arity k, then* $m(p)$ *satisfies the typing condition in item 4 of Def. 11.*

Definition 13. *An* ω-CCFA$^+$ *model* $\langle \mathfrak{A}, m \rangle$ *is* full *if* $\mathfrak{A}$ *has domain* $\mathcal{P}(S \times S)$ *for some nonempty set S, and the operations have their intended set-theoretical meaning.*

In the following paragraphs we will define a function mapping $For(\Sigma)$ to terms in ω-CCFA$^+$. The definition proceeds in two steps, because terms (as well as formulas) must be translated.

Notation 2. *We will denote by $IT(S)$ the set of terms over individuals from DL that can be built from constant and function symbols belonging to S.*

Definition 14. *The function $\delta_\sigma : IT(F) \to T(F)$, mapping individual terms whose variables have indices in σ to relational terms, is defined inductively by the conditions:*

$$\delta_\sigma(v_i) = \begin{cases} \rho^{Pos(i,\sigma)-1};\pi & \text{if } i \text{ is not the last index in } \sigma \,, \\ \rho^{Length(\sigma)-1} & \text{if } i \text{ is the last index in } \sigma \,. \end{cases}$$

$$\delta_\sigma(f(t_1,\ldots,t_m)) = (\delta_\sigma(t_1)\nabla\cdots\nabla\delta_\sigma(t_m));f \text{ for each } f\in F \,.$$

Before defining the mapping T_σ translating $For(\Sigma)$, we need to define some auxiliary terms. For σ of length l, $n \in \mathbb{N}$ and $k = Pos(n, \sigma\oplus n)$, we define the term $\Delta_{\sigma,n}$ by the condition[1]

$$\Delta_{\sigma,n} = \begin{cases} \delta_\sigma(v_{\sigma(1)})\nabla\cdots\nabla\delta_\sigma(v_{\sigma(k-1)})\nabla 1_s\nabla\delta_\sigma(v_{\sigma(k)})\nabla\cdots\nabla\delta_\sigma(v_{\sigma(l)}) & \text{if } k\le l, \\ \delta_\sigma(v_{\sigma(1)})\nabla\cdots\nabla\delta_\sigma(v_{\sigma(l)})\nabla 1_s & \text{if } k=l+1. \end{cases}$$

The term $\Delta_{\sigma,n}$ can be understood as a cylindrification [7, 8] in the k–th coordinate of an l-dimensional space.

Given a subsequence, τ, of σ, of length k, by $\Pi_{\sigma,\tau}$ we denote the term

$$\delta_\sigma(v_{\tau(1)})\nabla\cdots\nabla\delta_\sigma(v_{\tau(k)}) \,.$$

This term, given an object storing values for the variables whose indices occur in σ, builds an object storing values for the variables occurring in τ.

In the next definition we present functions M_σ and T_σ mapping action terms and formulas, respectively, to relational terms. The general assumption in the definition, is that whenever we apply M_σ to a program p, σ includes all the indices of state variables occurring in p. Similarly, if we apply T_σ to a formula α, then σ includes all the indices of variables with free occurrences in α.

Definition 15. *The functions M_σ and T_σ are mutually defined by*

1. $M_\sigma(a) = (\Pi_{\sigma,\sigma_a};a \;\nabla\; \Pi_{\sigma,\sigma-\sigma_a});Merge_{\sigma_a,\sigma-\sigma_a}$, *for each* $a\in A$,
2. $M_\sigma(R^*) = M_\sigma(R)^*$,
3. $M_\sigma(R\cup S) = M_\sigma(R)+M_\sigma(S)$,
4. $M_\sigma(R;S) = M_\sigma(R);M_\sigma(S)$,
5. $M_\sigma(\alpha?) = T_\sigma(\alpha)\cdot 1'$,
6. $T_\sigma(p(t_1,\ldots,t_n)) = (\delta_\sigma(t_1)\nabla\cdots\nabla\delta_\sigma(t_n));p$,
7. $T_\sigma(\neg\alpha) = \overline{T_\sigma(\alpha)}$,
8. $T_\sigma(\alpha\vee\beta) = T_\sigma(\alpha)+T_\sigma(\beta)$,
9. $T_\sigma((\exists v_i)\,\alpha) = \Delta_{\sigma,i};T_{\sigma\oplus i}(\alpha)$,

[1] By 1_s we denote the relation $1;1'_s$.

10. $T_\sigma\left(\langle p\rangle\,\alpha\right) = M_\sigma(p);T_\sigma(\alpha)$.

Notation 3. *We will denote by* $\vdash_{\omega\text{-CCFA}}$ *the entailment relation in the calculus* ω-CCFA.

As usual, a formula α will be called a *sentence* if it does not contain any free occurrences of variables. The next theorem states the interpretability of presentations of theories from DL as equational theories in ω-CCFA. A detailed sketch of the proof is given in the Appendix, Thm. 8.

Theorem 2. *Let* $\Gamma \cup \{\varphi\}$ *be a set of sentences. Then,*

$$\Gamma \models_{DL} \varphi \qquad \Longleftrightarrow \qquad \{\, T_{\langle\rangle}(\gamma) = 1 : \gamma \in \Gamma \,\} \vdash_{\omega\text{-CCFA+}} T_{\langle\rangle}(\varphi) = 1\ .$$

5 Conclusions

Since the interpretability of first-order dynamic logic allows one to reason about UML–like diagrams with dynamic content in an equational calculus, the results in this paper are at the heart of relational methods in computer science.

The contributions of this paper can be summarized as follows:

1. Definition of the class of omega closure fork algebras.
2. Definition of the class of proper closure fork algebras.
3. Proof of a representation theorem of ω-CFA in PCFA.
4. Definition of a complete calculus for PCFA.
5. Proof of an interpretability theorem of first-order dynamic logic in the calculus ω-CCFA.

Reasoning about properties of UML–like diagrams with dynamic content can be done in two ways. One way is by using formulas from dynamic logic to represent diagrams, and then translate the formulas to relational equations. The other way is by skipping the passage through dynamic logic and using relational equations directly.

References

1. Burris, S. and Sankappanavar, H.P., *A Course in Universal Algebra*, Graduate Texts in Mathematics 78, Springer–Verlag, 1981.
2. Frias M. F., Baum G. A. and Haeberer A. M., *Fork Algebras in Algebra, Logic and Computer Science*, Fundamenta Informaticae Vol. 32 (1997), pp. 1–25.
3. Frias, M. F., Haeberer, A. M. and Veloso, P. A. S., *A Finite Axiomatization for Fork Algebras*, Logic Journal of the IGPL, Vol. 5, No. 3, 311–319, 1997.
4. Frias, M. F. and Orlowska E., *Equational Reasoning in Non–Classical Logics*, Journal of Applied Non Classical Logic, Vol. 8, No. 1–2, 1998.
5. Gyuris, V., *A Short Proof for Representability of Fork Algebras*, Theoretical Computer Science, vol. 188, 1–2, pp. 211–220, 1997.
6. Harel, D., Kozen, D. and Tiuryn, J., *Dynamic Logic*, MIT Press, 2000.

7. Henkin, L., Monk, D. and Tarski, A., *Cylindric Algebras Part I*, Studies in Logic and the Foundations of Mathematics, vol.64, North–Holland, 1971.
8. Henkin, L., Monk, D. and Tarski, A., *Cylindric Algebras Part II*, Studies in Logic and the Foundations of Mathematics, vol.115, North–Holland, 1985.
9. Lyndon, R., *The Representation of Relational Algebras*, Annals of Mathematics (Series 2) vol.51 (1950), 707–729.
10. Maddux, R.D., *Topics in Relation Algebras*, Doctoral Dissertation, Univ. California, Berkeley, 1978, pp. iii+241.
11. Maddux, R.D., *Finitary Algebraic Logic*, Zeitschr. f. math. Logik und Grundlagen d. Math. vol. 35, pp. 321–332, 1989.
12. Maddux, R.D., *Pair-Dense Relation Algebras*, Transactions of the A.M.S., vol. 328, No. 1, pp. 83–131, 1991.
13. Maddux, R.D., *Relation Algebras*, Chapter 2 of Relational Methods in Computer Science, Springer Wien–New York, 1997.
14. Orlowska, E., *Relational proof systems for modal logics.* In: Wansing,H. (ed) Proof Theory of Modal Logics. Kluwer, pp. 55–77, 1996.
15. Rasiowa, H. and Sikorski, R. *The Mathematics of Metamathematics.* Polish Science Publishers, Warsaw, 1963.
16. Tarski, A., *On the Calculus of Relations*, Journal of Symbolic Logic, Vol. 6, 73–89, 1941.
17. Tarski, A. and Givant, S.,*A Formalization of Set Theory without Variables*, A.M.S. Coll. Pub., vol. 41, 1987.

A On ω-CCFA

Theorem 3. *The class* ω-CFA *is representable in* PCFA.

Proof. Let $\mathfrak{A} \in \omega$-CFA. Let us consider the fork algebra reduct $\mathfrak{A}'$ of the algebra $\mathfrak{A}$. By the representation theorem for fork algebras [3, 5] there exists a representation $\langle \mathfrak{C}', h \rangle$ of $\mathfrak{A}'$. Let us consider the structure $\mathfrak{C} = \langle \mathfrak{C}', \underline{^\diamond}, \underline{^*} \rangle$, where $\underline{^\diamond}$ and $\underline{^*}$ are defined by the conditions:

$$x^{\underline{\diamond}} = h\left(\left(h^{-1}(x)\right)^{\diamond}\right) \text{ and } x^{\underline{*}} = h\left(\left(h^{-1}(x)\right)^{*}\right) \quad \text{for each } x \in \mathfrak{C}'.$$

It is easy to check that $h : \mathfrak{A} \to \mathfrak{C}$ is an ω-CFA isomorphism. It is also easy to check using the infinitary rule that for $x \in \mathfrak{A}$, $x^* = \mathbf{lub}\{ x^i : i \in \omega \}$. Thus, using the isomorphism h, we can prove that for each $x \in \mathfrak{C}$,

$$x^{\underline{*}} = \mathbf{lub}\{ x^i : i \in \omega \}. \tag{1}$$

Notice that it is not necessarily the case that for $x \in \mathfrak{C}$, $x^{\underline{*}}$ equals the reflexive-transitive closure of x, since infinite sums may not correspond to the infinite union of the binary relations[2]. Let $\mathfrak{D}$ be $\mathfrak{C}$'s relation algebra reduct. Since $\mathfrak{D}$ is point–dense (see [11, 12] for details on point-density), by [11, Thm. 8] $\mathfrak{D}$ has a complete representation[3] $\langle \mathfrak{D}', g \rangle$. Since the representation is complete, g satisfies

$$g(\mathbf{lub}\{ x_i : i \in I \}) = \bigcup_{i \in I} g(x_i) \ .$$

Let $\mathfrak{B} = \langle \mathfrak{D}', \nabla, ^\diamond, ^* \rangle$, where ∇, $^\diamond$ and * are defined as follows:

$$\begin{aligned} x \nabla y &= g\left(g^{-1}(x) \underline{\nabla} g^{-1}(y)\right), \\ x^\diamond &= g\left(\left(g^{-1}(x)\right)^{\underline{\diamond}}\right), \\ x^* &= g\left(\left(g^{-1}(x)\right)^{\underline{*}}\right). \end{aligned}$$

Then, $g : \mathfrak{C} \to \mathfrak{B}$ is also an isomorphism. That ∇ as defined is a fork operation on $\mathfrak{B}$ follows from the fact relations $g(\pi)$ and $g(\rho)$ are a pair of quasi-projections [11, 17] in $\mathfrak{B}$. Since $^\diamond$ satisfies Ax. 4–Ax. 6, it is a choice operator. Finally, let us check that x^* is the reflexive-transitive closure of x, for each $x \in \mathfrak{B}$.

2 Examples of proper relation algebras in which $\mathbf{lub}A$ does not agree with $\bigcup A$ (for some subset A of the universe) are given in [9, 10] and elsewhere.

3 A representation $\langle \mathfrak{A}, h \rangle$ is complete if h maps infinite suprema into the corresponding infinite set unions.

$$
\begin{aligned}
x^* &= g\left(\left(g^{-1}(x)\right)^{\underline{*}}\right) && \text{(by Def. } x^*\text{)}\\
&= g\left(\mathbf{lub}\left\{\left(g^{-1}(x)\right)^i : i \geq 0\right\}\right) && \text{(by (1))}\\
&= \bigcup_{i\geq 0} g\left(\left(g^{-1}(x)\right)^i\right) && (g \text{ complete representation})\\
&= \bigcup_{i\geq 0} \left(g\left(g^{-1}(x)\right)\right)^i && (g \text{ homomorphism})\\
&= \bigcup_{i\geq 0} x^i . && (g \text{ one-to-one})
\end{aligned}
$$

The last union computes the reflexive-transitive closure of the relation x, as was to be proved, and therefore $\mathfrak{B} \in$ PCFA. Finally, $\langle \mathfrak{B}, g \circ h \rangle$ is a representation of the ω-CFA $\mathfrak{A}$ in PCFA. □

B On the Interpretability of DL in the ω-CCFA$^+$

Notation 4. *Given a structure $\mathcal{A} = \langle \mathbf{s}, m \rangle$, a valuation of the individual variables ν, and $\mathfrak{A} \in$ PCFA such that $Urel_{\mathfrak{A}} \supseteq \mathbf{s}$, by $s_{\nu,\sigma}$ we denote the element $a_1 \star \cdots \star a_i \star \cdots \star a_n \in U_{\mathfrak{A}}$ such that:*

1. *$n = Length(\sigma)$,*
2. *$a_i = \nu(v_{\sigma(i)})$ for all i, $1 \leq i \leq n$.*

In case $\sigma = \langle\rangle$, $s_{\nu,\sigma}$ denotes an arbitrary element from $U_{\mathfrak{A}}$.

Definition 16. *By the FullPCFA with set of urelements U (U a nonempty set), we refer to the algebra $\mathfrak{A}$ constructed as follows:*

1. *Let $\langle U^\star, \star \rangle$ be the totally free groupoid with set of free generators U.*
2. *Let $\mathfrak{B}$ be the full algebra of binary relations with field $U^\star$.*
3. *Let $\mathfrak{A} = \langle \mathfrak{B}, \nabla, {}^\diamond, {}^* \rangle$ where ∇, ${}^\diamond$ and * are defined by:*

$$
\begin{aligned}
R \nabla S &= \{ \langle x, y \star z \rangle : x R y \ \wedge \ x S z \} \ .\\
R^\diamond &= \{ \langle a, b \rangle \}, \textit{with } \langle a, b \rangle \in R, \ \textit{arbitrary} \ .\\
R^* &= \textit{the reflexive-transitive closure of } R \ .
\end{aligned}
$$

Definition 17. *A ω-CCFA$^+$ model $\langle \mathfrak{A}, m \rangle$ is* full *if $\mathfrak{A}$ is a FullPCFA.*

Theorem 4. *Let $\alpha \in For(S)$ and let $\mathcal{A}$ be a structure for Σ. Then there exists a full ω-CCFA$^+$ model $\mathcal{B} = \langle \mathfrak{B}, m' \rangle$ such that*[4]

$$
\mathcal{A}, \nu \models_{DL} \alpha \qquad \Longleftrightarrow \qquad s_{\nu,\sigma_\alpha} \in \mathsf{dom}\left(m'\left(T_{\sigma_\alpha}(\alpha)\right)\right) \ .
$$

[4] Given a binary relation R, by $\mathsf{dom}(R)$ we denote the domain set of R.

Sketch of the Proof. Assume $\mathcal{A} = \langle \mathbf{s}, m \rangle$. Let us define $\mathcal{B}$ as follows.

1. Let $\mathfrak{B}$ be the FullPCFA with set of urelements $\mathbf{s}$,
2. $m'(a) = \{ \langle s_{\nu,\sigma_a}, s_{\nu',\sigma_a} \rangle : \langle \nu, \nu' \rangle \in m(a) \}$, for each $a \in A$,
3. $m'(f) = \{ \langle a_1 \star \cdots \star a_i, b \rangle : m(f)(a_1, \ldots, a_i) = b \}$, for each $f \in F$,
4. $m'(p) = \{ \langle a_1 \star \cdots \star a_i, b \rangle : m(p)(a_1, \ldots, a_i)$ and $b \in U_{\mathfrak{B}} \}$, for each $p \in P$.

The proof continues by proving by simultaneous induction on the structure of programs and formulas the following properties. Let σ be an arbitrary sequence of indices. Then:

a) for each $R \in RT(\Sigma)$ with σ extending σ_R,

$$m'(M_\sigma(R)) = \{ \langle s_{\nu,\sigma}, s_{\nu',\sigma} \rangle : \langle \nu, \nu' \rangle \in m(R) \} .$$

b) for each $\beta \in For(S)$ and σ extending σ_β,

$$\mathsf{dom}\,(m'\,(T_\sigma(\beta))) = \{ s_{\nu,\sigma} : \mathcal{A}, \nu \models_{DL} \beta \} .$$

Finally, the theorem follows from b) choosing σ to be σ_α. □

Corollary 1. *Let $\alpha \in For(\Sigma)$ be a sentence, and let $\mathcal{A}$ be a structure for Σ. Then there exists a full ω-CCFA$^+$ model $\mathcal{B} = \langle \mathfrak{B}, m \rangle$ such that*

$$\mathcal{A} \models_{DL} \alpha \qquad \Longleftrightarrow \qquad m\,(T_{\langle\rangle}\,(\alpha)) = 1 .$$

Proof.

$$\begin{aligned} \mathcal{A} \models_{DL} \alpha &\iff \mathcal{A}, \nu \models_{DL} \alpha \text{ for each } \nu && \text{(by Def. } \models_{DL}) \\ &\iff s_{\nu,\langle\rangle} \in \mathsf{dom}\left(m\left(T_{\langle\rangle}(\alpha)\right)\right) \text{ for each } \nu . && \text{(by Thm. 4)} \end{aligned}$$

Since $s_{\nu,\langle\rangle}$ denotes an arbitrary element from $U_{\mathfrak{B}}$,

$$\mathcal{A} \models_{DL} \alpha \qquad \Longleftrightarrow \qquad e \in \mathsf{dom}\,(m\,(T_{\langle\rangle}(\alpha))) \text{ for each } e \in U_{\mathfrak{B}} .$$

Finally, since $m\left(T_{\langle\rangle}(\alpha)\right)$ is a right-ideal relation,

$$\mathcal{A} \models_{DL} \alpha \qquad \Longleftrightarrow \qquad m\,(T_{\langle\rangle}(\alpha)) = 1 .$$

□

Notation 5. *Let $\mathfrak{A} \in$ PCFA, $s = a_1 \star \cdots \star a_k$ ($a_j \in Urel_{\mathfrak{A}}$ for all j, $1 \leq j \leq k$), and i such that $0 \leq i \leq k$. By $\nu_{s,i}$ we denote the set of valuations of individual variables ν satisfying $\nu(v_j) = a_j$ for all $1 \leq j \leq i$. In case $i = 0$, for each $s \in U_{\mathfrak{A}}$ $\nu_{s,i}$ denotes the set of all valuations.*

Theorem 5. *Let $\alpha \in For(\Sigma)$ and $\mathcal{A} = \langle \mathfrak{A}, m \rangle$ be a full ω-CCFA$^+$ model. Then there exists a structure $\mathcal{B}$ such that*

$$s \in \mathsf{dom}\,(m\,(T_{\sigma_\alpha}(\alpha))) \qquad \Longleftrightarrow \qquad \mathcal{B}, \nu \models_{DL} \alpha \text{ for all } \nu \in \nu_{s,\sigma_\alpha} .$$

Sketch of the Proof. Let $\mathcal{B} = \langle \mathbf{s}, m' \rangle$ with

1. $\mathbf{s} = Urel_{\mathfrak{A}}$,
2. (a) For each $a \in A$, $m'(a) = \{\langle \nu, \nu' \rangle : \langle s_{\nu,\sigma_\alpha}, s_{\nu',\sigma_\alpha} \rangle \in m(a)\}$.
   (b) For each $f : s^k \to s \in F$,
$$m'(f)(a_1, \ldots, a_k) = b \quad \text{iff} \quad \langle a_1 \star \cdots \star a_k, b \rangle \in m(f) \ .$$
   (c) For each $p \subseteq s^k \in P$,
$$\langle a_1, \ldots, a_k \rangle \in m'(p) \quad \text{iff} \quad a_1 \star \cdots \star a_k \in \mathsf{dom}\,(m(p)) \ .$$

The remaining part of the proof follows by induction on the structure of terms and formulas and is analogous to the proof of Thm. 4. □

Corollary 2. *Let $\alpha \in For(S)$ be a sentence, and let $\mathcal{A} = \langle \mathfrak{A}, m \rangle$ be a full ω-CCFA$^+$ model. Then there exists a structure $\mathcal{B}$ such that*
$$\mathcal{B} \models_{DL} \alpha \quad \iff \quad m\left(T_{\langle\rangle}(\alpha)\right) = 1 \ .$$

Proof. Let $\mathcal{B}$ as in Thm. 5. Then
$$\begin{aligned}
\mathcal{B} \models_{DL} \alpha &\iff \mathcal{B}, \nu \models_{DL} \alpha \text{ for each } \nu && \text{(by Def. } \models_{DL}) \\
&\iff \text{for each } s \in U_{\mathfrak{A}}, \mathcal{B}, \nu \models_{DL} \alpha \text{ for each } \nu \in \nu_{s,\langle\rangle} && \text{(by Def. } \nu_{s,\langle\rangle}) \\
&\iff s \in \mathsf{dom}\left(m\left(T_{\langle\rangle}(\alpha)\right)\right) \text{ for each } s \in U_{\mathfrak{A}} && \text{(by Thm. 5)} \\
&\iff \mathsf{dom}\left(m\left(T_{\langle\rangle}(\alpha)\right)\right) = U_{\mathfrak{A}} && \text{(by Def. dom)} \\
&\iff m\left(T_{\langle\rangle}(\alpha)\right) = 1 \ . && \text{(by } T_{\langle\rangle}(\alpha) \text{ right-ideal)}
\end{aligned}$$
□

Notation 6. *We will denote by $\models_{Full}$ the semantic consequence relation on the class of full ω-CCFA models.*

Theorem 6. *Let $\Gamma \cup \{\varphi\}$ be a set of sentences. Then,*
$$\Gamma \models_{DL} \varphi \quad \iff \quad \{T_{\langle\rangle}(\gamma) = 1 : \gamma \in \Gamma\} \models_{\text{Full}} T_{\langle\rangle}(\varphi) = 1 \ .$$

Proof. In order to prove the theorem, it suffices to show that
$$\Gamma \not\models_{DL} \varphi \quad \iff \quad \{T_{\langle\rangle}(\gamma) = 1 : \gamma \in \Gamma\} \not\models_{\text{Full}} T_{\langle\rangle}(\varphi) = 1 \ . \tag{2}$$
Formula (2) follows from Cors. 1 and 2. □

Theorem 7. *Let $\mathcal{V}$ be the variety generated by* FullPCFA. *Then,* $\mathcal{V} = \omega$-CFA.

Proof. From Thm. 1, every ω-CFA is isomorphic to some PCFA. A proof similar to the one showing that proper relation algebras are isomorphic to subalgebras of products of full proper relation algebras, shows that[5] PCFA $\subseteq$ **ISP**FullPCFA. Thus, ω-CFA $\subseteq$ **ISP**FullPCFA $\subseteq$ **HSP**FullPCFA.

Since FullPCFA satisfies the equations defining ω-CFA and equations are preserved by **H**, **S** and **P**, **HSP**FullPCFA $\subseteq \omega$-CFA. □

[5] Given a class of algebras K, by $\mathbf{I}K$ we denote the closure of K under isomoprhisms. By $\mathbf{H}K$, the closure of K under homomorphic images. By $\mathbf{S}K$, the closure of K under subalgebras. By $\mathbf{P}K$, the closure of K under direct products.

Notation 7. *Given a class of algebras K, by $EqTh(K)$ we denote the equational theory of K, i.e., the set of all valid equations in K.*

Theorem 8. *Let $\Gamma \cup \{\varphi\}$ be a set of sentences. Then,*

$$\Gamma \models_{DL} \varphi \iff \{T_{\langle\rangle}(\gamma) = 1 : \gamma \in \Gamma\} \vdash_{\omega\text{-CCFA}} T_{\langle\rangle}(\varphi) = 1 .$$

Proof. By Thm. 6,

$$\Gamma \models_{DL} \varphi \iff \{T_{\langle\rangle}(\gamma) = 1 : \gamma \in \Gamma\} \models_{\text{Full}} T_{\langle\rangle}(\varphi) = 1 .$$

It then suffices to show that given a set of equations $E \cup \{e\}$,

$$E \models_{\text{Full}} e \iff E \vdash_{\omega\text{-CCFA}} e .$$

$\Rightarrow$)

$$\begin{aligned} E \models_{\text{Full}} e &\Rightarrow E \models_{\omega\text{-CFA}} e && \text{(by Thm. 7)} \\ &\Rightarrow E \cup EqTh\,(\omega\text{-CFA}) \models e && \text{(by Def. } \omega\text{-CFA)} \\ &\Rightarrow E \cup EqTh\,(\omega\text{-CFA}) \vdash e . && \text{(by completeness of eq. logic)} \end{aligned}$$

Notice now that an equational proof of e from $E \cup EqTh\,(\omega\text{-CFA})$ will be a finite height tree whose leaves are equations in $EqTh\,(\omega\text{-CFA})$. In order to show that $E \vdash_{\omega\text{-CCFA}} e$, it suffices to replace each leaf in $EqTh\,(\omega\text{-CFA})$ by its corresponding proof in ω-CCFA.

$\Leftarrow$) Let us proceed by induction on the height of proof trees.

- If the tree has height 1, then $e \in E$ or e is an axiom of ω-CCFA. Clearly, $E \models_{\text{Full}} e$.
- Assume the implication holds if there is a proof of an equation e' from E of height n, and let Π be a proof of e from E of height $n+1$. The only nontrivial case is when Π is of the form

$$\frac{\begin{array}{cccc} \Pi_0 & & \Pi_i & \Pi_{i+1} \\ \vdash 1' \leq y & \cdots & \vdash x^i \leq y & \vdash x^{i+1} \leq y \quad \cdots \end{array}}{\vdash x^* \leq y}$$

By inductive hypothesis,

$$E \models_{\text{Full}} 1' \leq y,\ \ldots, E \models_{\text{Full}} x^i \leq y,\ E \models_{\text{Full}} x^{i+1} \leq y, \ldots ,$$

which implies $E \models_{\text{Full}} x^* \leq y$.

□

Relations in GUHA Style Data Mining

Petr Hájek

Institute of Computer Science AS CR,
Pod Vodárenskou věží 2, Prague 8, Czech Republic

Abstract. The formalism of GUHA style data mining is confronted with the approach of relational structures of Orlowska and others. A computational complexity result on tautologies with implicational quantifiers is presented.

1 Introduction

The first aim of the present paper is to compare the logic of observational calculi, presented is depth in the monograph [5] and serving as foundation for the GUHA method of automated generation of hypotheses, (see e.g. [6,7]) with the formalism of relational structures as presented in [11] (see e.g. Chapter 16 by Düntsch and Orlowska). Our main aim is to contribute to "building bridges" between these two approaches by showing some possiblities of mutual influence and application. For this purpose we survey, in Section 2,various kinds of relations emerging in this context and present, as an illustration, the notion of an i-algebra (Boolean algebra with an implicational relation) and formulate a corresponding representability problem. The second aim is to present a result on the computational complexity of an important class of logical formulas called implicational tautologies, i.e. formulas built using a binary *quantifier* $\Rightarrow^*$ and logically true for each interpretation of $\Rightarrow^*$ as an implicational quantifier (in the sense of GUHA): the set of all such tautologies is shown to be co-NP-complete.

The two parts of the paper can be read independently of each other; the reader not familiar with GUHA finds in the first part all definitions needed for the second part. Interestingly, the second part makes non-trivial use of mathematical fuzzy logic.

Support of COST Action 274 TARSKI (Theory and Applications of Relational Structures as Knowledge Instruments) is acknowledged. The second part of the paper is also relevant for the grant project A13004/00 of the Grant Agency of the Academy of Sciences of the Czech Republic. Thanks are due to an anonymous referee, whose remarks helped to improve the paper.

2 Relations – Where Are They from

One basic notion common to both approaches is that called *information system* **U** in the terminology of [9,10,12,13] and *data matrix* or *observational model* in GUHA. It is given by a finite non-empty set U of objects, a finite non-empty set A of attributes, each $a \in A$ having a finite domain V_a and an evaluation

H. de Swart (Ed.): RelMiCS 2001, LNCS 2561, pp. 81–87, 2002.

function f_a assigning to each $u \in U$ an element $f_a(u) \in V_a$. More generally, one may assume $f_a(u) \subseteq V_a$; less generally, one may assume $f_a(u) \in \{0,1\}$, i.e. $V_a = \{0,1\}$ for all a. Obviously, the most general notion is reducible to particular case of $\{0,1\}$-valued attributes by an appropriate coding; for simplicity we shall restrict ourselves to this particular case.[1] Without loss of generality assume $U = \{1,\ldots,m\}$ and $A = \{1,\ldots,n\}$. Then the system $\mathbf{U}$ is obviously coded by a matrix $\{u_{i,j}\}_{i=1,\ldots,m}^{j=1,\ldots,n}$ of zeros and ones where $u_{i,j} = f_j(i)$. Name the j-th column of the matrix by a predicate P_j.

GUHA uses monadic predicate calculus with generalized quantifiers. Open formulas are built from atomic formulas $P_j(x)$ (where P_j is a unary predicate and x is a variable) using logical connectives; for open formulas φ, ψ containing just one free variable x, the formula $\varphi \sim \psi$ is built using the quantifier $\sim$ (read $\varphi \sim \psi$ "φ is associated with ψ").

Given such a matrix $\mathbf{U}$ with columns named by predicates $P_1, \ldots, P_n$, the i-th object *satisfies* the formula $P_j(x)$ iff $u_{i,j} = 1$, i.e. the value in the i-th row and j-th column is 1. For each data matrix $\mathbf{U}$ let a, b, c, d be the number of objects satisfying $\varphi\&\psi, \varphi\&\neg\psi, \neg\varphi\&\psi, \neg\varphi\&\neg\psi$ respectively. The quadruple (a,b,c,d) is commonly called the *four-fold table* of φ, ψ in $\mathbf{U}$) and can be presented as follows:

	ψ	$\neg\psi$	
φ	a	b	r
$\neg\varphi$	c	d	s
	k	l	m

(Here $r = a+b$, $s = c+d$, $k = a+c$, $l = b+d$ are the marginal sums; $r+s = k+l = a+b+c+d = m$.)

The semantics of $\sim$ is given by a truth function $tr_\sim$ assigning to each (a,b,c,d) the truth value $tr_\sim(a,b,c,d) \in \{0,1\}$. A quantifier $\sim$ is *associational* if

$$a_1 \geq a_2, b_1 \leq b_2, c_1 \leq c_2, d_1 \geq d_2 \text{ and } tr_\sim(a_1,b_1,c_1,d_1) = 1 \text{ implies } tr_\sim(a_2,b_2,c_2,d_2) = 1.$$

It is *implicational* if

$$a_1 \geq a_2, b_1 \leq b_2 \text{ and } tr_\sim(a_1,b_1,c_1,d_1) = 1 \text{ implies } tr_\sim(a_2,b_2,c_2,d_2) = 1.$$

(Thus each implicational quantifier is associational.) Some examples are presented in Sect. 3. The GUHA software works with particular associational/implicational quantifiers making use of their theory. It generates formulas $\varphi \sim \psi$ true in the data.

Orlowska's approach uses information systems to define binary relations on U, particularly *information relations* of similarity and diversity, e.g.

[1] But note that this is equivalent to the assumption of having just one attribute $\hat{a}$ and $f_{\hat{a}}(u) \subseteq V_{\hat{a}}$ is the *item set* in the terminology of [1] - the set of attributes for which u has the value 1.

uRv iff u, v have the same attributes,
$uR'v$ iff u, v have no attribute in common.

For deep modal and algebraic aspects of this approach see [11]. Our question now reads: does the GUHA approach lead to some interesting relations? Let us offer three possibilities.

First, note that each information system, i.e. data matrix ($\{0,1\}$-valued) can be seen as a binary relation, subset of $U \times A$, namely $(u,a) \in R$ iff $f_a(u) = 1$. Evidently, this relation uniquely extends to a relation on $U \times Form$, where $Form$ is the set of all open formulas built from atoms $P_1(x), P_2(x), \ldots$. Each n-tuple $\{\varphi_1, \ldots, \varphi_n\}$ of such open formulas determines the corresponding relation on $U \times \{\varphi_1, \ldots, \varphi_n\}$ – the truth table of $\varphi_1, \ldots, \varphi_n$ given by $\mathbf{U}$. Note that this representation is very near to that used in Agrawal's kind of *mining associational rules* [1] mentioned above. See also [14,2].

Second, assume a data matrix $\mathbf{U}$ and a quantifier $\sim$ to be fixed; this defines a binary relation HYP on $Form$, namely of all pairs φ, ψ such that $\varphi \sim \psi$ is true in $\mathbf{U}$. If $\sim$ is associational then HYP represents all pairs (φ, ψ) of open formulas (composed properties) such that φ is associated with ψ. GUHA generates subrelations of this relation given by syntactical conditions on φ, ψ (think e.g. on φ as a combination of symptoms and ψ a combination of diseases or so). Let us stress the importance of *logical rules* (deduction rules) for an optimal representation of true hypotheses; for example if $\sim$ is symmetric ($\varphi \sim \psi$ logically implies $\psi \sim \varphi$) one can make use of it.

Third, let $\mathbf{U}$ and $\sim$ be as before; instead of pairs of formulas let us think of pairs of *subsets* of U definable by open formulas. We get a binary relation on the Boolean algebra $\mathbf{B}_\mathbf{U}$ of definable subsets of $\mathbf{U}$; properties of the quantifier $\sim$ (for example $\sim$ being an implicational quantifier) determine properties of the relation. In more details, for $u, v \in \mathbf{B}_\mathbf{U}$ let
$a_{uv} = \mathrm{card}(u \cap v)$,
$b_{uv} = \mathrm{card}(u \cap -v)$,
$c_{uv} = \mathrm{card}(-u \cap v)$,
$d_{uv} = \mathrm{card}(-u \cap -v)$.
Let $tr_\sim$ be the truth function of an implicational quantifier $\sim$ and put

$$R_\mathbf{U} = \{(u,v) | tr_\sim(a_{uv}, b_{uv}, c_{uv}, d_{uv}) = 1\}.$$

This leads us to the following

Definition 1. *A Boolean algebra with an implicational relation (briefly, an i-Boolean algebra) is a structure* $\mathbf{B} = (B, \wedge, \vee, -, R)$ *where* $(B, \wedge, \vee, -)$ *is a Boolean algebra and* $R \subseteq B \times B$ *is a relation such that for each* $u_1, u_2, v_1, v_2 \in B$, *if* $u_1 \wedge v_1 \le u_2 \wedge v_2, u_1 \wedge -v_1 \le u_2 \wedge -v_2$ *and* $(u_1, v_1) \in R$ *then* $(u_2, v_2) \in R$.

Clearly, for each $\mathbf{U}$ and each implicational quantifier $\sim$, the algebra of definable subsets of $\mathbf{U}$ together with the relation HYP as above is an i-Boolean algebra. This leads us to the following *problem:*

Is each (finite) i-Boolean algebra isomorphic to the algebra $\mathbf{B}_\mathbf{U}$ *given by a data matrix* $\mathbf{U}$ *and an implicational quantifier* $\sim$*?* This may be easy to decide; but it seems at least to show that our approach may offer some possibly interesting purely algebraic problems. The methods of the next section seem to be useful for solution of the present problem.

3 Complexity of GUHA-Implicational Quantifiers

Implicational quantifiers[2] were defined in the preceding section; for full treatment see [5]. Note that implicational quantifiers are usually denoted by the symbol $\Rightarrow^*$ and by similar symbols. Just recall the following examples:

- classical implicational quantifier – $\varphi \Rightarrow \psi$ is $(\forall x)(\varphi(x) \to \psi(x))$, i.e. $\varphi \Rightarrow \psi$ is true iff $b = 0$;
- founded implication $\varphi \Rightarrow_{p,s} \psi$ is true iff $a \geq s$ and $a \geq p(a+b)$ where s is a positive natural number and $0 < p \leq 1$,
- lower critical implication: $\varphi \Rightarrow^{L}_{p,\alpha} \psi$ is true iff $LIMPL_p(a,b) = \sum_{i=a}^{a+b} \binom{a+b}{i} p^i (1-p)^{a+b-i} \leq \alpha$

For example, $LIMPL_{0.9}(240,3) = 2.76.10^{-8}$.

Consider the language with unary predicates $P_1, P_2, \ldots$ and a binary quantifier $\Rightarrow^*$ (as well as object variables and logical connectives). Let x be an object variable called the *designated variable.* A *pure prenex formula* has the form $\varphi \Rightarrow^* \psi$, where φ and ψ are quantifier-free open formulas containing no variable except x (i.e. Boolean combinations of formulas $P_1(x), P_2(x), \ldots$). It is understood that $\Rightarrow^*$ binds the variable x; the pedantical writing would be $(\Rightarrow^* x)(\varphi(x), \psi(x))$. The *normal form theorem* (see [5] 3.1.30) says that each closed formula of our language is logically equivalent to a Boolean combination of pure prenex formulas. Note the the proof of this fact is constructive and uniform: given a closed formula Φ one finds a normal form $NF(\Phi)$ which is a (possibly empty) disjunction of elementary conjunctions of pure prenex formulas and is logically equivalent to Φ for *any* semantics of the quantifier $\Rightarrow^*$. It follows that if you have only finitely many predicates (atributes) $P_1, \ldots, P_n$ then there is a finite set **NF** of normal form formulas such that for each Φ, $NF(\Phi)$ can be taken from **NF.**

Call Φ an *implicational tautology* (or a tautology with implicational quantifiers) if Φ is true in each data matrix **U** (of appropriate arity) for each implicational quantifier $\Rightarrow^*$.

Φ is *implicationally satisfiable* (briefly, i-satisfiable) if there is a data matrix **U** (interpreting predicates occuring in Φ) and an implicational quantifier $\Rightarrow^*$ such that Φ is true in **U** w.r.t $\Rightarrow^*$.

Examples of implicational tautologies (easy to verify):

$$((\varphi \& \psi) \Rightarrow^* \chi) \to (\varphi \Rightarrow^* (\chi \vee \neg\psi)),$$

$$(\varphi \Rightarrow^* (\psi \& \chi)) \to ((\varphi \& \psi) \Rightarrow^* \chi).$$

(Compute the corresponding four-fold tables.)

[2] An alternative name is *multitudinal quantifiers*, cf. [4].

The normal form theorem gives a cheap decidability result: For each fixed n, the set of all implicational tautologies containing no predicate except $P_1, \ldots, P_n$, is decidable (since the set of the corresponding normal forms is finite, cf. [5] Chapter III, Problem 9). We are going to show decidability and determine the computational complexity of the set of implicational tautologies having arbitrary many predicates. Since a formula is an implicational tautology iff its negation is not i-satisfiable, everything is solved by the following

Theorem 1. *The set all of implicationally satisfiable formulas in normal form is NP-complete.*

Proof. The reader is assumed to know basics notions and facts of polynomial complexity theory as: a set X of formulas (or words in an alphabet) is in NP (accepted by a non-deterministic Turing machine working in polynomial time); X is NP-complete (is in NP and each Y in NP is reducible to X in polynomial time); the set of satisfiable formulas of propositional logic is NP-complete and, in more details, the function assigning to each propositional formula A and each evaluation of its atoms by zeros and ones the corresponding truth value of A is computable in polynomial time. See e.g. [3] for details.

Formulas in question result from propositional formulas $A(p_1, \ldots, p_n)$ and pure prenex formulas $\Phi_1, \ldots \Phi_n$ (Φ_i being $\varphi_i \Rightarrow^* \psi_i$) as $A(\Phi_1, \ldots, \Phi_n)$ (substituting Φ_i for p_i in $A(p_1, \ldots p_n)$). We describe a non-deterministic algorithm accepting exactly all implicationally satisfiable formulas; checking that it works in polynomial time is routine. This will show that the set of all implicationally satisfiable formulas are in NP; then we show NP-completeness.

Given $A(p_1, \ldots, p_n)$ and $\Phi_1, \ldots \Phi_n$, first guess an evaluation e assigning to each p_i a truth value $e(p_i) \in \{0, 1\}$ and check if e makes $A(p_1, \ldots p_n)$ true. (If not, fail.) If it does, we want to find a data matrix (information structure) $\mathbf{U}$ and an implicational quantifier $\Rightarrow^*$ such that $\mathbf{U}$ *respects* e for $\Rightarrow^*$, i.e. for $i = 1, \ldots, n$, Φ_i is true in $\mathbf{U}$ iff $e(p_i) = 1$. If all this succeeds then $A(\Phi_1, \ldots, \Phi_n)$ is implicationally satisfiable and vice versa.

Call Φ_i *positive* if $e(p_i) = 1$, and *negative* if $e(p_i) = 0$. Recall that Φ_i is $\varphi_i \Rightarrow^* \psi_i$. If $\mathbf{U}$ and $\Rightarrow^*$ are given then only the frequency of $\varphi_i \& \psi_i$ and of $\varphi_i \& \neg\psi_i$ (number of objects satisfying the respective formula) decide of Φ_i is true in $\mathbf{U}$ or not. Call all the formulas $\varphi_i \& \psi_i$, $\varphi_i \& \neg\psi_i$ ($i = 1, \ldots, n$) *critical* formulas. Their frequencies (in $\mathbf{U}$) define a linear preorder $\leq_{\mathbf{U}}$ on them: for any critical γ_1, γ_2, we put $\gamma_1 \leq_{\mathbf{U}} \gamma_2$ if $freq_{\mathbf{U}}(\gamma_1) \leq freq_{\mathbf{U}}(\gamma_2)$. For an arbitrary linear preorder $\leq$ of the critical formulas, call $\leq$ *acceptable* for e of the following holds for any $1 \leq i, j \leq n$: whenever $\varphi_i \Rightarrow^* \psi_i$ is positive (with respect to e), $\varphi_j \& \psi_j \geq \varphi_i \& \psi_i$ and $\varphi_j \& \neg\psi_j \leq \varphi_i \& \neg\psi_i$ then $\varphi_j \Rightarrow^* \psi_j$ is also positive. From the definition of an implicational quantifier it follows immediately that the above preorder $\leq_{\mathbf{U}}$ is acceptable (for any $\mathbf{U}, \Rightarrow^*$). Thus we continue as follows:

Having e, guess an acceptable linear preorder $\leq$ of critical formulas and ask whether there is an $\mathbf{U}$ such that $\leq$ is $\leq_{\mathbf{U}}$ (say then that $\leq$ is *realizable*). We show that this can be done in non-deterministic polynomial time. To do this we use the result of [8] on the complexity of fuzzy probabilistic logic over Łukasiewicz propositional calculus. (This calculus is also described in [4].) For each critical formula γ, let $\mathcal{P}(\gamma)$ be the formula saying "γ is probable". Let

$\gamma_1, \ldots, \gamma_n$ be a sequence of all critical formulas non-decreasing with respect to $\leq$. Let ε be a new atomic formula. For each $1 \leq i < k$, if $\gamma_i < \gamma_k$ let Λ_i be the formula $(\mathcal{P}(\gamma_i) \oplus \mathcal{P}(\varepsilon) \to \mathcal{P}(\gamma_{i+1})$ (where $\oplus$ is Łukasiewicz strong disjunction, also denoted by $\underline{\vee}$); if $\gamma_i \leq \gamma_{i+1}$ and $\gamma_{i+1} \leq \gamma_i$ then Λ_i is $\mathcal{P}(\gamma_i) \equiv \mathcal{P}(\gamma_{i+1})$. Let T be the finite theory over FPŁ whose axioms are all the Λ_i ($i = 1, \ldots k-1$) and also $P(\gamma_{i_0}) \to P(\neg\varepsilon)$ where i_0 is the largest index such that $\gamma_{i_o} < \gamma_k$. (The singular case that all γ_i are $\leq$-equivalent is left to the reader as an exercise.) Observe that there is a probability on open formulas built from predicates occuring the critical formulas and coherent with $\leq$ iff the theory T has a model over FPŁ in which $\mathcal{P}(\varepsilon)$ has a non-extremal value, i.e. $\mathcal{P}(\varepsilon) \vee \neg\mathcal{P}(\varepsilon)$ has not value 1. Using the method of [8] and [4] one easily reduces this problem to a Mixed Integer Programming problem, showing that the last problem is in NP.

Three things remain: First, to show that if such probability exists then we may assume it has rational values on all open formulas from our predicates – this is done as in [4] 8.4.16. Thus multiplying by the common denominator we may get a finite model alias data matrix alias information system $\mathbf{U}$ such that frequencies of objects satisfying critical formulas order them in accordance with $\leq$. Second, let (a_i, b_i, c_i, d_i) be four-fold tables of pairs (φ_i, ψ_i) of formulas occuring in the sentences $\varphi_i \Rightarrow^* \psi_i$; define a quantifier $\Rightarrow^*$ by letting $tr_{\Rightarrow^*}(a, b, c, d) = 1$ iff for some i, $a \geq a_i, b \leq b_i, c \leq c_i$ and $d \geq d_i$. This is an implicational quantifier and $\mathbf{U}$ respects e for this $\Rightarrow^*$. This shows that our problem of satisfiability of Φ) is in NP.

The last (third) thing is to show NP-hardness. To this end we reduce the satisfiability problem of propositional logic (equivalently, satisfiability of open formulas built from atom $P_1(x), \ldots, P_n(x), \ldots$) to our problem. Let ε be as before; for each open formula φ as above let φ^* be $(\varepsilon \Rightarrow^* \varepsilon) \to (\varepsilon \Rightarrow^* \varphi)$. Let us show that φ is a Boolean tautology logic iff φ^* is an implicational tautology; thus φ is satisfiable in Boolean logic iff $\neg(\neg\varphi)^*$ is satisfiable in our logic with an implicational quantifier.

Indeed, if φ is a Boolean tautology and (a, b, c, d) is a four-fold table of (ε, φ) given by an $\mathbf{U}$; then $b = d = 0$; if in the same $\mathbf{U}$ the formula $\varepsilon \Rightarrow^* \varepsilon$ is true for an implicational quantifier $\Rightarrow^*$ then $tr_{\Rightarrow^*}(a, 0, 0, c) = 1$ (note that $(a, 0, 0, c)$ is the four-fold table of ϵ, ϵ in $\mathbf{U}$) and hence also $tr_{\Rightarrow^*}(a, 0, c, 0) = 1$. Then $\epsilon \Rightarrow^* \varphi$ is true and hence φ^* is true in our $\mathbf{U}$. Conversely, if φ is not a Boolean tautology then take an $\mathbf{U}$ in which all objects satisfy $\neg\varphi$ and all satisfy ε, thus $\varepsilon \Rightarrow^* \varepsilon$ is true and $\varepsilon \Rightarrow^* \varphi$ is false for the classical implicational quantifier $\Rightarrow^*$ ($\gamma \Rightarrow^* \delta$ being $(\forall x)(\gamma \to \delta)$). This completes the proof.

Corollary 1. *The set of all implicational tautologies is co-NP-complete.*

Remark 1. The same idea can be used to show that the problem of associational satisfiability (more precisely, the problem of showing that a formula Φ of the language with one binary quantifier $\sim$ in normal form, is satisfiable in a $\mathbf{U}$ by an associational quantifier) is in NP. The reader may try to show NP-completeness (possibly easy).

Conclusion. We hope to have shown that the logic of observational calculi as calculi speaking on data (alias information systems) is interesting and relevant for the study of relational structures and have contributed to the study of its computational complexity.

References

1. Agrawal R., Mannila H., Srikant R., Toivonen H., Verkamo A. I.: Fast discovery of association rules. In: (Fayyad V. et. al, ed.) Advances in knowledge discovery and data mining, AAA Press/MIT Press 1996, pp. 307–328.
2. Düntsch I., Gediga G., Orłowska E.: Relational attribute systems, International Journal of Human Computer Studies, 55 (2001) pp. 293-309.
3. Garrey M. R., Johnson D. S: Computers and intractability, New York, W. J. Freeman and Co., 1979.
4. Hájek P.: Metamathematics of fuzzy logic, Kluwer 1998.
5. Hájek P., Havránek T.: Mechanizing hypothesis formation – Mathematical foundations for a general theory, Springer Verlag 1978. Internet version (free): www.cs.cas.cz/~hajek/guhabook
6. Hájek P., Holeňa M.: Formal logics of discovery and hypothesis formation by machine. In: (Arikawa et al, ed.) Discovery Science, Springer Verlag 1998, 291-302
7. Hájek P., Sochorová A., Zvárová J.: GUHA for personal computers. Comp. Statistics and Data Analysis 19(1995), 149–153.
8. Hájek P., Tulipani S.: Complexity of fuzzy probability logic. Fundamenta Informaticae 45 (2001), 207–213.
9. Lipski W.: Informational systems with incomplete information. In Michaelson S., Milner R., editors. Third International Colloquium on Automata, Languages and Programming, University of Edinburgh, Edinburgh University Press, 1976, pp. 120-130.
10. Orlowska E., Pawlak Z: Logical foundationas of knowledge representation, ICS Research Report 573, Polish Academy of Sciences, 1984.
11. Orlowska E., Szalas A. (ed.): Relational methods in computer science applications, Springer-Physica Verlag 2001, 263–285.
12. Pawlak Z.: Mathematical foundations of information retrieval, ICS Research Report 101, Polish Academy of Sciences, 1973.
13. Pawlak Z.: Rough sets. Internat. Journal Comput. Inform. Sci., 11 (1982) pp. 341-356.
14. Vakarelov D.: Information systems, similarity relations and modal logics. In: Orlowska E., editor, Incomplete Information – Rough Set Analysis, Heidelberg, Physica-Verlag 1997, pp. 492-550.

Groups in Allegories

Yasuo Kawahara

Department of Informatics, Kyushu University 33, Fukuoka 812-8581, Japan
kawahara@i.kyushu-u.ac.jp

Abstract. Groups are one of the most fundamental notions in mathematics. This paper provides a foundation of group theory in allegories. Almost all results in the paper can be applied to theory of fuzzy groups.

1 Introduction

The motivation of the paper arose from the following three fundamental exercises in group theory:

(a) Let $G = (G, \cdot)$ be a semigroup, that is, a binary operation $\cdot : G \times G \to G$ is associative. Show that if $\forall x \in G : xe = x$ and $\forall x \in G \exists y \in G : xy = e$ for some element $e \in G$, then $\forall x \in G : ex = x$ and $\forall x \in G \exists y \in G : yx = e$.
(b) Let H be a subgroup of a group $G = (G, \cdot, e, \cdot^{-1})$. Prove that the (binary) relation $x \equiv y$ on G, defined by $x^{-1}y \in H$, is an equivalence relation.
(c) Show that the set of all normal subgroups of a group G forms a modular lattice. In other words, the following modular law holds for three normal subgroups S, T and U of G:

$$S \subseteq U \implies ST \cap U \subseteq S(T \cap U).$$

To define group objects in categories [9] is not new, but it is difficult to directly treat algebraic relations, such as residual relations (b) induced by subgroups, in the ordinary category theory. Allegories [4], as a kind of relation categories, give a natural and suitable setting for manipulating algebraic binary relations in group theory, lattice theory [8] and so on. This paper provides a foundation of group theory in allegories, and solves the above three questions (a), (b) and (c).

The paper is organised as follows: In section 2 we recall the definition of allegories [4] and remark some fundamentals on relational products in allegories. In section 3 we review a simple sharpness property [5,10,6] on relational products of relations, which was initiated by Schmidt and will play an important rôle in the proof of the main results. In section 4 we explore some fundamental properties of relational binary operations and show a suitable formalisation (Theorem 2) of the associative law, as well as the commutative law, using with relational global elements. The inverse law and the absorption laws (in lattice theory) are of course not the case. In section 5 we mention unitary and inverse operations for binary operations in allegories. We also give a relational version (Theorem 4) for

H. de Swart (Ed.): RelMiCS 2001, LNCS 2561, pp. 88–103, 2002.

the first question (a). In section 6 we define notions of (functional) groups and subgroups, and prove an elementary fact (b) that the residual relation induced by a subgroup is an equivalence relation. In section 7 we describe a notion of normal subgroups in allegories, and answer the final question (c) that the set of all normal subgroups of a group forms a modular lattice.

2 Allegories

In this section we recall the fundamentals on relation categories, called allegories [4].

Throughout this paper, a morphism α from an object X into an object Y in an allegory (which will be defined below) will be denoted by a half arrow $\alpha : X \rightharpoondown Y$, and the composite of a morphism $\alpha : X \rightharpoondown Y$ followed by a morphism $\beta : Y \rightharpoondown Z$ will be written as $\alpha\beta : X \rightharpoondown Z$. Also we will denote the identity morphism on X as id_X.

Definition 1. An allegory $\mathcal{A}$ is a category satisfying the following:
D1. [Meet Semi-Lattice] For all pairs of objects X and Y the hom-set $\mathcal{A}(X,Y)$ consisting of all morphisms of X into Y is a meet semi-lattice with the greatest morphism ∇_{XY}. Its semi-lattice structure will be denoted by

$$\mathcal{A}(X,Y) = (\mathcal{A}(X,Y), \sqsubseteq, \sqcap, \nabla_{XY}).$$

D2. [Converse] There is given a converse operation $^\sharp : \mathcal{A}(X,Y) \to \mathcal{A}(Y,X)$. That is, for all morphisms $\alpha, \alpha' : X \rightharpoondown Y$, $\beta : Y \rightharpoondown Z$, the following converse laws hold:
(a) $(\alpha\beta)^\sharp = \beta^\sharp\alpha^\sharp$, (b) $(\alpha^\sharp)^\sharp = \alpha$, (c) If $\alpha \sqsubseteq \alpha'$, then $\alpha^\sharp \sqsubseteq \alpha'^\sharp$.

D3. [Dedekind Formula] For all morphisms $\alpha : X \rightharpoondown Y$, $\beta : Y \rightharpoondown Z$ and $\gamma : X \rightharpoondown Z$ the Dedekind formula $\alpha\beta \sqcap \gamma \sqsubseteq \alpha(\beta \sqcap \alpha^\sharp\gamma)$ holds.

D4. [Sub-Distributivity] The composition preserves order: If $\alpha \sqsubseteq \alpha'$ and $\beta \sqsubseteq \beta'$, then $\alpha\beta \sqsubseteq \alpha'\beta'$. □

The fundamental properties of relational categories is referred to [1,4,11,7]. The following is a basic property of allegories.

Proposition 1. *Let $\alpha : X \rightharpoondown Y$ and $\gamma : Y \rightharpoondown X$ be morphisms in an allegory $\mathcal{A}$. If $\alpha\gamma = \mathrm{id}_X$ and $\gamma\alpha = \mathrm{id}_Y$, then $\alpha = \gamma^\sharp$.* □

A *unit* I in an allegory $\mathcal{A}$ is an object such that $\mathrm{id}_I = \nabla_{II}$. A morphism $\alpha : X \rightharpoondown Y$ is *total* if $\mathrm{id}_X \sqsubseteq \alpha\alpha^\sharp$ (or equivalently, $\alpha\nabla_{YX} = \nabla_{XX}$). A morphism $f : X \rightharpoondown Y$ such that $f^\sharp f \sqsubseteq \mathrm{id}_Y$ (*univalent*) is called a *function* and may be introduced as $f : X \to Y$. In what follows the word *relation* is used as synonym for morphisms in allegories.

Definition 2. A pair (A, B) of objects in an allegory $\mathcal{A}$ *has a relational product* if there exists a pair $(p : T \to A, q : T \to B)$ of total functions such that $p^\sharp q = \nabla_{AB}$ and $pp^\sharp \sqcap qq^\sharp = \mathrm{id}_T$. The pair $(p : T \to A, q : T \to B)$ is called a pair of projections for (A, B). An allegory $\mathcal{A}$ *has a relational product* if every pair of objects in $\mathcal{A}$ has a relational product. □

Let $(p : T \to A, q : T \to B)$ be a pair of projections for a pair (A, B) of objects. For each pair of relations $\alpha : X \rightharpoondown A$ and $\beta : X \rightharpoondown B$, we define a relation $\alpha \top \beta : X \rightharpoondown T$ by $\alpha \top \beta = \alpha p^\sharp \sqcap \beta q^\sharp$. It is trivial that $p \top q = \mathrm{id}_T$.

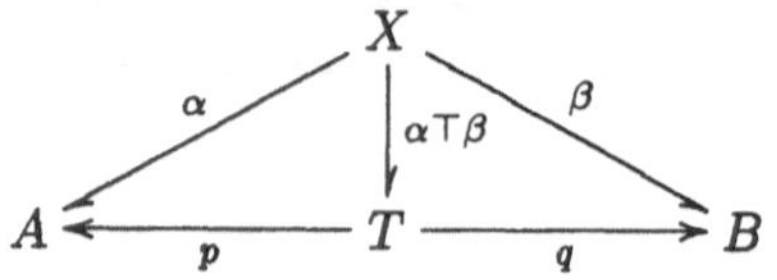

The following proposition is a list of elementary properties of relational products in allegories. The proof is trivial and so omitted.

Proposition 2. *Let $(p : T \to A, q : T \to B)$ be a pair of projections for (A, B) and let $\alpha, \alpha' : X \rightharpoondown A$, $\beta, \beta' : X \rightharpoondown B$, $\gamma : Y \rightharpoondown A$, $\delta : Y \rightharpoondown B$ and $\xi : Z \rightharpoondown X$ be relations.*

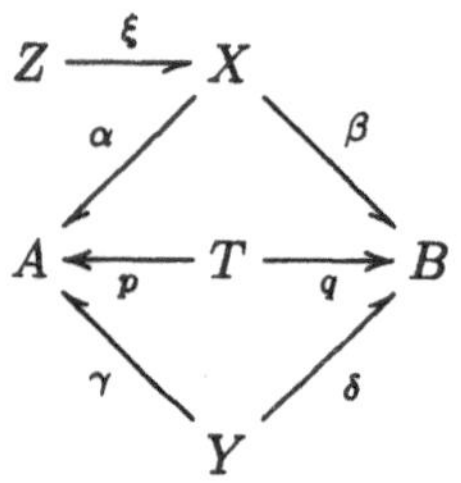

Then the following statements hold:

(a) *If $\alpha \sqsubseteq \alpha'$ and $\beta \sqsubseteq \beta'$, then $\alpha \top \beta \sqsubseteq \alpha' \top \beta'$,*
(b) $\xi(\alpha \top \beta) \sqsubseteq \xi\alpha \top \xi\beta$,
(c) $\xi(\alpha \top \nabla_{XY}\delta) = \xi\alpha \top \nabla_{ZY}\delta$,
(d) $\xi(\nabla_{XY}\gamma \top \beta) = \nabla_{ZY}\gamma \top \xi\beta$,
(e) *If $\xi : Z \rightharpoondown X$ is a function, then $\xi(\alpha \top \beta) = \xi\alpha \top \xi\beta$.*
(f) *If α and β are total functions, then $\alpha \top \beta$ is a unique total function such that $(\alpha \top \beta)p = \alpha$ and $(\alpha \top \beta)q = \beta$,*
(g) $(\alpha \sqcap \alpha')\top\beta = (\alpha \top \beta) \sqcap (\alpha' \top \beta)$ *and* $\alpha \top (\beta \sqcap \beta') = (\alpha \top \beta) \sqcap (\alpha \top \beta')$. □

As in ordinary category theory, the common domain T of projections $p : T \to A$ and $q : T \to B$ is uniquely determined up to isomorphism by the virtue of the last Proposition 2(f). This enables us to write the object T as $A \times B$ and the pair of projections for (A, B) as (p, q), if it exists.

Let (p, q) be a pair of projections for (A, A). The twisting function $t : A \times A \to A \times A$ is defined by $t = q \top p \; (= qp^\sharp \sqcap pq^\sharp)$. That is, t is a unique total function such that $tp = q$ and $tq = p$. Then $tt = t(q \top p) = tq \top tp = p \top q = \mathrm{id}_{A \times A}$.

In addition we assume that (p_1, q_1) and (p_2, q_2) are pairs of projections for $(A \times A, A)$ and $(A, A \times A)$, respectively.

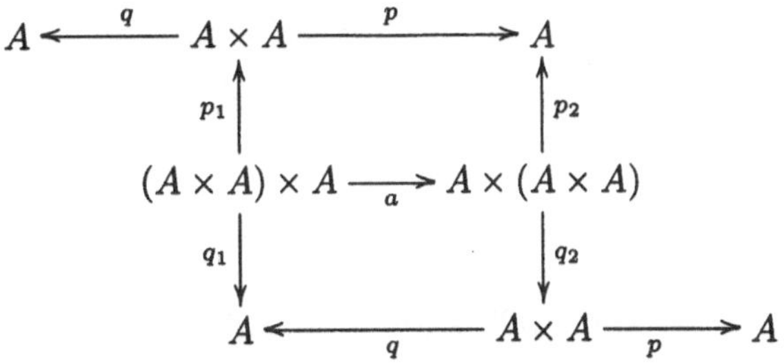

The associative function $a : (A \times A) \times A \to A \times (A \times A)$ is defined by

$$a = p_1 p \top (p_1 q \top q_1).$$

That is, a is a unique total function such that $ap_2 = p_1 p$, $aq_2 p = p_1 q$ and $aq_2 q = q_1$. Another associative function $b : A \times (A \times A) \to (A \times A) \times A$ is defined by

$$b = (p_2 \top q_2 p) \top q_2 q.$$

That is, b is a unique total function such that $bp_1 p = p_2$, $bp_1 q = q_2 p$ and $bq_1 = q_2 q$. It is trivial that a and b are mutually inverses, that is, $ab = \mathrm{id}_{(A \times A) \times A}$ and $ba = \mathrm{id}_{A \times (A \times A)}$, and consequently $b = a^\sharp$ by Proposition 1. It is easy to see that a pair of total functions $p_2' = p_1 p$ and $q_2' = p_1 q \top q_1$ is a pair of projections for $(A, A \times A)$. This fact indicates that if a relational product for $(A \times A, A)$ exists iff a relational product for $(A, A \times A)$ exists. Obviously $a = b = \mathrm{id}_{(A \times A) \times A}$ when $p_2 = p_2'$ and $q_2 = q_2'$.

The following proposition can be readily seen by a simple computation:

Proposition 3. *Let $\alpha, \beta, \gamma : X \rightharpoonup A$ be relations. Then the following identities hold:*

(a) $\{(\alpha \top \beta) \top \gamma\} a = \alpha \top (\beta \top \gamma)$,
(b) $\{\alpha \top (\beta \top \gamma)\} b = (\alpha \top \beta) \top \gamma$. □

Let (p, q) and (p_0, q_0) be pairs of relational products for (A, B) and (X, Y), respectively. For each pair of relations $\xi : A \rightharpoonup X$ and $\eta : B \rightharpoonup Y$ we define a relation $\xi \times \eta : A \times B \rightharpoonup X \times Y$ by $\xi \times \eta = p\xi \top q\eta = p\xi p_0^\sharp \sqcap q\eta q_0^\sharp$.

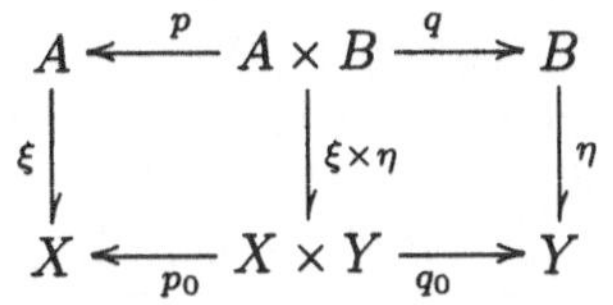

If $f : A \to X$ and $g : B \to Y$ are total functions, then $f \times g$ is a unique function such that $(f \times g)p_0 = pf$ and $(f \times g)q_0 = qg$. Remark that an equality

$(f \times g)(f' \times g') = ff' \times gg'$ always holds for all total functions f, f', g and g'. Schmidt (Cf. [10,6]) initially suggested that the so-called sharpness property

$$(\xi \times \eta)(\xi' \times \eta') = \xi\xi' \times \eta\eta'$$

does not hold for relations ξ, ξ', η and η' in general. In the rest of the section we review a simple sharpness property [5] needed in the later disscusion.

Theorem 1 (Sharpness).

(a) *If two conditions* (1) $\alpha^\sharp\beta \sqcap \gamma^\sharp\delta \sqsubseteq \rho^\sharp\tau$ *and* (2) $\alpha\alpha^\sharp\beta \sqsubseteq \beta$ *hold, then*

$$\alpha\gamma^\sharp \sqcap \beta\delta^\sharp \sqsubseteq (\alpha\rho^\sharp \sqcap \beta\tau^\sharp)(\rho\gamma^\sharp \sqcap \tau\delta^\sharp).$$

(b) *If* $p : T \to A$ *and* $q : T \to B$ *is a pair of total functions with* $p^\sharp q = \nabla_{AB}$, *and if there exists a relation* $\eta : T \rightharpoonup X$ *such that* $\eta^\sharp\eta \sqsubseteq \mathrm{id}_X$ *and* $\eta^\sharp p = \nabla_{XA}$, *then*

$$\alpha\gamma^\sharp \sqcap \beta\delta^\sharp = (\alpha p^\sharp \sqcap \beta q^\sharp)(p\gamma^\sharp \sqcap q\delta^\sharp).$$

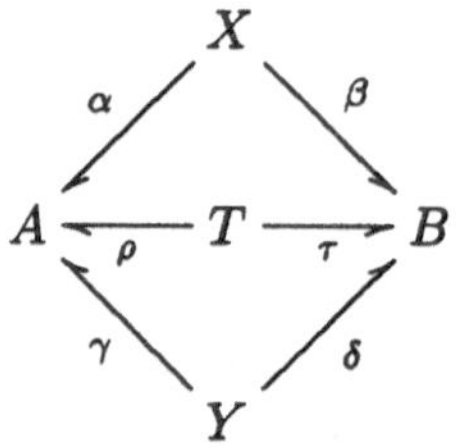

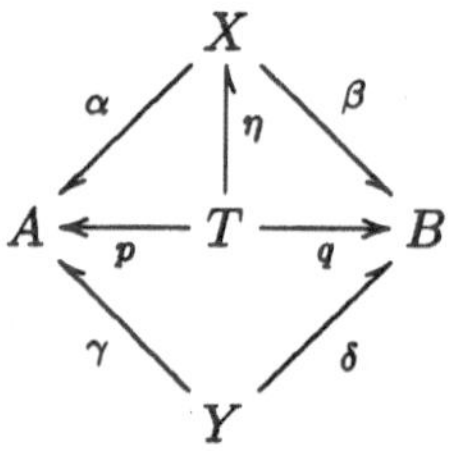

Proof. (a) It is direct from the following computation.

$$\begin{aligned}
\alpha\gamma^\sharp \sqcap \beta\delta^\sharp &\sqsubseteq \alpha(\gamma^\sharp \sqcap \alpha^\sharp\beta\delta^\sharp) && \{\text{ Dedekind Formula }\}\\
&= \alpha\{\gamma^\sharp \sqcap (\gamma^\sharp\delta \sqcap \alpha^\sharp\beta)\delta^\sharp\} && \{\text{ DF }\}\\
&\sqsubseteq \alpha\{\gamma^\sharp \sqcap (\rho^\sharp\tau \sqcap \alpha^\sharp\beta)\delta^\sharp\} && \{\text{ (1) }\}\\
&\sqsubseteq \alpha\{\gamma^\sharp \sqcap (\rho^\sharp \sqcap \alpha^\sharp\beta\tau^\sharp)\tau\delta^\sharp\} && \{\text{ DF }\}\\
&\sqsubseteq \alpha(\rho^\sharp \sqcap \alpha^\sharp\beta\tau^\sharp)\{(\rho \sqcap \tau\beta^\sharp\alpha)\gamma^\sharp \sqcap \tau\delta^\sharp\} && \{\text{ DF }\}\\
&\sqsubseteq (\alpha\rho^\sharp \sqcap \alpha\alpha^\sharp\beta\tau^\sharp)(\rho\gamma^\sharp \sqcap \tau\delta^\sharp) && \{\text{ Sub-distributive }\}\\
&\sqsubseteq (\alpha\rho^\sharp \sqcap \beta\tau^\sharp)(\rho\gamma^\sharp \sqcap \tau\delta^\sharp). && \{\text{ (2) }\}
\end{aligned}$$

(b) An inclusion $(\alpha p^\sharp \sqcap \beta q^\sharp)(p\gamma^\sharp \sqcap q\delta^\sharp) \sqsubseteq \alpha\gamma^\sharp \sqcap \beta\delta^\sharp$ is trivial from the univalency $p^\sharp p \sqsubseteq \mathrm{id}_T$ and $q^\sharp q \sqsubseteq \mathrm{id}_T$. To see the converse inclusion, we set $\rho = pp^\sharp$, $\tau = q$, $\hat{\alpha} = \alpha p^\sharp \sqcap \eta^\sharp$ and $\hat{\gamma} = \gamma p^\sharp$. We verify that ρ, τ, $\hat{\alpha}$, β, $\hat{\gamma}$ and δ satisfy two conditions (1) and (2) of (a): (1) $\hat{\alpha}^\sharp\beta \sqcap \hat{\gamma}^\sharp\delta \sqsubseteq \nabla_{TB} \sqsubseteq p\nabla_{AB} = pp^\sharp q = \rho^\sharp\tau$ by the totality of p and $p^\sharp q = \nabla_{AB}$. (2) $\hat{\alpha}\hat{\alpha}^\sharp\beta \sqsubseteq \eta^\sharp\eta\beta \sqsubseteq \beta$ by the univalency $\eta^\sharp\eta \sqsubseteq \mathrm{id}_Y$. Therefore we have

$$\begin{aligned}
\alpha\gamma^\sharp \sqcap \beta\delta^\sharp &\sqsubseteq (\alpha \sqcap \beta\delta^\sharp\gamma)\gamma^\sharp \sqcap \beta\delta^\sharp && \{\text{ DF }\}\\
&\sqsubseteq (\alpha \sqcap \eta^\sharp p)\gamma^\sharp \sqcap \beta\delta^\sharp && \{\ \nabla_{XA} = \eta^\sharp p\ \}\\
&\sqsubseteq (\alpha p^\sharp \sqcap \eta^\sharp)p\gamma^\sharp \sqcap \beta\delta^\sharp && \{\text{ DF }\}\\
&= \hat{\alpha}\hat{\gamma}^\sharp \sqcap \beta\delta^\sharp\\
&\sqsubseteq (\hat{\alpha}\rho^\sharp \sqcap \beta\tau^\sharp)(\rho\hat{\gamma}^\sharp \sqcap \tau\delta^\sharp) && \{\text{(a)}\}\\
&\sqsubseteq (\alpha p^\sharp \sqcap \beta q^\sharp)(p\gamma^\sharp \sqcap q\delta^\sharp). && \{\ p^\sharp p \sqsubseteq \mathrm{id}_T\ \}
\end{aligned}$$

□

Corollary 1. *Let (p, q) and (p_0, q_0) be pairs of projections for (A, B) and (Y, Z), respectively.*

(a) *If X or Y is identical with one of three objects A, B and I, then*

$$\alpha\gamma^\sharp \sqcap \beta\delta^\sharp = (\alpha \top \beta)(\gamma \top \delta)^\sharp.$$

(b) *If X or $Y \times Z$ is identical with one of three objects A, B and I, then*

$$\alpha\xi \top \beta\eta = (\alpha \top \beta)(\xi \times \eta).$$

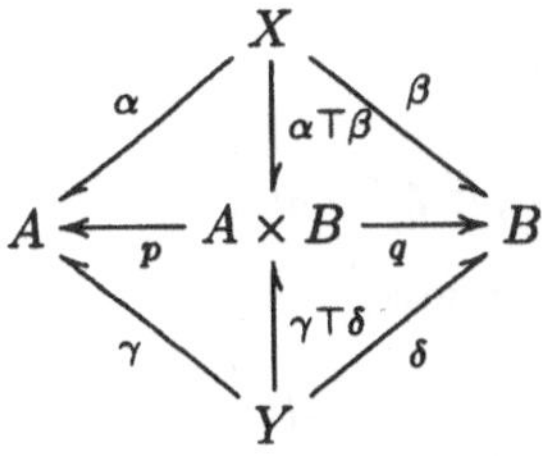

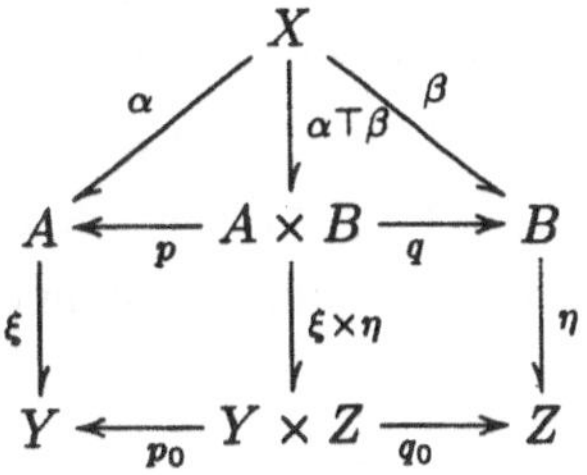

Proof. It is a direct corollary of Theorem 1(b). □

3 Binary Operations

In this section we will study fundamental properties of binary operations in allegories.

Throughout of the rest of the paper we assume that G is an object in an allegory $\mathcal{A}$ and $\mathcal{A}$ has at least two pairs (p, q) and (p_1, q_1) of projections for (G, G) and $(G \times G, G)$, respectively. Under this condition a pair (p_2, q_2) of projections for $(G, G \times G)$ exists, as stated in the section 2. Also we assume that all relations are in $\mathcal{A}$ unless othewise stated.

Definition 3. A binary operation on G is a relation $\mu : G \times G \rightharpoondown G$. □

Let $\mu : G \times G \rightharpoondown G$ be a binary operation. Then we define $\alpha \odot \beta = (\alpha \top \beta)\mu$ for a pair of relations $\alpha, \beta : X \rightharpoondown G$. Remark that $p \odot q = \mu$.

Lemma 1. *Assume $\mu : G \times G \rightharpoondown G$ is a binary operation on G. Let $\alpha, \alpha', \beta, \beta' : X \rightharpoondown G$, $\gamma, \delta : Y \rightharpoondown G$ and $\xi : Z \rightharpoondown X$ be relations in $\mathcal{A}$.*

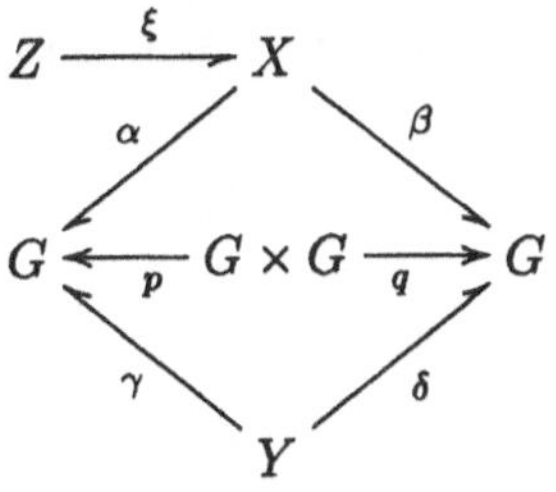

Then the following hold:

(a) *If* $\alpha \sqsubseteq \alpha'$ *and* $\beta \sqsubseteq \beta'$, *then* $\alpha \odot \beta \sqsubseteq \alpha' \odot \beta'$,
(b) $\xi(\alpha \odot \beta) \sqsubseteq \xi\alpha \odot \xi\beta$,
(c) $\xi(\alpha \odot \nabla_{XY}\delta) = \xi\alpha \odot \nabla_{ZY}\delta$,
(d) $\xi(\nabla_{XY}\gamma \odot \beta) = \nabla_{ZY}\gamma \odot \xi\beta$,
(e) *If* $\xi : Z \rightharpoondown X$ *is a function, then* $\xi(\alpha \odot \beta) = \xi\alpha \odot \xi\beta$,
(f) $(\alpha \sqcap \alpha') \odot \beta \sqsubseteq (\alpha \odot \beta) \sqcap (\alpha' \odot \beta)$ *and* $\alpha \odot (\beta \sqcap \beta') \sqsubseteq (\alpha \odot \beta) \sqcap (\alpha \odot \beta')$,

Proof. It is just a corollary of Proposition 2. □

The next proposition is a simple result from the sharpness Corollary 1, but it will play an important rôle in the proof of Theorems 5 and 6.

Proposition 4. *Assume that a binary operation* $\mu : G \times G \rightharpoondown G$ *is total. If* $X = G$ *or* $X = I$, *then* $\alpha\gamma^\sharp \sqcap \beta\delta^\sharp \sqsubseteq (\alpha \odot \beta)(\gamma \odot \delta)^\sharp$ *holds.*

Proof. It is immediate from Corollary 1 (sharpness):

$$\begin{aligned} \alpha\gamma^\sharp \sqcap \beta\delta^\sharp &= (\alpha\top\beta)(\gamma\top\delta)^\sharp \quad \{\text{ Corollary 1(a) }\} \\ &\sqsubseteq (\alpha\top\beta)\mu\mu^\sharp(\gamma\top\delta)^\sharp \ \{ \text{ id}_{G\times G} \sqsubseteq \mu\mu^\sharp \ (\mu : \text{total}) \} \\ &= (\alpha \odot \beta)(\gamma \odot \delta)^\sharp. \end{aligned}$$

□

Definition 4. A (relational) semigroup $G = (G, \mu)$ in $\mathcal{A}$ is a pair of an object G in $\mathcal{A}$ and a binary operation $\mu : G \times G \rightharpoondown G$, satisfying the associative law $(\mu \times \text{id}_G)\mu = a(\text{id}_G \times \mu)\mu$, where $a : (G \times G) \times G \to G \times (G \times G)$ is the associative function. □

The associative law is an indispensable property on binary operations to simplify iterations of operations. There are often difficulties when one manipulates the associative law in terms of morphisms. So it is convenient to use the traditional form of the associative law

$$(x \cdot y) \cdot z = x \cdot (y \cdot z).$$

The following theorem guarantees that the associative law in terms of morphisms and the traditional form of the associative law using global elements are equivalent.

Theorem 2. *A binary operation* $\mu : G \times G \rightharpoondown G$ *satisfies the associative law* $(\mu \times \text{id}_G)\mu = a(\text{id}_G \times \mu)\mu$ *if and only if* $(\alpha \odot \beta) \odot \gamma = \alpha \odot (\beta \odot \gamma)$ *for all relations* $\alpha, \beta, \gamma : X \rightharpoondown G$.

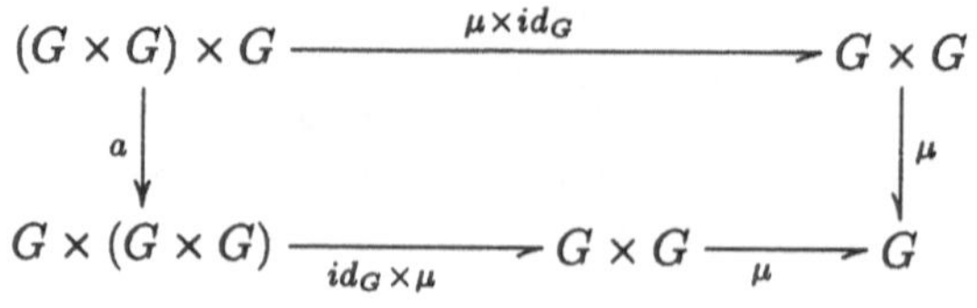

Proof. (⇒)

$$\begin{aligned}(\alpha \odot \beta) \odot \gamma &= \{(\alpha \top \beta)\mu \top \gamma\}\mu \\ &= \{(\alpha \top \beta) \top \gamma\}(\mu \times \mathrm{id}_G)\mu && \{ \text{Corollary 1(b) (sharpness)} \} \\ &= \{(\alpha \top \beta) \top \gamma\}a(\mathrm{id}_G \times \mu)\mu && \{ \text{associative law} \} \\ &= \{\alpha \top (\beta \top \gamma)\}(\mathrm{id}_G \times \mu)\mu && \{ \text{Proposition 3(a)} \} \\ &= \{\alpha \top (\beta \top \gamma)\mu\}\mu && \{ \text{Corollary 1(b) (sharpness)} \} \\ &= \alpha \odot (\beta \odot \gamma)\end{aligned}$$

(⇐)

$$\begin{aligned}(\mu \times \mathrm{id}_G)\mu &= (p_1\mu \top q_1)\mu && \{ \mu \times \mathrm{id}_G = p_1\mu \top q_1 \} \\ &= p_1(p \odot q) \odot q_1 && \{ \mu = p \odot q \} \\ &= (p_1 p \odot p_1 q) \odot q_1 && \{ \text{Lemma 1(e)} \} \\ &= p_1 p \odot (p_1 q \odot q_1) && \{ \text{(associative law)} \} \\ &= a p_2 \odot (a q_2 p \odot a q_2 q) && \{ a p_2 = p_1 p, a q_2 p = p_1 q, a q_2 q = q_1 \} \\ &= a(p_2 \top q_2 \mu)\mu && \{ \text{Lemma 1(e)},\ \mu = p \odot q \} \\ &= a(\mathrm{id}_G \times \mu)\mu && \{ \mathrm{id}_G \times \mu = p_2 \top q_2 \mu \}\end{aligned}$$

□

Remark. It is trivial that a binary operation $\mu : G \times G \rightharpoondown G$ satisfies the commutative law $t\mu = \mu$ (where $t : G \times G \to G \times G$ is the twisting function defined by $t = q \top p$) if and only if $\alpha \odot \beta = \beta \odot \alpha$ for all relations $\alpha, \beta : X \rightharpoondown G$.

4 Unitary and Inverse Operations

A total function $x : I \to X$ is called an *I-point* of an object X. A relation $\rho : I \rightharpoondown X$ is *nonempty* if there is some I-point $x : I \to X$ such that $x \sqsubseteq \rho$.

In this section we assume $G = (G, \mu)$ is a semigroup in $\mathcal{A}$.

Proposition 5. *Let $\varepsilon, \varepsilon' : I \rightharpoondown G$ be relations. Then*

(a) $\mathrm{id}_G \odot \nabla_{GI}\varepsilon = \mathrm{id}_G$ *iff* $\alpha \odot \nabla_{XI}\varepsilon = \alpha$ *for all relations* $\alpha : X \rightharpoondown G$.
(b) $\nabla_{GI}\varepsilon \odot \mathrm{id}_G = \mathrm{id}_G$ *iff* $\nabla_{XI}\varepsilon \odot \alpha = \alpha$ *for all relations* $\alpha : X \rightharpoondown G$,
(c) *If* $\mathrm{id}_G \odot \nabla_{GI}\varepsilon = \mathrm{id}_G$ *and* $\nabla_{GI}\varepsilon' \odot \mathrm{id}_G = \mathrm{id}_G$, *then* $\varepsilon = \varepsilon'$.
(d) *If* $\mathrm{id}_G \odot \nabla_{GI}\varepsilon = \mathrm{id}_G$ *or* $\nabla_{GI}\varepsilon \odot \mathrm{id}_G = \mathrm{id}_G$, *then* $\mu^\sharp$ *is total.*

Proof. (a) (⇒)

$$\begin{aligned}\alpha &= \alpha\,\mathrm{id}_G \\ &= \alpha(\mathrm{id}_G \odot \nabla_{GI}\varepsilon) && \{ \mathrm{id}_G \odot \nabla_{GI}\varepsilon = \mathrm{id}_G \} \\ &= \alpha\,\mathrm{id}_G \odot \nabla_{XI}\varepsilon && \{ \text{Lemma 1(c)} \} \\ &= \alpha \odot \nabla_{XI}\varepsilon\end{aligned}$$

($\Leftarrow$) Set $\alpha = \mathrm{id}_G$. Then $\mathrm{id}_G \odot \nabla_{GI}\varepsilon = \mathrm{id}_G$ simply follows from $\alpha = \alpha \odot \nabla_{XI}\varepsilon$.
(b) It is similar to (a).
(c)

$$\begin{aligned} \varepsilon' &= \varepsilon' \odot \nabla_{II}\varepsilon \quad \{\, (a) : \mathrm{id}_G \odot \nabla_{GI}\varepsilon = \mathrm{id}_G \,\} \\ &= \varepsilon' \odot \varepsilon \quad \{\, \nabla_{II} = \mathrm{id}_I \,\} \\ &= \nabla_{II}\varepsilon' \odot \varepsilon \quad \{\, \mathrm{id}_I = \nabla_{II} \,\} \\ &= \varepsilon \quad \{\, (b) : \nabla_{GI}\varepsilon' \odot \mathrm{id}_G = \mathrm{id}_G \,\} \end{aligned}$$

(d) Assume $\mathrm{id}_G \odot \nabla_{GI}\varepsilon = \mathrm{id}_G$. Then $\nabla_{GG} = \nabla_{GG}(\mathrm{id}_G \top \nabla_{GI}\varepsilon)\mu \sqsubseteq \nabla_{GG\times G}\mu$ and so $\nabla_{GG\times G}\mu = \nabla_{GG}$, which is equivalent to $\mathrm{id}_G \sqsubseteq \mu^\sharp\mu$. □

A relation $\varepsilon : I \rightharpoonup G$ is called a *unitary* operation for μ if it satisfies two conditions $\mathrm{id}_G \odot \nabla_{GI}\varepsilon = \mathrm{id}_G$ and $\nabla_{GI}\varepsilon \odot \mathrm{id}_G = \mathrm{id}_G$. As in the ordinary group theory a unitary operation for a binary operation is unique by Proposition 5(c).

Corollary 2. *Let $\varepsilon : I \rightharpoonup G$ a unitary operation for μ, and let $\alpha, \beta, \gamma : X \rightharpoonup G$ be relations. If $\alpha \odot \beta = \beta \odot \gamma = \nabla_{XI}\varepsilon$, then $\alpha = \gamma$.*

Proof.

$$\begin{aligned} \alpha &= \alpha \odot \nabla_{XI}\varepsilon \quad \{\, \text{Proposition 5(a)} : \mathrm{id}_G = \mathrm{id}_G \odot \nabla_{GI}\varepsilon \,\} \\ &= \alpha \odot (\beta \odot \gamma) \quad \{\, \beta \odot \gamma = \nabla_{XI}\varepsilon \,\} \\ &= (\alpha \odot \beta) \odot \gamma \quad \{\, \text{(associative)} \,\} \\ &= \nabla_{XI}\varepsilon \odot \gamma \quad \{\, \alpha \odot \beta = \nabla_{XI}\varepsilon \,\} \\ &= \gamma. \quad \{\, \text{Proposition 5(b)} : \mathrm{id}_G = \nabla_{GI}\varepsilon \odot \mathrm{id}_G \,\} \end{aligned}$$

□

The following states that a nonempty unitary operation for a total binary operation is an I-point.

Theorem 3. *Let $\varepsilon : I \rightharpoonup G$ be a unitary operation for μ. If μ is total and ε is nonempty, then ε is an I-point.*

Proof. As ε is nonempty there is an I-point $e : I \to G$ such that $e \sqsubseteq \varepsilon$. Then $\mathrm{id}_G \odot \nabla_{GI}e = (\mathrm{id}_G \top \nabla_{GI}e)\mu$ is total (since μ is total by the assumtion) and $\mathrm{id}_G \odot \nabla_{GI}e \sqsubseteq \mathrm{id}_G \odot \nabla_{GI}\varepsilon = \mathrm{id}_G$. Hence $\mathrm{id}_G \odot \nabla_{GI}e = \mathrm{id}_G$ (remarking that $\mathrm{id}_G \odot \nabla_{GI}e$ is total and id_G is a function), and so $\varepsilon = e$ by Proposition 5(c). □

Remark. Every total relation $\varepsilon : I \rightharpoonup G$ is nonempty under the *relational axiom of choice*: For all relations $\alpha : A \rightharpoonup B$ there exists a function $f : A \to B$ such that $f \sqsubseteq \alpha$ and $f\nabla_{BA} = \alpha\nabla_{BA}$. □

The notion of inverse operations of course depends on unitary operations. Here we only mention a few general properties on inverse-like relations.

Corollary 3. *Let $\varepsilon : I \rightharpoonup G$ be a unitary operation for μ. If two relations $\iota, \iota' : G \rightharpoonup G$ satisfy $\mathrm{id}_G \odot \iota = \iota' \odot \mathrm{id}_G = \nabla_{GI}\varepsilon$, then $\iota = \iota'$.*

Proof. It directly follows from Corollary 2. □

Recall a fundamental excercise in the ordinary group theory: If $\forall x \in G : xe = x$ and $\forall x \in G \exists y \in G : xy = e$ for some element $e \in G$, then $\forall x \in G : ex = x$ and $\forall x \in G \exists y \in G : yx = e$. (Answer. Assume $xy = e$ and $yz = e$. Then $yx = (yx)e = (yx)(yz) = (y(xy))z = (ye)z = yz = e$ and $ex = (xy)x = x(yx) = xe = x$.)

The above fact reflects the following theorem on our framework:

Theorem 4. *Let* $\varepsilon : I \rightharpoonup G$ *and* $\iota : G \rightharpoonup G$ *be relations satisfying* $\mathrm{id}_G \odot \nabla_{GI}\varepsilon = \mathrm{id}_G$ *and* $\mathrm{id}_G \odot \iota = \nabla_{GI}\varepsilon$. *Then the following four conditions are equivalent:*

(a) ι *is a total function,*
(b) $\iota \odot \iota^2 = \nabla_{GI}\varepsilon$,
(c) $\nabla_{GI}\varepsilon \odot \iota^2 = \mathrm{id}_G$ *and* $\nabla_{GI}\varepsilon \odot \mathrm{id}_G = \mathrm{id}_G$,
(d) $\iota^2 = \mathrm{id}_G$.

Proof. (a)⇒(b)

$$\begin{aligned} \iota \odot \iota^2 &= \iota(\mathrm{id}_G \odot \iota) && \{ \text{ Lemma 1(e) } \iota : \text{function } \} \\ &= \iota\nabla_{GI}\varepsilon && \{ \mathrm{id}_G \odot \iota = \nabla_{GI}\varepsilon \} \\ &= \nabla_{GI}\varepsilon. && \{ \iota : \text{total} \} \end{aligned}$$

(b)⇒(c) (i)

$$\begin{aligned} \nabla_{GI}\varepsilon \odot \iota^2 &= (\mathrm{id}_G \odot \iota) \odot \iota^2 && \{ \mathrm{id}_G \odot \iota = \nabla_{GI}\varepsilon \} \\ &= \mathrm{id}_G \odot (\iota \odot \iota^2) && \{ \text{(associative)} \} \\ &= \mathrm{id}_G \odot \nabla_{GI}\varepsilon && \{ \text{(b)} \} \\ &= \mathrm{id}_G, && \{ \mathrm{id}_G \odot \nabla_{GI}\varepsilon = \mathrm{id}_G \} \end{aligned}$$

(ii)

$$\begin{aligned} \iota \odot \mathrm{id}_G &= \iota \odot (\nabla_{GI}\varepsilon \odot \iota^2) && \{ \text{(i)} \} \\ &= (\iota \odot \nabla_{GI}\varepsilon) \odot \iota^2 && \{ \text{(associative)} \} \\ &= \iota \odot \iota^2 && \{ \mathrm{id}_G \odot \nabla_{GI}\varepsilon = \mathrm{id}_G \} \\ &= \nabla_{GI}\varepsilon, && \{ \text{(b} \} \end{aligned}$$

(iii)

$$\begin{aligned} \nabla_{GI}\varepsilon \odot \mathrm{id}_G &= (\mathrm{id}_G \odot \iota) \odot \mathrm{id}_G && \{ \mathrm{id}_G \odot \iota = \nabla_{GI}\varepsilon \} \\ &= \mathrm{id}_G \odot (\iota \odot \mathrm{id}_G) && \{ \text{(associative)} \} \\ &= \mathrm{id}_G \odot \nabla_{GI}\varepsilon && \{ \text{(ii)} \} \\ &= \mathrm{id}_G. && \{ \mathrm{id}_G \odot \nabla_{GI}\varepsilon = \mathrm{id}_G \} \end{aligned}$$

(c)⇒(d)

$$\begin{aligned} \iota^2 &= \iota^2(\nabla_{GI}\varepsilon \odot \mathrm{id}_G) && \{ \text{(c)} \} \\ &= \nabla_{GI}\varepsilon \odot \iota^2 && \{ \text{ Lemma 1(d) } \} \\ &= \mathrm{id}_G. && \{ \text{(c)} \} \end{aligned}$$

(d)⇒(a) It is trivial by Proposition 1. □

Remark. Supposed the relational axiom of choice. If $\mathrm{id}_G \odot \nabla_{GI}\varepsilon = \mathrm{id}_G$ and $\mathrm{id}_G \odot \iota = \nabla_{GI}\varepsilon$, and if $\mu : G \times G \rightharpoonup G$ and $\iota : G \rightharpoonup G$ are total and $\varepsilon : I \rightharpoonup G$ is a function, then ι is a (unique) total function. (By the relational axiom of choice there exists a total function $i : G \to G$ such that $i \sqsubseteq \iota$. Then $\mathrm{id}_G \odot i \sqsubseteq \mathrm{id}_G \odot \iota = \nabla_{GI}\varepsilon$ and so $\mathrm{id}_G \odot i = \nabla_{GI}\varepsilon$, since $\mathrm{id}_G \odot i$ is total and $\nabla_{GI}\varepsilon$ is a function. Hence, by the virtue of the last Theorem 4, i is a right and left inverse, and finally we have $\iota = i$ by Corollary 3.)

5 Groups

In this section we define the notion of subgroups and residual relations induced by subgroups, and show a fundamental fact that the residual relations are equivalence relations in allegories. However we limit all the operations of group structures to be total functions for the sake of simplicity.

Definition 5. A group $G = (G, m, e, i)$ in $\mathcal{A}$ is a quartet of an object G and three total functions $m : G \times G \to G$, $e : I \to G$ and $i : G \to G$ satisfying the following conditions:
(Associative Law) $(m \times \mathrm{id}_G)m = a(\mathrm{id}_G \times m)m$,
(Right Unitary) $(\mathrm{id}_G \top \nabla_{GI} e)m = \mathrm{id}_G$,
(Right Inverse) $(\mathrm{id}_G \top i)m = \nabla_{GI} e$.
□

The following proposition is trivial from the arguments in the previous sections:

Proposition 6. *Let $G = (G, m, e, i)$ be a group in $\mathcal{A}$ and $\alpha, \beta, \gamma : X \rightharpoondown G$ relations. Then the following identities hold:*

(a) $(\alpha \odot \beta) \odot \gamma = \alpha \odot (\beta \odot \gamma)$ (associative),
(b) $\alpha \odot \nabla_{XI} e = \nabla_{XI} e \odot \alpha = \alpha$ (unitary),
(c) $\mathrm{id}_G \odot i = i \odot \mathrm{id}_G = \nabla_{GI} e$, (inverse),
(d) $i^2 = \mathrm{id}_G$,
(e) $ei = e$ *and* $e \odot e = e$,
(f) $mi = qi \odot pi$.

Proof. The statement (a) has already been seen in Theorem 2. The statements (b), (c) and (d) follow from Proposition 5 and Theorem 4.
(e) An identity $e \odot e = e$ is trivial by (b) and it is easily seen that $ei = ei \odot e = e(i \odot \mathrm{id}_G) = e\nabla_{GI} e = e$.
(f) It follows from Corollary 2, since we have $mi \odot m = (qi \odot pi) \odot m = \nabla_{G\times GI} e$ as follows:

$$\begin{aligned} mi \odot m &= m(i \odot \mathrm{id}_G) && \{ \text{ Lemma 1(e) } \} \\ &= m\nabla_{GI} e && \{ i \odot \mathrm{id}_G = \nabla_{GI} \} \\ &= \nabla_{G\times GI} e && \{ m : \text{total} \} \end{aligned}$$

and

$$\begin{aligned} m \odot (qi \odot pi) &= (p \odot q) \odot (qi \odot pi) && \{ m = p \odot q \} \\ &= \{p \odot (q \odot qi)\} \odot pi && \{ \text{ (associative) } \} \\ &= \{p \odot q(\mathrm{id}_G \odot i)\} \odot pi && \{ \text{ Lemma 1(e) } \} \\ &= (p \odot q\nabla_{GI} e) \odot pi && \\ &= (p \odot \nabla_{G\times GI} e) \odot pi && \{ q : \text{total} \} \\ &= p \odot pi && \\ &= p(\mathrm{id}_G \odot i) && \{ \text{ Lemma 1(e) } \} \\ &= p\nabla_{GI} e && \\ &= \nabla_{G\times GI} e. && \{ p : \text{total} \} \end{aligned}$$

Note that $qi \odot pi = t(i \times i)m$, where $t = q \top p : G \times G \to G \times G$ is the twisting function. □

In what follows we assume $G = (G, m, e, i)$ is a group in $\mathcal{A}$. In relational calculus there are a few different ways how to specify subobjects. For example, Schmidt and Ströhlein [11] made use of "vectors", to represent subobjects in relation algebras. We are going to use relations from a unit I into some object G; these do in fact satisfy the vector equation $\nabla_{II}\rho = \rho$.

Definition 6. A nonempty relation $\rho : I \rightharpoondown G$ is a *subgroup* of G if $\rho i \sqsubseteq \rho$ and $\rho \odot \rho \sqsubseteq \rho$. □

Note that every subgroup $\rho : I \rightharpoondown G$ contains the unitary operation $e : I \to G$, that is, $e \sqsubseteq \rho$. Assume $x \sqsubseteq \rho$ for some I-point $x : I \to G$. Then $e = x\nabla_{GI}e = x(i \odot \mathrm{id}_G) = xi \odot x \sqsubseteq \rho i \odot \rho \sqsubseteq \rho \odot \rho \sqsubseteq \rho$. Thus a relation $\rho : I \rightharpoondown G$ is a subgroup of G iff $e \sqsubseteq \rho$, $\rho i = \rho$ and $\rho \odot \rho = \rho$. The unitary operation e and the universal relation ∇_{IG} are trivial subgroups of G.

Proposition 7. *If $\rho, \rho' : I \rightharpoondown G$ are two subgroups of G, then so is $\rho \sqcap \rho'$.*

Proof. It is trivial. □

Now let us go back to an exercise (that is, the second question (b) in the introduction) in classical group theory: Let S be a subgroup of a group G. Prove that the right residual relation $x \equiv y$ on G, defined by $x^{-1}y \in S$, is an equivalence relation. (Answer. The reflexive law: $x^{-1}x = e \in S$, the symmetric law: If $x^{-1}y \in S$, then $y^{-1}x = (x^{-1}y)^{-1} \in S$, and the transitive law: If $x^{-1}y \in S$ and $y^{-1}z \in S$, then $x^{-1}z = (x^{-1}y)(y^{-1}z) \in S$.)

The following theorem is a generalization of this fundamental fact.

Theorem 5. *If $\rho : I \rightharpoondown G$ is a subgroup, then $\theta = \mathrm{id}_G \odot \nabla_{GI}\rho$ is an equivalence relation on G.*

Proof. Reflexivity $\mathrm{id}_G \sqsubseteq \theta$ is direct from $\mathrm{id}_G = \mathrm{id}_G \odot \nabla_{GI}e \sqsubseteq \mathrm{id}_G \odot \nabla_{GI}\rho = \theta$. Next we can show transitivity $\theta\theta \sqsubseteq \theta$ by the following computation:

$$
\begin{aligned}
\theta\theta &= \theta(\mathrm{id}_G \odot \nabla_{GI}\rho) \\
&= \theta \odot \nabla_{GI}\rho && \{ \text{ Lemma 1(c) } \} \\
&= (\mathrm{id}_G \odot \nabla_{GI}\rho) \odot \nabla_{GI}\rho \\
&= \mathrm{id}_G \odot (\nabla_{GI}\rho \odot \nabla_{GI}\rho) && \{ \text{ (associative) } \} \\
&= \mathrm{id}_G \odot \nabla_{GI}(\rho \odot \rho) && \{ \nabla_{GI} : \text{total function, Lemma 1(e)} \} \\
&\sqsubseteq \mathrm{id}_G \odot \nabla_{GI}\rho && \{ \rho \odot \rho \sqsubseteq \rho \} \\
&= \theta.
\end{aligned}
$$

Finally we will see symmetry $\theta^\sharp \sqsubseteq \theta$. The proof is somewhat complicated as follows:

$$\begin{aligned}
\theta &= (p^\sharp \sqcap \nabla_{GI}\rho q^\sharp)m \\
&= \{p^\sharp \sqcap \nabla_{GI}\rho(q^\sharp \sqcap \rho^\sharp \nabla_{IG\times G})\}m && \{\ \rho q^\sharp = \rho(q^\sharp \sqcap \rho^\sharp \nabla_{IG\times G}) : \text{DF}\ \} \\
&= \{p^\sharp \sqcap \nabla_{GI}\rho(q^\sharp \sqcap i\rho^\sharp \nabla_{IG\times G})\}m && \{\ \rho i = \rho,\ i^\sharp = i\ \} \\
&\sqsubseteq \{p^\sharp \sqcap \nabla_{GI}\rho(\mathrm{id}_G \odot i)(q \odot \nabla_{G\times GI}\rho)^\sharp\}m && \{\ \text{Proposition 4}\ \} \\
&= \{p^\sharp \sqcap \nabla_{GI}\rho\nabla_{GI}e(q \odot \nabla_{G\times GI}\rho)^\sharp\}m && \{\ \mathrm{id}_G \odot i = \nabla_{GI}e\ \} \\
&= \{p^\sharp \sqcap \nabla_{GI}e(q \odot \nabla_{G\times GI}\rho)^\sharp\}m && \{\ \rho\nabla_{GI} = \mathrm{id}_I\ \} \\
&\sqsubseteq (\mathrm{id}_G \odot \nabla_{GI}e)\{p \odot (q \odot \nabla_{G\times GI}\rho)\}^\sharp m && \{\ \text{Proposition 4}\ \} \\
&= \{(p \odot q) \odot \nabla_{G\times GI}\rho\}^\sharp m && \{\ \mathrm{id}_G \odot \nabla_{GI}e = \mathrm{id}_G\ \} \\
&= (m \odot \nabla_{G\times GI}\rho)^\sharp m && \{\ p \odot q = m\ \} \\
&= \{m(\mathrm{id}_G \odot \nabla_{GI}\rho)\}^\sharp m && \{\ \text{Lemma 1(c)}\ \} \\
&= \theta^\sharp m^\sharp m \\
&\sqsubseteq \theta^\sharp. && \{\ m : \text{function}\ \}
\end{aligned}$$

□

6 Normal Subgroups

In the final section we define normal subgroups of groups in allegories and show the main result that normal subgroups of a group form a modular lattice.

In classical group theory a subgroup of a group is normal iff the right and the left residual relations coincide. We adopt this property to define normal subgroups in our framework:

Definition 7. A subgroup $\rho : I \rightharpoondown G$ of G is called *normal* if $\mathrm{id}_G \odot \nabla_{GI}\rho = \nabla_{GI}\rho \odot \mathrm{id}_G$. □

It is trivial that a subgroup $\rho : I \rightharpoondown G$ is normal iff $\alpha \odot \nabla_{XI}\rho = \nabla_{XI}\rho \odot \alpha$ for all relations $\alpha : X \rightharpoondown G$.

The following is also an analogy from the classical case:

Proposition 8. *If $\rho, \sigma : I \rightharpoondown G$ are normal subgroups of G, then so is $\rho \odot \sigma$.*

Proof. It is easy to see that $\rho \odot \sigma$ is a subgroup of G. We only show the normality:

$$\begin{aligned}
\mathrm{id}_G \odot \nabla_{GI}(\rho \odot \sigma) &= \mathrm{id}_G \odot (\nabla_{GI}\rho \odot \nabla_{GI}\sigma) && \{\ \text{Lemma 1(e)}\ \} \\
&= (\mathrm{id}_G \odot \nabla_{GI}\rho) \odot \nabla_{GI}\sigma \\
&= (\nabla_{GI}\rho \odot \mathrm{id}_G) \odot \nabla_{GI}\sigma && \{\ \rho : \text{normal}\ \} \\
&= \nabla_{GI}\rho \odot (\mathrm{id}_G \odot \nabla_{GI}\sigma) \\
&= \nabla_{GI}\rho \odot (\nabla_{GI}\sigma \odot \mathrm{id}_G) && \{\ \sigma : \text{normal}\ \} \\
&= (\nabla_{GI}\rho \odot \nabla_{GI}\sigma) \odot \mathrm{id}_G \\
&= \nabla_{GI}(\rho \odot \sigma) \odot \mathrm{id}_G. && \{\ \text{Lemma 1(e)}\ \}
\end{aligned}$$

□

Let $\rho, \sigma : I \rightharpoondown G$ be normal subgroups of G. Then it is obvious that the normal subgroup $\rho \odot \sigma$ is the supremum (or, join) of ρ and σ with respect to the relational inclusion $\sqsubseteq$ of normal subgroups.

Definition 8. Let $m : X \times X \to X$ be a binary operation on X. A relation $\theta : X \rightharpoondown X$ is a *congruence* with respect to m if it is an equivalence relation such that $(\theta \times \theta)m \sqsubseteq m\theta$. □

Proposition 9. *If $\rho : I \rightharpoondown G$ is a normal subgroup of G, then $\theta = \mathrm{id}_G \odot \nabla_{GI}\rho$ is a congruence with respect to m.*

Proof.

$$\begin{aligned}
(\theta \times \theta)m &= (p\theta \top q\theta)m \\
&= p\theta \odot q\theta \\
&= p(\mathrm{id}_G \odot \nabla_{GI}\rho) \odot q(\mathrm{id}_G \odot \nabla_{GI}\rho) \\
&= (p \odot \nabla_{G\times GI}\rho) \odot (q \odot \nabla_{G\times GI}\rho) && \{\text{ Lemma 1(e) }\} \\
&= p \odot (\nabla_{G\times GI}\rho \odot q) \odot \nabla_{G\times GI}\rho && \{\text{ (associative) }\} \\
&= p \odot (q \odot \nabla_{G\times GI}\rho) \odot \nabla_{G\times GI}\rho && \{\ \rho : \text{normal }\} \\
&= (p \odot q) \odot (\nabla_{G\times GI}\rho \odot \nabla_{G\times GI}\rho) && \{\text{ (associative) }\} \\
&= m \odot \nabla_{G\times GI}(\rho \odot \rho) && \{\ m = p \odot q,\ \text{Lemma 1(e) }\} \\
&= m \odot \nabla_{G\times GI}\rho && \{\ \rho : \text{subgroup }\} \\
&= m(\mathrm{id}_G \odot \nabla_{GI}\rho) && \{\text{ Lemma 1(c) }\} \\
&= m\theta
\end{aligned}$$

□

In the ordinary group theory it is well-known that the set of all normal subgroups of a group forms a modular lattice. In other words, for three normal subgroups S, T and U of a group G the following modular law holds:

$$S \subseteq U \implies ST \cap U \subseteq S(T \cap U).$$

The proof of the fact is fundamental: Assume $S \subseteq U$ and $x \in ST \cap U$. Then $x \in U$ and $x \in ST$, so $x = st$ for some $s \in S$ and $t \in T$. Hence $t = s^{-1}x \in U$ holds since $x \in U$ and $s \in S \subseteq U$ and U is a subgroup. That is, $x = st$ with $s \in S$ and $t \in T \cap U$. Hence $x \in S(T \cap U)$.

To see the above fact in allegories we need the next lemma:

Lemma 2. *If $\tau : I \rightharpoondown G$ is a subgroup of G, then $\tau p^\sharp \sqcap \tau m^\sharp \sqsubseteq \tau q^\sharp$.*

Proof. First recall that $\tau = \tau i$. Applying Proposition 4

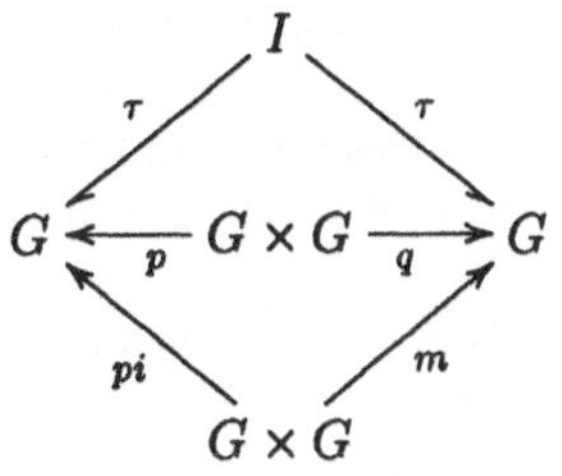

we have

$$\begin{aligned}
\tau p^\sharp \sqcap \tau m^\sharp &= \tau i^\sharp p^\sharp \sqcap \tau m^\sharp && \{\ \tau = \tau i^\sharp\ \} \\
&\sqsubseteq (\tau \odot \tau)(pi \odot m)^\sharp && \{\text{ Proposition 4 }\} \\
&\sqsubseteq \tau(pi \odot m)^\sharp. && \{\ \tau \odot \tau \sqsubseteq \tau\ \}
\end{aligned}$$

On the other hand we can see the identity $pi \odot m = q$ from

$$\begin{aligned} pi \odot m &= pi \odot (p \odot q) && \{\, m = p \odot q \,\} \\ &= (pi \odot p) \odot q && \{\, \text{(associative)} \,\} \\ &= p(i \odot \mathrm{id}_G) \odot q && \{\, \text{Lemma 1(e)} \,\} \\ &= p\nabla_{GI} e \odot q && \{\, i \odot \mathrm{id}_G = \nabla_{GI} e \,\} \\ &= \nabla_{G\times GI} e \odot q && \{\, p : \text{total} \,\} \\ &= q. && \{\, \text{Proposition 6(b)} \,\} \end{aligned}$$

□

Finally we prove a main result that the set of all normal subgroups in allegories forms a modular lattice, as in the classical case:

Theorem 6. *Let $\rho, \sigma, \tau : I \rightharpoondown G$ be normal subgroups of a group G. If $\rho \sqsubseteq \tau$, then $(\rho \odot \sigma) \sqcap \tau \sqsubseteq \rho \odot (\sigma \sqcap \tau)$.*

Proof. It follows from the simple computation:

$$\begin{aligned} (\rho \odot \sigma) \sqcap \tau &= (\rho p^\sharp \sqcap \sigma q^\sharp) m \sqcap \tau \\ &\sqsubseteq (\rho p^\sharp \sqcap \sigma q^\sharp \sqcap \tau m^\sharp) m && \{\, \text{DF} \,\} \\ &= \{\rho p^\sharp \sqcap \sigma q^\sharp \sqcap (\tau p^\sharp \sqcap \tau m^\sharp)\} m && \{\, \rho \sqsubseteq \tau \,\} \\ &\sqsubseteq (\rho p^\sharp \sqcap \sigma q^\sharp \sqcap \tau q^\sharp) m && \{\, \text{Lemma 2} : \tau p^\sharp \sqcap \tau m^\sharp \sqsubseteq \tau q^\sharp \,\} \\ &= \{\rho p^\sharp \sqcap (\sigma \sqcap \tau) q^\sharp\} m \\ &= \rho \odot (\sigma \sqcap \tau). \end{aligned}$$

□

7 Conclusion

Group theory is an important field of mathematics with applications in many other science. In this paper the author has tried to demonstrate some fundamental exercises of group theory entirely at the point-free relation-algebraic level. The paper identifies allegories with a unit and relational products as the appropriate relational setting for this aim and presents purely calculational proofs for group theoretic theorems that cannot be easily be shown in the ordinary category-theoretical approach to group theory. Of course, since allegories are sufficiently abstract idea for the relational methods, all the results in the paper can be applicable to theory of fuzzy or L-fuzzy groups, for example.

It is interesting to contrast these proofs with graphical approach to relational reasoning due to Curtis and Lowe [2] and Dougherty and Gutierrez [3]. We just remark that by using the graphical rewriting laws 1, 3, 5, 6, 7 and 9 in [2] one can easily prove that $\alpha\gamma^\sharp \sqcap \beta\delta^\sharp \sqsubseteq \rho^\sharp\tau$ implies $\alpha\gamma^\sharp \sqcap \beta\delta^\sharp \sqsubseteq (\alpha\rho^\sharp \sqcap \beta\tau^\sharp)(\rho\gamma^\sharp \sqcap \tau\delta^\sharp)$. But, as stated in the section 2, Maddux [10] gave a counterexample for the sharpness problem in finitely generated relation algebras. Also the law 2 and 10 seems to show the uniformity condition $\nabla_{XY}\nabla_{YZ} = \nabla_{XZ}$ for all objects

X, Y, Z. These examples indicate that the graphical calculus is based on theory of (homogeneous) binary relations on sets, and the laws of graphical calculus form a stronger setting than those of relation algebras, allegories and Dedekind categories in some sense.

Finally the author would like to thank the anonymous referee for his valuable comments and suggestions to the paper.

References

1. C. Brink, W. Kahl and G. Schmidt (eds.), Relational methods in computer science. Advances in Computing Science, (Springer, Wien, New York, 1997).
2. S. Curtis and G. Lowe, *A graphical calculus*, Lecture Notes in Computer Science **947** (1995), 214-231.
3. D. Dougherty and C. Gutiérrez, *Normal forms and reduction for theories of binary relations*, Lecture Notes in Computer Science, **1833**(2000), 95–109.
4. P. Freyd and A. Scedrov, Categories, allegories (North-Holland, Amsterdam, 1990).
5. J. Desharnais, *Monomorphic characterization of n-ary direct products*, Information Sciences **119**(1999), 275–288.
6. W. Kahl and G. Schmidt, *Exploring (finite) relation algebras using tools written in Haskell*, Technical Report 2000-02, Fakultät für Informatik, Universität der Bundeswehr München, October 2000.
7. Y. Kawahara, *Relational set theory*, Lecture Notes in Computer Science, **953**(1995), 44–58.
8. Y. Kawahara, *Lattices in Dedekind categories*, In: Orlowska, E. and Szalas, A. (Eds), Relational Methods for Computer Science Applications, Physica-Verlag, 2001, 247–260.
9. S. Mac Lane, *Categories for the working mathematician*, (Springer-Verlag, 1972).
10. R. Maddux, *On the derivation of identities invloving projection functions*, Logic Colloquium '92, ed. Csirmaz, Gabbay, de Rijke, Center for the Study of Language and Information Publications, Stanford, 1995, 145–163.
11. G. Schmidt and T. Ströhlein, Relations and graphs – Discrete Mathematics for Computer Scientists – (Springer-Verlag, 1993).
12. A. Rosenfeld, *Fuzzy groups*, J. Math. Anal. Appl. **35**(1971), 512–517.

Distributed Conceptual Structures

Robert E. Kent

The Ontology Consortium
rekent@ontologos.org

Abstract. The theory of distributed conceptual structures, as outlined in this paper, is concerned with the distribution and conception of knowledge. It rests upon two related theories, Information Flow and Formal Concept Analysis, which it seeks to unify. Information Flow (IF) [2] is concerned with the distribution of knowledge. The foundations of Information Flow are explicitly based upon the Chu Construction in *-autonomous categories [1] and implicitly based upon the mathematics of closed categories [6]. Formal Concept Analysis (FCA) [3] is concerned with the conception and analysis of knowledge. In this paper, we connect these two studies by categorizing the basic theorem of Formal Concept Analysis, thus extending it to the distributed realm of Information Flow. The main result is the representation of the basic theorem as a categorical equivalence at three different levels of functional and relational constructs. This representation accomplishes a rapprochement between Information Flow and Formal Concept Analysis.

1 Introduction

Figure 1 illustrates distributed conceptual knowledge as a two-dimensional structure. The first dimension is along the distribution/conception distinction. Information Flow exists on the distributional side, whereas Formal Concept Analysis extends this toward the conceptual direction. The second dimension is along a functional/relational distinction. All of the development of Information Flow has taken place on the functional level, but some of Formal Concept Analysis

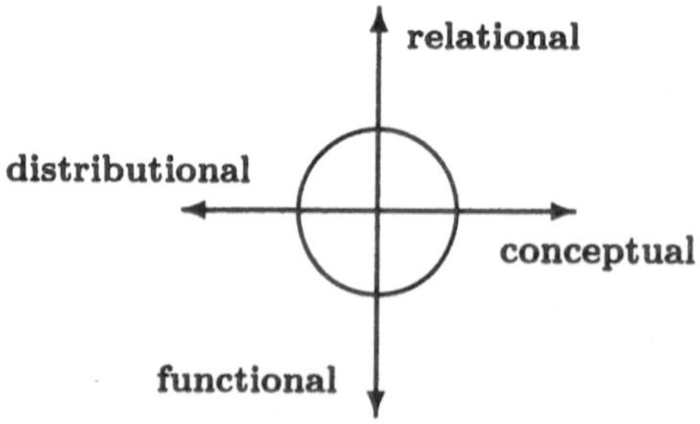

Fig. 1. Two Dimensions of Distributed Conceptual Structures

H. de Swart (Ed.): RelMiCS 2001, LNCS 2561, pp. 104–123, 2002.

extends this into the relational direction, which still might be considered *terra incognita.*

To a large extent the foundation of distributed conceptual structures is based upon binary relations (or matrices) and centered upon the axiom of adjointness between composition and residuation. This composition/residuation adjointness axiom is similar to the axiom of adjointness between conjunction and implication. Since composition and residuation are binary, the axiom has two statements: (1) Left composition is (left) adjoint to left residuation: $r \circ s \subseteq t$ iff $s \subseteq r \backslash t$, for any compatible binary relations r, s and t. (2) Right composition is (left) adjoint to right residuation: $r \circ s \subseteq t$ iff $r \subseteq t/s$, for any compatible binary relations r, s and t. Some derived properties are that residuation preserves composition: $(r_1 \circ r_2) \backslash t = r_2 \backslash (r_1 \backslash t)$ and $t/(s_1 \circ s_2) = (t/s_2)/s_1$ and that residuation preserves identity: $Id_A \backslash t = t$ and $t/Id_B = t$. The involutions of transpose and negation are of secondary importance. The axiom for transpose states that transpose dualizes residuation: $(r \backslash t)^{\propto} = t^{\propto}/r^{\propto}$ and $(t/s)^{\propto} = s^{\propto} \backslash t^{\propto}$.

There are two important associative laws – one unconstrained the other constrained. There is an unconstrained associative law: $(r \backslash t)/s = r \backslash (t/s)$, for all $t \subseteq A \times B$, $r \subseteq A \times C$ and $s \subseteq D \times B$. There is also an associative law constrained by closure: if t is an endorelation and r and s are closed with respect to t, $r = t/(r \backslash t)$ and $s = (t/s) \backslash t$, then $(t/s) \backslash r = s/(r \backslash t)$, for all $t \subseteq A \times A$, $r \subseteq A \times B$ and $s \subseteq C \times A$. Functions have a special behavior with respect to derivation. If function f and relation r are composable, then $f^{\propto} \backslash r = f \circ r$. If relation s and the opposite of function g are composable, then $s/g = s \circ g^{\propto}$.

The paper consists of five sections: Introduction, Basic Notions, Architecture, Limit/Colimit Constructions, and Summary and Future Work. The section on Basic Notions is a review of some of the basic ideas of Information Flow and Formal Concept Analysis (classifications, concept lattices, and functional infomorphisms), and an introduction of some new ideas (relational infomorphisms, bonds, and bonding pairs). The section on Architecture, the central section of the paper, is principally concerned with the three categorical equivalences between the distributional pole and the conceptual pole that represent the fundamental theorem of Formal Concept Analysis. The section on Limit/Colimit Constructions, the closest to applications, gives an enhanced fibrational description of the architecture, thereby situating various basic constructions. The final section gives a summary and points out future plans to apply the logic of distributed conceptual structures to actual distributed representational frameworks on the Internet.

2 Basic Notions

There are six basic notions in distributed conceptual structures: classifications, concept lattices, functional infomorphisms, relational infomorphisms, bonds, and bonding pairs. The first two are object-like structures, with classifications being the central object-like notion of Information Flow, and concept lattices being the central object-like notion of Formal Concept Analysis. The last four are morphism-like, with functional infomorphisms being the central morphism notion of Information Flow, relational infomorphisms being newly defined in this paper, bonds being used as an analytic tool in Formal Concept Analysis, and bonding pairs being equivalent (in a categorical sense) to complete homomorphisms, the central morphism notion in FCA.

2.1 Objects

Classifications. According to the theory of Information Flow [2], information presupposes a system of classification. Classifications have been important in library science for the last 2,000 years. Major classification systems in library science include the Dewey Decimal System (DDS) and the Library of Congress (LC). However, the library science classification system most in accord with the philosophy and techniques of IF is the Colon classification system invented by the library scientist Ranganathan. A domain-neutral notion of classification is given by the following abstract mathematical definition. A *classification* $\boldsymbol{A} = \langle inst(\boldsymbol{A}), typ(\boldsymbol{A}), \models_{\boldsymbol{A}} \rangle$ consists of (1) a set, $inst(\boldsymbol{A})$, of things to be classified, called the *instances* of $\boldsymbol{A}$, (2) a set, $typ(\boldsymbol{A})$, of things used to classify the instances, called the *types* of $\boldsymbol{A}$, (3) and a binary *classification* relation, $\models_{\boldsymbol{A}}$, between $inst(\boldsymbol{A})$ and $typ(\boldsymbol{A})$. The notation $a \models_{\boldsymbol{A}} \alpha$ is read "instance a is of type α in $inst(\boldsymbol{A})$". A classification has an alternate expression in either a suitable extension of the theory of closed categories [6] or *-autonomous categories [1]: a *classification* is a triple $\boldsymbol{A} = \langle inst(\boldsymbol{A}), typ(\boldsymbol{A}), \models_{\boldsymbol{A}} \rangle$, where $inst(\boldsymbol{A})$ and $typ(\boldsymbol{A})$ are sets and $\models_{\boldsymbol{A}}: typ(\boldsymbol{A}) \times inst(\boldsymbol{A}) \to 2$ is a function, representing a 2-valuated matrix. Classifications are known as *formal contexts* in Formal Concept Analysis [3]. In FCA, types are called *formal attributes* and instances are called *formal objects.*

Define the following pair of *derivation operators*: $A^{\boldsymbol{A}} = \{\alpha \in typ(\boldsymbol{A}) \mid a \models_{\boldsymbol{A}} \alpha \text{ for all } a \in A\}$ for any instance subset $B \subseteq inst(\boldsymbol{A})$, and $\Gamma^{\boldsymbol{A}} = \{a \in inst(\boldsymbol{A}) \mid a \models_{\boldsymbol{A}} \alpha \text{ for all } \alpha \in \Gamma\}$ for any type subset $\Gamma \subseteq typ(\boldsymbol{A})$. When $A : inst(\boldsymbol{A}) \to 1$ and $\Gamma : typ(\boldsymbol{A}) \leftarrow 1$ are regarded as relations, derivation is seen to be residuation, $A^{\boldsymbol{A}} = A \backslash \boldsymbol{A}$ and $\Gamma^{\boldsymbol{A}} = \boldsymbol{A}/\Gamma$, where the classification relation is represented as $\boldsymbol{A}$. An *extent* of $\boldsymbol{A}$ is a subset of instances of the form $\Gamma^{\boldsymbol{A}}$, and an *intent* of $\boldsymbol{A}$ is a subset of types of the form $A^{\boldsymbol{A}}$. For any instance $a \in inst(\boldsymbol{A})$, the *intent* or *type set* of a is the set $typ(a) = a^{\boldsymbol{A}} = \{\alpha \in typ(\boldsymbol{A}) \mid a \models_{\boldsymbol{A}} \alpha\}$. Intent induces a preorder on the instances $inst(\boldsymbol{A})$ defined by: $a \leq_{\boldsymbol{A}} a'$ when $a^{\boldsymbol{A}} \supseteq a'^{\boldsymbol{A}}$. For any type $\alpha \in typ(\boldsymbol{A})$, the *extent* or *instance set* of α is the set $inst(\alpha) = \alpha^{\boldsymbol{A}} = \{a \in inst(\boldsymbol{A}) \mid a \models_{\boldsymbol{A}} \alpha\}$. Extent induces a preorder on types $typ(\boldsymbol{A})$ defined by: $\alpha \leq_{\boldsymbol{A}} \alpha'$ when $\alpha^{\boldsymbol{A}} \subseteq \alpha'^{\boldsymbol{A}}$. Note that $a \in \alpha^{\boldsymbol{A}}$ iff $\alpha \in a^{\boldsymbol{A}}$ for any instance $a \in inst(\boldsymbol{A})$ and any type $\alpha \in typ(\boldsymbol{A})$.

As befitting such an important and generic notion, classifications abound. Organisms (instances) are classified by scientists into categories (types), such as Plant, Animal, Fungus, Bacterium, Alga, Eukaryote, Prokaryote, etc. Words (instances) are classified in a dictionary by parts of speech (types), such as Noun, Verb, Adjective, Adverb, etc. The following is a motivating example in Barwise and Seligman [2]: Given a first-order language $\boldsymbol{L}$, the *truth classification* of $\boldsymbol{L}$ has $\boldsymbol{L}$-structures as instances, sentences of $\boldsymbol{L}$ as types, and satisfaction as the classification relation: $M \models \phi$ when sentence ϕ is true in structure M. An ontology forms a classification with either explicit subtyping or subsumption being the classification relation. Any preorder $\boldsymbol{P} = \langle P, \leq_P \rangle$ is a classification $\boldsymbol{P} = \langle P, P, \leq_P \rangle$, where the preorder elements function as both types and instances, and the ordering relation is the classification.

Systemic examples of classifications also abound. Given any set A (of instances), the *instance powerset classification* $\wp A = \langle A, \wp A, \in \rangle$ associated with A, has elements of A as instances, subsets of A as types, with the membership relation serving as the classification relation. Given any classification $\boldsymbol{A} = \langle inst(\boldsymbol{A}), typ(\boldsymbol{A}), \models_{\boldsymbol{A}} \rangle$, the *dual classification* $\boldsymbol{A}^{\propto} = \langle typ(\boldsymbol{A}), inst(\boldsymbol{A}), \models_{\boldsymbol{A}}{}^{\propto} \rangle$ is the involution of $\boldsymbol{A}$. This is the classification, whose instances are types of $\boldsymbol{A}$, whose types are instances of $\boldsymbol{A}$, and whose classification is the transpose of the $\boldsymbol{A}$ classification. The involution operator applies also to morphisms of classifications and limiting constructions on classifications.

Concept Lattices. The basic notion of FCA is the notion of a formal concept. A *formal concept* is a pair $\boldsymbol{a} = (A, \Gamma)$ of subsets, $A \subseteq inst(\boldsymbol{A})$, called the *extent* of the concept $\boldsymbol{a}$ and denoted $ext_{\boldsymbol{A}}(\boldsymbol{a})$, and $\Gamma \in typ(\boldsymbol{A})$, called the *intent* of the concept $\boldsymbol{a}$ and denoted $int_{\boldsymbol{A}}(\boldsymbol{a})$, that satisfies the closure properties $A = \Gamma^{\boldsymbol{A}}$ and $\Gamma = A^{\boldsymbol{A}}$. There is a naturally defined concept order: $\boldsymbol{a}_1 \leq \boldsymbol{a}_2$ when $\boldsymbol{a}_1$ is more specific than $\boldsymbol{a}_2$ or dually when $\boldsymbol{a}_2$ is more generic than $\boldsymbol{a}_1$: $ext_{\boldsymbol{A}}(\boldsymbol{a}_1) \subseteq ext_{\boldsymbol{A}}(\boldsymbol{a}_2)$, or equivalently, $int_{\boldsymbol{A}}(\boldsymbol{a}_1) \supseteq int_{\boldsymbol{A}}(\boldsymbol{a}_2)$. This partial order is part of a complete lattice $\mathbf{L}(\boldsymbol{A}) = \langle L(\boldsymbol{A}), \leq_{\boldsymbol{A}}, \bigwedge_{\boldsymbol{A}}, \bigvee_{\boldsymbol{A}} \rangle$ underlying the concept lattice of $\boldsymbol{A}$, where the meet and join are defined by: $\bigwedge_{\boldsymbol{A}}(C) = \langle \bigcap_{c \in C} ext_{\boldsymbol{A}}(c), (\bigcup_{c \in C} int_{\boldsymbol{A}}(c))^{\boldsymbol{AA}} \rangle$ and $\bigvee_{\boldsymbol{A}}(C) = \langle (\bigcup_{c \in C} ext_{\boldsymbol{A}}(c))^{\boldsymbol{AA}}, \bigcap_{c \in C} int_{\boldsymbol{A}}(c) \rangle$ for any collection of concepts $C \subseteq L(\boldsymbol{A})$.

Define the *instance embedding relation* $\iota_{\boldsymbol{A}} : inst(\boldsymbol{A}) \rightarrow L(\boldsymbol{A})$, as follows: for every instance $a \in inst(\boldsymbol{A})$ and every formal concept $\boldsymbol{a} \in L(\boldsymbol{A})$ the relationship $a \iota_{\boldsymbol{A}} \boldsymbol{a}$ holds when $a \in ext(\boldsymbol{a})$, a is in the extent of $\boldsymbol{a}$. This relation is closed on the right with respect to lattice order. Instances are mapped into the lattice by the *instance embedding function* $\iota_{\boldsymbol{A}} : inst(\boldsymbol{A}) \rightarrow L(\boldsymbol{A})$, where $\iota_{\boldsymbol{A}}(a) = (a^{\boldsymbol{AA}}, a^{\boldsymbol{A}})$ for each instance $a \in inst(\boldsymbol{A})$. This function is expressed in terms of the relation as the meet $\iota_{\boldsymbol{A}}(a) = \bigwedge_{\boldsymbol{A}} a \iota_{\boldsymbol{A}}$. Concepts in $\iota_{\boldsymbol{A}}[inst(\boldsymbol{A})]$ are called *instance concepts.* Any concept $\boldsymbol{a} \in L(\boldsymbol{A})$ can be expressed as the join $\boldsymbol{a} = \bigvee a{\in}ext(\boldsymbol{a}) \iota_{\boldsymbol{A}}(a)$ of a subset of instance concepts – $\iota_{\boldsymbol{A}}[inst(\boldsymbol{A})]$ is join-dense in $L(\boldsymbol{A})$. Dually, define the *type embedding relation* $\tau_{\boldsymbol{A}} : L(\boldsymbol{A}) \rightarrow typ(\boldsymbol{A})$, as follows: for every type $\alpha \in typ(\boldsymbol{A})$ and every formal concept $\boldsymbol{a} \in L(\boldsymbol{A})$ the relationship $\boldsymbol{a} \tau_{\boldsymbol{A}} \alpha$ holds when $\alpha \in int(\boldsymbol{a})$, α is in the intent of $\boldsymbol{a}$. This relation is closed on the left respect to lattice order. Types are mapped into the lattice by the *type embedding function* $\tau_{\boldsymbol{A}} : typ(\boldsymbol{A}) \rightarrow L(\boldsymbol{A})$, where $\tau_{\boldsymbol{A}}(\alpha) = (\alpha^{\boldsymbol{A}}, \alpha^{\boldsymbol{AA}})$ for each type $\alpha \in typ(\boldsymbol{A})$. This function is expressed in terms of the relation as the join $\tau_{\boldsymbol{A}}(\alpha) = \bigvee_{\boldsymbol{A}} \tau_{\boldsymbol{A}} \alpha$. Concepts in $\tau_{\boldsymbol{A}}[typ(\boldsymbol{A})]$ are called *type concepts.* Any concept $\boldsymbol{a} \in L(\boldsymbol{A})$ can be expressed as the meet $\boldsymbol{a} = \bigwedge_{\alpha \in int(\boldsymbol{a})} \tau_{\boldsymbol{A}}(\alpha)$ of a subset of type concepts – $\tau_{\boldsymbol{A}}[typ(\boldsymbol{A})]$ is meet-dense in $L(\boldsymbol{A})$. Any classification $a \models_{\boldsymbol{A}} \alpha$ can be expressed in terms of the instance/type mappings as: $\iota_{\boldsymbol{A}}(a) \leq_{\boldsymbol{A}} \tau_{\boldsymbol{A}}(\alpha)$ – the classification relation decomposes as the relational composition: $\models_{\boldsymbol{A}} = \iota_{\boldsymbol{A}} \circ \leq_{\boldsymbol{A}} \circ \tau_{\boldsymbol{A}}{}^{\propto}$. The quintuple $\mathsf{L}(\boldsymbol{A}) = \langle \mathsf{L}(\boldsymbol{A}), inst(\boldsymbol{A}), typ(\boldsymbol{A}), \iota_{\boldsymbol{A}}, \tau_{\boldsymbol{A}} \rangle$, is called the *concept lattice* of the classification $\boldsymbol{A}$. More abstractly, a *concept lattice* $\boldsymbol{L} = \langle L, inst(\boldsymbol{L}), typ(\boldsymbol{L}), \iota_{\boldsymbol{L}}, \tau_{\boldsymbol{L}} \rangle$ consists of a complete lattice L, two sets $inst(\boldsymbol{L})$ and $typ(\boldsymbol{L})$ called the *instance set* and *type set* of $\boldsymbol{L}$, respectively; along with two functions mapping to the lattice, the *instance embedding* $\iota_{\boldsymbol{L}} : inst(\boldsymbol{L}) \to L$ and the *type embedding* $\tau_{\boldsymbol{L}} : typ(\boldsymbol{L}) \to L$, such that $\iota_{\boldsymbol{L}}[inst(\boldsymbol{L})]$ is join-dense in L and $\tau_{\boldsymbol{L}}[typ(\boldsymbol{L})]$ is meet-dense in L. For each classification $\boldsymbol{A}$ the associated quintuple $\mathsf{L}(\mathbf{A})$ is a concept lattice.

Let $\boldsymbol{A}$ be any classification and let A be any (indexing) set. A *collective $\boldsymbol{A}$-instance* indexed by A is any relation $a : inst(\boldsymbol{A}) \to A$, and a *collective $\boldsymbol{A}$-type* indexed by A is any relation $a : A \to typ(\boldsymbol{A})$. Let $[inst(\boldsymbol{A}), A]$ denote the preorder of A-indexed $\boldsymbol{A}$-instances, and let $[A, typ(\boldsymbol{A})]$ denote the preorder of A-indexed $\boldsymbol{A}$-types. The equivalence $\alpha \leq a \backslash \boldsymbol{A}$ iff $a \circ \alpha \leq \boldsymbol{A}$ iff $a \leq \boldsymbol{A}/\alpha$ states that derivation with respect to $\boldsymbol{A}$ forms an adjoint pair (Galois connection) $(\ )\backslash \boldsymbol{A} \dashv \boldsymbol{A}/(\ ) : [inst(\boldsymbol{A}), A] \rightleftharpoons [A, typ(\boldsymbol{A})]^{op}$. This observation can be extended to the statement that derivation forms a natural transformation $\S_{\boldsymbol{A}} : \mathbf{typ}_{\boldsymbol{A}} \Rightarrow \mathbf{inst}_{\boldsymbol{A}} : \mathbf{Relation}^{op} \to \mathbf{Adjoint}$ between two functors between $\mathbf{Relation}^{op}$ the opposite of the category of sets and binary relations and **Adjoint** the category of preorders and adjoint pairs of monotonic functions. Define the functor $\mathbf{typ}_{\boldsymbol{A}}$ to map sets A to the preorder of collective $\boldsymbol{A}$-types $[A, typ(\boldsymbol{A})]^{\propto}$ and relations $r : A \leftarrow B$ to the adjoint pair $r\backslash(\ ) \dashv r\circ(\ )$, and define the functor $\mathbf{inst}_{\boldsymbol{A}}$ to map sets A to the preorder of collective $\boldsymbol{A}$-instances $[inst(\boldsymbol{A}), A]$ and relations $r : A \leftarrow B$ to the adjoint pair $(\ )\circ r \dashv (\ )/r$. For each set X let $\S_{\boldsymbol{A},X}$ denote the derivation adjoint pair $(\ )\backslash \boldsymbol{A} \dashv \boldsymbol{A}/(\ )$. Then $\S_{\boldsymbol{A}}$ is a natural transformation with X^{th} component $\S_{\boldsymbol{A},X}$.

An *A-indexed collective $\boldsymbol{A}$-concept* is a pair (a, α), where $a : inst(\boldsymbol{A}) \to A$ is a collective $\boldsymbol{A}$-instance, $\alpha : A \to typ(\boldsymbol{A})$ is a collective $\boldsymbol{A}$-type, which satisfy the closure conditions $a = \boldsymbol{A}/\alpha$ and $\alpha = a\backslash\boldsymbol{A}$. Let $\boldsymbol{A} \rhd A$ denote the collection (lattice) of all A-indexed collective $\boldsymbol{A}$-concepts. The basic example of a collective $\boldsymbol{A}$-concept is the pair $(\iota_{\boldsymbol{A}}, \tau_{\boldsymbol{A}})$ consisting of the instance relation $\iota_{\boldsymbol{A}} : inst(\boldsymbol{A}) \to \mathsf{L}(\boldsymbol{A})$ and the type relation $\tau_{\boldsymbol{A}} : \mathsf{L}(\boldsymbol{A}) \to typ(\boldsymbol{A})$, since the closure conditions $\iota_{\boldsymbol{A}}\backslash\boldsymbol{A} = \tau_{\boldsymbol{A}}$ and $\boldsymbol{A}/\tau_{\boldsymbol{A}} = \iota_{\boldsymbol{A}}$ hold. This collective concept is indexed by the concept lattice $\mathsf{L}(\boldsymbol{A})$. Any A-indexed collective $\boldsymbol{A}$-concept (a, α) induces a unique *mediating function* $f : A \to \mathsf{L}(\boldsymbol{A})$ satisfying the constraints $\alpha = f \circ \tau_{\boldsymbol{A}} = f^{\propto}\backslash\tau_{\boldsymbol{A}}$ and $a = \iota_{\boldsymbol{A}} \circ f^{\propto} = \iota_{\boldsymbol{A}}/f$. The definition $f(x) \doteq (ax, x\alpha)$ is well-defined, since the closure conditions are equivalent to the (pointwise) fact that $(ax, x\alpha) \in \mathsf{L}(\boldsymbol{A})$ is a formal concept for each indexing element $x \in A$. Conversely, for any function $f : A \to \mathsf{L}(\boldsymbol{A})$, the pair $(\iota_{\boldsymbol{A}} \circ f^{\propto}, f \circ \tau_{\boldsymbol{A}})$ is an A-indexed collective $\boldsymbol{A}$-concept, since the conditions $\iota_{\boldsymbol{A}} \circ f^{\propto} = \boldsymbol{A}/(f \circ \tau_{\boldsymbol{A}}) = (\boldsymbol{A}/\tau_{\boldsymbol{A}})/f = \iota_{\boldsymbol{A}}/f$ and $f \circ \tau_{\boldsymbol{A}} = (\iota_{\boldsymbol{A}} \circ f^{\propto})\backslash\boldsymbol{A} = f^{\propto}\backslash(\iota_{\boldsymbol{A}}\backslash\boldsymbol{A}) = f^{\propto}\backslash\tau_{\boldsymbol{A}}$ hold. So, we have the isomorphism $\boldsymbol{A} \rhd A \cong \mathsf{L}(\boldsymbol{A})^{A}$, representing the fact that any A-indexed collective $\boldsymbol{A}$-concept can equivalently be define as a function $f : A \to \mathsf{L}(\boldsymbol{A})$.

The left adjoint $r\backslash(\ )$ preserves intents: if (b, β) is a B-indexed collective $\boldsymbol{A}$-concept, then $(\boldsymbol{A}/((b \circ r)\backslash\boldsymbol{A}), r\backslash\beta)$ is an A-indexed collective $\boldsymbol{A}$-concept, since $r\backslash\beta = r\backslash(b\backslash\boldsymbol{A}) = (b \circ r)\backslash\boldsymbol{A}$, an A-indexed $\boldsymbol{A}$-intent; in particular, if β is a B-indexed collective $\boldsymbol{A}$-intent, then $r\backslash\beta$ is an A-indexed collective $\boldsymbol{A}$-intent. The right adjoint $(\ )/r$ preserves extents: if (a, α) is an A-indexed collective $\boldsymbol{A}$-concept, then $(a/r, (\boldsymbol{A}/(r \circ \alpha))\backslash\boldsymbol{A})$ is a B-indexed collective $\boldsymbol{A}$-concept, since $a/r = (\boldsymbol{A}/\alpha)/r = \boldsymbol{A}/(r \circ \alpha)$, a B-indexed $\boldsymbol{A}$-extent; in particular, if a is an A-indexed collective $\boldsymbol{A}$-extent, then a/r is a B-indexed collective $\boldsymbol{A}$-extent. These two observations are concentrated in an adjoint pair of monotonic functions, $\langle \phi_r, \psi_r \rangle : \boldsymbol{A} \triangleright B \rightleftharpoons \boldsymbol{A} \triangleright A$, where the left adjoint $\phi_r : \boldsymbol{A} \triangleright B \rightarrow \boldsymbol{A} \triangleright A$ is defined by $\phi_r((b, \beta)) = (\boldsymbol{A}/((b \circ r)\backslash\boldsymbol{A}), r\backslash\beta)$ and the right adjoint $\psi_r : \boldsymbol{A} \triangleright B \leftarrow \boldsymbol{A} \triangleright A$ is defined by $\psi_r((a, \alpha)) = (a/r, (\boldsymbol{A}/(r \circ \alpha))\backslash\boldsymbol{A})$. Adjointness follows from the expressions $\alpha \leq r\backslash\beta$ iff $r \circ \alpha \leq \beta$ implies $b \leq a/r$ and the expressions $b \leq a/r$ iff $b \circ r \leq a$ implies $\alpha \leq r\backslash\beta$.

2.2 Morphisms

Functional Infomorphisms. Classifications are connected through infomorphisms. Infomorphisms connect classifications and provide a way to move information back and forth between classifications. There exist two generality levels of infomorphism: functional infomorphisms and relational infomorphisms. Functional infomorphisms are the infomorphisms that are used in the literature on Information Flow [2] and *-autonomous categories and the Chu construction [1]. Relational infomorphisms are newly defined in this paper. This subsection gives the definition of functional infomorphism. The next subsection will define the relational version.

A *(functional) infomorphism* $\boldsymbol{f} = \langle inst(\boldsymbol{f}), typ(\boldsymbol{f}) \rangle : \boldsymbol{A} \rightleftharpoons \boldsymbol{B}$ from classification $\boldsymbol{A}$ to classification $\boldsymbol{B}$ is a contravariant pair of functions, a function $\hat{f} = typ(\boldsymbol{f}) : typ(\boldsymbol{A}) \rightarrow typ(\boldsymbol{B})$ in the forward direction between types and a function $\check{f} = inst(\boldsymbol{f}) : inst(\boldsymbol{A}) \leftarrow inst(\boldsymbol{B})$ in the reverse direction between instances, satisfying the following fundamental property $\check{f}(b) \models_{\boldsymbol{A}} \alpha$ iff $b \models_{\boldsymbol{B}} \hat{f}(\alpha)$ for each instance $b \in inst(\boldsymbol{B})$ and each type $\alpha \in typ(\boldsymbol{A})$. In the theory of the Chu construction [1] an infomorphism (a morphism in the category $Chu(\mathbf{Set}, 2)$) is known as a Chu transformation.

Systemic examples of functional infomorphisms abound. For any two sets (of instances) A and B, any function $f : A \leftarrow B$ and its inverse image function $f^{-1} : \wp A \rightarrow \wp B$ form a *instance powerset infomorphism* $\wp f = \langle f, f^{-1} \rangle : \wp A \rightleftharpoons \wp B$ between the instance powerset classifications. Given any classification $\boldsymbol{A} = \langle inst(\boldsymbol{A}), typ(\boldsymbol{A}), \models_{\boldsymbol{A}} \rangle$, the *intent infomorphism* $\eta_{\boldsymbol{A}} : \boldsymbol{A} \rightleftharpoons \wp(inst(\boldsymbol{A}))$ from $\boldsymbol{A}$ to the powerset classification of the instance set of $\boldsymbol{A}$, is the identity function on instances and the intent function $\eta_{\boldsymbol{A}}(\alpha) = int_{\boldsymbol{A}}(\alpha)$ on types.

Given any two infomorphisms $\boldsymbol{f} = \langle inst(\boldsymbol{f}), typ(\boldsymbol{f}) \rangle : \boldsymbol{A} \rightleftharpoons \boldsymbol{B}$ and $\boldsymbol{g} = \langle inst(\boldsymbol{g}), typ(\boldsymbol{g}) \rangle : \boldsymbol{B} \rightleftharpoons \boldsymbol{C}$, there is a *composite* infomorphism $\boldsymbol{f} \circ \boldsymbol{g} = \langle inst(\boldsymbol{g}) \cdot inst(\boldsymbol{f}), typ(\boldsymbol{f}) \cdot typ(\boldsymbol{g}) \rangle : \boldsymbol{A} \rightleftharpoons \boldsymbol{C}$ defined by composing the type and instance functions. Given any classification $\boldsymbol{A} = \langle inst(\boldsymbol{A}), typ(\boldsymbol{A}), \models_{\boldsymbol{A}} \rangle$, the pair of identity functions on types and instances forms an *identity* infomorphism

$id_A : \langle id_{inst(A)}, id_{typ(A)} \rangle : A \rightleftharpoons A$ (with respect to composition). Given any infomorphism $f = \langle inst(f), typ(f) \rangle : A \rightleftharpoons B$, the *dual* infomorphism is the infomorphism $f^{\propto} : \langle typ(f), inst(f) \rangle : B^{\propto} \rightleftharpoons A^{\propto}$, whose source classification is the dual of the target of f, whose target classification is the dual of the source of f, whose instance function is the type function of f, and whose type function is the instance function of f. Classifications and functional infomorphisms form a category **Classification** with involution $(\ )^{\propto}$.

For any infomorphism $f = \langle inst(f), typ(f) \rangle : A \rightleftharpoons B$, the instance function is a monotonic function between instance preorders $inst(f) : inst(A) \leftarrow inst(B)$, and the type function is a monotonic function between type preorders $typ(f) : typ(A) \rightarrow typ(B)$. These facts are represented as two (projection) functors: the functor **inst** (contravariant) from **Classification** to $\mathbf{Preorder}^{op}$, the opposite of the category of preorders and monontonic functions, maps each classification to its instance preorder and each infomorphism to its instance monotonic function; and the functor **typ** from **Classification** to **Preorder** maps each classification to its type preorder and each infomorphism to its type monotonic function.

An *adjoint pair* of monotonic functions $f = \langle left(f), right(f) \rangle : P \rightleftharpoons Q$ from preorder $P = \langle P, \leq_P \rangle$ to preorder $Q = \langle Q, \leq_Q \rangle$ is a contravariant pair of functions, a monotonic function $left(f) : P \rightarrow Q$ in the forward direction and a monotonic function $right(f) : P \leftarrow Q$ in the reverse direction, satisfying the following fundamental adjointness property $left(f)(p) \leq_Q q$ iff $p \leq_P right(f)(q)$ for each element $p \in P$ and each element $q \in Q$. The fundamental property of functional infomorphisms is clearly related to the notion of adjointness. Preorders and adjoint pairs of monotonic functions form the category **Adjoint**. Projecting out to the left and right adjoint monotonic functions give rise to two (projection) functors: the functor **left** from **Adjoint** to **Preorder**, and the functor **right** from **Adjoint** to $\mathbf{Preorder}^{op}$. Any adjoint pair of monotonic functions is a functional infomorphism between the associated classifications. This fact is expressed as the functor **incl** from **Adjoint** to $\mathbf{Classification}^{op}$ that commutes with projection functors: $\mathbf{incl} \circ \mathbf{inst}^{op} = \mathbf{left}$ and $\mathbf{incl} \circ \mathbf{typ}^{op} = \mathbf{right}$.

Relational Infomorphisms. A (*relational*) *infomorphism* $r = \langle inst(r), typ(r) \rangle : A \rightleftharpoons B$ from classification A to classification B is a pair of binary relations, a *type* relation $\hat{r} = typ(r) : typ(A) \rightarrow typ(B)$ between types and an *instance* relation $\check{r} = inst(r) : inst(A) \rightarrow inst(B)$ between instances, satisfying the following fundamental property: $\check{r} \backslash A = B / \hat{r}$. This property is equivalent to either of the properties in Table 1. The common relation $\check{r} \backslash A = B / \hat{r} : inst(B) \rightarrow typ(A)$ in the fundamental property is called the *bond* of r.

Given any two relational infomorphisms $r = \langle inst(r), typ(r) \rangle : A \rightleftharpoons B$ and $s = \langle inst(s), typ(s) \rangle : B \rightleftharpoons C$, there is a *composite* infomorphism $r \circ s = \langle inst(r) \circ inst(s), typ(r) \circ typ(s) \rangle : A \rightleftharpoons C$ defined by composing the instance and type relations of the component infomorphisms, and whose fundamental property follows from the preceding composition and associative laws. Given any classification $A = \langle inst(A), typ(A), \models_A \rangle$, the pair of identity rela-

Table 1. Relational Infomorphism

$$\alpha \in (\check{r}b)^{A} \text{ iff } b \in (\alpha\hat{r})^{B} \qquad \forall_{b \in inst(B)} \text{ and } \forall_{a \in inst(A)}$$

$$\check{r}b \subseteq \alpha^{A} \text{ iff } \alpha\hat{r} \subseteq b^{B}$$

$$\check{r}B \subseteq \Gamma^{A} \text{ iff } \forall_{b \in B, \alpha \in \Gamma}\, \check{r}b \subseteq \alpha^{A} \text{ iff } \forall_{b \in B, \alpha \in \Gamma}\, \alpha\hat{r} \subseteq b^{B} \text{ iff } \Gamma\hat{r} \subseteq B^{B} \qquad \forall_{B \subseteq inst(B)} \text{ and } \forall_{\Gamma \subseteq typ(A)}$$

$$\check{r} \circ B \subseteq A/\Gamma \text{ iff } B \circ \Gamma \subseteq \check{r}\backslash A \text{ iff } B \circ \Gamma \subseteq B/\hat{r} \text{ iff } \Gamma \circ \hat{r} \subseteq B\backslash B \qquad \forall_{B : inst(B) \to 1} \text{ and } \forall_{\Gamma : 1 \to typ(A)}$$

	$typ(A)$	$typ(B)$
$inst(A)$	$\models_A$	
$inst(B)$	$\check{r}\backslash A = B/\hat{r}$	$\models_B$

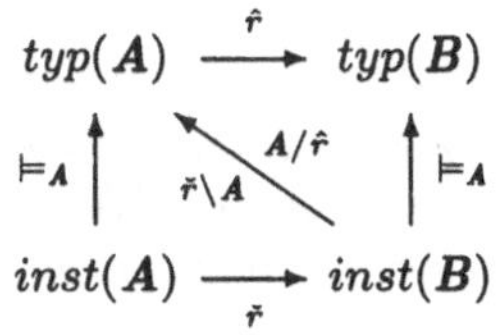

tions on instances and types, with the bond being A, forms an *identity* infomorphism $id_A = \langle id_{inst(A)}, id_{typ(A)} \rangle : A \rightleftharpoons A$ (with respect to composition). For any given infomorphism $r = \langle inst(r), typ(r) \rangle : A \rightleftharpoons B$ the *dual* infomorphism $r^{\propto} : \langle typ(r), inst(r) \rangle : B^{\propto} \rightleftharpoons A^{\propto}$ is the infomorphism, whose source classification is the dual of the target of r, whose target classification is the dual of the source of r, whose instance relation is the transpose of the type relation of r, and whose type relation is the transpose of the instance relation of r. The fundamental property of relational infomorphisms for composition, identity and involution follow from basic properties of residuation. Classifications and relational infomorphisms form a category **Classification$_{\mathsf{rel}}$** with involution $(\ )^{\propto}$.

Any functional infomorphism $f = \langle \check{f}, \hat{f} \rangle = \langle inst(f), typ(f) \rangle : A \rightleftharpoons B$ has an associated relational infomorphism **fn2rel**$(f) = \langle \check{f}_{\bullet}, \hat{f}^{\bullet} \rangle : A \rightleftharpoons B$, whose bond is the relation in the fundamental property $[\check{f}(b) \models_A \alpha \text{ iff } b \models_B \hat{f}(\alpha)] : inst(B) \to typ(A)$. The definition of the instance and type relations uses the induced orders on instances and types: the type relation $\hat{f}^{\bullet} : typ(A) \to typ(B)$ is defined by $\hat{f}^{\bullet}(\alpha, \beta) = \hat{f}(\alpha) \leq_B \beta$, and the instance relation $\check{f}_{\bullet} : inst(A) \to inst(B)$ is defined by $\check{f}_{\bullet}(a, b) = a \leq_A \check{f}(b)$. The operator **fn2rel**, which maps classifications to themselves, is a functor from **Classification** to **Classification$_{\mathsf{rel}}$**.

The theory of classifications with relational infomorphisms can profitably be regarded as a theory of Boolean matrices, with classifications being matrices, infomorphisms being matrix pairs, composition involving matrix multiplication in the two dual senses of relational composition and residuation, and involution using matrix transpose.

Bonds. A *bond* [3] $F : A \to B$ between two classifications A and B is a classification $F = \langle inst(B), typ(A), \models_F \rangle$, sharing types with A and instances with B, that is compatible with A and B in the sense of closure: type sets $\{bF \mid b \in inst(B)\}$ are intents of A and instance sets $\{F\alpha \mid \alpha \in typ(A)\}$ are extents of B. Closure can be expressed categorically as $(A/F)\backslash A = F$ and $B/(F\backslash B) = F$. The first expression says that $(A/F, F)$ is an $inst(B)$-indexed collective A-concept, whereas the second says that $(F, F\backslash B)$ is a $typ(A)$-indexed collective $\boldsymbol{B}$-concept. A bond is order-closed on left and right: $b' \leq_{\boldsymbol{B}} b, b\boldsymbol{F}\alpha$ imply $b'\boldsymbol{F}\alpha$, and $b\boldsymbol{F}\alpha, \alpha \leq_{\boldsymbol{A}} \alpha'$ imply $b\boldsymbol{F}\alpha'$. Or, $b' \leq_{\boldsymbol{B}} b$ implies $b'\boldsymbol{F} \supseteq b\boldsymbol{F}$, and $\alpha \leq_{\boldsymbol{A}} \alpha'$ implies $\boldsymbol{F}\alpha \subseteq \boldsymbol{F}\alpha'$.

For any two bonds $\boldsymbol{F} : \boldsymbol{A} \rightarrow \boldsymbol{B}$ and $\boldsymbol{G} : \boldsymbol{B} \rightarrow \boldsymbol{C}$, the *composition* is the bond $\boldsymbol{F} \diamond \boldsymbol{G} \doteq (\boldsymbol{B}/\boldsymbol{G})\backslash\boldsymbol{F} : \boldsymbol{A} \rightarrow \boldsymbol{C}$ defined using left and right residuation. Since both $\boldsymbol{F}$ and $\boldsymbol{G}$ being bonds are closed with respect to $\boldsymbol{B}$, an equivalent expression for the composition is $\boldsymbol{F} \diamond \boldsymbol{G} \doteq \boldsymbol{G}/(\boldsymbol{F}\backslash\boldsymbol{B}) : \boldsymbol{A} \rightarrow \boldsymbol{C}$. Pointwise, the composition is $\boldsymbol{F} \diamond \boldsymbol{G} = \{(c, \alpha) \mid \boldsymbol{F}\alpha \supseteq (c\boldsymbol{G})^{\boldsymbol{B}}\}$. To check closure, $(\boldsymbol{A}/(\boldsymbol{F} \diamond \boldsymbol{G}))\backslash\boldsymbol{A} = (\boldsymbol{A}/((\boldsymbol{B}/\boldsymbol{G})\backslash\boldsymbol{F}))\backslash\boldsymbol{A} = (\boldsymbol{B}/\boldsymbol{G})\backslash\boldsymbol{F}$, since $\boldsymbol{F}$ being a collective $\boldsymbol{A}$-intent means that $(\boldsymbol{B}/\boldsymbol{G})\backslash\boldsymbol{F}$ is also a collective $\boldsymbol{A}$-intent. With respect to bond composition, the *identity* bond at any classification $\boldsymbol{A}$ is the classification relation $\models_{\boldsymbol{A}}: \boldsymbol{A} \rightarrow \boldsymbol{A}$; its closed relational infomorphism is the instance-type order pair $\langle \leq_{\boldsymbol{A}}, \leq_{\boldsymbol{A}} \rangle : \boldsymbol{A} \rightarrow \boldsymbol{A}$. With bond composition and bond identities, classifications and bonds form the category **Bond**.

For any relational infomorphism $\boldsymbol{r} = \langle \check{r}, \hat{r} \rangle = \langle inst(\boldsymbol{r}), typ(\boldsymbol{r}) \rangle : \boldsymbol{A} \rightleftharpoons \boldsymbol{B}$, the common relation bond $\mathbf{bond}(\boldsymbol{r}) = \check{r}\backslash\boldsymbol{A} = \boldsymbol{B}/\hat{r}$ in the fundamental property is a bond: for any instance $b \in inst(\boldsymbol{B})$ the bth row of the bond $\check{r}\backslash\boldsymbol{A}$ is an intent of $\boldsymbol{A}$, since $b(\check{r}\backslash\boldsymbol{A}) = \{\alpha \in typ(\boldsymbol{A}) \mid \check{r}b \in \alpha^{\boldsymbol{A}}\} = (\check{r}b)^{\boldsymbol{A}}$, and for any type $\alpha \in typ(\boldsymbol{A})$, the αth column of the bond $\boldsymbol{B}/\hat{r}$ is an extent of $\boldsymbol{B}$, since $(\boldsymbol{B}/\hat{r})\alpha = \{b \in inst(\boldsymbol{B}) \mid \alpha\hat{r} \in b^{\boldsymbol{B}}\} = (\alpha\hat{r})^{\boldsymbol{B}}$. Any bond $\boldsymbol{F} : \boldsymbol{A} \rightarrow \boldsymbol{B}$ is the bond of some (closed) relational infomorphism $\boldsymbol{r} : \boldsymbol{A} \rightleftharpoons \boldsymbol{B}$ – just make the definitions: $\check{r} \doteq \boldsymbol{A}/\boldsymbol{F}$ and $\hat{r} \doteq \boldsymbol{F}\backslash\boldsymbol{B}$, or pointwise $\check{r}b \doteq (b\boldsymbol{F})^{\boldsymbol{A}}$ and $\alpha\hat{r} \doteq (\boldsymbol{F}\alpha)^{\boldsymbol{B}}$. There is a naturally defined equivalence relation on the collection of relational infomorphisms between any two classifications $\boldsymbol{A}$ and $\boldsymbol{B}$: two infomorphisms are equivalent when they have the same bond. Since the bond of a composition is the composition of the bonds and the bond of the identity is the identity bond, this defines a bond quotient functor from relational infomorphisms to bonds, that makes the category **Bond** a quotient category of $\mathbf{Classification}_{\mathbf{rel}}$.

The notion of a bond of a relational infomorphism is (ignoring orientation) the same notion of a bond as defined in [3]. The notion of a relational infomorphism, defined for the first time in this paper, supports a categorical rendering of the notion of bond as defined in [3].

Bonding Pairs. A *complete* (*lattice*) *homomorphism* $\psi : \boldsymbol{L} \rightarrow \boldsymbol{K}$ between complete lattices $\boldsymbol{L}$ and $\boldsymbol{K}$ is a (monotonic) function that preserves both joins and meets. Being meet-preserving, ψ has a left adjoint $\phi : \boldsymbol{K} \rightarrow \boldsymbol{L}$, and being join-preserving ψ has a right adjoint $\theta : \boldsymbol{K} \rightarrow \boldsymbol{L}$. So, a complete homomorphism is the middle monotonic function in two adjunctions $\phi \dashv \psi \dashv \theta$. Let **Complete Lattice** denote the category of complete lattices and complete homomorphisms.

The bond equivalent to a complete homomorphism would seem to be given by two bonds $\boldsymbol{F} : \boldsymbol{A} \rightarrow \boldsymbol{B}$ and $\boldsymbol{G} : \boldsymbol{B} \rightarrow \boldsymbol{A}$, where the right adjoint $\psi_{\boldsymbol{F}} : \mathsf{L}(\boldsymbol{A}) \rightarrow \mathsf{L}(\boldsymbol{B})$ of the complete adjoint $\mathsf{A}(\boldsymbol{F}) = \langle \phi_{\boldsymbol{F}}, \psi_{\boldsymbol{F}} \rangle : \mathsf{L}(\boldsymbol{B}) \rightleftharpoons \mathsf{L}(\boldsymbol{A})$ of one bond (say $\boldsymbol{F}$, without loss of generality) is equal to the left adjoint $\phi_{\boldsymbol{G}} : \mathsf{L}(\boldsymbol{A}) \rightarrow \mathsf{L}(\boldsymbol{B})$ of the complete adjoint $\mathsf{A}(\boldsymbol{G}) = \langle \phi_{\boldsymbol{G}}, \psi_{\boldsymbol{G}} \rangle : \mathsf{L}(\boldsymbol{A}) \rightleftharpoons \mathsf{L}(\boldsymbol{B})$ of the other bond $\boldsymbol{G}$ with the resultant adjunctions, $\phi_{\boldsymbol{F}} \dashv \psi_{\boldsymbol{F}} = \phi_{\boldsymbol{G}} \dashv \psi_{\boldsymbol{G}}$, where the middle adjoint is the complete homomorphism. This is indeed the case, but the question is what constraint to place on $\boldsymbol{F}$ and $\boldsymbol{G}$ in order for this to hold. The simple answer is to identify the actions of the two monotonic functions $\psi_{\boldsymbol{G}}$ and $\phi_{\boldsymbol{F}}$, and this is exactly the solution given in [3]. Let $(A, \Gamma) \in \mathbf{L}(\boldsymbol{A})$ be any formal concept in $\mathbf{L}(\boldsymbol{A})$. The action of the left adjoint $\phi_{\boldsymbol{G}}$ on this concept is $(A, \Gamma) \mapsto (A^{GB}, A^{G})$, whereas the action of the right adjoint $\psi_{\boldsymbol{F}}$ on this concept is $(A, \Gamma) \mapsto (\Gamma^{F}, \Gamma^{FB})$. So the appropriate pointwise constraints are: $A^{GB} = \Gamma^{F}$ and $\Gamma^{FB} = A^{G}$, for every concept $(A, \Gamma) \in \mathbf{L}(\boldsymbol{A})$. We now give these pointwise constraints a categorical rendition.

A *bonding pair* $\langle \boldsymbol{F}, \boldsymbol{G} \rangle : \boldsymbol{A} \rightleftharpoons \boldsymbol{B}$ between two classifications $\boldsymbol{A}$ and $\boldsymbol{B}$ is a contravariant pair of bonds, a bond $\boldsymbol{F} : \boldsymbol{A} \rightarrow \boldsymbol{B}$ in the forward direction and a bond $\boldsymbol{G} : \boldsymbol{B} \rightarrow \boldsymbol{A}$ in the reverse direction, satisfying the following *pairing constraints*: (1) $\boldsymbol{F}/\tau_{\boldsymbol{A}} = \boldsymbol{B}/(\iota_{\boldsymbol{A}} \backslash \boldsymbol{G})$ and (2) $\iota_{\boldsymbol{A}} \backslash \boldsymbol{G} = (\boldsymbol{F}/\tau_{\boldsymbol{A}}) \backslash \boldsymbol{B}$, which state that $(\boldsymbol{F}/\tau_{\boldsymbol{A}}, \iota_{\boldsymbol{A}} \backslash \boldsymbol{G})$ is an $\mathbf{L}(\boldsymbol{A})$-indexed collective $\boldsymbol{B}$-concept. The definitions of the relations $\boldsymbol{F}/\tau_{\boldsymbol{A}}$ and $\iota_{\boldsymbol{A}} \backslash \boldsymbol{G}$ are given as follows: $\boldsymbol{F}/\tau_{\boldsymbol{A}} = \{(b, \boldsymbol{a}) \mid int(\boldsymbol{a}) \subseteq b\boldsymbol{F}\} = \{(b, \boldsymbol{a}) \mid ((b\boldsymbol{F})^{\boldsymbol{A}}, b\boldsymbol{F}) \leq_{\boldsymbol{B}} \boldsymbol{a}\}$ and $\iota_{\boldsymbol{A}} \backslash \boldsymbol{G} = \{(\boldsymbol{a}, \beta) \mid ext(\boldsymbol{a}) \subseteq \boldsymbol{G}\beta\} = \{(\boldsymbol{a}, \beta) \mid \boldsymbol{a} \leq_{\boldsymbol{B}} (\boldsymbol{G}\beta, (\boldsymbol{G}\beta)^{\boldsymbol{A}})\}$. Any concept $\boldsymbol{a} = (A, \Gamma) \in \mathbf{L}(\boldsymbol{A})$ is mapped by the relations as: $(\boldsymbol{F}/\tau_{\boldsymbol{A}})((A, \Gamma)) = \{b \mid \Gamma \subseteq b\boldsymbol{F}\} = \Gamma^{\boldsymbol{F}}$ and $(\iota_{\boldsymbol{A}} \backslash \boldsymbol{G})((A, \Gamma)) = \{\beta \mid A \subseteq \boldsymbol{G}\beta\} = A^{\boldsymbol{G}}$. Hence, pointwise the constraints are $\Gamma^{\boldsymbol{F}} = B^{\boldsymbol{GB}}$ and $A^{\boldsymbol{G}} = \Gamma^{\boldsymbol{FB}}$. These are the original pointwise constraints of [3] discussed above.

The pointwise constraints can be lifted to a collective setting – any bonding pair $\langle \boldsymbol{F}, \boldsymbol{G} \rangle : \boldsymbol{A} \rightleftharpoons \boldsymbol{B}$ preserves collective concepts: for any A-indexed collective $\boldsymbol{A}$-concept $(a, \alpha), a \backslash \boldsymbol{A} = \alpha$ and $\boldsymbol{A}/\alpha = a$, the conceptual image $(\boldsymbol{F}/\alpha, a \backslash \boldsymbol{G})$ is an A-indexed collective $\boldsymbol{B}$-concept, $\boldsymbol{B}/(a \backslash \boldsymbol{G}) = \boldsymbol{F}/\alpha$ and $(\boldsymbol{F}/\alpha) \backslash \boldsymbol{B} = a \backslash \boldsymbol{G}$. An important special case is the $L(\boldsymbol{A})$-indexed collective $\boldsymbol{A}$-concept $(\iota_{\boldsymbol{A}}, \tau_{\boldsymbol{A}})$. To state that the $\langle \boldsymbol{F}, \boldsymbol{G} \rangle$-image $(\boldsymbol{F}/\tau_{\boldsymbol{A}}, \iota_{\boldsymbol{A}} \backslash \boldsymbol{F})$ is an $L(\boldsymbol{A})$-indexed collective $\boldsymbol{B}$-concept, is to assert the pairing constraints $\boldsymbol{F}/\tau_{\boldsymbol{A}} = \boldsymbol{B}/(\iota_{\boldsymbol{A}} \backslash \boldsymbol{G})$ and $\iota_{\boldsymbol{A}} \backslash \boldsymbol{G} = (\boldsymbol{F}/\tau_{\boldsymbol{B}}) \backslash \boldsymbol{B}$. So, the concise definition in terms of pairing constraints, the original pointwise definition of [3], and the assertion that $\langle \boldsymbol{F}, \boldsymbol{G} \rangle$ preserves all collective concepts, are equivalent versions of the notion of a bonding pair.

Two other special cases are the collective $\boldsymbol{B}$-concepts $(\boldsymbol{F}, \boldsymbol{A} \backslash \boldsymbol{G})$ and $(\boldsymbol{F}/\boldsymbol{A}, \boldsymbol{G})$ with pairing constraints $\boldsymbol{B}/(\boldsymbol{A} \backslash \boldsymbol{G}) = \boldsymbol{F}$ and $(\boldsymbol{F}/\boldsymbol{A}) \backslash \boldsymbol{B} = \boldsymbol{G}$, which are $\langle \boldsymbol{F}, \boldsymbol{G} \rangle$-images of the collective $\boldsymbol{A}$-concepts $(\models_{\boldsymbol{A}}, \leq_{\boldsymbol{A}})$ and $(\leq_{\boldsymbol{A}}, \models_{\boldsymbol{A}})$. Since any collective concept uniquely factors in terms of its mediating function, the image of any collective concept can be computed in two steps: (1) Factor an A-indexed collective $\boldsymbol{A}$-concept (a, α) in terms of the mediating function $f : A \rightarrow L(\boldsymbol{A})$ and the basic collective $\boldsymbol{A}$-concept $(\iota_{\boldsymbol{A}}, \tau_{\boldsymbol{A}})$; and (2) Compose the $\langle \boldsymbol{F}, \boldsymbol{G} \rangle$-image $(\boldsymbol{F}/\tau_{\boldsymbol{A}}, \iota_{\boldsymbol{B}} \backslash \boldsymbol{G})$ with the mediating function f, resulting in the $\langle \boldsymbol{F}, \boldsymbol{G} \rangle$-image A-indexed collective $\boldsymbol{A}$-concept $(\boldsymbol{F}/\alpha, a \backslash \boldsymbol{G})$.

Let $\langle \boldsymbol{F}, \boldsymbol{G} \rangle : \boldsymbol{A} \rightleftharpoons \boldsymbol{B}$ and $\langle \boldsymbol{M}, \boldsymbol{N} \rangle : \boldsymbol{B} \rightleftharpoons \boldsymbol{C}$ be two bonding pairs. Define the bonding pair *composition* $\langle \boldsymbol{F}, \boldsymbol{G} \rangle \diamond \langle \boldsymbol{M}, \boldsymbol{N} \rangle \doteq \langle \boldsymbol{F} \diamond \boldsymbol{M}, \boldsymbol{N} \diamond \boldsymbol{G} \rangle : \boldsymbol{A} \rightleftharpoons \boldsymbol{C}$ in terms of bond composition. We can check, either categorically or pointwise, that bonding pair composition is well defined. Let **Bonding Pair** denote the category, whose objects are classifications and whose morphisms are bonding pairs.

3 Architecture

The architecture of the distribution/conception distinction is a categorical equivalence at both the functional and the relational poles of the other (scope) distinction. This architecture is in one sense a categorical expression of the basic theorem of Formal Concept Analysis. The central architecture is revealed at the functional level to be the equivalence between **Classification** and **Concept Lattice**, at the relational level to be the equivalence between **Bond** and **Complete Adjoint**^{op},and at the complete relational level to be the equivalence between **Bonding Pair** and **Complete Lattice**. Not to be forgotten is the fact from Information Flow, that **Classification** is also equivalent to **Regular Theory**, the category of regular theories and theory morphisms.

3.1 Functional Equivalence

Information Flow (IF) and Formal Concept Analysis (FCA) are intimately connected. Every classification supports, and is equivalent to, an associated complete lattice called its concept lattice. Every infomorphism defines an adjoint pair of monotonic functions between the concept lattices of its source and target classifications. This section formalizes these observations in a theorem on categorical equivalence.

The Concept Lattice Functor. Let $\boldsymbol{f} = \langle inst(\boldsymbol{f}), typ(\boldsymbol{f})\rangle : \boldsymbol{A} \rightleftharpoons \boldsymbol{B}$ be any functional infomorphism between two classifications with instance function $\check{f} = inst(\boldsymbol{f}) : inst(\boldsymbol{B}) \rightarrow inst(\boldsymbol{A})$ and type function $\hat{f} = typ(\boldsymbol{f}) : typ(\boldsymbol{A}) \rightarrow typ(\boldsymbol{B})$. How are the two concept lattices $\mathbf{L}(\boldsymbol{A})$ and $\mathbf{L}(\boldsymbol{B})$ related?

Since for any concept $(A, \Gamma) \in L(\boldsymbol{A})$ the equality $\check{f}^{-1}[A] = \hat{f}[\Gamma]^{\boldsymbol{B}}$ holds between direct and inverse images, the mapping $(A, \Gamma) \mapsto (\check{f}^{-1}[A], (\check{f}^{-1}[A])^{\boldsymbol{B}})$ is a well-defined monotonic function $\mathbf{L}(\check{f}) : \mathbf{L}(\boldsymbol{B}) \leftarrow \mathbf{L}(\boldsymbol{A})$ that preserves type concepts in the sense that: $\tau_{\boldsymbol{A}} \cdot \mathbf{L}(\check{f}) = \hat{f} \cdot \tau_{\boldsymbol{B}}$. Since it is always true that meet-irreducible concepts are type concepts, if $\mathbf{L}(\boldsymbol{B})$ is type reduced, then $\mathbf{L}(\check{f})$ preserves meet-irreducibility – it maps meet-irreducible concepts to meet-irreducible concepts. Dually, since for any concept $(B, \Delta) \in L(\boldsymbol{B})$ the equality $\check{f}[B]^{\boldsymbol{A}} = \hat{f}^{-1}[\Delta]$ holds, the mapping $(B, \Delta) \mapsto (\hat{f}^{-1}[\Delta]^{\boldsymbol{A}}, \hat{f}^{-1}[\Delta])$ is a well-defined monotonic function $\mathbf{L}(\hat{f}) : \mathbf{L}(\boldsymbol{B}) \rightarrow \mathbf{L}(\boldsymbol{A})$ that preserves instance concepts in the sense that: $\iota_{\boldsymbol{B}} \cdot L(\hat{f}) = \check{f} \cdot \iota_{\boldsymbol{A}}$. Since it is always true that join-irreducible concepts are instance concepts, if $\mathbf{L}(\boldsymbol{A})$ is instance reduced, then $\mathbf{L}(\hat{f})$ preserves join-irreducibility – it maps join-irreducible concepts to join-irreducible concepts. Moreover, $\langle \mathbf{L}(\hat{f}), \mathbf{L}(\check{f})\rangle : \mathbf{L}(\boldsymbol{B}) \rightleftharpoons \mathbf{L}(\boldsymbol{A})$ is a pair of adjoint monotonic functions between concept lattices, with $\mathbf{L}(\hat{f})$ left adjoint and $\mathbf{L}(\check{f})$ right adjoint.

The quadruple $\mathbf{L}(\boldsymbol{f}) = \langle \mathbf{L}(\hat{f}), \mathbf{L}(\check{f}), \check{f}, \hat{f}\rangle$ is called the *concept lattice morphism* of the infomorphism $\boldsymbol{f} = \langle \check{f}, \hat{f}\rangle = \langle inst(\boldsymbol{f}), typ(\boldsymbol{f})\rangle : \boldsymbol{A} \rightleftharpoons \boldsymbol{B}$. More abstractly, a *concept lattice morphism* $\boldsymbol{h} = \langle \acute{h}, \grave{h}, \check{h}, \hat{h}\rangle = \langle left(\boldsymbol{h}), right(\boldsymbol{h}), inst(\boldsymbol{h}), typ(\boldsymbol{h})\rangle : \boldsymbol{L} \rightleftharpoons \boldsymbol{K}$ between two concept lattices $\boldsymbol{L}$ and $\boldsymbol{K}$ consists of a pair of ordinary functions $\check{h} = inst(\boldsymbol{h}) : inst(\boldsymbol{L}) \leftarrow inst(\boldsymbol{K})$ and $\hat{h} = typ(\boldsymbol{h}) : typ(\boldsymbol{L}) \rightarrow typ(\boldsymbol{K})$

between instance sets and type sets, respectively; and an adjoint pair $adj(\boldsymbol{h}) = \langle \acute{h}, \grave{h} \rangle = \langle left(\boldsymbol{h}), right(\boldsymbol{h}) \rangle : \boldsymbol{K} \rightleftharpoons \boldsymbol{L}$ of monotonic functions, where the right adjoint $\grave{h} = right(\boldsymbol{h}) : \boldsymbol{L} \to \boldsymbol{K}$ is a monotonic function in the forward direction that preserves types $\tau_{\boldsymbol{L}} \cdot \grave{h} = \hat{h} \cdot \tau_{\boldsymbol{K}}$, and the left adjoint $\acute{h} = left(\boldsymbol{h}) : \boldsymbol{L} \leftarrow \boldsymbol{K}$ is a monotonic function in the reverse direction that preserves instances $\iota_{\boldsymbol{K}} \cdot \acute{h} = \check{h} \cdot \iota_{\boldsymbol{L}}$.

Let **Concept Lattice** denote the category of concept lattices and concept lattice morphisms. For each infomorphism $\boldsymbol{f} = \boldsymbol{A} \rightleftharpoons \boldsymbol{B}$ the quadruple $\mathsf{L}(\boldsymbol{f}) : \mathsf{L}(\boldsymbol{A}) \rightleftharpoons \mathsf{L}(\boldsymbol{B})$ is a concept lattice morphism from $\mathsf{L}(\boldsymbol{A})$, the concept lattice of classification $\boldsymbol{A}$, to $\mathsf{L}(\boldsymbol{B})$, the concept lattice of classification $\boldsymbol{B}$. The operator L is a functor called *concept lattice* from **Classification**, the category of classifications and functional infomorphisms, to **Concept Lattice**, the category of concept lattices and concept lattice morphisms.

Classification Functor. Any concept lattice $\boldsymbol{L} = \langle L, inst(\boldsymbol{L}), typ(\boldsymbol{L}), \iota_{\boldsymbol{L}}, \tau_{\boldsymbol{L}} \rangle$ has an associated classification $\mathsf{C}(\boldsymbol{L}) = \langle inst(\boldsymbol{L}), typ(\boldsymbol{L}), \models_{\boldsymbol{L}} \rangle$, which has $\boldsymbol{L}$-instances as its instance set, $\boldsymbol{L}$-types as its type set, and the relational composition $\models_{\boldsymbol{L}} = \iota_{\boldsymbol{L}} \circ \leq_{\boldsymbol{L}} \circ \tau_{\boldsymbol{L}}{}^{op}$ as its classification relation. Associated with any concept lattice morphism $\boldsymbol{h} = \langle left(\boldsymbol{h}), right(\boldsymbol{h}), inst(\boldsymbol{h}), typ(\boldsymbol{h}) \rangle : \boldsymbol{L} \rightleftharpoons \boldsymbol{K}$, from concept lattice $\boldsymbol{L}$ to concept lattice $\boldsymbol{K}$, is the infomorphism $\mathsf{C}(\boldsymbol{h}) = \langle inst(\boldsymbol{h}), typ(\boldsymbol{h}) \rangle : \mathsf{C}(\boldsymbol{L}) \rightleftharpoons \mathsf{C}(\boldsymbol{K})$. The fundamental property of infomorphisms is an easy translation of the adjointness condition for $adj(\boldsymbol{h}) : \boldsymbol{K} \rightleftharpoons \boldsymbol{L}$ and the commutativity of the instance/type functions with the adjoint pair of monotonic functions (left/right functions). The operator C is a functor called *classification* from **Concept Lattice**, the category of concept lattices and their morphisms, to **Classification**, the category of classifications and infomorphisms.

Equivalence. The functor composition $\mathsf{C} \circ \mathsf{L}$ is naturally isomorphic to the identity functor $Id_{\textsf{Concept Lattice}}$. To see this, let $\boldsymbol{L} = \langle L, inst(\boldsymbol{L}), typ(\boldsymbol{L}), \iota_{\boldsymbol{L}}, \tau_{\boldsymbol{L}} \rangle$ be a concept lattice with associated classification $\mathsf{C}(\boldsymbol{L}) = \langle inst(\boldsymbol{L}), typ(\boldsymbol{L}), \models_{\boldsymbol{L}} \rangle$. Part of the fundamental theorem of concept lattices [3] asserts the isomorphism $\boldsymbol{L} \cong \mathsf{L}(\mathsf{C}(\boldsymbol{L}))$. In particular, define the map $\mathsf{L}(\mathsf{C}(\boldsymbol{L})) \to \boldsymbol{L}$ by $(A, \Gamma) \mapsto \bigvee_{\boldsymbol{L}} \iota_{\boldsymbol{L}}(A) = \bigwedge_{\boldsymbol{L}} \tau_{\boldsymbol{L}}(\Gamma)$ for every formal concept (A, Γ) in $\mathsf{L}(\mathsf{C}(\boldsymbol{L}))$, and the map $\boldsymbol{L} \to \mathsf{L}(\mathsf{C}(\boldsymbol{L}))$ by $x \mapsto (\{a \in inst(\boldsymbol{L}) \mid \iota_{\boldsymbol{L}}(a) \leq_{\boldsymbol{L}} x\}, \{\alpha \in typ(\boldsymbol{L}) \mid x \leq_{\boldsymbol{L}} \tau_{\boldsymbol{L}}(\alpha)\})$ for every element $x \in \boldsymbol{L}$. These are inverse monotonic functions. Let $\boldsymbol{h} = \langle left(\boldsymbol{h}), right(\boldsymbol{h}), inst(\boldsymbol{h}), typ(\boldsymbol{h}) \rangle : \boldsymbol{L} \rightleftharpoons \boldsymbol{K}$ be a concept lattice morphism from concept lattice $\boldsymbol{L}$ to concept lattice $\boldsymbol{K}$, with associated infomorphism $\mathsf{C}(\boldsymbol{h}) = \langle inst(\boldsymbol{h}), typ(\boldsymbol{h}) \rangle : \mathsf{C}(\boldsymbol{L}) \rightleftharpoons \mathsf{C}(\boldsymbol{K})$. Then, up to isomorphism, $\mathsf{L}(\mathsf{C}(\boldsymbol{h})) = \boldsymbol{h}$. This defines the natural isomorphism: $\mathsf{C} \circ \mathsf{L} \cong Id_{\textsf{Concept Lattice}}$.

The functor composition $\mathsf{L} \circ \mathsf{C}$ is equal to the identity functor $Id_{\textsf{Classification}}$. To see this, consider whether $\boldsymbol{A} = \mathsf{C}(\mathsf{L}(\boldsymbol{A}))$ for any classification $\boldsymbol{A}$. Obviously, the type and instance sets are the same. What about the classification relations? The classification relation in $\mathsf{C}(\mathsf{L}(\boldsymbol{A}))$ is defined in terms of the lattice order and instance/type embeddings by $\models_{\boldsymbol{L}} = \iota_{\boldsymbol{L}} \circ \leq_{\boldsymbol{L}} \circ \tau_{\boldsymbol{L}}{}^{op}$; which is easily seen to be equal $\models_{\boldsymbol{L}}$. Hence, $\boldsymbol{A} = \mathsf{C}(\mathsf{L}(\boldsymbol{A}))$. What about infomorphisms? The functional infomorphisms $\boldsymbol{f}$ and $\mathsf{C}(\mathsf{L}(\boldsymbol{f}))$ are equal.

3.2 Relational Equivalence

Relational infomorphisms are more general than functional infomorphisms. This increased flexibility and expressiveness must be balanced with a decreased number of properties. However, the property of (categorical) equivalence between the distributional side and the conceptual side still holds. This section formalizes these observations in a theorem on categorical equivalence.

The Complete Adjoint Functor. Let $\boldsymbol{F} : \boldsymbol{A} \to \boldsymbol{B}$ be any bond and let $\boldsymbol{r} = \langle \check{r}, \hat{r} \rangle = \langle inst(\boldsymbol{r}), typ(\boldsymbol{r}) \rangle : \boldsymbol{A} \rightleftharpoons \boldsymbol{B}$ be any relational infomorphism having $\boldsymbol{F}$ as its bond. We again ask how the two concept lattices $\mathsf{L}(\boldsymbol{A})$ and $\mathsf{L}(\boldsymbol{B})$ are related, but now in terms of relational infomorphisms. More particularly, how are $\mathsf{L}(\boldsymbol{A})$ and $\mathsf{L}(\boldsymbol{B})$ related to $\mathsf{L}(\boldsymbol{F})$, the concept lattice of the bond itself?

Since for any concept $(B, \Gamma) \in L(\boldsymbol{F})$ the intent $\Gamma = B^{\boldsymbol{F}} = \bigcap_{b \in B} b\boldsymbol{F}$ is an intent of $L(\boldsymbol{A})$, the mapping $(B, \Gamma) \mapsto (\Gamma^{\boldsymbol{A}}, \Gamma)$ is a well-defined monotonic function $\partial_0 : \mathsf{L}(\boldsymbol{F}) \to \mathsf{L}(\boldsymbol{A})$ that has a right-adjoint-right-inverse $\tilde{\partial}_0 : \mathsf{L}(\boldsymbol{A}) \to \mathsf{L}(\boldsymbol{F})$ defined as the mapping $(A, \Gamma) \mapsto (\Gamma^{\boldsymbol{F}}, \Gamma^{\boldsymbol{FF}})$. Let $int_{\boldsymbol{F}} = \langle \partial_0, \tilde{\partial}_0 \rangle : \mathsf{L}(\boldsymbol{F}) \rightleftharpoons \mathsf{L}(\boldsymbol{A})$ denote this adjoint pair (coreflection). Dually, since for any concept $(B, \Gamma) \in L(\boldsymbol{F})$ the extent $B = \Gamma^{\boldsymbol{F}} = \bigcap_{\alpha \in \Gamma} \boldsymbol{F}\alpha$ is an extent of $L(\boldsymbol{B})$, the mapping $(B, \Gamma) \mapsto (B, B^{\boldsymbol{B}})$ is a well-defined monotonic function $\partial_1 : \mathsf{L}(\boldsymbol{F}) \to \mathsf{L}(\boldsymbol{B})$ that has a left-adjoint-right-inverse $\tilde{\partial}_1 : \mathsf{L}(\boldsymbol{B}) \to \mathsf{L}(\boldsymbol{F})$ defined as the mapping $(B, \Delta) \mapsto (B^{\boldsymbol{FF}}, B^{\boldsymbol{F}})$. Let $ext_{\boldsymbol{F}} = \langle \tilde{\partial}_1, \partial_1 \rangle : \mathsf{L}(\boldsymbol{B}) \rightleftharpoons \mathsf{L}(\boldsymbol{F})$ denote this adjoint pair (reflection).

Since for any concept $(A, \Gamma) \in L(\boldsymbol{A})$ the equality $\Gamma^{\boldsymbol{F}} = \boldsymbol{F}/\Gamma = (\boldsymbol{B}/\hat{r})/\Gamma = \boldsymbol{B}/(\Gamma \circ \hat{r}) = (\Gamma\hat{r})^{\boldsymbol{B}}$ holds, the mapping $(A, \Gamma) \mapsto (\Gamma^{\boldsymbol{F}}, \Gamma^{\boldsymbol{FB}})$ is a well-defined monotonic function $\mathsf{L}(\hat{r}) : \mathsf{L}(\boldsymbol{A}) \to \mathsf{L}(\boldsymbol{B})$, which is the composition $\mathsf{L}(\hat{r}) = \tilde{\partial}_0 \cdot \partial_1$. Dually, since for any concept $(B, \Delta) \in L(\boldsymbol{B})$ the equality $B^{\boldsymbol{F}} = B \backslash \boldsymbol{F} = B \backslash (\check{r} \backslash \boldsymbol{A}) = (\check{r} \circ B) \backslash \boldsymbol{A} = (\check{r}B)^{\boldsymbol{A}}$ holds, the mapping $(B, \Delta) \mapsto (B^{\boldsymbol{FA}}, B^{\boldsymbol{F}})$ is a well-defined monotonic function $\mathsf{L}(\check{r}) : \mathsf{L}(\boldsymbol{A}) \leftarrow \mathsf{L}(\boldsymbol{B})$, which is the composition $\mathsf{L}(\check{r}) = \tilde{\partial}_1 \cdot \partial_0$. Basic properties of residuation show that this is an adjoint pair of monotonic functions $\langle \mathsf{L}(\check{r}), \mathsf{L}(\hat{r}) \rangle = int_{\boldsymbol{F}} \circ ext_{\boldsymbol{F}} : \mathsf{L}(\boldsymbol{B}) \rightleftharpoons \mathsf{L}(\boldsymbol{A})$, since $B^{\boldsymbol{F}} \supseteq \Gamma$ iff $B \backslash \boldsymbol{F} \supseteq \Gamma$ iff $B \circ \Gamma \subseteq \boldsymbol{F}$ iff $B \subseteq \boldsymbol{F}/\Gamma$ iff $B \subseteq \Gamma^{\boldsymbol{F}}$. The pair $\mathsf{A}(\boldsymbol{F}) \doteq \langle \mathsf{L}(\check{r}), \mathsf{L}(\hat{r}) \rangle$ is called the *complete adjoint* of the bond $\boldsymbol{F} : \boldsymbol{A} \to \boldsymbol{B}$.

More abstractly, a *complete adjoint* $\boldsymbol{h} = \langle left(\boldsymbol{h}), right(\boldsymbol{h}) \rangle : \boldsymbol{L} \rightleftharpoons \boldsymbol{K}$ between complete lattices $\boldsymbol{L}$ and $\boldsymbol{K}$ is an adjoint pair of monotonic functions. Let **Complete Adjoint** denote the category of complete lattices and adjoint pairs of monotonic functions. For each bond $\boldsymbol{F} : \boldsymbol{A} \to \boldsymbol{B}$ the pair $\mathsf{A}(\boldsymbol{F}) : \mathsf{L}(\boldsymbol{B}) \rightleftharpoons \mathsf{L}(\boldsymbol{A})$ is a complete adjoint between the concept lattices of the target and source classifications $\boldsymbol{B}$ and $\boldsymbol{A}$. The operator A is a functor called *complete adjoint* from **Bond** the category of classifications and bonds to **Complete Adjoint**^{op} the opposite of the category of complete lattices and adjoint pairs.

The Bond Functor. Associated with any complete lattice $\boldsymbol{L} = \langle L, \leq_{\boldsymbol{L}}, \bigwedge_{\boldsymbol{L}}, \bigvee_{\boldsymbol{L}} \rangle$ is the classification $\mathsf{B}(\boldsymbol{L}) = \langle L, L, \leq_{\boldsymbol{L}} \rangle$, which has $\boldsymbol{L}$-elements as its instances and types, and the lattice order as its classification relation. Associated with any complete adjoint $\boldsymbol{h} = \langle \phi, \psi \rangle = \langle left(\boldsymbol{h}), right(\boldsymbol{h}) \rangle : \boldsymbol{K} \rightleftharpoons \boldsymbol{L}$ is the bond

$\mathbf{B}(h) : \mathbf{B}(L) \to \mathbf{B}(K)$ defined by its adjointness property: $y\mathbf{B}(h)x$ iff $\phi(y) \leq_L x$ iff $y \leq_K \psi(x)$ for all elements $x \in L$ and $y \in K$. The closure property of bonds is obvious, since $y\mathbf{B}(h) = \uparrow_L \phi(y)$ for all elements $y \in K$ and $\mathbf{B}(h)x = \downarrow_K \psi(x)$ for all elements $x \in L$. The operator $\mathbf{B}$ is a functor called *bond* from **Complete Adjoint**^{op} the category of complete lattices and adjoint pairs to **Bond** the category of classifications and bonds.

Equivalence. The functor composition $\mathbf{B} \circ \mathbf{A}$ is naturally isomorphic to the identity functor $Id_{\mathbf{Complete\ Adjoint}^{op}}$. To see this, let $L = \langle L, \leq_L, \bigwedge_L, \bigvee_L \rangle$ be a complete lattice with associated classification $\mathbf{B}(L) = \langle L, L, \leq_L \rangle$. Part of the fundamental theorem of concept lattices [3] asserts the isomorphism $L \cong \mathbf{A}(\mathbf{B}(L))$. In particular, formal concepts of $\mathbf{A}(\mathbf{B}(L))$ are of the form $(\downarrow_L x, \uparrow_L x)$ for elements $x \in L$. So, define the obvious maps $\mathbf{A}(\mathbf{B}(L)) \to L$ by $(\downarrow_L x, \uparrow_L x) \mapsto x$ and $L \to \mathbf{A}(\mathbf{B}(L))$ by $x \mapsto (\downarrow_L x, \uparrow_L x)$, for every element $x \in L$. Let $h = \langle \phi, \psi \rangle = \langle left(h), right(h) \rangle : K \rightleftharpoons L$ be a complete adjoint with associated bond $\mathbf{B}(h) : \mathbf{B}(L) \to \mathbf{B}(K)$. Then, the right adjoint of $\mathbf{A}(\mathbf{B}(h))$ maps $(\downarrow_L x, \uparrow_L x) \mapsto (\downarrow_L \psi(x), \uparrow_L \psi(x))$ and the left adjoint of $\mathbf{A}(\mathbf{B}(h))$ maps $(\downarrow_K y, \uparrow_K y) \mapsto (\downarrow_K \phi(y), \uparrow_K \phi(y))$. So, up to isomorphism, $\mathbf{A}(\mathbf{B}(h))$ is the same as h. This defines the natural isomorphism: $\mathbf{B} \circ \mathbf{A} \cong Id_{\mathbf{Complete\ Adjoint}^{op}}$.

The basic theorem of Formal Concept Analysis [3] can be framed in terms of two fundamental bonds (relational infomorphisms) between any classification and its associated concept lattice. For any classification A the instance embedding relation is a bond $\iota_A : L(A) \to A$ from the concept lattice to A itself, and the type embedding relation is a bond $\tau_A : A \to L(A)$ in the reverse direction. The pair $\langle L(A)/\iota_A, \tau_A \rangle : L(A) \rightleftharpoons A$ is a relational infomorphism whose bond is the instance embedding relation. The pair $\langle \iota_A, \tau_A \backslash L(A) \rangle : A \rightleftharpoons L(A)$ is a relational infomorphism whose bond is the type embedding relation. The instance and type embedding bonds are inverse to each other: $\iota_A \diamond \tau_A = Id_{L(A)}$ and $\tau_A \diamond \iota_A = Id_A$.

The functor composition $\mathbf{A} \circ \mathbf{B}$ is naturally isomorphic to the identity functor $Id_{\mathbf{Bond}}$. To see this, let A be a classification with associated concept lattice $\mathbf{A}(A)$. The comments above demonstrate the isomorphism $A \cong \mathbf{B}(\mathbf{A}(A))$. Let $F : A \to B$ be a bond between classifications A and B with associated complete adjoint $\mathbf{A}(F) : \mathbf{A}(B) \rightleftharpoons \mathbf{A}(A)$. The bond $\mathbf{B}(\mathbf{A}(F)) : \mathbf{B}(\mathbf{A}(A)) \to \mathbf{B}(\mathbf{A}(B))$ contains a conceptual pair (b, a) of the form $b = (B, \Delta)$ and $a = (A, \Gamma)$ iff $B \circ \Gamma \subseteq F$, where $B \circ \Gamma = B \times \Gamma$ a Cartesian product or rectangle, iff $B \subseteq \Gamma^F$ iff $\Gamma \subseteq B^F$. So, $(\iota_A \diamond F) \diamond \tau_B = (F/(\iota_A \backslash A)) \diamond \tau_B = (F/\tau_A) \diamond \tau_B = (B/\tau_B) \backslash (F/\tau_A) = \iota_B \backslash (F/\tau_A) = \mathbf{B}(\mathbf{A}(F))$, by bond composition and properties of the instance and type relations. Hence, $\mathbf{B}(\mathbf{A}(F)) \diamond \iota_B = \iota_A \diamond F$. This proves the required naturality condition.

3.3 Complete Relational Equivalence

In the section on relational equivalence, we have seen how bonds are categorically equivalent to the opposite of complete adjoints, adjoint pairs between complete lattices. Unfortunately, these are not the best morphisms for making structural comparisons between complete lattices. Complete homomorphisms are best for this [3]. Since complete homomorphisms are special cases of complete adjoints on the conceptual side, we are interested in what constraints to place on bonds on the distributional side.

The Complete Lattice Functor. Let $\langle F, G\rangle : A \rightleftharpoons B$ be any bonding pair. Then $F : A \to B$ is a bond in the forward direction from classification A to classification B, and $G : A \to B$ is a bond in the reverse direction to classification A from classification B. Applying the complete adjoint functor, we get two adjoint pairs in opposite directions: $\langle \phi_F, \psi_F\rangle : \mathsf{L}(B) \rightleftharpoons \mathsf{L}(A)$ in the forward direction and $\langle \phi_G, \psi_G\rangle : \mathsf{L}(A) \rightleftharpoons \mathsf{L}(B)$ in the reverse direction. It was shown above that for bonding pairs the meet-preserving monotonic function $\psi_F : \mathsf{L}(A) \to \mathsf{L}(B)$ is equal to the join-preserving monotonic function $\phi_G : \mathsf{L}(A) \to \mathsf{L}(B)$, giving a complete homomorphism. This function is the unique mediating function for the $\mathsf{L}(A)$-indexed collective B-concept $(F/\tau_A, \iota_A\backslash G)$, the $\langle F, G\rangle$-image of the $\mathsf{L}(A)$-indexed collective A-concept (ι_A, τ_A), whose closure expressions define the pairing constraints.

The *complete lattice functor* A^2 : **Bonding Pair** $\to$ **Complete Lattice** is the operator that maps a classification A to its concept lattice $\mathsf{A}^2(A) \doteq L(A)$ regarded as a complete lattice only, and maps a bonding pair $\langle F, G\rangle : A \rightleftharpoons B$ to its complete homomorphism $\mathsf{A}^2(\langle F, G\rangle) \doteq \psi_F = \phi_G : L(A) \to L(B)$.

The Bonding Pair Functor. Let $\psi : L \to K$ be a complete homomorphism between complete lattices L and K with associated adjunctions $\phi \dashv \psi \dashv \theta$. Then $\langle \phi, \psi\rangle : K \rightleftharpoons L$ and $\langle \psi, \theta\rangle : L \rightleftharpoons K$ are morphisms in **Complete Adjoint**. Application of the bond functor B produces the bonds $F = \mathsf{B}(\langle \phi, \psi\rangle) : \mathsf{B}(L) \to \mathsf{B}(K)$ and $G = \mathsf{B}(\langle \psi, \theta\rangle) : \mathsf{B}(K) \to \mathsf{B}(L)$ between the classifications $\mathsf{B}(L) = \langle L, L, \leq_L\rangle$ and $\mathsf{B}(K) = \langle K, K, \leq_K\rangle$. Note that for any complete lattice L the instance and type relations for the classification $\mathsf{B}(L)$ are both equal to the order relation $\iota_L = \leq_L = \tau_L$. Since $F/\tau_L = F/\leq_L = F$, $\mathsf{B}(K)/(\iota_L\backslash G) = \leq_K/(\leq_L\backslash G) = \leq_K/G = F$, $\iota_L\backslash G = \leq_L\backslash G = G$ and $(F/\tau_L)\backslash\mathsf{B}(K) = (F/\leq_L)\backslash\leq_K = F\backslash\leq_K = G$, the pair of bonds (F, G) is a bonding pair $(F, G) = (\mathsf{B}(\langle \phi, \psi\rangle), \mathsf{B}(\langle \psi, \theta\rangle)) : \mathsf{B}(L) \rightleftharpoons \mathsf{B}(K)$. Let $\mathsf{B}^2(\psi)$ denote this pair.

The *bonding pair functor* B^2 : **Complete Lattice** $\to$ **Bonding Pair** is the operator that maps a complete lattice L to its classification $\langle L, L, \leq_L\rangle$ and maps a complete homomorphism to its bonding pair as above. Since the bond functor B is functorial, so is B^2.

Equivalence. The functor composition $\mathsf{A}^2 \circ \mathsf{B}^2$ is naturally isomorphic to the identity functor $Id_{\textbf{Bonding Pair}}$. Consider any classification A. The type and instance embedding relations form bonding pairs in two different ways, $\langle \tau_A, \iota_A\rangle : A \rightleftharpoons \mathsf{L}(A) = \mathsf{A}^2\circ\mathsf{B}^2(A)$ and $\langle \iota_A, \tau_A\rangle : \mathsf{A}^2\circ\mathsf{B}^2(A) = \mathsf{L}(A) \rightleftharpoons A$, and these are inverse to each other: $\langle \tau_A, \iota_A\rangle \diamond \langle \iota_A, \tau_A\rangle = Id_A$ and $\langle \iota_A, \tau_A\rangle \diamond \langle \tau_A, \iota_A\rangle = Id_{\mathsf{L}(A)}$. So that each classification is isomorphic in **Bonding Pair** to its concept lattice: $A \cong \mathsf{B}^2(\mathsf{L}(A)) = \mathsf{A}^2\circ\mathsf{B}^2(A)$. Let $\langle F, G\rangle : A \rightleftharpoons B$ be a bonding pair. As shown above, the naturality conditions for bonds F and G are expressed as

$\iota_A \diamond F \diamond \tau_B = \mathsf{A}{\circ}\mathsf{B}(F)$ and $\iota_B \diamond G \diamond \tau_A = \mathsf{A}{\circ}\mathsf{B}(G)$. So $\langle \iota_A, \tau_A \rangle \diamond \langle F, G \rangle \diamond \langle \tau_B, \iota_B \rangle = \langle \iota_A \diamond F \diamond \tau_B, \iota_B \diamond G \diamond \tau_A \rangle = \langle \mathsf{A}{\circ}\mathsf{B}(F), \mathsf{A}{\circ}\mathsf{B}(G) \rangle = \mathsf{A}^2{\circ}\mathsf{B}^2(\langle F, G \rangle)$.

The functor composition $\mathsf{B}^2 \circ \mathsf{A}^2$ is naturally isomorphic to the identity functor $Id_{\textsf{Complete Lattice}}$. Consider any complete lattice L. As we have seen in studying relational equivalence, any complete lattice L is isomorphic to the complete lattice $(\mathsf{B}^2{\circ}\mathsf{A}^2)(L) = \mathsf{L}(\langle L, L, \leq_L \rangle)$ via the bijection $x \mapsto (\downarrow_L x, \uparrow_L x)$. Now consider any complete homomorphism $\psi : L \rightarrow K$ between complete lattices L and K with associated adjunctions $\phi \dashv \psi \dashv \theta$. The bonding pair functor maps this to the bonding pair $\mathsf{B}^2(\psi) = (\mathsf{B}(\langle \phi, \psi \rangle), \mathsf{B}(\langle \psi, \theta \rangle))$, and the complete lattice functor maps this to the complete homomorphism $\mathsf{A}^2(\mathsf{B}^2(\psi)) = \tilde{\psi} : \mathsf{L}(\langle L, L, \leq_L \rangle) \rightarrow \mathsf{L}(\langle K, K, \leq_K \rangle)$ with associated adjunctions $\tilde{\phi} \dashv \tilde{\psi} \dashv \tilde{\theta}$, where $\tilde{\phi}(\downarrow_K y, \uparrow_K y) = (\downarrow_L \phi(y), \uparrow_L \phi(y))$, $\tilde{\psi}(\downarrow_L x, \uparrow_L x) = (\downarrow_K \psi(x), \uparrow_K \psi(x))$ and $\tilde{\theta}(\downarrow_K y, \uparrow_K y) = (\downarrow_L \theta(y), \uparrow_L \theta(y))$. Clearly, the naturality condition holds between ψ and $(\mathsf{B}^2{\circ}\mathsf{A}^2)(\psi)$.

3.4 Theorem and Architectural Diagram

Figure 2 contains a commuting diagram of functors, categories and equivalences that represents the architecture of distributed conceptual structures. This is the central contribution of this paper. Many of the details that support this diagram are known from the literature ([2], [3]). This paper seeks to bring these facts together into a coherent view. Figure 2 is two-dimensional, having the same orientation as Fig. 1. The vertical dimension contains the functional-relational distinction. The morphisms of the equivalent categories **Classification** and **Concept Lattice** are function-based, whereas the equivalent categories **Bond** and **Complete Adjoint**^{op} are relation-based. The horizontal dimension contains the distributional-conceptual distinction. This is represented as the three categorical equivalences expressed in Theorem 1. The equivalences between the categories **Classification** and **Concept Lattice** at the functional level and the categories **Bond** and **Complete Adjoint**^{op} at the relational level are expressions of the basic theorem of Formal Concept Analysis [3]. The equivalence between the categories **Classification** and **Regular Theory** is one of the main contributions of the theory of Information Flow [2].

Theorem 1. *The distributional/conceptual distinction in the architecture of distributed conceptual structures is represented by three categorical equivalences.*

- *The category of classifications and functional infomorphisms is categorically equivalent to the category of concept lattices and concept lattice morphisms*

  $$\textsf{Classification} \equiv \textsf{Concept Lattice}$$

  via the lattice functor and classification functor, which are generalized inverses: $\mathsf{L} \circ \mathsf{C} = Id_{\textsf{Classification}}$ *and* $\mathsf{C} \circ \mathsf{L} \cong Id_{\textsf{ConceptLattice}}$.
- *The category of classifications and bonds is categorically equivalent to the category of complete lattices and complete adjoints*

  $$\textsf{Bond} \equiv \textsf{Complete Adjoint}^{op}$$

  via the complete adjoint functor and bond functor, which are generalized inverses: $\mathsf{A} \circ \mathsf{B} \cong Id_{\textsf{Bond}}$ *and* $\mathsf{B} \circ \mathsf{A} \cong Id_{\textsf{Complete Adjoint}^{op}}$.

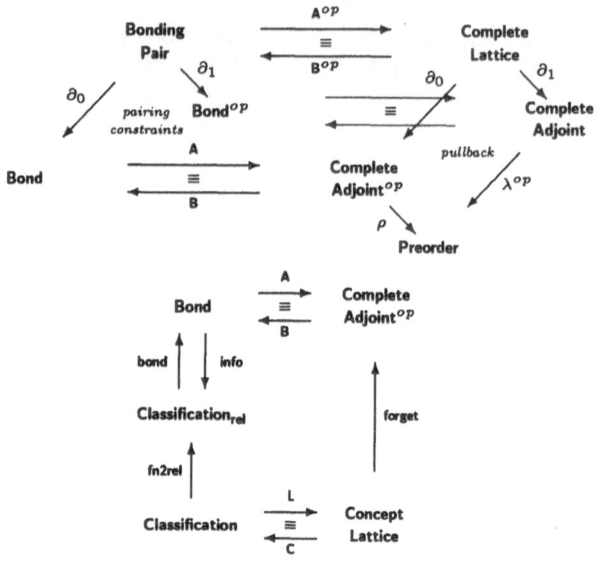

Fig. 2. Architectural Diagram of Distributed Conceptual Structures

- *The category of classifications and bonding pairs is categorically equivalent to the category of complete lattices and complete homomorphisms*

$$\textsf{\textbf{Bonding Pair}} \equiv \textsf{\textbf{Complete Lattice}}$$

via the complete lattice functor and bonding pair functor, which are generalized inverses: $\mathbf{A}^2 \circ \mathbf{B}^2 \cong Id_{\textsf{Bonding Pair}}$ *and* $\mathbf{B}^2 \circ \mathbf{A}^2 \cong Id_{\textsf{Complete Lattice}}$.

Proof. The bulk of this paper provides the proof for this theorem. □

4 Limit/Colimit Constructions

Limit/colimit constructions are very important in applications, often harboring the central semantics. Categorical equivalence is more general and more useful than categorical isomorphism. Fact 1 in the appendix expresses and proves the well-known observation that equivalent categories have the same limit/colimit structures. Since the distributional-conceptual distinction, the horizontal dimension illustrated in Fig. 2 and expressed in Theorem 1, has three categorical equivalences, limit/colimit structures will give added meaning to distributive conceptual structures. In brief comments below, although we describe limit/colimit constructions on one side of the equivalence, we are assured of their preservation when mapped to the other side of the equivalence.

A true conception of the limit/colimit architecture requires an understanding of the two-dimensional fibrational nature of distributive conceptual structures. Types and instances form fibered spans. The fibrational instance dimension is defined in terms of the instance forgetful functor **inst** : **Classification** → **Set**^{op}, here simplified by ignoring instance order. For any set (of instances) A, the Ath fiber category of **inst**, denoted $\mathbf{inst}^{-1}(A)$, is the subcategory of classifications $\boldsymbol{A}$ with instance set $inst(\boldsymbol{A}) = A$ and infomorphisms $\boldsymbol{f} : \boldsymbol{A} \rightleftharpoons \boldsymbol{B}$ with instance function $inst(\boldsymbol{f}) = Id_A$. The fibrational type dimension is defined dually in terms of the type forgetful functor $\mathbf{typ} : \mathbf{Classification} \to \mathbf{Set}$. The initial object in the A-th fiber category $\mathbf{inst}^{-1}(A)$ is the empty classfication $\boldsymbol{0}_A = \langle A, \emptyset, \emptyset \rangle$, the terminal object in $\mathbf{inst}^{-1}(A)$ is the instance powerset classification $\wp A = \langle A, \wp A, \in \rangle$ – for any classification $\boldsymbol{A}$, the extent infomorphism $\eta_{\boldsymbol{A}} : \boldsymbol{A} \rightleftharpoons \wp(inst(\boldsymbol{A}))$ is the unique infomorphism in $\mathbf{inst}^{-1}(A)$ from $\boldsymbol{A}$ to $\wp(inst(\boldsymbol{A}))$. Given two classifications $\boldsymbol{A}_0$ and $\boldsymbol{A}_1$ in the A-th fiber category $\mathbf{inst}^{-1}(A)$, the apposition $\boldsymbol{A}_0 \mid \boldsymbol{A}_1$ is the coproduct of $\boldsymbol{A}_0$ and $\boldsymbol{A}_1$ in $\mathbf{inst}^{-1}(A)$. Dual comments hold for the type fibrational dimension, where for example subposition is the product construction.

The full category **Classification** contains all limits and colimits. Given two classifications $\boldsymbol{A}$ and $\boldsymbol{B}$, the *sum* (*semiproduct*) $\boldsymbol{A}+\boldsymbol{B}$ is the coproduct of $\boldsymbol{A}$ and $\boldsymbol{B}$ in **Classification**. The dual notion provides the product. All limits/colimits exist when not only products/coproducts but also quotient constructions exist. Given a classification $\boldsymbol{A}$, a *dual invariant* is a pair $J = (A, R)$ consisting of a set $A \subseteq inst(\boldsymbol{A})$ of instances of $\boldsymbol{A}$ and a binary relation R on types of $\boldsymbol{A}$ satisfying the constraint: if $\alpha R \beta$, then for each $a \in A$, $a \models_{\boldsymbol{A}} \alpha$ if and only if $a \models_{\boldsymbol{A}} \beta$. The *dual quotient* of $\boldsymbol{A}$ by J, written $\boldsymbol{A}/J$, is the classification with instances A, whose types are the R-equivalence classes of types of $\boldsymbol{A}$, and whose classification is $a \models_{\boldsymbol{A}/J} [\alpha]$ if and only if $a \models_{\boldsymbol{A}} \alpha$. Dual quotients include kernel image factorizations, coequalizors and pushouts. Suitable colimit constructions in **Classification** have been used by the author to define the semantics of ontology sharing [4,5].

5 Summary and Future Work

This paper has had as its goal the formulation of a framework for conceptual knowledge representation. For this it uses the language of category theory in order to represent some of the essence of Information Flow and Formal Concept Analysis, thereby unifying these two studies. This has culminated in the recognition of the two-dimensional nature of distributed conceptual structures (Fig. 1), whose particulars, described as a commuting diagram (Fig. 2), represent the fundamental theorem of Formal Concept Analysis in terms of the three categorical equivalences expressed in Theorem 1. Information Flow has initiated development of the distributed nature of the logic of information, principally represented by the morphisms on the distributional side of the diagram. Formal Concept Analysis has developed the conceptual nature of knowledge, principally represented by the objects on the conceptual side of the diagram.

The author is currently engaged in the development of the Information Flow Framework (IFF), whose mission is to further the development of the theory of Information Flow and Formal Concept Analysis and to apply these to distributed logic, ontologies, and knowledge representation. The Information Flow Framework is manifest as the IFF Foundation Ontology[1], which is being formulated in a new version of the Knowledge Interchange Format (KIF). The IFF

[1] Located at: URL: `http://suo.ieee.org/IFF/`

Foundation Ontology represents metalogic. It provides a principled foundation for the metalevel (structural level) of the Standard Upper Ontology (SUO). The SUO metalevel can be used as a logical framework for manipulating collections of object level ontologies. The IFF Foundation Ontology is rather large, but highly structured. It is partitioned into three metalevels: top, upper and lower. These metalevels correspond to the set-theoretic distinction in foundations between the generic, the large and the small. Each metalevel is partitioned into numerous namespaces and services the level below it. The top metalevel provides an interface between the KIF logical language and the upper metalevel of the IFF Foundation Ontology. The upper metalevel is partitioned into three sub-ontologies: the IFF Upper Core Ontology, the IFF Upper Classification Ontology, and the IFF Category Theory Ontology. The lower metalevel contains, amongst other modules, the IFF Model Theory Ontology, the IFF Algebraic Theory Ontology, and the IFF Ontology Ontology. The lower metalevel represents object-level ontologies by providing terminology and axiomatization for (1) the formal or axiomatic semantics of IFF theories, (2) the interpretative semantics of IFF model-theoretic structures, and (3) the combined semantics of IFF logics. Guided by the Upper Classification Ontology, the lower metalevel also offers a complete framework for a conceptual "lattice of theories" as concentrated in the fundamental truth meta-classification and truth concept meta-lattice[2]. The IFF Upper Classification Ontology, which enables metalevel reasoning about truth, formalizes the terminological and axiomatic content of this paper.

References

1. Barr, M.: *-Autonomous categories and linear logic. Mathematical Structures in Computer Science **1** (1991) 159–178.
2. Barwise, J., Seligman, J.: Information Flow: the Logic of Distributed Systems. Cambridge University Press, Cambridge (1997).
3. Ganter, B., Wille, R.: Formal Concept Analysis: Mathematical Foundations. Springer, Heidelberg (1999).
4. Kent, R.: The Information Flow Foundation for Conceptual Knowledge Organization. Beghtol, C., Howarth, C., Williamson, N. (eds.): Dynamism and Stability in Knowledge Organization. Proceedings of the Sixth International ISKO Conference. Advances in Knowledge Organization **7** (2000) 111–117. Ergon Verlag, Würzburg.
5. Kent, R.: The IFF Foundation for Ontological Knowledge Organization. Beghtol, C., Williamson, N. (eds.): Knowledge Organization and Classification in International Information Retrieval. Cataloging and Classification Quarterly. The Haworth Press Inc., Binghamton, New York (2003).

[2] The truth classification of a first-order language $\boldsymbol{L}$ is the meta-classification, whose instances are $\boldsymbol{L}$-structures, whose types are $\boldsymbol{L}$-sentences, and whose classification relation is satisfaction. The truth concept lattice, the concept lattice of the truth classification, functions as the appropriate "lattice of ontological theories" for the SUO. A formal concept in this lattice has an intent that is a closed theory and an extent that is the collection of all models for that theory. The theory (intent) of the join or supremum of two concepts is the intersection of the theories (conceptual intents), and the theory (intent) of the meet or infimum of two concepts is the theory of the common models.

6. Lawvere, W.: Metric Spaces, Generalized Logic and Closed Categories. Rendiconti del Seminario Matematico e Fisico di Milano **43** (1973) 135–166.
7. Mac Lane, S.: Categories for the Working Mathematician. Springer, New York (1971).

A Appendix

According to Saunders Mac Lane [7], equivalences between categories are more general, and more useful, than isomorphisms between categories. We emphatically concur and we argue that the main reason for their usefulness is derived from the well-known observation that equivalent categories have the same limit/colimit structures. We make this precise by the following fact.

Fact 1 *If categories* $\mathbf{A}$ *and* $\mathbf{B}$ *are equivalent, then they have the same limit/colimit structures. More particularly, if* $\mathbf{B}$ *has limits (colimits) for all* $\mathbf{C}$*-shaped diagrams, then* $\mathbf{A}$ *also has limits (colimits) for all* $\mathbf{C}$*-shaped diagrams, and vice-versa. In particular,* $\mathbf{A}$ *is complete (co-complete) iff* $\mathbf{B}$ *is complete (co-complete).*

Proof. Assume $\mathbf{F} : \mathbf{A} \to \mathbf{B}$ and $\mathbf{G} : \mathbf{B} \to \mathbf{A}$ are functors that mediate the equivalence through natural isomorphisms $\eta : Id_{\mathbf{A}} \Rightarrow \mathbf{F} \circ \mathbf{G}$ and $\epsilon : \mathbf{G} \circ \mathbf{F} \Rightarrow Id_{\mathbf{B}}$. This means the following.

- η is the unit and ϵ is the counit of adjunction $\mathbf{F} \dashv \mathbf{G}$:
  $\eta\mathbf{F} \bullet \mathbf{F}\epsilon = Id_{\mathbf{F}}$ and $\mathbf{G}\eta \bullet \epsilon\mathbf{G} = Id_{\mathbf{G}}$.
- The natural transformations η are η^{-1} are inverse:
  $\eta \bullet \eta^{-1} = Id_{Id_{\mathbf{A}}}$ and $\eta^{-1} \bullet \eta = Id_{\mathbf{F}\circ\mathbf{G}}$.
- The natural transformations ϵ are ϵ^{-1} are inverse:
  $\epsilon \bullet \epsilon^{-1} = Id_{\mathbf{G}\circ\mathbf{F}}$ and $\epsilon^{-1} \bullet \epsilon = Id_{Id_{\mathbf{B}}}$.
- ϵ^{-1} is the unit and η^{-1} is the counit of adjunction $\mathbf{G} \dashv \mathbf{F}$:
  $\epsilon^{-1}\mathbf{G} \bullet \mathbf{G}\eta^{-1} = Id_{\mathbf{G}}$ and $\mathbf{F}\epsilon^{-1} \bullet \eta^{-1}\mathbf{F} = Id_{\mathbf{F}}$.

Let $\mathbf{D} : \mathbf{C} \to \mathbf{A}$ be any $\mathbf{C}$-shaped diagram in $\mathbf{A}$. Diagram $\mathbf{D}$ is mapped by $\mathbf{F}$ to $\mathbf{D} \circ \mathbf{F} : \mathbf{C} \to \mathbf{B}$, a $\mathbf{C}$-shaped diagram in $\mathbf{B}$. By assumption, there is a limiting cone $\lambda : \mathbf{B} \Rightarrow \mathbf{D} \circ \mathbf{F}$ for $\mathbf{D} \circ \mathbf{F}$ in $\mathbf{B}$ with limit object B. We will show that $\lambda\mathbf{G} \bullet \mathbf{D}\eta^{-1} : \mathbf{G}(B) \Rightarrow \mathbf{D}$ is a limiting cone for $\mathbf{D}$ in $\mathbf{A}$ with limit object $\mathbf{G}(B)$.

Let $\gamma : A \Rightarrow \mathbf{D}$ be any cone for $\mathbf{D}$ in $\mathbf{A}$. Then $\gamma\mathbf{F} : \mathbf{F}(A) \Rightarrow \mathbf{D} \circ \mathbf{F}$ is a cone for $\mathbf{D} \circ \mathbf{F}$ in $\mathbf{B}$. Since λ is a limiting cone, there is a unique $\mathbf{B}$-arrow $g : \mathbf{F}(A) \to B$ with $g \cdot \lambda = \gamma\mathbf{F}$. Define $\mathbf{A}$-arrow $f \doteq \eta_A \cdot \mathbf{G}(g) : A \to \mathbf{G}(\mathbf{F}(A)) \to \mathbf{G}(B)$. We will show that f is the unique mediating $\mathbf{A}$-arrow for cone $\lambda\mathbf{G} \bullet \mathbf{D}\eta^{-1}$; that is, f is the unique $\mathbf{A}$-arrow satisfying the constraint $f \cdot \lambda\mathbf{G} \bullet \mathbf{D}\eta^{-1} = \gamma$.

[Existence] $f \cdot \mathbf{G}(\lambda_i) \cdot {\eta^{-1}}_{\mathbf{D}_i} = \eta_A \cdot \mathbf{G}(g) \cdot \mathbf{G}(\lambda_i) \cdot {\eta^{-1}}_{\mathbf{D}_i} = \eta_A \cdot \mathbf{G}(g \cdot \lambda_i) \cdot {\eta^{-1}}_{\mathbf{D}_i} = \eta_A \cdot \mathbf{G}(\mathbf{F}(\gamma_i)) \cdot {\eta^{-1}}_{\mathbf{D}_i} = \eta_A \cdot {\eta_A}^{-1} \cdot \gamma_i = \gamma_i$. [Uniqueness] Suppose that $\tilde{f} : A \to \mathbf{G}(B)$ is any $\mathbf{A}$-arrow satisfying $\tilde{f} \cdot (\lambda\mathbf{G} \bullet \mathbf{D}\eta^{-1}) = \gamma$. Applying $\mathbf{F}$, get $\mathbf{F}(\tilde{f}) \cdot {\eta^{-1}}_B \cdot \lambda_i = \mathbf{F}(\tilde{f}) \cdot \mathbf{F}(\mathbf{G}(\lambda_i)) \cdot {\eta^{-1}}_{\mathbf{F}(\mathbf{D}_i)} = \mathbf{F}(\tilde{f}) \cdot \mathbf{F}(\mathbf{G}(\lambda_i)) \cdot \mathbf{F}({\eta^{-1}}_{\mathbf{D}_i}) = \mathbf{F}(\tilde{f} \cdot \mathbf{G}(\lambda_i) \cdot {\eta^{-1}}_{\mathbf{D}_i}) = \mathbf{F}(\gamma_i) = g \cdot \lambda_i$. By uniqueness $\mathbf{F}(\tilde{f}) \cdot {\eta_B}^{-1} = \mathbf{F}(\tilde{f}) \cdot {\eta^{-1}}_B = g$. Applying $\mathbf{G}$, get $\tilde{f} = \eta_A \cdot \mathbf{G}(\mathbf{F}(\tilde{f})) \cdot \mathbf{G}({\eta_B}^{-1}) = \eta_A \cdot \mathbf{G}(g)$. □

A Computer Algebra Approach to Relational Systems Using Gröbner Bases

L.M. Laita[1], E. Roanes-Lozano[2], L. de Ledesma[1], T. Calvo[3], and L. Gozález-Sotos[3]

[1] Universidad Politécnica de Madrid, Dept. Artificial Intelligence, Campus de Montegancedo, Boadilla del Monte, 28660-Madrid, Spain
[2] Universidad Complutense de Madrid, Dept. Algebra, Edificio "Almudena", c/ Rector Royo Villanova s/n, 28040-Madrid, Spain
[3] Universidad de Alcalá, Dept. de Ciencias de la Computación, Escuela Politécnica, Campus Universitario, 28871-Alcalá de Henares, Madrid, Spain

Abstract. The aim of this paper is to suggest that Gröbner bases can be applied to the study of reflexive and transitive relations. To illustrate our suggestion, we consider two examples. The first example is a Rule-Based-Expert-System: the reflexive and transitive relation is the implication arrow. The second example deals with railway interlocking systems: the reflexive and transitive relation is the accessibility to sections in a railway network. The study is generic; it can be applied to any reflexive and transitive relation.

1 Introduction

The aim of this paper is to suggest that some Computer Algebra theoretical results and their implementation using Gröbner bases, can be applied to reflexive and transitive relations.

To illustrate our suggestion, we consider two examples.

The first example is a Rule-Based-Expert-System (denoted RBES) dealing with what in medical circles is known as "appropriateness criteria" for the application of certain techniques in medicine [5]. Considering the RBES as a RAS (Relational Attribute System) [4], the reflexive and transitive relation is the implication arrow.

The second example deals with railway interlocking systems. Railway interlocking systems are designed to prevent conflicting actions (related to the position of switches and signals) during railway exploitation. A decision model (independent from the topology of the station) based on the use of polynomial ideals and Gröbner bases is presented [9]. The reflexive and transitive relation is the accessibility to sections in a railway network.

2 The RAS-RBES "Appropriateness Criteria"

The next Computer Algebra theoretical result links "tautological consequence" in logic with an ideal membership problem in algebra using Gröbner bases [3,6,7].

H. de Swart (Ed.): RelMiCS 2001, LNCS 2561, pp. 124–133, 2002.

Theorem 1. *A formula A_0 is a tautological consequence of a set of formulae $\{A_1, A_2, ..., A_m\}$, in any p-valued logic for p a prime number, iff the polynomial translation of (the negation of) A_0 belongs to the ideal $J+I$, where J is the ideal generated by the polynomial translations of (the negations of) $A_1, A_2, ..., A_m$, and I is the ideal generated by the polynomials $x_1^p - x_1, x_2^p - x_2, ..., x_n^p - x_n$. Intuitively written:*

$$NEG(A_0) \in < NEG(A_1), NEG(A_2), ..., NEG(A_m), x_1^p - x_1, x_2^p - x_2, ..., x_n^p - x_n) >$$

which can be denoted as

$$NEG(A_0) \in J + I \ .$$

Particularly interesting is the application of this theorem to a type of RAS whose relations are reflexive and transitive: RBES represented in multi-valued and modal logic. In this case the formulae $A_1, A_2, ..., A_m$, are the production rules and facts (and possibly any other information as integrity constraints).

We have been able to apply the theorem to verification and extraction of new knowledge in RBES containing up to 200 propositional variables under bi-valued logic and up to 150 propositional variables under three-valued and modal logic. The process of verification of consistency of a certain ideal takes a few seconds. The same holds for the process of extraction of consequences. Next we give an outline of a particular example of RAS-RBES dealing with what is known in Medicine as "appropriateness criteria".

2.1 Table Description

A set of medical data -effort test proof, one, two, or three blood vessels diseased, LVEF (Left Ventricle Ejection Fraction) value- were presented to a panel of ten experts on coronary diseases. They were asked about the appropriateness of performing or not the action of revascularization, which could have been done by means of two techniques: PTCA (Percutaneous Transluminal Coronary Angioplasty) and CABG (Coronary Artery Bypass Grafting).

The experts' opinions were set out in a table containing 260 information items. Items 1 to 24 are transcribed afterwards as illustration.

```
(The patients are asymptomatic)
1. Effort test positive.
1.1. Left common trunk diseased
1.1.1. Surgical risk low-moderate
```

% LVEF (F)	Revascularization	PTCA and CABG
$F > 50$	1:1 2 3 4 5 6 7 $8^1 *^9 +A$	2:1 2 3 4 5 6 7 $8^1 *^9 - + A$
$50 \geq F > 30$	5:1 2 3 4 5 6 7 $8 *^{10} +A$	6:1 2 3 4 5 6 7 $8\ ^1 *^9 - + A$
$30 \geq F \geq 20$	9:1 2 3 4 5 6 7 $8^1 *^9 +A$	10:1 2 3 4 5 6 $7^1 8^1 *^8 - + A$

1.1.2. Surgical risk high.

% LVEF (F)	Revascularization	PTCA and CABG
$F > 50$	3:1 2 3 4 5 6 7 $8^4 *^6$ $+A$	4:1 2 3 4 $5^1 6$ $7^1 8^1$ $*^7$ $-+A$
$50 \geq F > 30$	7:1 2 3 4 5 6 7 $8^3 *^7$ $+A$	8:1 2 3 4 5 $6^1 7^1 8$ $^2 *^6$ $-+A$
$30 \geq F \geq 20$	11:1 2 3 4 5 6 $7^1 8^2$ $*^7$ $+A$	12:1 2 3 4 5 $6^2 7$ 8^2 $*^6$ $-+A$

1.2. Three blood vessels diseased

1.2.1. Surgical risk low-moderate

% LVEF (F)	Revascularization	PTCA and CABG
$F > 50$	13:1 2 3 4 $5^1 6^1 7^1 8^2 * 9^5 + A$	14:1 2 3 4 $5^1 6^1 7^1 8^2 * 9^5 ? + A$
$50 \geq F > 30$	17:1 2 3 4 5 6 $7^1 8^2$ $*^7$ $+A$	18:1 2 3 4 $5^1 6$ $7^2 8^2 * 9^5 ? + A$
$30 \geq F \geq 20$	21:1 2 3 $4^1 5$ 6 7 8^2 $*^7$ $+A$	22:1 2 3 4 5^1 $6^1 7^2$ $*^2 9^4 ? + A$

1.2.2. Surgical risk high.

% LVEF (F)	Revascularization	PTCA and CABG
$F > 50$	15:$1^1 2$ 3 4 $5^1 6$ 7^2 $*^2 9^4 + A$	16:1 2 3^3 4^2 5 $*7$ $8^1 9^4$ $++I$
$50 \geq F > 30$	19:1 2 3 4 5 6 $7^1 8^3$ $*^6$ $+A$	20:1 2 3^3 4^2 5 $*$ 7 8 $^2 9^3$ $++D$
$30 \geq F \geq 20$	23:1 2 3 $4^1 5$ 6 7 8^3 $*^6$ $+A$	24:1 $2^2 3^1 4^1 5^1 *$ $6^1 7^1 8^1 9^2$ $++D$

A bold-type digit (**1, 2**,..., **260**) is assigned to each row of digits, symbols $+, -, *$, and letters A, D, I of the table. The rules R1, R2,..., R260 of the RBES that translates the table, are numbered according to the digits mentioned above.

The experts were informed, for instance, in the six cases **1, 2, 5, 6, 9, 10** that, for a certain patient, the data were: effort test positive, suffers from left common trunk disease, surgical risk is low/moderate and LVEF is in a given percentage bracket.

Once a set of data is given, the panelists are asked about the appropriateness of a therapeutic action: revascularization, which can be performed by means of two different techniques, PTCA and CABG. The panelists' response is transcribed as a row of digits with or without superscripts, symbols $+, -, ?, *$, and letters A, D, I.

The superscripts express the number of experts that have assigned a value from 1 to 9 to the appropriateness of an action. "$*$" stands for the median.

The symbols $+$, $-$ (and ? used in other parts of the table) respectively mean that an action is, respectively, appropriate, inappropriate (and of undecided appropriateness). If the panelists refer to revascularization, only one of these symbols is transcribed. PTCA and CABG are referred to simultaneously by a pair of these symbols.

The letter A at the end of a row means "agreement" about appropriateness, inappropriateness or undecided appropriateness. D means "disagreement"and I (which appears in other parts of the table) means "undecided agreement". There is disagreement D if the ratings of three panelists rank between 1 and 3, and the ratings of another three rank between 7 and 9. There is agreement when there are no more than two opinions outside an interval of *[1,3]*, *[4,6]* or *[7,9]* that contains the median. In any other case, there is undecided agreement I (i. e. big standard deviation).

2.2 Translation of the Tables into a Set of Production Rules

The tables can be translated into a set of production rules as follows.

- Assign a propositional variable, denoted $x[i]$ ($i = 1, ..., 9, 13$), $y[1], y[2]$ (possibly preceded by a permissible combination of symbols $\neg$, $\Box$, $\Diamond$), to each symptom and therapeutic action. Let us refer as illustration to a few of them, for instance:
  - Surgical risk high: $\neg x[1]$
  - Effort test positive: $x[2]$
  - ...
  - Effort test not done or not decisive: $\Diamond\neg x[2]$
  - LVEF > 50%: $x[7]$
  - ...
  - Left common trunk disease: $x[3]$
  - ...
  - Anterior proximal descendent affected: $x[13]$
  - ...
  - PTCA: $y[1]$
  - CABG: $y[2]$

  The relation between a given set of symptoms and the respective experts' response can be reinterpreted as a production rule subject to the following convention.

- In production rules based on the three-valued and modal logic here illustrated, the propositional variables $x[i]$ ($i = 1, .., 9$), $x[13]$, $y[1]$ and $y[2]$ can be preceded by $\neg$, or/and the modal logic connectives $\Box$, $\Diamond$ or by any permissible combinations of these symbols. For reasons given in [5], the symbols $\Box$ and $\Diamond$ do not precede, except for the case "effort test not done or not decisive" (represented as said above as $\Diamond\neg x[2]$), the variables $x[i]$, $i = 1, .., 9$ and $x[13]$ or negations of variables that represent symptoms.

The relationship between the symptoms and the respective list of digits and symbols in **6**, for instance, can be reinterpreted as the statement: "*IF* the effort test of a patient is positive *AND* he/she suffers from left common trunk disease *AND* his/her surgical risk is low/moderate *AND* his/her LVEF is less-equal 50% and over 30%, *THEN* the experts have assessed that PTCA is inappropriate but CABG is appropriate"; moreover, "there is agreement (A) on this assessment". These "*IF-THEN*" assertions can be translated into production rules as follows.

- We write the symbol $\wedge$ in the conclusion when at the end of the row of digits the pair of symbols composed of $+$, $-$ and ? are $-+$, $+-$, $--$, $-?$, $?-$, $+?$ and $?+$ (see item **4** below for instance). We write the symbol $\vee$ when the above pair are $++$ (as in item **20** below) and ??. The reason for choosing these translations are given in [5].

- The different degrees of appropriateness are translated as follows (for instance, if the experts assess that a therapeutic action y, is not appropriate under agreement, $y\ -A$, we propose the translation, "necessarily not y": $\Box\neg y$) :

$$\begin{array}{l} y\ -A \leftrightarrow \Box\neg y \\ y\ -I \leftrightarrow \neg y \\ y\ -D \leftrightarrow \Diamond\neg y \\ y\ ? \leftrightarrow \mathit{true\ or\ false},\ (\mathit{see\ below}) \\ y\ +D \leftrightarrow \Diamond y \\ y\ +I \leftrightarrow y \\ y\ +A \leftrightarrow \Box y \end{array}$$

We translate the information items that contain "?" as follows. If the consequent refers to both PTCA and CABG, "?" is translated as "true" if the consequent is a conjunction and as false if it is a disjunction, except in the case where two "?"s appear, in which case the whole consequent is translated as "true". The reason is that in this last case, the whole rule becomes a trivially true implication, whereas in the first and second cases, the consequent becomes PTCA and CABG whichever is not marked with "?".

For example, information item **4** is translated into a production rule as follows.

R4: $\neg x[1] \wedge x[2] \wedge x[3] \wedge x[7] \rightarrow \Box\neg y[1] \wedge \Box y[2]$).

Information **40** is translated as:

R40: $\neg x[1] \wedge x[2] \wedge x[5] \wedge \neg x[13] \wedge x[7] \rightarrow \mathit{true} \wedge \Box\neg y[2]$.

An example of a rule that expresses disagreement (using $\Diamond$), is the one that translates item **20**:

R20: $\neg x[1] \wedge x[2] \wedge x[4] \wedge x[8] \rightarrow \Diamond y[1] \vee \Diamond y[2]$.

Note that $\rightarrow$ is a reflexive and transitive relation. If denoted **R**, production rule R4 can be written as follows

$$(\neg x[1], x[2], x[3], x[7])\mathbf{R}(\Box\neg y[1], \Box y[2]) \ .$$

The set U is a set of tuples of symptoms that correspond to a set of classes of patients. The set of values V_a is a set of ordered pairs of assessments of degrees of appropriateness $y[...]$. The relation admits a representation using three-valued and modal logic operators. The symbol $\vee$ is sometimes required, as in the relation that would correspond to R20.

2.3 CoCoA Implementation

The method follows the next three steps.

(i) To traduce IF-THEN statements into production rules. This has been already illustrated above.

(ii) To traduce production rules to polynomials. The theoretical process is described in [3,6,7]. We just give the results for three-valued logic with modal operators next:

$\neg x$ (NOT-x) is translated as $2 + 2x$
$\Diamond x$ (POSSIBLY-x) is translated as $2x^2$
$\Box x$ (NECESSARILY-x) is translated as $x^2 + 2x$
$x \vee y$ (x-OR-y) is translated as $x^2y^2 + x^2y + xy^2 + 2xy + x + y$
$x \wedge y$ (x-AND-y) is translated as $2x^2y^2 + 2x^2y + 2xy^2 + xy$
$x \to y$ (IF x-THEN-y) is translated as $x^2y^2 + x^2y + xy^2 + 2x + 2$.

(iii) To apply Theorem 1. As advanced above, this theorem allows to perform the following two important steps. The foundations are in [1,10] and the tool used is the computer algebra language CoCoA [1][2].

(1) To check automatically that the given set of logical formulae does not lead to inconsistencies.
(2) To extract new information from the information contained in the mentioned set.

Step (1) requires just to type a command (Gröbner basis) in CoCoA .

```
GBasis(I + J);
```

If the output is [1], there is inconsistency. The inconsistencies can be

- purely logical, as when one obtains, say, $\Box y[2] \wedge \Box\neg y[2]$ ("necessarily $y[2]$" and "necessarily-not $y[2]$") from firing R4 and R40
- due to having found in the process of obtaining inferences, say some $z[9]$ and some $w[27]$, being these two pieces of information incompatible because the experts have determined so. This incompatibility is called an integrity constraint.

Step (2) requires the use of Normal Forms. For instance, in order to check that, say, a propositional variable $v[6]$ (or some formula, say $x[8] \to \Box\neg y[1]$) follows from the RBES, it is enough to type, respectively:

```
NF(NEG(v[6]), J + I);
```

```
NF(NEG(IMP(x[8], NEC(NEG(y[1])))), J + I);
```

Depending on whether the output is 0 or not, $v[6]$ (resp. $x[8] \to \Box\neg y[1]$), follows or does not follow from the RBES.

We have also implemented a program that exactly locates the inconsistencies.

[1] CoCoA, a system for doing Computations in Commutative Algebra. Authors: A. Capani, G. Niesi, L. Robbiano. Available via anonimous ftp from: `cocoa.dima.unige.it`

3 Railway Interlocking System

Clearance can be given simultaneously to more than one train. This action should not allow two trains to collide at any point (this is not a scheduling problem but a problem of compatibility of permissions). This is not a trivial problem if the network is not very small (Fig. 1). We shall consider that all trains are allowed to move at the same time in any direction at any speed, unless there is a signal forbidding the movement.

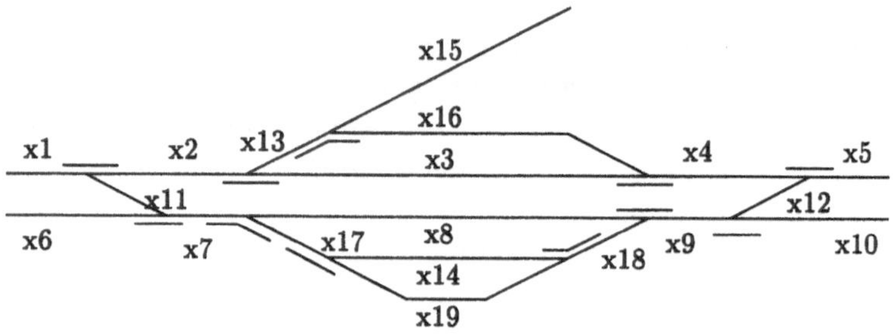

Fig. 1. Example of small station

The way the safety of the logical problem is treated is very similar to that used above to check consistency of RBES. In both cases what has to be checked is the degeneracy of an ideal of a polynomial residue class ring into the whole ring (what is done using Gröbner bases). In this case the steps are

$$\text{railway situation} \rightarrow \text{graph} \rightarrow \text{(algebraic) decision model} .$$

Let us detail the model.

Four oriented graphs (GD, GS, G^*, G) are considered. In all four cases the vertices of the graph are the sections of the line, but the edges will be different, as explained below.

Graph GD corresponds to the layout (and the position of the switches of the turnouts). There is an edge connecting section xi and section xj iff one of the following conditions holds:

- sections xi and xj are consecutive in the line (i.e., they are two consecutive sections of a block-system)
- there is a turnout connecting sections xi and xj and the switch is in the position that connects sections xi and xj
- there is a turnout connecting sections xi and xj and the switch is in the position such that it is possible to pass from section xi to section xj trailing through this switch set against.

Graph GS corresponds to semaphores. There is an edge connecting section xi with section xj iff there is a semaphore controlling the pass from section xi to section xj and it forbids such movement.

Therefore, a next section is accessible from another iff the layout (and the position of the switches of the turnouts) make it possible and the semaphores do not forbid it, i.e., iff there is an edge connecting them in $G^* = GD - GS$. General accessibility is given by the transitive closure of the graph G^* above, to be denoted G.

Let us denote the sections by $x_1, x_2, ..., x_n$ (polynomial variables). The graph G will be interpreted as a polynomial ideal

$$I \subseteq \mathbb{Q}[x_1, x_2, ..., x_n]$$

that is initialized as $\{0\}$. That it is possible to move from section x_i to a (next) section x_j will be represented by including the polynomial

$$x_i \cdot (x_i - x_j)$$

in the ideal I.

Trains will be denoted by positive integers. If train α is in section x_i, $x_i - \alpha$ will be added to the set where the position of trains is stored (PT).

Theorem 2. *Safety: A situation of the switches and signals given by the ideal I and a position of trains given by the ideal $<PT>$ is safe iff*

$$I + <PT> \neq <1>$$

(what can be checked using Gröbner Bases).

The idea in the background is that if, for example, train 11 is in section x_3 and trains can move from section x_3 to section x_4, then $x_3 - 11 = 0$ and $x_3 \cdot (x_3 - x_4) = 0$. From both equations, $x_4 - 11 = 0$, that is, section x_4 is reachable by train 11.

The whole package is implemented in the computer algebra language Maple [8] (the implementation is similar to that of section 3.3).

4 General Reflexive and Transitive Relations

Both models can be used to perform computations in the reflexive and transitive closure of general relations.

4.1 First Approach

The reflexive and transitive closure of a relation $\mathcal{R}$, defined in a set C, can be approached through a polynomial model, the same way as a Boolean logic RBES was simulated using Theorem 1. Note that it is enough:

- to treat the elements of C as the propositional variables of the RBES
- to identify *to be related to* in C with the *if-then* statements of the RBES.

Then the reflexive and transitive closure of $\mathcal{R}$ will behave exactly the same way as *to be a tautological consequence* in the RBES (although $\wedge$ and $\vee$ do not have any use or translation).

Moreover, in case the relation is an equivalence one, the quotient Boolean algebra can be identified with the corresponding (propositional) Lindenbaum algebra.

4.2 Second Approach

The reflexive and transitive closure of a relation $\mathcal{R}$, defined in a set C, can also be approached through a polynomial model the same way as the railway interlocking system of the previous section. In this case it is enough:

- to treat the elements of C as the sections of the line (identifying them with polynomial variables)
- to identify *to be related to* in C with the edges of the main (and now only) oriented graph (G), by including the adequate $x_i \cdot (x_i - x_j)$ polynomial in the corresponding ideal.

Then reflexive and transitive closure of $\mathcal{R}$ will behave exactly the same way as *to be accesible.* For instance, to check if $a, b \in C$ are such that a is related to b in the reflexive and transitive closure of $\mathcal{R}$, it is enough to add $a - 1$ to the ideal corresponding to G and to check whether $b - 1$ is also in it or not.

5 Conclusions

It has been illustrated, by using two examples, that Gröbner Bases can be applied to the study of consistence of sets of reflexive and transitive relations. Automated extraction of knowledge, using Normal Forms can be performed in these sets. This information is related to appropriateness criteria in medicine in the first example, and to transit security in railway stations.

Gröbner Bases can be applied to any equivalence and partial order relations.

In the case of RBES considered as RAS, it can be possible to extend the idea to the multi-valued information systems of [4] by using production rules as, say:

$$x[1] \wedge x[2] \wedge x[3] \wedge ... \wedge \ x[k] \rightarrow v[1] \wedge ... \wedge v[l]$$

(allowing also the connective $\vee$ when necessary), where U is a set of tuples of single facts (that correspond to a class of objects, say sick persons) and $\{v[1], ..., v[l]\} \in 2^{V_a}$.

The symbols for necessity and possibility can be used to translate the relations I and B.

References

1. V. Adams and P. Loustaunau, *An Introduction to Gröbner Bases.* Graduate Studies in Mathematics 3. American Mathematical Society Press, Providence, RI (1994).
2. A. Capani, and G. Niesi, *CoCoA User's Manual (v. 3.0b)*, Dept. of Mathematics University of Genova (1996).
3. J. Chazarain, A. Riscos, J. A. Alonso and E. Briales, Multivalued Logic and Gröbner Bases with Applications to Modal Logic. *Journal of Symbolic Computation,* **11** (1991) 181-194.
4. I. Düntsch, G. Gediga, E. Orlowska, *Relational attribute systems.* Preprint.
5. L. M. Laita, E. Roanes-Lozano, V. Maojo, L. de Ledesma, L. Laita, An Expert System for Managing Medical Appropriateness Criteria based on Computer Algebra Techniques. *Computers and Mathematics with Applications,* **42/12** (2001) 1505-1522.
6. L. M. Laita, E. Roanes-Lozano, L. de Ledesma and J. A. Alonso, A Computer Algebra Approach to Verification and Deduction in Many-Valued Knowledge Systems, *Soft Computing* **3-1** (1999) 7-19.
7. E. Roanes-Lozano, L. M. Laita and E. Roanes-Macías, A Polynomial Model for Multivalued Logics with a Touch of Algebraic Geometry and Computer Algebra, *Mathematics and Computers in Simulation* **45/1** (1998) 83-99.
8. E. Roanes-Macías and E. Roanes-Lozano, *Cálculos matemáticos por ordenador con Maple V.5.* Ed. RubiñOS 1860, Madrid (1999).
9. E. Roanes-Lozano, E. Roanes-Macías, L.M. Laita, Railway Interlocking Systems and Gröbner Bases, *Mathematics and Computers in Simulation* **51/5** (2000) 473-482.
10. F. Winkler, *Polynomial Algorithms in Computer Algebra.* Springer, Wien (1996).

Fuzzy Relational Images in Computer Science

Mike Nachtegael, Martine De Cock, Dietrich Van der Weken, and Etienne E. Kerre

Fuzziness and Uncertainty Modelling Research Unit
Department of Applied Mathematics and Computer Science
Ghent University, Krijgslaan 281 (S9), 9000 Gent, Belgium
{mike.nachtegael,martine.decock}@rug.ac.be,
{dietrich.vanderweken,etienne.kerre}@rug.ac.be,
http://fuzzy.rug.ac.be

Abstract. Relations appear in many fields of mathematics and computer science. In classical mathematics these relations are usually crisp, i.e. two objects are related or they are not. However, many relations in real-world applications are intrinsically fuzzy, i.e. objects can be related to each other to a certain degree.
With each fuzzy relation, different kinds of fuzzy relational images can be associated, all with a very practical interpretation in a wide range of application areas. In this paper we will explicite the formal link between well known direct and inverse images of fuzzy sets under fuzzy relations on one hand, and different kinds of compositions of fuzzy relations on the other. Continuing from this point of view we are also able to define a new scale of so-called double images. The wide applicability in mathematics and computer science of all these fuzzy relational images is illustrated with several examples.

1 Introduction

For many relations in the real world it is very difficult to obtain a justifiable crisp partition of the universe into those objects satisfying the relationship and those who do not. For example [18] suppose we have some set of patients $P = \{p_1, p_2, ..., p_n\}$ and some set of symptoms $S = \{s_1, s_2, ..., s_m\}$. The first step in a diagnosis problem consists in identifying the set of symptoms related to each patient, i.e. in establishing a relation R from P to S. One of the symptoms may be for example fever (denoted f). Classically only two situations may appear: a given patient p has fever f or not, depending on the temperature of his body. This presupposes that the concept of fever has been defined in a very crisp way, for example: p has f if and only if temperature(p) ≥ 38, and hence

$$(p, f) \in R \text{ if and only if temperature}(p) \geq 38$$
$$(p, f) \notin R \text{ if and only if temperature}(p) < 38$$

Using the same symbol R to denote the characteristic mapping of R, we can write $(p, f) \in R$ as $R(p, f) = 1$, while $(p, f) \notin R$ can be written as $R(p, f) = 0$. Now we know that there is a great difference between a body temperature of

H. de Swart (Ed.): RelMiCS 2001, LNCS 2561, pp. 134–151, 2002.

38.2°C and one of 41°C. Nevertheless both situations are handled in the same way according to the model above. This unrealistic black-or-white distinction can be made more realistic by introducing a continuous scale for degrees of relationship. We may define for instance $R(p, f) = A(\text{temperature}(p))$ with

$$A(x) = \begin{cases} 0 & \text{if } x < 37 \\ 2\frac{(x-37)^2}{9} & \text{if } 37 \leq x \leq 38.5 \\ 1 - 2\frac{(x-40)^2}{9} & \text{if } 38.5 \leq x \leq 40 \\ 1 & \text{if } 40 < x \end{cases} \tag{1}$$

for all x in the universe of temperatures T.

I.e. we may define R as a $P \times S - [0, 1]$ mapping that associates with every patient-symptom couple (p, s) a number between 0 and 1, interpreted as the degree to which p and s are related. Such a mapping is called a fuzzy relation from P to S. The mapping A defined in (1) associates with every temperature x the degree between 0 and 1 to which x is considered to be "feverous". Therefore this mapping is called the membership function of the fuzzy set "feverous".

Like relations are fundamental tools in classical set theory (also called *crisp* set theory), fuzzy relations are a basic concept of fuzzy set theory [28]. They appear in a wide range of fuzzy systems: for instance various kinds of fuzzy relations are used in fuzzy relational database systems (see e.g. [4], [27]); fuzzy tolerance relations, similarity relations and document description relations are used in information retrieval systems (see e.g. [5], [32]); preference relations are used in multicriteria decision support systems (see e.g. [14]); in approximate reasoning systems IF-THEN rules are modelled by means of fuzzy relations (see e.g. [24]), etc.

Since fuzzy relations are so widely used it is not surprising that most tutorials on fuzzy set theory include a definition of the concept of a fuzzy relation as well as of some kind of composition of fuzzy relations (see e.g. [16], [18], [21], [33]). For instance if X, Y and Z are three universes, the sup-min composition of a fuzzy relation R_1 from X to Y, and a fuzzy relation R_2 from Y to Z is a fuzzy relation from X to Z denoted by $R_1 \circ R_2$ and defined by

$$(R_1 \circ R_2)(x, z) = \sup_{y \in Y} \min(R_1(x, y), R_2(y, z)), \tag{2}$$

for all (x, z) in $X \times Z$. However only few textbooks also mention some concept of image of a fuzzy set under a fuzzy relation [16], [18]. Since, as we will show, fuzzy relational images can and are used all over the place, it is quite remarkable that they are not more formalized and studied as a concept an sich. One of the oldest applications of fuzzy relational images is the compositional rule of inference [29], [30]. If A is a fuzzy set in X, R is a fuzzy relation from X to Y, and v_1 and v_2 are variables on X and Y respectively, this rule dictates

$$\begin{array}{ll} v_1 \text{ is } A & (*) \\ (v_1, v_2) \text{ is } R & (**) \\ \hline v_2 \text{ is } B & (***) \end{array}$$

i.e. from fact (*) and fact (**) we can derive fact (***) in which B is the fuzzy set in Y defined by

$$B(y) = \sup_{x \in X} \min(A(x), R(x, y)), \tag{3}$$

for all y in Y. In [29], [30] B is quite ad hoc called "the composition of a unary fuzzy relation A and a binary fuzzy relation R", somehow referring to (2). But rather than some composition, formula (3) is a straightforward generalization of the classical direct image $R(A)$ of a crisp set A under a crisp relation R which is defined by

$$R(A) = \{y \in Y | (\exists x \in X)(x \in A \wedge (x, y) \in R)\}$$

Although meanwhile a fuzzification of this image has been realized (see [6], [16], [17]), in recent works regarding approximate reasoning many authors (e.g. [1], [26], [31], [33]) still hang on to the compositional notation $A \circ R$ introduced in [29], [30].

In this paper we will explicite a mathematically correct connection between the direct image of A under R and the composition of fuzzy relations. It is inspired by an idea proposed in [15] on one hand, and the concept of cylindrical extension [30] on the other. We will do the same exercise for three other kinds of direct images, as well as for four kinds of inverse images. Continuing in the same line, we will also introduce several new kinds of double images of fuzzy sets under fuzzy relations. We will show that all these images are very powerful tools that can be used in many applications of computer science. The study of their definition and properties is therefore more than worth the while.

This paper is structured as follows: in Section 2 we recall the definition of four direct and four inverse fuzzy relational images and some of their properties. In Section 3 we recall the definition of four kinds of fuzzy relational compositions and we are able to draw the mathematically correct link with the images of Section 2 due to the newly introduced concepts left and right extension of a fuzzy set. These extensions will also allow us to define several kinds of double images in Section 4. We end with examples from application fields in Section 5.

2 Fuzzy Relational Images

Throughout this paper X, Y and Z denote universes of discourse, i.e. classical (or "crisp") sets containing the objects we want to say something about. In real-world applications, relations between sets of 'objects' are not always crisp by nature. For example, as clearly illustrated in the introduction of this paper, the patient-fever relation is an intrinsically vague relation. In order to cope with the vagueness involved in these kind of relations, the notion of fuzzy relations has been introduced, as a special kind of fuzzy set [28].

Definition 1 (Fuzzy set). *A fuzzy set A in a universe X is characterized by a $X - [0, 1]$ mapping, also denoted by A, that associates with every element x in X the degree to which x belongs to A. $A(x)$ is therefore called the membership* *degree of x in the fuzzy set A. The class of all fuzzy sets in X is denoted by $\mathcal{F}(X)$.*

Given two fuzzy sets A and B in X, their (Zadeh)-union $A \cup B$ and (Zadeh)-intersection $A \cap B$ are fuzzy sets in X, defined by, for all x in X:

$$(A \cup B)(x) = \max(A(x), B(x))$$
$$(A \cap B)(x) = \min(A(x), B(x)).$$

The algebraic structure $(\mathcal{F}(X), \cap, \cup)$ is a lattice with the ordering defined by

$$A \subseteq B \text{ if and only if } (\forall x \in X)(A(x) \leq B(x))$$

for all A and B in $\mathcal{F}(X)$. A fuzzy set A is normal if there exists an x in X such that $A(x) = 1$. A fuzzy set in which 0 and 1 are the only membership values, is called a crisp set or a classical set.

Definition 2 (Fuzzy relation). *A fuzzy relation R from X to Y is a fuzzy set in $X \times Y$. It is characterized by a $X \times Y - [0,1]$ mapping, also denoted by R, that associates with every pair (x, y) in $X \times Y$ the degree to which x and y are related by R. If $X = Y$, then R is called a binary fuzzy relation in X.*

The inverse fuzzy relation of R is the fuzzy relation R^{-1} from Y to X defined by $R^{-1}(y, x) = R(x, y)$ for all x in X and y in Y.

Definition 3 (Foreset, afterset). *[3] Let R be a fuzzy relation from X to Y. For all x in X, the R-afterset xR of x is the fuzzy set on Y defined as $xR(y) = R(x, y)$, for all y in Y. Likewise, for all y in Y, the R-foreset Ry of y is the fuzzy set on X defined as $Ry(x) = R(x, y)$, for all x in X.*

An important role in the definitions of fuzzy relational images is played by logical operators on $[0,1]$ which are an extension of logical conjunction and implication on $\{0,1\}$.

Definition 4 (Conjunctor). *A conjunctor on $[0,1]$ is a $[0,1] \times [0,1] - [0,1]$ mapping $\mathcal{T}$ that satisfies $\mathcal{T}(0,0) = \mathcal{T}(0,1) = \mathcal{T}(1,0) = 0$ and $\mathcal{T}(1,1) = 1$, and that is increasing in both arguments. It is called a semi-norm if $\mathcal{T}(a,1) = \mathcal{T}(1,a) = a$, for all a in $[0,1]$. A semi-norm is called a* **t-norm** *if it is also commutative and associative.*

The minimum $\mathcal{T}_M$, the algebraic product $\mathcal{T}_P$ and the Łukasiewicz t-norm $\mathcal{T}_W$ (also called bounded sum) are very popular t-norms. They are given by $\mathcal{T}_M(a,b) = \min(a,b)$, $\mathcal{T}_P(a,b) = a \cdot b$ and $\mathcal{T}_W(a,b) = \max(0, a+b-1)$, respectively.

Definition 5 (Implicator). *An implicator on $[0,1]$ is a $[0,1] \times [0,1] - [0,1]$ mapping $\mathcal{I}$ that satisfies $\mathcal{I}(0,0) = \mathcal{I}(0,1) = \mathcal{I}(1,1) = 1$ and $\mathcal{I}(1,0) = 0$, and that is decreasing in the first and increasing in the second argument. It is called a border implicator if $\mathcal{I}(1,a) = a$, for all a in $[0,1]$.*

The Łukasiewicz implicator $\mathcal{I}_L$, the Kleene-Dienes implicator $\mathcal{I}_{KD}$ and the Reichenbach implicator $\mathcal{I}_R$ are among the most popular border implicators. They

are respectively given by $\mathcal{I}_L(a,b) = \min(1, 1-a+b)$, $\mathcal{I}_{KD}(a,b) = \max(1-a,b)$ and $\mathcal{I}_R(a,b) = 1-a+a\cdot b$.
The first condition in the definitions of conjunctors and implicators guarantees that these operators are extensions of their classical counterparts. The monotonicity conditions guarantee that the operators have a logical interpretation. Using a conjunctor $\mathcal{T}$, the notions of reflexivity, symmetry and $\mathcal{T}$-transitivity can be defined for binary fuzzy relations in the following way:

Definition 6. *Let R be a binary fuzzy relation in X. R is called*

- *reflexive iff $R(x,x) = 1$, for all x in X*
- *symmetric iff $R(x_1,x_2) = R(x_2,x_1)$, for all x_1 and x_2 in X*
- *sup-$\mathcal{T}$ transitive iff $R(x_1,x_2) \geq \sup\limits_{z\in X} \mathcal{T}(R(x_1,z), R(z,x_2))$, for all x_1 and x_2 in X*

Definition 7 (Degree of overlap). *Let $\mathcal{T}$ be a conjunctor. For two fuzzy sets A and B in X, the degree of $\mathcal{T}$-overlap of A and B is given by*

$$\sup_{x\in X} \mathcal{T}(A(x), B(x))$$

Definition 8 (Degree of inclusion). *[2] Let $\mathcal{I}$ be an implicator. For two fuzzy sets A and B in X, the degree of $\mathcal{I}$-inclusion of A in B is given by*

$$\inf_{x\in X} \mathcal{I}(A(x), B(x))$$

Definition 9 (Degree of equality). *[2] Let $\mathcal{T}$ be a conjunctor and $\mathcal{I}$ an implicator. For two fuzzy sets A and B in X, the degree of $(\mathcal{T},\mathcal{I})$-equality of A and B is given by*

$$\mathcal{T}(\inf_{x\in X} \mathcal{I}(A(x), B(x)), \inf_{x\in X} \mathcal{I}(B(x), A(x)))$$

If R is a crisp relation from X to Y, it can map a crisp subset A of X onto a crisp subset B of Y in many different ways. Interesting ones are:

- The direct image of A under R, which is the set of objects in Y that are related to at least one object of A.
- The subdirect image of A under R, which is the set of objects in Y that are related to all objects of A.
- The superdirect image of A under R, which is the set of objects in Y that are related only to objects of A.
- The square direct image of A under R, which is the set of objects in Y that are related to all and only to objects of A. In other words it is the set of objects typically related to A.

Fuzzification of these images leads to the following definition [6], [17]:

Definition 10 (Direct images). *For $\mathcal{T}$ a conjunctor, $\mathcal{I}$ an implicator, R a fuzzy relation from X to Y, and A a fuzzy set in X, the direct image $R_{\mathcal{T}}(A)$, the subdirect image $R_{\mathcal{I}}^{\triangleleft}(A)$, the superdirect image $R_{\mathcal{I}}^{\triangleright}(A)$ and the square direct image $R_{\mathcal{T},\mathcal{I}}^{\diamond}(A)$ of A under R are the fuzzy sets in Y defined by, for all y in Y,*

- $R_{\mathcal{T}}(A)(y) = \sup\limits_{x\in X} \mathcal{T}(A(x), R(x,y))$ *(direct image)*
- $R_{\mathcal{I}}^{\triangleleft}(A)(y) = \inf\limits_{x\in X} \mathcal{I}(A(x), R(x,y))$ *(subdirect image)*
- $R_{\mathcal{I}}^{\triangleright}(A)(y) = \inf\limits_{x\in X} \mathcal{I}(R(x,y), A(x))$ *(superdirect image)*
- $R_{\mathcal{T},\mathcal{I}}^{\diamond}(A)(y) = \mathcal{T}(R_{\mathcal{I}}^{\triangleleft}(A)(y), R_{\mathcal{I}}^{\triangleright}(A)(y))$ *(square direct image)*

That Definition 10 corresponds to the semantics of the images described above becomes more clear when viewing $R(x,y)$ as $Ry(x)$, i.e. the degree to which x belongs to the R-foreset of y. Indeed Ry is actually the fuzzy set of objects related to y. So when determining the degree to which y belongs to the direct image of A under R, i.e. the degree to which y is related to at least one object of A, one looks at the degree of overlap of A and Ry. Likewise when determining the degree to which y belongs to the subdirect image of A under R, i.e. the degree to which y is related to all objects of A, one looks at the degree to which A is included in Ry etc.

Note that it is a changement in the position of the arguments of $\mathcal{I}$ in the definition of subdirect image which gives rise to that of superdirect image. A similar thing can be done for the direct image, leading to another image $R'_{\mathcal{T}}(A)$ defined by

$$R'_{\mathcal{T}}(A)(y) = \sup_{x\in X} \mathcal{T}(R(x,y), A(x))$$

for all y in Y. This image was included in the study in [11]. Most common conjunctors $\mathcal{T}$ are however commutative. One can easily verify that for these kind of conjunctors $R_{\mathcal{T}}(A) = R'_{\mathcal{T}}(A)$.

If R is a relation from X to Y and B is a set in Y it also makes sense to talk about the set of objects of X related to at least one object of B, related to all objects of B, only related to objects of B, etc. This gives rise to the definition of several inverse images [6].

Definition 11 (Inverse images). *For $\mathcal{T}$ a conjunctor, $\mathcal{I}$ an implicator, R a fuzzy relation from X to Y, and B a fuzzy set in Y, we say that*

- $(R^{-1})_{\mathcal{T}}(B)$ *is the inverse image of B under R*
- $(R^{-1})_{\mathcal{I}}^{\triangleleft}(B)$ *is the subinverse image of B under R*
- $(R^{-1})_{\mathcal{I}}^{\triangleright}(B)$ *is the superinverse image of B under R*
- $(R^{-1})_{\mathcal{T},\mathcal{I}}^{\diamond}(B)$ *is the square inverse image of B under R*

The above definitions have a certain degree of freedom: for every choice of the conjunctor $\mathcal{T}$ and the implicator $\mathcal{I}$, a fuzzy relational image is obtained. This freedom of choice is reflected in the notation. In theoretical studies one can focus on the direct fuzzy relational images, since the inverse images w.r.t. R can be regarded as direct images w.r.t. R^{-1}.

The fuzzy relational images defined in Definition 10 and 11 have an important characteristic which was pointed out in [6], namely the lack of non-emptiness conditions. Indeed if the fuzzy set A is almost empty, i.e. all the membership degrees in A are small, then $R_{\mathcal{I}}^{\triangleleft}(A)(y)$ tends to be high. In the extreme case when $A = \emptyset$, the subdirect image of A under every imaginable fuzzy relation

R will be the universe for most common implicators $\mathcal{I}$. A similar phenomenon occurs with the superdirect image, and of course then also with the square direct image. To avoid this the following images have been defined [6]:

Definition 12 (Kerre direct images). *For $\mathcal{T}$ a t-norm, $\mathcal{I}$ an implicator, R a fuzzy relation from X to Y, and A a fuzzy set in X, the Kerre subdirect image $R^{\triangleleft k}_{\mathcal{T},\mathcal{I}}(A)$, the Kerre superdirect image $R^{\triangleright k}_{\mathcal{T},\mathcal{I}}(A)$ and the Kerre square direct image $R^{\diamond k}_{\mathcal{T},\mathcal{I}}(A)$ of A under R are the fuzzy sets in Y defined by*

- $R^{\triangleleft k}_{\mathcal{T},\mathcal{I}}(A) = R^{\triangleleft}_{\mathcal{I}}(A) \cap R_{\mathcal{T}}(A)$ *(Kerre subdirect image)*
- $R^{\triangleright k}_{\mathcal{T},\mathcal{I}}(A) = R^{\triangleright}_{\mathcal{I}}(A) \cap R_{\mathcal{T}}(A)$ *(Kerre superdirect image)*
- $R^{\diamond k}_{\mathcal{T},\mathcal{I}}(A) = R^{\diamond}_{\mathcal{T},\mathcal{I}}(A) \cap R_{\mathcal{T}}(A)$ *(Kerre square direct image)*

The definition in [6] requires $\mathcal{T}$ to be a t-norm. If $\mathcal{T}$ would have been a conjunctor in general it would be more natural to involve the alternative direct image $R'_{\mathcal{T}}(A)$ defined above in the definition of the Kerre subdirect image and square direct image.

Proposition 1. *Let $\mathcal{T}$ be a t-norm and $\mathcal{I}$ a border implicator. Furthermore let R be a fuzzy relation from X to Y and A a fuzzy set in X.*

1. *If A is normal then $R^{\triangleleft k}_{\mathcal{T},\mathcal{I}}(A) = R^{\triangleleft}_{\mathcal{I}}(A)$.*
2. *If all R-foresets are normal then $R^{\triangleright k}_{\mathcal{T},\mathcal{I}}(A) = R^{\triangleright}_{\mathcal{I}}(A)$.*
3. *If A is normal and all R-foresets are normal then $R^{\diamond k}_{\mathcal{T},\mathcal{I}}(A) = R^{\diamond}_{\mathcal{T},\mathcal{I}}(A)$*

Proof. If A is normal there is an x_0 in X such that $A(x_0) = 1$. For all y in Y:

$$\begin{aligned} R^{\triangleleft}_{\mathcal{I}}(A)(y) &= \inf_{x\in X} \mathcal{I}(A(x), R(x,y)) \leq \mathcal{I}(A(x_0), R(x_0,y)) \\ &\leq \mathcal{T}(A(x_0), R(x_0,y)) \leq \sup_{x\in X} \mathcal{T}(A(x), R(x,y)) = R_{\mathcal{T}}(A)(y) \end{aligned}$$

Therefore $R^{\triangleleft}_{\mathcal{I}}(A) \subseteq R_{\mathcal{T}}(A)$ and hence 1. If all R-foresets are normal likewise we can prove that $R^{\triangleright}_{\mathcal{I}}(A) \subseteq R_{\mathcal{T}}(A)$ and hence 2. If both the conditions of 1 and 2 are fulfilled we have for all y in Y:

$$\mathcal{T}(R^{\triangleleft}_{\mathcal{I}}(A)(y), R^{\triangleright}_{\mathcal{I}}(A)(y)) \leq \min(R^{\triangleleft}_{\mathcal{I}}(A)(y), R^{\triangleright}_{\mathcal{I}}(A)(y)) \leq R_{\mathcal{T}}(A)(y)$$

and hence 3.

□

The conditions stated in Proposition 1 are rather weak. For example, the condition on R will be satisfied if R is a reflexive binary fuzzy relation in X, and most common conjunctors and implicators are indeed t-norms and border implicators. As we illustrate in Section 5, many applications live by the conditions of Proposition 1, so throughout the remainder of this paper we will focus on the fuzzy relational images defined in Definition 10. The properties of these images are studied in detail in [6], [11], [12]. We summarize some of the results with interesting interpretation for the applications.

Proposition 2 (Monotonicity). *Let A be a fuzzy set in X, R a fuzzy relation from X to Y, $\mathcal{T}$ a conjunctor and $\mathcal{I}$ an implicator.*

– *The direct and the square direct image are increasing w.r.t. the conjunctor. The subdirect, the superdirect and the square direct image are increasing w.r.t the implicator. I.e. if $\mathcal{T}_1$ and $\mathcal{T}_2$ are conjunctors such that $\mathcal{T}_1 \le \mathcal{T}_2$ and $\mathcal{I}_1$ and $\mathcal{I}_2$ are implicators such that $\mathcal{I}_1 \le \mathcal{I}_2$, then:*

$$\begin{aligned} R_{\mathcal{T}_1}(A) &\subseteq R_{\mathcal{T}_2}(A) \\ R^{\triangleleft}_{\mathcal{I}_1}(A) &\subseteq R^{\triangleleft}_{\mathcal{I}_2}(A) \\ R^{\triangleright}_{\mathcal{I}_1}(A) &\subseteq R^{\triangleright}_{\mathcal{I}_2}(A) \\ R^{\diamond}_{\mathcal{T}_1,\mathcal{I}}(A) &\subseteq R^{\diamond}_{\mathcal{T}_2,\mathcal{I}}(A) \\ R^{\diamond}_{\mathcal{T},\mathcal{I}_1}(A) &\subseteq R^{\diamond}_{\mathcal{T},\mathcal{I}_2}(A). \end{aligned}$$

– *The direct and the subdirect image are increasing w.r.t. the fuzzy relation, while the superdirect image is decreasing w.r.t. the fuzzy relation. I.e. if R_1 and R_2 are fuzzy relations from X to Y such that $R_1 \subseteq R_2$, then:*

$$\begin{aligned} (R_1)_{\mathcal{T}}(A) &\subseteq (R_2)_{\mathcal{T}}(A) \\ (R_1)^{\triangleleft}_{\mathcal{I}}(A) &\subseteq (R_2)^{\triangleleft}_{\mathcal{I}}(A) \\ (R_1)^{\triangleright}_{\mathcal{I}}(A) &\supseteq (R_2)^{\triangleright}_{\mathcal{I}}(A). \end{aligned}$$

– *The direct and the superdirect image are increasing w.r.t. the fuzzy set, while the subdirect image is decreasing w.r.t the fuzzy set. I.e. if A_1 and A_2 are fuzzy sets in X such that $A_1 \subseteq A_2$, then:*

$$\begin{aligned} R_{\mathcal{T}}(A_1) &\subseteq R_{\mathcal{T}}(A_2) \\ R^{\triangleright}_{\mathcal{I}}(A_1) &\subseteq R^{\triangleright}_{\mathcal{I}}(A_2) \\ R^{\triangleleft}_{\mathcal{I}}(A_1) &\supseteq R^{\triangleleft}_{\mathcal{I}}(A_2). \end{aligned}$$

Proposition 2 expresses that we can strengthen or weaken the direct images by adjusting the logical operators. It also expresses that more specific relations or more specific fuzzy sets lead to less or more specific direct images. For example, let A_1 be the set of male corpulent patients, and let A_2 be the set of corpulent patients (male or female). The superdirect image $R^{\triangleright}_{\mathcal{I}}(A_1)$ is the fuzzy set of symptoms that only occur with the male corpulent patients, while $R^{\triangleright}_{\mathcal{I}}(A_2)$ is the fuzzy set of symptoms that only occur with male or female corpulent patients. It is clear that the latter set will not be smaller than the first one.

From the proof of Proposition 1 we know that the superdirect image is a subset of the direct image if all R-foresets are normal. As we mentioned above this is the case for every reflexive binary fuzzy relation R. The direct images of a fuzzy set A in X under such a fuzzy relation R again are fuzzy sets in X, which makes it meaningful to study a possible ordering between the original fuzzy set A and its images. In particular we have the following proposition:

Proposition 3 (Expansiveness and restrictiveness). *Let $\mathcal{T}$ be a semi-norm and $\mathcal{I}$ a border implicator. Furthermore let R be a binary fuzzy relation in X and A a fuzzy set in X. If R is reflexive, then:*

$$R^{\diamond}_{\mathcal{T},\mathcal{I}}(A) \subseteq R^{\triangleright}_{\mathcal{I}}(A) \subseteq A \subseteq R_{\mathcal{T}}(A)$$

Proposition 3 shows that the square and the superdirect images are smaller (i.e. more specific) than the original fuzzy set, but that the direct image is larger (i.e. less specific) than the original fuzzy set.

Proposition 4 (Interaction with union of fuzzy sets). *Let $\mathcal{T}$ be a conjunctor and $\mathcal{I}$ be an implicator. For a fuzzy relation R from X to Y and fuzzy sets A_1 and A_2 in X it holds that:*

$$R_{\mathcal{T}}(A_1 \cup A_2) = R_{\mathcal{T}}(A_1) \cup R_{\mathcal{T}}(A_2)$$
$$R_{\mathcal{I}}^{\triangleleft}(A_1 \cup A_2) \subseteq R_{\mathcal{I}}^{\triangleleft}(A_1) \cup R_{\mathcal{I}}^{\triangleleft}(A_2)$$
$$R_{\mathcal{I}}^{\triangleright}(A_1 \cup A_2) \supseteq R_{\mathcal{I}}^{\triangleright}(A_1) \cup R_{\mathcal{I}}^{\triangleright}(A_2)$$

The direct image of the union of two fuzzy sets can be expressed as the union of the direct images of the separate fuzzy sets. For the other types of direct images only a containment relation w.r.t. the union of the separate direct images holds.

The property $R_{\mathcal{T}}(A_1 \cup A_2) = R_{\mathcal{T}}(A_1) \cup R_{\mathcal{T}}(A_2)$ is very interesting from a practical and computational point of view. For example, let A_1 be the set of corpulent male patients, and let A_2 be the set of corpulent female patients. One can then compute the direct images $R_{\mathcal{T}}(A_1)$ and $R_{\mathcal{T}}(A_2)$ as respectively the fuzzy set of symptoms that occur with corpulent male patients and the fuzzy set of symptoms that occur with corpulent female patients. Without many other calculations, one can now easily compute the direct image $R_{\mathcal{T}}(A_1 \cup A_2)$, which is the fuzzy set of symptoms that occur with corpulent patients (both male or female).

3 Fuzzy Relational Compositions

Definition 13 (Compositions). *[3], [6], [28] For R in $\mathcal{F}(X \times Z)$ and S in $\mathcal{F}(Z \times Y)$, the sup-$\mathcal{T}$-composition and the Bandler-Kohout compositions of R and S are fuzzy relations from X to Y defined as, for all x in X and all y in Y:*

- $R \circ_{\mathcal{T}} S\ (x,y) = \sup_{z \in Z} \mathcal{T}(R(x,z), S(z,y))$ *(sup-$\mathcal{T}$-composition)*
- $R \triangleleft_{\mathcal{I}} S\ (x,y) = \inf_{z \in Z} \mathcal{I}(R(x,z), S(z,y))$ *(subproduct)*
- $R \triangleright_{\mathcal{I}} S\ (x,y) = \inf_{z \in Z} \mathcal{I}(S(z,y), R(x,z))$ *(superproduct)*
- $R \diamond_{\mathcal{T},\mathcal{I}} S\ (x,y) = \mathcal{T}(R \triangleleft_{\mathcal{I}} S(x,y), R \triangleright_{\mathcal{I}} S(x,y))$ *(squareproduct)*

Following the principle of cylindrical extension we define the extension operators $\overset{\rightarrow}{\cdot}$ and $\overset{\leftarrow}{\cdot}$.

Definition 14 (Left extension, right extension). *The left extension operator $\overset{\rightarrow}{\cdot}$ and the right extension operator $\overset{\leftarrow}{\cdot}$ are $\mathcal{F}(X) - \mathcal{F}(X \times X)$ mappings that turn every fuzzy set A in X into the binary fuzzy relations $\overrightarrow{A}$ and $\overleftarrow{A}$ in X defined by*

$$\overleftarrow{A}(x,y) = A(y), \text{ for all } (x,y) \text{ in } X^2$$

$$\overrightarrow{A}(x,y) = A(x), \text{ for all } (x,y) \text{ in } X^2$$

$\overleftarrow{A}$ and $\overrightarrow{A}$ are called the left and the right extension of A respectively.

Characterization 1 The binary fuzzy relations R_1 and R_2 in X are respectively the left and the right extensions of the fuzzy set A in X if and only if

$$xR_1 = A, \text{ for all } x \text{ in } X$$

$$R_2y = A, \text{ for all } y \text{ in } X$$

Since the left extension of a fuzzy set A in X is a binary fuzzy relation in X, it makes sense to study its compositions with a fuzzy relation R from X to Y. Likewise we can study the compositions of R with the right extension of a fuzzy set B in Y.

Proposition 5. *Let $\mathcal{T}$ be a conjunctor and $\mathcal{I}$ an implicator, R in $\mathcal{F}(X \times Y)$, A in $\mathcal{F}(X)$ and B in $\mathcal{F}(Y)$. For all x in X and y in Y it holds that:*

(1) $\overleftarrow{A} \circ_{\mathcal{T}} R(x,y) = R_{\mathcal{T}}(A)(y)$ (5) $R \circ_{\mathcal{T}} \overrightarrow{B}(x,y) = (R^{-1})_{\mathcal{T}}(B)(x)$
(2) $\overleftarrow{A} \triangleleft_{\mathcal{I}} R(x,y) = R_{\mathcal{I}}^{\triangleleft}(A)(y)$ (6) $R \triangleleft_{\mathcal{I}} \overrightarrow{B}(x,y) = (R^{-1})_{\mathcal{I}}^{\triangleright}(B)(x)$
(3) $\overleftarrow{A} \triangleright_{\mathcal{I}} R(x,y) = R_{\mathcal{I}}^{\triangleright}(A)(y)$ (7) $R \triangleright_{\mathcal{I}} \overrightarrow{B}(x,y) = (R^{-1})_{\mathcal{I}}^{\triangleleft}(B)(x)$
(4) $\overleftarrow{A} \diamond_{\mathcal{T},\mathcal{I}} R(x,y) = R_{\mathcal{T},\mathcal{I}}^{\diamond}(A)(y)$ (8) $R \diamond_{\mathcal{T},\mathcal{I}} \overrightarrow{B}(x,y) = (R^{-1})_{\mathcal{T},\mathcal{I}}^{\diamond}(B)(x)$

Proof. The proposition above can be proven by applying the definitions of the concepts involved. As an example we prove (1) and (6).

$$\begin{aligned}
\overleftarrow{A} \circ_{\mathcal{T}} R(x,y) &= \sup_{z \in X} \mathcal{T}(\overleftarrow{A}(x,z), R(z,y)) \\
&= \sup_{z \in X} \mathcal{T}(A(z), R(z,y)) = R_{\mathcal{T}}(A)(y) \\
R \triangleleft_{\mathcal{I}} \overrightarrow{B}(x,y) &= \inf_{z \in X} \mathcal{I}(R(x,z), \overrightarrow{B}(z,y)) \\
&= \inf_{z \in X} \mathcal{I}(R(x,z), B(z)) = (R^{-1})_{\mathcal{I}}^{\triangleright}(B)(x)
\end{aligned}$$

□

Hence the compositions of the left extension of A and R give rise to the direct, the subdirect, the superdirect and the square direct image of A under R, while the compositions of R and the right extension of B result in the inverse, the superinverse, the subinverse and the square inverse image of B under R. Note that it is the subproduct of $\overrightarrow{B}$ and R that results in the superinverse image of B under R, and that it is the superproduct of $\overrightarrow{B}$ and R that results in the subinverse image of B under R. Furthermore note that while the left hand sides of the equalities in Proposition 5 seem to be depending on x and y the right hand sides only depend on either x or y.

4 Double Images

If A is a fuzzy set in X and B is a fuzzy set in Y their left and right extension will respectively be a binary fuzzy relation in X and a binary fuzzy relation in Y. The compositions of $\overleftarrow{A}$ with a fuzzy relation R from X to Y result in fuzzy relations from X to Y which we can once again compose with $\overrightarrow{B}$. Likewise the compositions of R and $\overrightarrow{B}$ lead to fuzzy relations from X to Y which can be composed with $\overleftarrow{A}$. This is precisely what we study in this section.

Proposition 6. *Let $\mathcal{T}$ be a conjunctor and $\mathcal{I}$ an implicator. Furthermore let R be a fuzzy relation from X to Y, A a fuzzy set in X and B a fuzzy set in Y. For all x in X and y in Y it holds that:*

$$(1)((\overleftarrow{A} \circ_\mathcal{T} R)\circ_\mathcal{T} \overrightarrow{B})(x,y) = \sup_{z\in Y} \mathcal{T}(R_\mathcal{T}(A)(z), B(z))$$

$$(2)(\overleftarrow{A} \circ_\mathcal{T} (R\circ_\mathcal{T} \overrightarrow{B}))(x,y) = \sup_{z\in X} \mathcal{T}(A(z), (R^{-1})_\mathcal{T}(B)(z))$$

$$(3)((\overleftarrow{A} \triangleleft_\mathcal{I} R)\triangleleft_\mathcal{I} \overrightarrow{B})(x,y) = \inf_{z\in Y} \mathcal{I}(R^{\triangleleft}_\mathcal{I}(A)(z), B(z))$$

$$(4)(\overleftarrow{A} \triangleleft_\mathcal{I} (R\triangleleft_\mathcal{I} \overrightarrow{B}))(x,y) = \inf_{z\in X} \mathcal{I}(A(z), (R^{-1})^{\triangleright}_\mathcal{I}(B)(z))$$

$$(5)((\overleftarrow{A} \triangleright_\mathcal{I} R)\triangleright_\mathcal{I} \overrightarrow{B})(x,y) = \inf_{z\in Y} \mathcal{I}(B(z), R^{\triangleright}_\mathcal{I}(A)(z))$$

$$(6)(\overleftarrow{A} \triangleright_\mathcal{I} (R\triangleright_\mathcal{I} \overrightarrow{B}))(x,y) = \inf_{z\in X} \mathcal{I}((R^{-1})^{\triangleleft}_\mathcal{I}(B)(z), A(z))$$

$$(7)((\overleftarrow{A} \diamond_{\mathcal{T},\mathcal{I}} R)\diamond_{\mathcal{T},\mathcal{I}} \overrightarrow{B})(x,y) = \mathcal{T}(\inf_{z\in Y} \mathcal{I}(B(z), R^{\diamond}_{\mathcal{T},\mathcal{I}}(A)(z)), \inf_{z\in Y} \mathcal{I}(R^{\diamond}_{\mathcal{T},\mathcal{I}}(A)(z), B(z)))$$

$$(8)(\overleftarrow{A} \diamond_{\mathcal{T},\mathcal{I}} (R\diamond_{\mathcal{T},\mathcal{I}} \overrightarrow{B}))(x,y) = \mathcal{T}(\inf_{z\in X} \mathcal{I}(A(z), (R^{-1})^{\diamond}_{\mathcal{T},\mathcal{I}}(B)(z)), \inf_{z\in Y} \mathcal{I}((R^{-1})^{\diamond}_{\mathcal{T},\mathcal{I}}(B)(z), A(z)))$$

Proof. All equalities can be proven using Definition 10 and 11 and Proposition 5. As an example we prove (2) and (7).

$$\begin{aligned}(\overleftarrow{A} \circ_\mathcal{T} (R\circ_\mathcal{T} \overrightarrow{B}))(x,y) &= (R\circ_\mathcal{T} \overrightarrow{B}))_\mathcal{T}(A)(y) \\ &= \sup_{z\in X} \mathcal{T}(A(z), R\circ_\mathcal{T} \overrightarrow{B}\,(z,y)) \\ &= \sup_{z\in X} \mathcal{T}(A(z), (R^{-1})_\mathcal{T}(B)(z))\end{aligned}$$

$$\begin{aligned}((\overleftarrow{A} \diamond_{\mathcal{T},\mathcal{I}} R)\diamond_{\mathcal{T},\mathcal{I}} \overrightarrow{B})(x,y) &= ((\overleftarrow{A} \diamond_{\mathcal{T},\mathcal{I}} R)^{-1})^{\diamond}_{\mathcal{T},\mathcal{I}}(B)(x) \\ &= \mathcal{T}(((\overleftarrow{A} \diamond_{\mathcal{T},\mathcal{I}} R)^{-1})^{\triangleleft}_\mathcal{I}(B)(x), ((\overleftarrow{A} \diamond_{\mathcal{T},\mathcal{I}} R)^{-1})^{\triangleright}_\mathcal{I}(B)(x)) \\ &= \mathcal{T}(\inf_{z\in Y} \mathcal{I}(B(z), \overleftarrow{A} \diamond_{\mathcal{T},\mathcal{I}} R(x,z)), \inf_{z\in Y} \mathcal{I}(\overleftarrow{A} \diamond_{\mathcal{T},\mathcal{I}} R(x,z), B(z))) \\ &= \mathcal{T}(\inf_{z\in Y} \mathcal{I}(B(z), R^{\diamond}_{\mathcal{T},\mathcal{I}}(A)(z)), \inf_{z\in Y} \mathcal{I}(R^{\diamond}_{\mathcal{T},\mathcal{I}}(A)(z), B(z)))\end{aligned}$$

□

All compositions presented above result in constant fuzzy relations. In [15]

$$(\overleftarrow{A} \circ_\mathcal{T} R)\circ_\mathcal{T} \overrightarrow{B}$$

is called the double image of A and B under R. It is "the extent to which A and B are related under R". The analogy with the preceding section inspires the following definition:

Definition 15 (Double images). *For $\mathcal{T}$ a conjunctor, $\mathcal{I}$ an implicator, R a fuzzy relation from X to Y, A a fuzzy set in X, and B a fuzzy set in Y we say that*

- $(\overleftarrow{A} \circ_{\mathcal{T}} R) \circ_{\mathcal{T}} \overrightarrow{B}$ *is the double direct image of A and B under R*
- $\overleftarrow{A} \circ_{\mathcal{T}} (R \circ_{\mathcal{T}} \overrightarrow{B})$ *is the double inverse image of A and B under R*
- $(\overleftarrow{A} \lhd_{\mathcal{I}} R) \lhd_{\mathcal{I}} \overrightarrow{B}$ *is the double subdirect image of A and B under R*
- $\overleftarrow{A} \lhd_{\mathcal{I}} (R \lhd_{\mathcal{I}} \overrightarrow{B})$ *is the double subinverse image of A and B under R*
- $(\overleftarrow{A} \rhd_{\mathcal{I}} R) \rhd_{\mathcal{I}} \overrightarrow{B}$ *is the double superdirect image of A and B under R*
- $\overleftarrow{A} \rhd_{\mathcal{I}} (R \rhd_{\mathcal{I}} \overrightarrow{B})$ *is the double superinverse image of A and B under R*
- $(\overleftarrow{A} \diamond_{\mathcal{T},\mathcal{I}} R) \diamond_{\mathcal{T},\mathcal{I}} \overrightarrow{B}$ *is the double square direct image of A and B under R*
- $\overleftarrow{A} \diamond_{\mathcal{T},\mathcal{I}} (R \diamond_{\mathcal{T},\mathcal{I}} \overrightarrow{B})$ *is the double square inverse image of A and B under R*

The following question which naturally comes to mind is whether it is possible to combine two kinds of fuzzy relational composition into one new kind of double image. As can be expected this leads to mixed double images such as the ones listed below, derived using Definition 10 and 11 as well as Proposition 5:

- $((\overleftarrow{A} \circ_{\mathcal{T}} R) \lhd_{\mathcal{I}} \overrightarrow{B})(x, y) = \inf_{z \in Y} \mathcal{I}(R_{\mathcal{T}}(A)(z), B(z))$
- $((\overleftarrow{A} \circ_{\mathcal{T}} R) \rhd_{\mathcal{I}} \overrightarrow{B})(x, y) = \inf_{z \in Y} \mathcal{I}(B(z), R_{\mathcal{T}}(A)(z))$
- $(\overleftarrow{A} \lhd_{\mathcal{I}} (R \rhd_{\mathcal{I}} \overrightarrow{B}))(x, y) = \inf_{z \in Y} \mathcal{I}(A(z), (R^{-1})^{\lhd}_{\mathcal{I}}(B)(z))$
- etc.

The main idea for all the double images is similar: A is a fuzzy set in X, B is a fuzzy set in Y, and R is a fuzzy relation from X to Y. If you want to know the degree to which A and B are related under R, first you transform either A into a fuzzy set on Y (using one of the direct images of Definition 10), or B into a fuzzy set on X (using one of the inverse images of Definition 11). Then you are left with two fuzzy sets on the same universe, which you can compare by looking to their degree of overlap, their degree of inclusion or their degree of equality.

5 A Look at Some Applications

So far we have illustrated the use of the fuzzy relational images in a medical data mining setting. In this section we will briefly discuss some other kinds of applications that can be described using fuzzy relational images, in particular fuzzy morphology, fuzzy rough sets, linguistic modifers and an application of the double images. An application in the context of modal logic has already been discussed in [13].

5.1 Fuzzy Morphology

Mathematical morphology is a theory for the processing and analysis of images, using operators based on geometrical concepts [25]. Different theories have been developed for binary images (represented as subsets of $\mathbb{R}^n$) and gray-scale images (represented as fuzzy subsets of $\mathbb{R}^n$). In the past years, fuzzy morphology has been introduced as an extension of binary morphology to gray-scale morphology, using techniques from fuzzy set theory [8], [19], [20]. The basic morphological operations are dilation and erosion. These operations transform an image A into a new image A' by means of a structuring element S.

Definition 16 (Fuzzy dilation and erosion). *Let $\mathcal{T}$ be a t-norm, $\mathcal{I}$ a border implicator, and A and S fuzzy sets (gray-scale images) in $\mathbb{R}^n$. The fuzzy dilation $D_{\mathcal{T}}(A,S)$ and the fuzzy erosion $E_{\mathcal{I}}(A,S)$ are defined by, for all $y \in \mathbb{R}^n$:*

$$D_{\mathcal{T}}(A,S)(y) = \sup_{x\in\mathbb{R}^n} \mathcal{T}(S(x-y), A(x))$$
$$E_{\mathcal{I}}(A,S)(y) = \inf_{x\in\mathbb{R}^n} \mathcal{I}(S(x-y), A(x))$$

In many image processing applications it is required that $S(\mathbf{0}) = 1$, i.e. the origin completely belongs to the structuring element. Now, if V is the substraction in $\mathbb{R}^n$, i.e. $V(x,y) = x - y$ for all x and y in $\mathbb{R}^n$, then this implies that the relation $R = V \circ S$ is a reflexive binary fuzzy relation in $\mathbb{R}^n$. In that case we have:

$$D_{\mathcal{T}}(A,S) = R_{\mathcal{T}}(A),$$
$$E_{\mathcal{I}}(A,S) = R^{\triangleright}_{\mathcal{I}}(A).$$

Note the nice interpretation of Proposition 4: the fuzzy dilation $D_{\mathcal{T}}(A,S)$ can be obtained by partitioning A into subsets A_i $(i = 1,...,n)$, by (simultaneously) computing the fuzzy dilations $D_{\mathcal{T}}(A_i,S)$ $(i = 1,...,n)$, and by combining the obtained results. This property can be used to decrease computation time.

5.2 Fuzzy Rough Sets

Rough set theory was introduced in the field of knowledge representation for modelling incomplete information [22]. Later on it was generalized to fuzzy rough set theory which can be used to model imprecise information as well (see e.g. [23]). Given a reflexive binary fuzzy relation R on a universe X of objects, the pair (X,R) is called a fuzzy approximation space. The basic operations in fuzzy rough set theory are the upper and lower fuzzy rough approximation.

Definition 17 (Upper and lower fuzzy rough approximation). *Let $\mathcal{T}$ be a t-norm, $\mathcal{I}$ a border implicator, (X,R) a fuzzy approximaton space and A a fuzzy set in X. The upper fuzzy rough approximation $\overline{\text{FAS}}^{\mathcal{T}}(A)$ and the lower fuzzy rough approximation $\underline{\text{FAS}}_{\mathcal{I}}(A)$ of A are given by, for all $y \in X$:*

$$\overline{\text{FAS}}^{\mathcal{T}}(A)(y) = \sup_{x\in X} \mathcal{T}(R(x,y), A(x)),$$
$$\underline{\text{FAS}}_{\mathcal{I}}(A)(y) = \inf_{x\in X} \mathcal{I}(R(x,y), A(x)).$$

$\overline{\text{FAS}}^{\mathcal{T}}(A)$ is the fuzzy set of objects that possibly belong to A while $\underline{\text{FAS}}_{\mathcal{I}}(A)$ is the fuzzy set of objects that necessarily belong to A. The link with the fuzzy direct and the fuzzy superdirect image is obvious, namely:

$$\overline{\text{FAS}}^{\mathcal{T}}(A) = R_{\mathcal{T}}(A),$$
$$\underline{\text{FAS}}_{\mathcal{I}}(A) = R^{\triangleright}_{\mathcal{I}}(A).$$

Proposition 3 guarantees that the lower fuzzy rough approximation of A is a fuzzy subset of A, while A is a fuzzy subset of its upper fuzzy rough approximation.

5.3 Linguistic Modifiers

If a linguistic term is represented by a fuzzy set A in a universe X, then a linguistic modifier (i.e. an adverb) can be modelled by a $\mathcal{F}(X) - \mathcal{F}(X)$ mapping, called a fuzzy modifier. Applying a fuzzy modifier to the fuzzy set modelling a linguistic term then yields the representation of the modified linguistic term; for instance applying a fuzzy modifier for very onto a fuzzy set for tall gives rise to the membership function for very tall. In fuzzy systems two kinds of interpretations of linguistic modifiers such extremely, very, more or less and roughly are common. In the first one, called the inclusive interpretation, it is assumed that

$$\text{extremely } A \subseteq \text{very } A \subseteq A \subseteq \text{more or less } A \subseteq \text{roughly } A \tag{4}$$

In [9] it is suggested to represent the linguistic modifiers by means of resemblance relations, i.e. reflexive binary fuzzy relations that model approximate equality. Let E_1 and E_2 be two resemblance relations in X and $E_1 \subset E_2$ (hence the approximate equality modelled by E_1 is stronger than the approximate equality modelled by E_2). For $\mathcal{T}$ a t-norm and $\mathcal{I}$ a border implicator the following representations are proposed (for all A in $\mathcal{F}(X)$) :

$$\text{more or less } A = (E_1)_{\mathcal{T}}(A) \quad \text{very } A = (E_1)^{\triangleright}_{\mathcal{I}}(A)$$
$$\text{roughly } A = (E_2)_{\mathcal{T}}(A) \quad \text{extremely } A = (E_2)^{\triangleright}_{\mathcal{I}}(A)$$

These representations can be interpreted like "y belongs to more or less A if y resembles to an object of A" and "y belongs to very A if y resembles only to objects of A". Propositions 2 and 3 guarantee that the inclusions in (4) hold.

In the non-inclusive interpretation which is often used in fuzzy control applications, a modified linguistic term does not denote a subset nor a superset of the original term. The original term and the modified term denote two different (possibly overlapping categories). If the universe X is numerical this situation can be modelled by shifting fuzzy modifiers that shift the original fuzzy set to the left or to the right. In [9] it was demonstrated that shifting modifiers can be seen as operators that take direct images under a fuzzy relation R_α defined as

$$R_\alpha(x,y) = \begin{cases} 1 \text{ if } x = y - \alpha \\ 0 \text{ otherwise} \end{cases}$$

for all x and y in X. The value of the numerical parameter α determines the size of the shift as well as its direction.

5.4 Applications of the Double Images

As far as we know until now the power of double images for applications in computer science has not yet been investigated. We suggest some examples:

- If X is a universe of people, A and B are fuzzy sets in X and the binary fuzzy relation R in X expresses the degree of friendship between two people, then the double direct image of A and B under R can be interpreted as the degree to which group A and group B are friends.
- If R is a fuzzy patient-symptom relation, A is the fuzzy set of people that live nearby a nuclear powerplant and B is the set of fuzzy set of symptoms for cancer then the subdirect double image of A and B under R expresses the degree to which the fuzzy set of symptoms that occur with all patients that live nearby a plant is included in the fuzzy set of symptoms of cancer.
- If R is a fuzzy relation between the universe of terms X and the universe of documents Y in a document retrieval system, if A is the fuzzy set of terms present in a profile for a research vacancy and B is the set of publications of a particular researcher, then the double direct image of A and B under R expresses the degree to which this researcher matches the vacancy.

A particularly interesting case arises when we denote the crisp equality on X by E, i.e. $E(x,y) = 1$ if $x = y$ and $E(x,y) = 0$ otherwise. Indeed then one can easily verify that $E_{\mathcal{T}}(A) = A$ for all A in $\mathcal{F}(X)$. For all A and B in $\mathcal{F}(X)$ it then holds that

- $(\overleftarrow{A} \circ_{\mathcal{T}} E) \circ_{\mathcal{T}} \overrightarrow{B}$ is the degree of $\mathcal{T}$-overlap of A and B
- $(\overleftarrow{A} \circ_{\mathcal{T}} E) \triangleleft_{\mathcal{I}} \overrightarrow{B}$ is the degree of $\mathcal{I}$-inclusion of A in B
- $(\overleftarrow{A} \circ_{\mathcal{T}} E) \diamond_{\mathcal{T},\mathcal{I}} \overrightarrow{B}$ is the degree of $(\mathcal{T},\mathcal{I})$-equality of A and B

By (slightly) changing the values of the membership degrees in E, new and refined measures of similarity and inclusion can be defined. One option is to define $E(x,x) = 0$ for the objects x in an uninteresting part of the universe. Another option is to use a fuzzy relation instead of a crisp one to model equality (see [10]). The advantages of such an approach become clear in the following example: suppose we want to compare the first two pictures depicted in Figure 1. When representing them as fuzzy sets A and B, all pixels that have membership degree 1 in A (namely the white pixels) have degree 0 in B (black pixels) and vice versa. Hence for all $\mathcal{T}$ and $\mathcal{I}$ the degree of $(\mathcal{T},\mathcal{I})$-equality of A and B will be 0. However if we model approximate equality on the pixel level by a fuzzy relation E_1 such that every pixel is equal to itself to degree 1 and neighbouring pixels are equal to each other to degree 0.5, using min as a t-norm, the direct image of picture a of Figure 1 under E_1 looks like picture c. The white pixels of the original picture remain white but the black pixels of the original picture have gotten a higher membership degree in the new one (namely 0.5) and therefore look gray. $(\overleftarrow{A} \circ_{\min} E_1) \diamond_{\mathcal{T},\mathcal{I}} \overrightarrow{B}$ might be greater than 0. For example, using the minimum operator as a t-norm and the Kleene-Dienes implicator, this mixed double image of A and B under E_1 is 0.5. If needed we can even use a fuzzy relation such that we obtain an equality degree of 1. Such a result is desirable in the context of measuring similarity between images (two images that have been shifted w.r.t. each other are similar images). To achieve this result we can choose a fuzzy relation such that the direct image acts like a shifting modifier as briefly discussed in the preceding subsection.

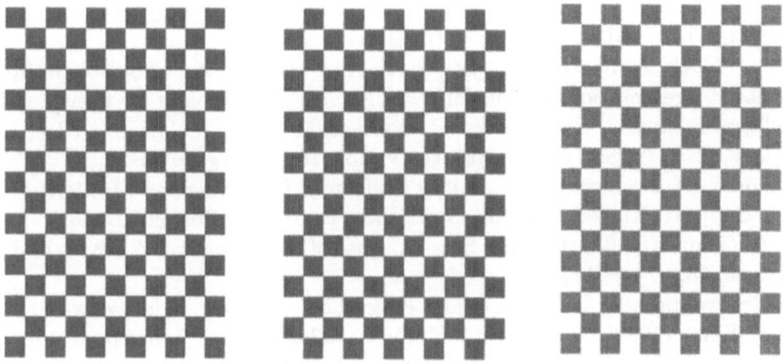

Fig. 1. Pictures a, b and c all containing 11 by 17 pixels

References

1. H. Bandemer, S. Gottwald (1995). Fuzzy Sets, Fuzzy Logic, Fuzzy Methods. John Wiley and Sons, Chisester.
2. W. Bandler, L. J. Kohout (1980). Fuzzy Power Sets and Fuzzy Implication Operators. Fuzzy Sets and Systems, 4, p. 13–30.
3. W. Bandler, L. J. Kohout (1980). Fuzzy Relational Products as a Tool for Analysis and Synthesis of the Behaviour of Complex Natural and Artificial Systems. In: Fuzzy Sets: Theory and Application to Policy Analysis and Information Systems S. K. Wang, P. P. Chang (eds.). Plenum Press, New York and London, p. 341–367.
4. G. Chen (1993). Fuzzy Data Modeling: Perspectives, Problems and Solutions. In: [18].
5. F. Crestani, G. Pasi (eds.) (2000). Soft Computing in Information Retrieval, Techniques and Applications. Studies in Fuzziness and Soft Computing Vol. 50. Physica-Verlag Heidelberg/New York.
6. B. De Baets (1995). Solving Fuzzy Relational Equations: an Order Theoretical Approach. Ph.D. Thesis, Ghent University (in Dutch).
7. B. De Baets, E. E. Kerre (1993). A Revision of Bandler-Kohout Compositions of Relations. Mathematica Pannonica, 4, p. 59–78.
8. B. De Baets (1997). Fuzzy Morphology: a Logical Approach. In: Uncertainty Analysis in Engineering and Sciences: Fuzzy Logic, Statistics, and Neural Network Approach. Kluwer Academic Publishers, Boston, p. 53–67.
9. M. De Cock, E. E. Kerre (2000). A New Class of Fuzzy Modifiers. In: Proceedings of 30th IEEE International Symposium on Multiple-Valued-Logic (ISMVL'2000), IEEE Computer Society, p. 121-126.

10. M. De Cock, E. E. Kerre (2002). On (Un)suitable Fuzzy Relations to Model Approximate Equality. To appear in: Fuzzy Sets and Systems.
11. M. De Cock, M. Nachtegael, E. E. Kerre (2000). Images under Fuzzy Relations: a Master-Key to Fuzzy Applications. In: Intelligent Techniques and Soft Computing in Nuclear Science and Engineering (D. Ruan, H. A. Abderrahim, P. D'hondt, E. E. Kerre (eds.)). World Scientific Publishing, Singapore, p. 47–54.
12. M. De Cock, M. Nachtegael, E. E. Kerre (2000). Images under Fuzzy Relations: Technical Report FUM-2000-02 (22 pages). Fuzziness and Uncertainty Modelling Research Unit, Ghent University (in Dutch).
13. M. De Cock, A. M. Radzikowska, E. E. Kerre (2002). A Fuzzy-Rough Approach to the Representation of Linguistic Hedges. In: Technologies for Constructing Intelligent Systems, Part 1: Tasks (B. Bouchon-Meunier, J. Gutierrez-Rios, L. Magdalena, R. R. Yager (eds.)). Springer-Verlag, Heidelberg, p. 33–42.
14. J. Fodor, M. Roubens (1994). Fuzzy Preference Modelling and Multicriteria Decision Support. Kluwer Academic Publishers Dordrecht/Boston/London.
15. J. A. Goguen (1967). L-Fuzzy Sets. Journal of Mathematical Analysis and Applications, 18, p. 145–174.
16. S. Gottwald (1993). Fuzzy Sets and Fuzzy Logic. Vieweg, Braunschweig.
17. E. E. Kerre (1992). A Walk through Fuzzy Relations and their Applications to Information Retrieval, Medical Diagnosis and Expert Systems. In: Analysis and Management of Uncertainty: Theory and Applications (B. M. Ayyub, M. M. Gupta, L. N. Kanal (eds.)). Elsevier Science Publishers, p. 141–151.
18. E. E. Kerre (ed.) (1993). Introduction to the Basic Principles of Fuzzy Set Theory and Some of its Applications. Communication and Cognition, Gent.
19. M. Nachtegael, E. E. Kerre (1999). Different Approaches towards Fuzzy Mathematical Morphology. In: Proceedings of EUFIT'99 (on CD-ROM), Aachen.
20. M. Nachtegael, E. E. Kerre (2000). Classical and Fuzzy Approaches towards Mathematical Morphology. In: Fuzzy Techniques in Image Processing (E. E. Kerre, M. Nachtegael (eds.)). Springer-Verlag, Heidelberg, p. 3–57.
21. V. Novak (1986). Fuzzy Sets and Their Applications. Adam Hilger, Bristol.
22. Z. Pawlak (1982). Rough sets. International Journal of Computer and Information Science, 11, p. 341–356.
23. A. M. Radzikowska, E. E. Kerre (2000). A Comparative Study of Fuzzy Rough Sets. To appear in Fuzzy Sets and Systems.
24. D. Ruan, E. E.Kerre (eds.) (2000). Fuzzy IF-THEN Rules in Computational Intelligence. Kluwer Academic Publishers Boston/Dordrecht/London.
25. J. Serra (1982). Image Analysis and Mathematical Morphology. Academic Press, London.
26. J. F.-F. Yao, J.-S. Yao (2001). Fuzzy Decision Making for Medical Diagnosis based on Fuzzy Number and Compositional Rule of Inference. Fuzzy Sets and Systems, 120, p. 351–366.
27. A. Yazici, R. George (1999). Fuzzy Database Modeling. Studies in Fuzziness and Soft Computing, Vol. 26. Physica-Verlag, Heidelberg.
28. L. A. Zadeh (1965). Fuzzy Sets, Information and Control, 8, p. 338–353.
29. L. A. Zadeh (1975). Calculus of Fuzzy Restrictions. In: Fuzzy Sets and Their Applications to Cognitive and Decision Processes L. A. Zadeh, K.-S. Fu, K. Tanaka, M. Shimura (eds.). Academic Press Inc., New York, p. 1–40.
30. L. A. Zadeh (1975). The Concept of a Linguistic Variable and its Application to Approximate Reasoning I, II, III. Information Sciences, 8, p. 199–249, p. 301–357, and 9, p. 43–80.

31. L.A. Zadeh, J. Kacprzyk (eds.) (1999). Computing with Words in Information/Intelligent Systems 1. Studies in Fuzziness and Soft Computing, Vol. 33. Physica-Verlag, Heidelberg.
32. R. Zenner, R. De Caluwe, E. E. Kerre (1984). Practical determination of document description relations in document retrieval systems. In: Proceedings of the Workshop on the Membership Function, EIASM, Brussels, p. 127–138.
33. H.-J. Zimmerman (1996). Fuzzy Set Theory and its Applications. Kluwer Academic Publishers, Boston.

A Completeness Theorem for Extended Order Dependencies on Relational Attribute Models in Dedekind Categories

Hitomi Okuma and Yasuo Kawahara

Department of Informatics, Kyushu University 33, Fukuoka 812-8581, Japan
{okuma, kawahara}@i.kyushu-u.ac.jp

Abstract. Order dependencies in relational database due to Ginsburg and Hull are relationships between attributes with domains of ordered values. The basic notions in this paper are comparison systems and relational attribute models in Dedekind categories. A comparison system constitutes a formal structure of possible orders for attribute domains. A relational attribute model is a system of relations on an object (of tuples or records) in Dedekind categories, which can be constructed by a suitable relational interpretation of comparing symbols. Generalizing order dependencies as well as functional dependencies, the paper introduces extended order dependencies, and their satisfactory relations for relational attribute models in Dedekind category. Then we give a simple proof that a revised set of inference rules is sound and complete.

1 Introduction

Relational database theory has been studied since Codd introduced relational database models. Dependency theory relates to relationships (constraints) between attributes in database relations. One of important questions with dependencies is that of (logical) implication: What other dependencies are necessarily satisfied by an instance when we know that a set of dependencies is satisfied by this instance? Inference rules are used in order to show implication between dependencies. The appropriateness of a set of inference rules is expressed by "soundness" and "completeness". Armstrong and Beeri et al. [2] have proved the completeness theorems for functional and multivalued dependencies.

Dependency theory has been studied with relational methods. Orłowska [10] proposed a relational formulation of functional, multivalued and other dependencies, and Buszkowski and Orłowska [3] developed an axiomatic relational calculus for dependency theory. Schmidt and Ströhlein [12] explained a basic relational feature of functional dependency for relational models of databases. Jaoua et al. [6,11] studied difunctional dependencies in relational databases and gave some inference rules. MacCaull [7,8] investigated a relational formulation for functional and multivalued dependencies and association rules, and proved soundness and completeness for the implication problem of these

H. de Swart (Ed.): RelMiCS 2001, LNCS 2561, pp. 152–170, 2002.

dependencies with a Rasiowa/Sikorski-style tableaux method. The foundations and recent applications of relational methods in computer science are excellently summarized in [1].

In [5], Ginsburg and Hull introduced order dependencies to incorporate semantic information involving order on attribute domains. For example, we consider the information concerning the checks written by a specific individual (see Example 1). We assume that the checks have been written in sequence in ascending order, an order dependency can describe the statement that a check having larger check number would be written later than the smaller one. They defined a notion of marked attributes to describe the order between attribute values, and presented a sound and complete set of inference rules for order dependencies. They also discussed the complexity of determining logical implication for order dependency.

The aim of this paper is to show the soundness and the completeness theorems of a set of inference rules for extended order dependencies, based on theory of Dedekind categories. To discuss order dependencies in our general framework, we define comparison systems, which generalize marked attributes in sense of [5], and comparators, which correspond to sets of marked attributes. Then we introduce a notion of relational attribute models, which is an extension of those in [9] or indiscernibility relations in [10]. A satisfactory relation for extended order dependencies on relational attribute models in Dedekind categories is defined with the same formalism as that for functional dependencies in [9]. The extended order dependencies succeed in extending order dependencies in [5]. Almost all arguments in this paper are point-free except for comparison systems and comparators. Finally we show the soundness and the completeness of a set of inference rules.

The remainder of this paper is organized as follows: In Section 2, we review the definition of Dedekind categories, a kind of relational category, and some basic properties of Heyting algebra. In Section 3, first we recall the original definition of marked attribute and order dependency by Ginsburg and Hull [5]. Then the foundation of comparison systems to describe the relationships between attribute values is given. We present examples of comparison systems for functional dependencies and for order dependencies. In Section 4, we introduce a notion of relational attribute models in order to discuss extended order dependencies in Dedekind categories. Then we present basic properties of relational attribute models. In Section 5, we define extended order dependency and its satisfactory relation in Dedekind categories, and show some properties related to the propriety and the logical implication for extended order dependencies by using properties of Heyting algebra. In Section 6, a modified set of inference rules for extended order dependencies is presented and shown to be sound. In Section 7, properties of univalent comparators are shown, and finally we give a completeness theorem for the set of inference rules.

2 Dedekind Categories

In this section we recall the definition of a kind of a relation category which we will call Dedekind categories following Olivier and Serrato (1980). Dedekind categories are equivalent to locally complete division allegories, in particular they are equivalent to locally complete distributive allegories introduced in [4].

Throughout this paper, a morphism α from an object X into an object Y in a Dedekind category (which will be defined below) will be denoted by a half arrow $\alpha : X \rightharpoondown Y$, and the composite of a morphism $\alpha : X \rightharpoondown Y$ followed by a morphism $\beta : Y \rightharpoondown Z$ will be written as $\alpha\beta : X \rightharpoondown Z$. Also we will denote the identity morphism on X as id_X.

Definition 1. A Dedekind category $\mathcal{D}$ is a category satisfying the following:
D1. [Complete Heyting Algebra] For all pairs of objects X and Y the hom-set $\mathcal{D}(X,Y)$ consisting of all morphisms of X into Y is a complete Heyting algebra (namely, a complete distributive lattice) with the least morphism 0_{XY} and the greatest morphism ∇_{XY}. Its algebraic structure will be denoted by

$$\mathcal{D}(X,Y) = (\mathcal{D}(X,Y), \sqsubseteq, \sqcup, \sqcap, 0_{XY}, \nabla_{XY}).$$

That is, (a) $\sqsubseteq$ is a partial order on $\mathcal{D}(X,Y)$, (b) $\forall \alpha \in \mathcal{D}(X,Y) :: 0_{XY} \sqsubseteq \alpha \sqsubseteq \nabla_{XY}$, (c) $\sqcup_{\lambda\in\Lambda}\alpha_\lambda \sqsubseteq \alpha$ iff $\alpha_\lambda \sqsubseteq \alpha$ for all $\lambda \in \Lambda$, (d) $\alpha \sqsubseteq \sqcap_{\lambda\in\Lambda}\alpha_\lambda$ iff $\alpha \sqsubseteq \alpha_\lambda$ for all $\lambda \in \Lambda$, and (e) $\alpha \sqcap (\sqcup_{\lambda\in\Lambda}\alpha_\lambda) = \sqcup_{\lambda\in\Lambda}(\alpha \sqcap \alpha_\lambda)$.
D2. [Converse] There is given a converse operation $^\sharp : \mathcal{D}(X,Y) \to \mathcal{D}(Y,X)$. That is, for all morphisms $\alpha, \alpha' : X \rightharpoondown Y$, $\beta : Y \rightharpoondown Z$, the following laws hold:
(a) $(\alpha\beta)^\sharp = \beta^\sharp\alpha^\sharp$, (b) $(\alpha^\sharp)^\sharp = \alpha$, and (c) If $\alpha \sqsubseteq \alpha'$, then $\alpha^\sharp \sqsubseteq \alpha'^\sharp$.
D3. [Dedekind Formula] For all morphisms $\alpha : X \rightharpoondown Y$, $\beta : Y \rightharpoondown Z$ and $\gamma : X \rightharpoondown Z$ the Dedekind formula $\alpha\beta \sqcap \gamma \sqsubseteq \alpha(\beta \sqcap \alpha^\sharp\gamma)$ holds.
D4. [Residue] For all morphisms $\beta : Y \rightharpoondown Z$ and $\gamma : X \rightharpoondown Z$ the residue (or division or weakest precondition) $\gamma \div \beta : X \rightharpoondown Y$ is a morphism such that $\alpha\beta \sqsubseteq \gamma$ if and only if $\alpha \sqsubseteq \gamma \div \beta$ for all morphisms $\alpha : X \rightharpoondown Y$. □

A morphism $f : X \rightharpoondown Y$ such that $f^\sharp f \sqsubseteq \mathrm{id}_Y$ (*univalent*) and $\mathrm{id}_X \sqsubseteq ff^\sharp$ (*total*) is called a *function* and may be introduced as $f : X \to Y$. In what follows the word *relation* is a synonym for morphism of a Dedekind category. Each hom-set $\mathcal{D}(X,Y)$ has pseudo-complement, that is, for any two relations α and β in $\mathcal{D}$ there is a relation $\alpha \Rightarrow \beta$ in $\mathcal{D}(X,Y)$ such that $\alpha \sqcap \gamma \sqsubseteq \beta$ iff $\gamma \sqsubseteq \alpha \Rightarrow \beta$ for all relations γ.

The next proposition lists some basic properties of relations in Dedekind categories related to Heyting algebra.

Proposition 1. *Let $\alpha, \alpha', \beta, \beta', \gamma : X \rightharpoondown Y$ be relations in a Dedekind category $\mathcal{D}$. Then the following hold:*

(a) $\beta \sqsubseteq \alpha \Rightarrow \beta$.
(b) $\alpha \sqcap (\alpha \Rightarrow \beta) \sqsubseteq \beta$.
(c) *If* $\alpha \sqsupseteq \alpha'$ *and* $\beta \sqsubseteq \beta'$, *then* $\alpha \Rightarrow \beta \sqsubseteq \alpha' \Rightarrow \beta'$.
(d) $\alpha \Rightarrow \beta \sqsubseteq (\gamma \sqcap \alpha) \Rightarrow (\gamma \sqcap \beta)$.
(e) $(\alpha \Rightarrow \beta)^\sharp = \alpha^\sharp \Rightarrow \beta^\sharp$.
(f) $(\alpha \Rightarrow \beta) \sqcap (\beta \Rightarrow \gamma) \sqsubseteq \alpha \Rightarrow \gamma$.
(g) *If* $\alpha \sqsubseteq \beta$, *then* $\alpha \sqcap (\beta \Rightarrow \gamma) = \alpha \sqcap \gamma$.
(h) $(\alpha \Rightarrow \gamma) \sqcap (\beta \Rightarrow \gamma) \sqsubseteq (\alpha \sqcup \beta) \Rightarrow \gamma$. □

An object I of a Dedekind category $\mathcal{D}$ is called a *unit* if $0_{II} \neq \mathrm{id}_I = \nabla_{II}$. In this paper we assume that a Dedekind category $\mathcal{D}$ has a unit I. An I-*point* x of X is a function $x : I \to X$.

3 Comparison Systems

Ginsburg and Hull [5] introduced a dependency which incorporate information involving order, called order dependency. They defined a notion of marked attributes to describe a certain relationship between attributes with ordered domains of values.

In this section, as a foundation of a relational treatment for order dependencies in Dedekind categories, we first introduce comparison systems, a generalization of the marked attribute [5]. Then we define a notion of comparators which correspond to sets of marked attributes. At the end of the section we list two examples of comparison systems for functional dependencies and for extended order dependencies.

Before presenting the notions of comparison systems and comparators, we recall the original definition of marked attribute and order dependency in [5].

Definition 2. Let U be a finite set of attributes and T_a a domain of values for each attribute $a \in U$. A tuple is a function u from U such that $u(a) \in T_a$ for each attribute $a \in U$. T is the set of all tuples over U and an instance R over U is a subset of T.

For each attribute $a \in U$, *marked attributes* are formal symbols $a^=$, $a^\leq$, $a^\geq$, $a^<$, $a^>$, a^{in}, used in the following way: Let u, v be tuples in T. Write: (a) $u[a^=]v$ if $u(a) =_a v(a)$; (b) $u[a^\leq]v$ if $u(a) \leq_a v(a)$; (c) $u[a^\geq]v$ if $u(a) \geq_a v(a)$; (d) $u[a^<]v$ if $u(a) <_a v(a)$; (e) $u[a^>]v$ if $u(a) >_a v(a)$; (f) $u[a^{in}]v$ if $u(a)$ and $v(a)$ are incomparable under $\leq_a$, where $=_a$ is an equality on T_a and $\leq_a$ is a partial ordering on T_a.

An *order dependency* is an expression of the form $X \to Y$, where X and Y are sets of marked attributes. We say that an instance R satisfies order dependency $X \to Y$ if for all tuples u and v in R, $u[a^*]v$ for each marked attribute $a^* \in X$ implies $u[b^*]v$ for each marked attribute $b^* \in Y$. □

We next present an example of order dependencies in Checking-account database [5].

Example 1. Consider the information concerning the checks written by a specific individual. Let U be a set of four attributes N, D, P and A. The domain of N is the set of natural numbers representing the number of a check. The domain of D is the set of dates representing the date the check was written. The domain of P is the set of names of individuals and corporations, representing to whom the check was paid. The domain of A is the set of dollar amounts representing the amount for which the check was paid. We assume that the checks have been written in sequence in ascending order. Then the order dependencies $N^{\leq} \to D^{\leq}$ and $D^{<} \to N^{<}$ hold.

As a foundation of the relational treatment for order dependency in Dedekind categories, we now define a notion of comparison systems, which generalize the marked attributes.

Definition 3. A *comparison system* L is a quartet $(U, C, {}^{t}, q)$ of a finite set U of attributes, a finite set C of comparing symbols, a transposal mapping ${}^{t} : C \to C$ and an attribute mapping $q : C \to U$ satisfying the following:

(a) $s^{tt} = s$ and $q(s^{t}) = q(s)$ for each $s \in C$,
(b) For each $a \in U$ there is a distinguished element $e_a \in C$ such that $e_a^t = e_a$ and $q(e_a) = a$. □

Set $C_a = \{s \in C \mid q(s) = a\}$ (the inverse image of a by q) for each $a \in U$. Note that q is surjective and each C_a is a nonempty set, because $q(e_a) = a$ and so $e_a \in C_a$.

Definition 4. Let $L = (U, C, {}^{t}, q)$ be a comparison system. A *comparator* over L is a binary relation $X : U \rightharpoonup C$ (namely, a subset $X \subseteq U \times C$) such that for each $a \in U$ and $s \in C$, if $(a, s) \in X$, then $q(s) = a$. □

Note that a binary relation $X : U \rightharpoonup C$ is a comparator iff $Xq \sqsubseteq \mathrm{id}_U$ iff $X \sqsubseteq q^{\sharp}$ in a relational notation.

For a comparator X over L we denote by $X(a)$ a subset $\{s \in C \mid (a, s) \in X\}$ of C_a. Clearly X uniquely corresponds to a set of subsets $\{X(a) \mid a \in U\}$. A comparator X over L is *univalent* if $X(a)$ has at most one element for every $a \in U$, and is *total* if $X(a)$ is nonempty for every $a \in U$. It is easy to see that the union $X \cup Y$ and the intersection $X \cap Y$ of comparators X and Y over L are also comparators over L. The transposal mapping ${}^{t} : C \to C$ can be extended for relations from U into C as follows: $\forall (a, s) : (a, s) \in X^{t} \Leftrightarrow (a, s^{t}) \in X$. It is trivial that if X is a comparator over L then so is X^{t}.

A comparator $U^{\cdot}$ over L is defined by $U^{\cdot}(a) = \{e_a\}$ for all attributes $a \in U$. For an attribute $a \in U$ and a subset $x \subseteq C_a$ we defined a comparator a^{x} over L by $a^{x}(b) = x$ if $b = a$, and $a^{x}(b) = C_b$ otherwise. The comparator a^{x} is a generalization of marked attributes in sense of [5]. Note that the comparator $U^{\cdot}$ over L is always total and univalent by the definition.

To maintain the consistency with the traditional notation in database theory we turn the natural inclusion of comparators upside down. That is, we write $X \subseteq Y$, $X \cap Y$ and $X \cup Y$ as $X \supseteq^{\cdot} Y$, $X \cup^{\cdot} Y$ and $X \cap^{\cdot} Y$, respectively.

We now straightforwardly have the following proposition by the definition of comparators.

Proposition 2. *Let $L = (U, C, {}^t, q)$ be a comparison system, $a \in U$, x, y subsets of C_a and X a comparator over L. Then the following hold:*

(a) $a^x \cup^{\cdot} a^y = a^{x \cap y}$ *and* $a^x \cap^{\cdot} a^y = a^{x \cup y}$.
(b) $X = \cup^{\cdot}_{a \in U} a^{X(a)}$.

Proof. (a) It is easy to see that $(a^x \cup^{\cdot} a^y)(b) = x \cap y$ if $b = a$, and $(a^x \cup^{\cdot} a^y)(b) = C_b$ otherwise, and $(a^x \cap^{\cdot} a^y)(b) = x \cup y$ if $b = a$, and $(a^x \cap^{\cdot} a^y)(b) = C_b$ otherwise.
(b) For all $b \in U$ we have

$$\begin{aligned}(\cup^{\cdot}_{a \in U} a^{X(a)})(b) &= \cap_{a \in U} a^{X(a)}(b) \\ &= X(b) \cap \{\cap_{b \neq a} a^{X(a)}(b)\} \\ &= X(b) \cap C_b \\ &= X(b).\end{aligned}$$

□

The following proposition gives a basic property of containment between comparators, which is useful result to show the completeness theorem for order dependencies.

Proposition 3. *Let $L = (U, C, {}^t, q)$ be a comparison system and X and M comparators over L. Then the following hold:*

(a) *If X is univalent and $M \supseteq^{\cdot} X$, then either $M = X$ or M is not total.*
(b) *If X is univalent and $X \not\supseteq^{\cdot} M$, then $X \cup^{\cdot} M$ is not total.*

Proof. (a) Assume M is total. Then both X and M are functions, since $M \supseteq^{\cdot} X$. Therefore $M = X$.
(b) Assume $X \cup^{\cdot} M$ is total. Then X and $X \cup^{\cdot} M$ are functions, since $X \supseteq X \cap M = X \cup^{\cdot} M$. Hence $X = X \cup^{\cdot} M$. Therefore $X \supseteq^{\cdot} M$, which is a contradiction. □

We now show a trivial example of comparison systems for functional dependencies.

Example 2. When we consider only functional dependencies in a database, the comparison system is simple, that is, a set C of comparing symbols is just a cartesian product $U \times \{=, \neq\}$ (where U is a given finite set of attributes), a mapping ${}^t : C \to C$ is the identity mapping on C and a mapping $q : C \to U$ is the first projection from C onto U (see Fig. 1). Set $e_a = a \times \{=\}$ for each $a \in U$. For a subset S of U we define a comparator $S^{\cdot} : U \rightharpoondown C$ by $S^{\cdot} = \cup^{\cdot}_{a \in S} a^{e_a}$. Then it is easily seen that $S^{\cdot} \subseteq^{\cdot} U^{\cdot}$ (which is equivalent to $U^{\cdot} \subseteq S^{\cdot}$) for all $S \subseteq U$, and that $S \subseteq T \Leftrightarrow S^{\cdot} \subseteq^{\cdot} T^{\cdot}$. The traditional functional dependency pays attention only to comparators X such that $X = S^{\cdot}$ for some subset $S \subseteq U$. This is a reason why we use the symbols $\subseteq^{\cdot}$, $\cup^{\cdot}$ and $\cap^{\cdot}$.

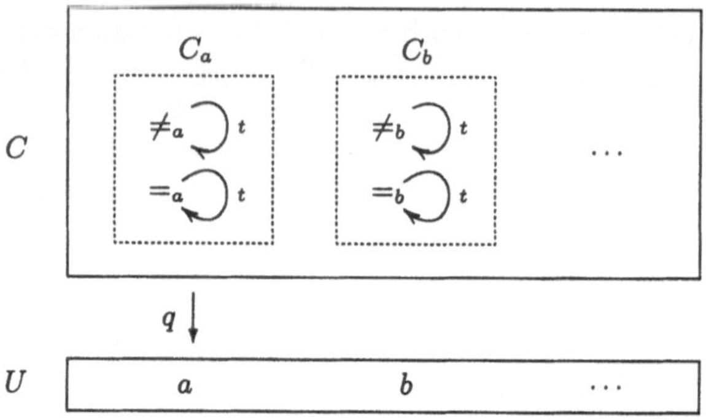

Fig. 1. A comparison system for functional dependency

Next we introduce an example of comparison systems for order dependencies.

Example 3. Consider the checking-account database mentioned in Example 1. Let C be a set of fifteen comparing symbols $<_N$, $>_N$, $=_N$, $<_D$, $>_D$, $=_D$, $\neq_P$, $=_P$, $<_{A2}$, $>_{A2}$, $<_{A1}$, $>_{A1}$, $<_{A0}$, $>_{A0}$, $=_A$ (see Fig. 2). The orders associated with each of the attribute domains for attributes N, D and P are obvious, and for example, for A it is the restriction of a divided-order for the real numbers: $r <_{A0} r'$ iff $r < r' < 10000$, $r <_{A1} r'$ iff $r < 10000 \leq r'$ and $r <_{A2} r'$ iff $10000 \leq r < r'$. This setting makes it possible to separate information of large payments, ten thousand dollars or more, from the others.

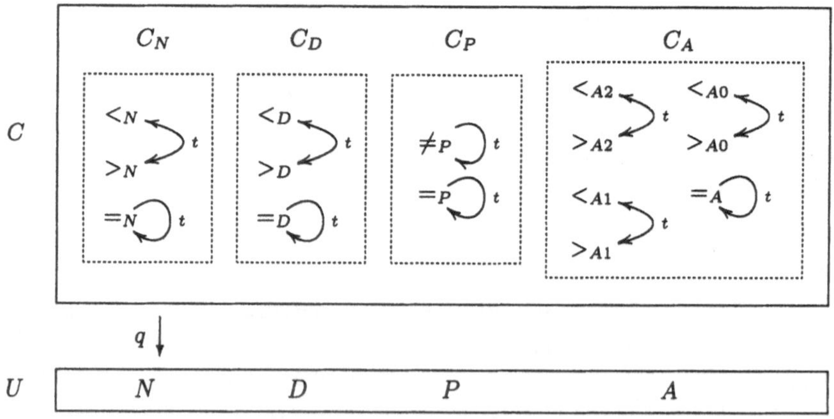

Fig. 2. A comparison system for checking-account database

4 Relational Attribute Models

In this section we introduce a notion of relational attribute models in Dedekind categories, which allow us to express a satisfactory relation of order dependencies, and present the basic properties of relational attribute models.

Throughout this paper we assume $L = (U, C, {}^t, q)$ is a comparison system and $\mathcal{D}$ is a Dedekind category with a unit I. The set of all comparators over L will be denoted by $Rel_L(U, C)$. The hom-set of all relations on an object T in $\mathcal{D}$ is denoted by $\mathcal{D}(T, T)$. Recall that the hom-set $\mathcal{D}(T, T)$ forms a complete Heyting algebra by Definition 1 (D1) of Dedekind categories.

Definition 5. A *relational attribute model* in $\mathcal{D}$ for L is a mapping

$$\theta : Rel_L(U, C) \to \mathcal{D}(T, T)$$

from the set $Rel_L(U, C)$ of all comparators over L into the hom-set $\mathcal{D}(T, T)$ for an object T in $\mathcal{D}$, satisfying the following: For all comparators X and Y over L, attributes a in U and subsets $x, y \subseteq C_a$,

(a) $\theta[X \cup^{\cdot} Y] = \theta[X] \sqcap \theta[Y]$,
(b) $\theta[X^t] = \theta[X]^\sharp$,
(c) $\theta[a^{x \cup y}] = \theta[a^x] \sqcup \theta[a^y]$,
(d) $\mathrm{id}_T \sqsubseteq \theta[U^{\cdot}]$,
(e) If X is not total, then $\theta[X] = 0_{TT}$. □

The following is the basic properties of relational attribute models.

Proposition 4. *Let X, Y, M and N be comparators over L, and $\theta : Rel_L(U, C) \to \mathcal{D}(T, T)$ a relational attribute model in $\mathcal{D}$ for L. Then the following hold:*

(a) *If $X \supseteq^{\cdot} Y$, then $\theta[X] \sqsubseteq \theta[Y]$.*
(b) *$\theta[X \cup^{\cdot} a^{x \cup y}] = \theta[X \cup^{\cdot} a^x] \sqcup \theta[X \cup^{\cdot} a^y]$ for all attributes a in U and subsets x,y of C_a.*
(c) *If $X \supseteq^{\cdot} M$, then $\theta[X] \sqcap (\theta[M] \Rightarrow \theta[N]) = \theta[X \cup^{\cdot} N]$.*
(d) *If X is univalent and $X \not\supseteq^{\cdot} M$, then $\theta[X] \sqsubseteq \theta[M] \Rightarrow \theta[N]$.*
(e) *If X is univalent, then $\theta[X] \sqsubseteq \theta[Y]$ iff $X \supseteq^{\cdot} Y$ or $\theta[X] = 0_{TT}$.*

Proof. (a) Assume $X \supseteq^{\cdot} Y$. Then we have $\theta[X] = \theta[X \cup^{\cdot} Y] = \theta[X] \sqcap \theta[Y] \sqsubseteq \theta[Y]$ by Definition 5(a) .
(b) It follows from Definition 5(a) and (c) that

$$\begin{aligned}
\theta[X \cup^{\cdot} a^x] \sqcup \theta[X \cup^{\cdot} a^y] &= (\theta[X] \sqcap \theta[a^x]) \sqcup (\theta[X] \sqcap \theta[a^y]) && \{\text{ Definition 5(a) }\} \\
&= \theta[X] \sqcap (\theta[a^x] \sqcup \theta[a^y]) && \{\text{ (distributive law) }\} \\
&= \theta[X] \sqcap \theta[a^{x \cup y}] && \{\text{ Definition 5(c) }\} \\
&= \theta[X \cup^{\cdot} a^{x \cup y}]. && \{\text{ Definition 5(a) }\}
\end{aligned}$$

(c) Assume $X \supseteq^{\cdot} M$. Then we have $\theta[X] \sqsubseteq \theta[M]$ by (a), and

$$\begin{aligned}
\theta[X] \sqcap (\theta[M] \Rightarrow \theta[N]) &= \theta[X] \sqcap \theta[N] && \{\text{ Proposition 1(g) }\} \\
&= \theta[X \cup^{\cdot} N]. && \{\text{ Definition 5(a) }\}
\end{aligned}$$

(d) Assume that X is univalent and $X \not\supseteq^{\cdot} M$. By Proposition 3(b) $X \cup^{\cdot} M$ is not total and so we have $\theta[X] \sqcap \theta[M] = 0_{TT}$ by Definition 5(a) and 5(e), which trivially means $\theta[X] \sqsubseteq \theta[M] \Rightarrow \theta[N]$.
(e) First assume that $X \supseteq^{\cdot} Y$ or $\theta[X] = 0_{TT}$. Then $\theta[X] \sqsubseteq \theta[Y]$ is trivial from (a). Conversely assume that X is univalent, $\theta[X] \sqsubseteq \theta[Y]$ and $X \not\supseteq^{\cdot} Y$. Then $X \cup^{\cdot} Y$ is not total by Proposition 3(b). Hence we have $\theta[X] = \theta[X] \sqcap \theta[Y] = \theta[X \cup^{\cdot} Y] = 0_{TT}$. □

Remark 1. In general an inclusion $\theta[X] \sqsubseteq \theta[Y]$ does not imply an inclusion $X \supseteq^{\cdot} Y$. For example, two comparators $X_a = U^{\cdot} \cup^{\cdot} a^{C_a - \{e_a\}}$ and $X_b = U^{\cdot} \cup^{\cdot} b^{C_b - \{e_b\}}$ $(a \neq b)$ give a counter example.

In the rest of this section we explain a construction of relational attribute models from information systems in Dedekind categories and interpretations of comparison systems into actual relations on attribute domains.

An *information system* in a Dedekind category $\mathcal{D}$ is a U-indexed set $\{f_a : T \to T_a \mid a \in U\}$ of functions $f_a : T \to T_a$ in $\mathcal{D}$ with a common domain T. An *interpretation* ξ of L into an information system $\{f_a : T \to T_a \mid a \in U\}$ is a U-indexed set $\{\xi_a : C_a \to \mathcal{D}(T_a, T_a) \mid a \in U\}$ of mappings such that

(a) $\xi_a[s] \sqcap \xi_a[s'] = 0_{T_a T_a}$ if $s, s' \in C_a$ and $s \neq s'$,
(b) $\xi_a[s^t] = \xi_a[s]^\sharp$,
(c) $\sqcup_{s \in C_a} \xi_a[s] = \nabla_{T_a T_a}$, and
(d) $\mathrm{id}_{T_a} \sqsubseteq \xi_a[e_a]$.

Then for a subset x of C_a we define a relation $\xi_a[x] : T_a \rightharpoonup T_a$ by

$$\xi_a[x] = \sqcup_{s \in x} \xi_a[s].$$

Of course, we set $\xi_a[x] = 0_{T_a T_a}$ for $x = \emptyset$. It is obvious that for all subsets x and y of C_a the following properties hold :
(i) $\xi_a[x] \sqcup \xi_a[y] = \xi_a[x \cup y]$, (ii) $\xi_a[x] \sqcap \xi_a[y] = \xi_a[x \cap y]$, and (iii) $\xi_a[C_a] = \nabla_{T_a T_a}$.

We now set $\theta[X] = \sqcap_{a \in U} f_a \xi_a[X(a)] f_a^\sharp$ for each comparator X over L. Then it is easy to see that for all comparators X and Y over L the following five properties of relational attribute models hold:
(a) It is a routine to compute the following:

$$\begin{aligned}
&\theta[X] \sqcap \theta[Y] \\
&= (\sqcap_{a \in U} f_a \xi_a[X(a)] f_a^\sharp) \sqcap (\sqcap_{a \in U} f_a \xi_a[Y(a)] f_a^\sharp) \\
&= \sqcap_{a \in U} f_a (\xi_a[X(a)] \sqcap \xi_a[Y(a)]) f_a^\sharp \\
&= \sqcap_{a \in U} f_a \xi_a[X(a) \cap Y(a)] f_a^\sharp \qquad \{ \text{(ii)} \} \\
&= \sqcap_{a \in U} f_a \xi_a[(X \cup^{\cdot} Y)(a)] f_a^\sharp \\
&= \theta[X \cup^{\cdot} Y].
\end{aligned}$$

(b) Clearly we have $\xi_a[X^t(a)] = \xi_a[X(a)]^\sharp$ by $\xi_a[s^t] = \xi_a[s]^\sharp$, and so

$$\begin{aligned}\theta[X^t] &= \sqcap_{a\in U} f_a\xi_a[X^t(a)]f_a^\sharp \\ &= \sqcap_{a\in U} f_a\xi_a[X(a)]^\sharp f_a^\sharp \\ &= (\sqcap_{a\in U} f_a\xi_a[X(a)]f_a^\sharp)^\sharp \\ &= \theta[X]^\sharp.\end{aligned}$$

(c) As $\theta[a^x] = f_a\xi_a[x]f_a^\sharp$ by (iii) we have

$$\begin{aligned}\theta[a^x] \sqcup \theta[a^y] &= f_a\xi_a[x]f_a^\sharp \sqcup f_a\xi_a[y]f_a^\sharp \\ &= f_a(\xi_a[x] \sqcup \xi_a[y])f_a^\sharp \\ &= f_a\xi_a[x \cup y]f_a^\sharp \qquad \{\text{ (i) }\} \\ &= \theta[a^{x\cup y}].\end{aligned}$$

(d) By $\mathrm{id}_{T_a} \sqsubseteq \xi_a[e_a]$ we have

$$\begin{aligned}\mathrm{id}_T &\sqsubseteq \sqcap_{a\in U} f_a f_a^\sharp \\ &\sqsubseteq \sqcap_{a\in U} f_a\xi_a[\{e_a\}]f_a^\sharp \\ &= \sqcap_{a\in U} f_a\xi_a[U^\cdot(a)]f_a^\sharp \\ &= \theta[U^\cdot].\end{aligned}$$

(e) If a comparator X over L is not total, then there is some attribute a in U such that $X(a) = \emptyset$, so $\theta[X] = 0_{TT}$ is clear from $\xi_a[X(a)] = 0_{T_aT_a}$.

5 Extended Order Dependencies

In this section we define extended order dependency and its satisfactory relation in Dedekind categories, and present some basic laws on extended order dependencies. We then show some properties of propriety and logical implication for extended order dependencies.

Throughout the rest of this paper, we assume $\theta : Rel_L(U, C) \to \mathcal{D}(T, T)$ is a fixed relational attribute model in a Dedekind category $\mathcal{D}$ for a comparison system L.

First we define extended order dependency and its satisfactory relation.

Definition 6. An *extended order dependency* is a formal expression of the form $X \to Y$, where X and Y are comparators over L. An *instance* (an abstraction of database or a set of tuples) on U is a relation $\rho : I \rightharpoonup T$ in $\mathcal{D}$. We say that an instance ρ *satisfies* $X \to Y$, denoted $\rho \models X \to Y$, if and only if $\rho^\sharp\rho \sqsubseteq \theta[X] \Rightarrow \theta[Y]$. An instance ρ on U satisfies a set Γ of extended order dependencies, denoted $\rho \models \Gamma$, if and only if $\rho \models X \to Y$ for each $X \to Y$ in Γ. □

The relational condition $\rho^\sharp\rho \sqsubseteq \theta[X] \Rightarrow \theta[Y]$ in the above definition is equivalent to the condition $(\rho^\sharp\rho \sqcap \mathrm{id}_T)\theta[X](\rho^\sharp\rho \sqcap \mathrm{id}_T) \sqsubseteq \theta[Y]$, that is the definition of functional dependencies in [9]. We use the former condition to express extended order dependencies, since it makes it easier to prove many properties of extended order dependencies by using the fundamental properties of Heyting algebra. Note that the set of all extended order dependencies is finite, since the set of all comparators over L is finite.

We now show the basic laws on extended order dependencies.

Lemma 1. *Let X, Y, Z, W be comparators over L, $a \in U$, $x, y \subseteq C_a$, and ρ an instance on U. Then the following hold:*

FD0. If $\theta[X] \sqsubseteq \theta[Y]$, then $\rho \models X \to Y$. (Reflexivity)
FD1. If $\rho \models X \to Y$ and $Z \supseteq^{\cdot} W$, then $\rho \models X \cup^{\cdot} Z \to Y \cup^{\cdot} W$. (Augmentation)
FD2. If $\rho \models X \to Y$ and $\rho \models Y \to Z$, then $\rho \models X \to Z$. (Transitivity)
OD1. If $\rho \models X \to Y$, then $\rho \models X^t \to Y^t$. (Reversal)
OD2. If $\rho \models X \cup^{\cdot} a^x \to Y$ and $\rho \models X \cup^{\cdot} a^y \to Y$, then $\rho \models X \cup^{\cdot} a^{x \cup y} \to Y$. (Generalized Disjunction)
OD3. If $U^{\cdot} \not\supseteq^{\cdot} N$ and $\rho \models U^{\cdot} \to N$, then $\rho \models X \to Y$ for all order dependencies $X \to Y$. (Generalized Impropriety)

Proof. FD0. Assume that $\theta[X] \sqsubseteq \theta[Y]$. Then $\theta[X] \Rightarrow \theta[Y] = \nabla_{TT}$ holds. Hence the claim is clear.
FD1. Assume $Z \supseteq^{\cdot} W$. Then we have $\theta[Z] \sqsubseteq \theta[W]$ and

$$
\begin{aligned}
\theta[X] \Rightarrow \theta[Y] &\sqsubseteq (\theta[X] \sqcap \theta[Z]) \Rightarrow (\theta[Y] \sqcap \theta[Z]) && \{ \text{ Proposition 1(d) } \} \\
&\sqsubseteq (\theta[X] \sqcap \theta[Z]) \Rightarrow (\theta[Y] \sqcap \theta[W]) && \{ \text{ Proposition 1(c) } \} \\
&= \theta[X \cup^{\cdot} Z] \Rightarrow \theta[Y \cup^{\cdot} W]. && \{ \text{ Definition 5(a) } \}
\end{aligned}
$$

Hence the claim is clear.
FD2. The claim directly follows from Proposition 1(f):

$$(\theta[X] \Rightarrow \theta[Y]) \sqcap (\theta[Y] \Rightarrow \theta[Z]) \sqsubseteq \theta[X] \Rightarrow \theta[Z].$$

OD1. Assume that $\rho^\sharp \rho \sqsubseteq \theta[X] \Rightarrow \theta[Y]$. Then we have

$$\rho^\sharp \rho \sqsubseteq (\theta[X] \Rightarrow \theta[Y])^\sharp = \theta[X]^\sharp \Rightarrow \theta[Y]^\sharp = \theta[X^t] \Rightarrow \theta[Y^t]$$

by Proposition 1(e) and Definition 5(b).
OD2. The claim is immediate from

$$
\begin{aligned}
&(\theta[X \cup^{\cdot} a^x] \Rightarrow \theta[Y]) \sqcap (\theta[X \cup^{\cdot} a^y] \Rightarrow \theta[Y]) \\
&\sqsubseteq (\theta[X \cup^{\cdot} a^x] \sqcup \theta[X \cup^{\cdot} a^y]) \Rightarrow \theta[Y] && \{ \text{ Proposition 1(h) } \} \\
&= \theta[X \cup^{\cdot} a^{x \cup y}] \Rightarrow \theta[Y] && \{ \text{ Proposition 4(b) } \}
\end{aligned}
$$

OD3. Assume $U^{\cdot} \not\supseteq^{\cdot} N$ and $\rho^\sharp \rho \sqsubseteq \theta[U^{\cdot}] \Rightarrow \theta[N]$. Then we have

$$
\begin{aligned}
\rho &= \rho \sqcap \rho \\
&\sqsubseteq \rho(\mathrm{id}_T \sqcap \rho^\sharp \rho) && \{ \text{ Dedekind Formula } \} \\
&\sqsubseteq \rho\{\theta[U^{\cdot}] \sqcap (\theta[U^{\cdot}] \Rightarrow \theta[N])\} && \{ \text{ Definition 5(d) } \} \\
&= \rho\theta[U^{\cdot} \cup^{\cdot} N] && \{ \text{ Proposition 4(c) } \} \\
&= 0_{IT}. && \{ \text{ Proposition 3(b) and Definition 5(e) } \}
\end{aligned}
$$

Therefore $\rho^\sharp\rho = 0_{TT} \sqsubseteq \theta[X] \Rightarrow \theta[Y]$ for all extended order dependencies $X \rightarrow Y$. □

A relation $\rho : I \rightharpoondown T$ in $\mathcal{D}$ is called *nonzero* if $\rho \neq 0_{IT}$. In what follows we assume T has at least one I-point.

The following defines the propriety and the impropriety of extended order dependencies (EODs, for short).

Definition 7. (a) A set Γ of EODs is *proper* if some nonzero instance on U satisfies Γ, and is *improper* otherwise.

$$\Gamma \text{ is proper} \Leftrightarrow \exists\rho\,(\text{nonzero instance on } U) : \rho \models \Gamma$$

$$\begin{aligned}\Gamma \text{ is improper} &\Leftrightarrow \forall\rho\,(\text{nonzero instance on } U) : \rho \not\models \Gamma \\ &\Leftrightarrow \forall\rho\,(\text{instance on } U) : \rho \models \Gamma \text{ implies } \rho = 0_{IT}\end{aligned}$$

(b) For a set Γ of EODs we define two relations $\theta[\Gamma], \theta^*[\Gamma] : T \rightharpoondown T$ as follows:

$$\theta[\Gamma] = \sqcap_{M\rightarrow N\in\Gamma}(\theta[M] \Rightarrow \theta[N]) \text{ and } \theta^*[\Gamma] = \theta[\Gamma] \sqcap \theta[\Gamma]^\sharp.$$

□

Note that for an instance ρ and a set Γ of EODs, $\rho \models \Gamma$ if and only if $\rho^\sharp\rho \sqsubseteq \theta[\Gamma]$ in a relational notation.

We now have the following equivalence on the propriety of EODs.

Proposition 5. *Let Γ be a set of EODs. Then the following three statements are equivalent:*

(a) *Γ is proper,*
(b) *$U^\cdot \not\supseteq^\cdot M$ or $U^\cdot \supseteq^\cdot N$ for all $M \rightarrow N$ in Γ,*
(c) $\mathrm{id}_T \sqsubseteq \theta[\Gamma]$.

Proof. (a)⇒(b) We will prove the contraposition ¬(b)⇒ ¬(a): If there exists an EOD $M \rightarrow N \in \Gamma$ such that $U^\cdot \supseteq^\cdot M$ and $U^\cdot \not\supseteq^\cdot N$, then Γ is improper. Let $M \rightarrow N$ be an EOD in Γ such that $U^\cdot \supseteq^\cdot M$ and $U^\cdot \not\supseteq^\cdot N$. Assume $\rho \models \Gamma$. Then $\rho \models M \rightarrow N$ by the definition. Then we have $\rho \models U^\cdot \rightarrow M$ by Proposition 4(a) and Lemma 1(FD0) and so $\rho \models U^\cdot \rightarrow N$ by Lemma 1(FD2). Hence $\rho = 0_{IT}$ follows from the proof of Lemma 1(OD3), and this means Γ is improper.
(b)⇒(c) Let $M \rightarrow N$ be an EOD in Γ. First assume $U^\cdot \not\supseteq^\cdot M$. Then $\mathrm{id}_T \sqsubseteq \theta[U^\cdot] \sqsubseteq \theta[M] \Rightarrow \theta[N]$ by Definition 5(d) and Proposition 4(d). Next suppose $U^\cdot \supseteq^\cdot N$. Then $\mathrm{id}_T \sqsubseteq \theta[U^\cdot] \sqsubseteq \theta[N] \sqsubseteq \theta[M] \Rightarrow \theta[N]$ by Proposition 4(a) and 1(a). Hence $\mathrm{id}_T \sqsubseteq \sqcap_{M\rightarrow N\in\Gamma}(\theta[M] \Rightarrow \theta[N]) = \theta[\Gamma]$.
(c)⇒(a) Recall that we assume T has at least one I-point. Let $u : I \rightarrow T$ be an I-point. Then u is a nonzero instance and $u^\sharp u \sqsubseteq \mathrm{id}_T \sqsubseteq \theta[\Gamma]$, that is $u \models \Gamma$. Hence Γ is proper. □

We now define the notion of logical implication for EODs.

Definition 8. Let Γ be a set of EODs, and $X \to Y$ an EOD. Then Γ *logically implies* $X \to Y$, denoted $\Gamma \models X \to Y$, if and only if $\rho \models \Gamma$ implies $\rho \models X \to Y$ for each instance ρ on U. □

Now we need the point-wise argument to get the following property related to logical implications. Suppose that every relation $\alpha : A \rightharpoondown B$ in $\mathcal{D}$ is equal to the supremum of all relations $u^\sharp v : A \rightharpoondown B$ such that $u^\sharp v \sqsubseteq \alpha$ for a pair $u : I \rightharpoondown A$ and $v : I \rightharpoondown B$ of univalent relations. It is easy to see that, under this hypothesis, for two relations $\alpha, \alpha' : A \rightharpoondown B$ in $\mathcal{D}$, $\alpha \sqsubseteq \alpha'$ if and only if $u^\sharp v \sqsubseteq \alpha$ implies $u^\sharp v \sqsubseteq \alpha'$ for all pairs of univalent relations $u : I \rightharpoondown A$ and $v : I \rightharpoondown B$. The hypothesis on $\mathcal{D}$ is in fact so week in computer science applications, because all Dedekind categories of so-called L-fuzzy relations (due to Goguen) trivially satisfy it.

Proposition 6. *Let a set Γ of EODs be proper and $X \to Y$ an EOD. Then $\Gamma \models X \to Y$ iff $\theta^*[\Gamma] \sqsubseteq \theta[X] \Rightarrow \theta[Y]$.*

Proof. Assume $\theta^*[\Gamma] \sqsubseteq \theta[X] \Rightarrow \theta[Y]$ and $\rho \models \Gamma$ for a nonzero instance ρ on U, that is, $\rho^\sharp \rho \sqsubseteq \theta[\Gamma]$. Then $\rho^\sharp \rho = (\rho^\sharp \rho)^\sharp \sqsubseteq \theta[\Gamma]^\sharp$ and so $\rho^\sharp \rho \sqsubseteq \theta^*[\Gamma] \sqsubseteq \theta[X] \Rightarrow \theta[Y]$. Hence $\rho \models X \to Y$. Therefore $\Gamma \models X \to Y$.
Conversely assume $\Gamma \models X \to Y$ and let $u^\sharp v \sqsubseteq \theta^*[\Gamma]$ for a pair of univalent relations $u : I \rightharpoondown T$ and $v : I \rightharpoondown T$. Then $v^\sharp u = (u^\sharp v)^\sharp \sqsubseteq \theta^*[\Gamma]^\sharp = \theta^*[\Gamma] \sqsubseteq \theta[\Gamma]$ and $u^\sharp v \sqsubseteq \theta[\Gamma]$. Set $\rho = u \sqcup v$. Then $\rho^\sharp \rho = u^\sharp u \sqcup v^\sharp v \sqcup u^\sharp v \sqcup v^\sharp u$, and by Proposition 5 $u^\sharp u, v^\sharp v \sqsubseteq \mathrm{id}_T \sqsubseteq \theta[\Gamma]$, since u and v are univalent. Hence $\rho^\sharp \rho \sqsubseteq \theta[\Gamma]$, that is, $\rho \models \Gamma$. By the assumption we have $\rho \models X \to Y$, which shows $u^\sharp v \sqsubseteq \rho^\sharp \rho \sqsubseteq \theta[X] \Rightarrow \theta[Y]$. Therefore we conclude $\theta^*[\Gamma] \sqsubseteq \theta[X] \Rightarrow \theta[Y]$ by the above hypothesis on $\mathcal{D}$. This completes the proof. □

6 Inference Rules

Ginsburg and Hull [5] introduced a sound and complete set of inference rules for order dependencies. In this section we modify the inference rules due to the generalization of marked attributes. We show the soundness and completeness with respect to a modified set of inference rules.

First we present the inference rules for EODs.

Definition 9. Let X, Y, Z and W be comparators over L and $a \in U$ and $x, y \subseteq C_a$. Inference rules for EODs are the following six rules:

$$\text{[FD0]} \qquad \{\theta[X] \sqsubseteq \theta[Y]\} \frac{}{X \to Y}$$

$$\text{[FD1]} \qquad \{Z \supseteq^{\cdot} W\} \frac{X \to Y}{X \cup^{\cdot} Z \to Y \cup^{\cdot} W}$$

$$\text{[FD2]} \qquad \frac{X \to Y \quad Y \to Z}{X \to Z}$$

$$\text{[OD1]} \qquad \frac{X \to Y}{X^t \to Y^t}$$

$$\text{[OD2]} \qquad \frac{X \cup^{\cdot} a^x \to Y \quad X \cup^{\cdot} a^y \to Y}{X \cup^{\cdot} a^{x \cup y} \to Y}$$

$$\text{[OD3]} \qquad \{U^{\cdot} \not\supseteq^{\cdot} N\} \frac{U^{\cdot} \to N}{X \to Y}.$$

The above inference rules [FD0], [FD1] and [OD3] can be used only when each assumption in the braces is satisfied. □

The rules [FD0]-[FD2] are analogous to ones for functional dependencies, [FD0]-[OD1] are the same as in [5], and [OD2] and [OD3] are brief modifications of the corresponding rules in [5].

Let Γ be a set of EODs and J a set of inference rules. A *proof* from Γ using J is a nonempty sequence

$$X_1 \to Y_1, \ldots, X_n \to Y_n$$

of EODs such that, for each $i = 1, \ldots, n$, either: (i) $X_i \to Y_i$ is in Γ, or (ii) $X_i \to Y_i$ follows from an application of some inference rule in J to some subset of $\{X_1 \to Y_1, \ldots, X_{i-1} \to Y_{i-1}\}$. An EOD $X \to Y$ is *provable* from Γ using J, we write $\Gamma \vdash_J X \to Y$, if there is a proof $X_1 \to Y_1, \ldots, X_n \to Y_n$ from Γ using J such that $X_n = X$ and $Y_n = Y$.

For example, it is well known that the union rule

$$\text{[FD3]} \qquad \frac{X \to Y \quad X \to Z}{X \to Y \cup^{\cdot} Z}$$

is proved from [FD1] and [FD2] as follows:

$$\frac{\{X \supseteq^{\cdot} X\} \dfrac{X \to Y}{X \to X \cup^{\cdot} Y}\text{[FD1]} \quad \{Y \supseteq^{\cdot} Y\} \dfrac{X \to Z}{X \cup^{\cdot} Y \to Y \cup^{\cdot} Z}\text{[FD1]}}{X \to Y \cup^{\cdot} Z}\text{[FD2]}.$$

The appropriateness of a set of inference rules is formally described by the following definition.

Definition 10. Let J be a set of inference rules. J is *sound* if for each set Γ of EODs and EOD $X \to Y$, $\Gamma \vdash_J X \to Y$ implies $\Gamma \models X \to Y$. Moreover, J is *complete* if for each set Γ of EODs and EOD $X \to Y$, $\Gamma \models X \to Y$ implies $\Gamma \vdash_J X \to Y$. □

In the following proposition we see the power of the inference rule $[OD2]$.

Proposition 7. *Let J be a set of inference rules containing $[OD2]$.*

(a) *Let $a \in U$. Then $\Gamma \vdash_J X \to Y$ if $\Gamma \vdash_J Z \to Y$ for all comparators Z such that $Z \supseteq^{\cdot} X$ and $|Z(a)| \leq 1$.*
(b) *If $\Gamma \vdash_J Z \to Y$ for all univalent comparators Z such that $Z \supseteq^{\cdot} X$, then $\Gamma \vdash_J X \to Y$.*

Proof. (a) Assume $|X(a)| \leq 1$. Then X itself can be a candidate of Z's and so the claim is trivial. Next assume $X(a) = \{s_1, \cdots, s_k\}$ $(k \geq 2)$. First note that $X = X \cup^{\cdot} a^{X(a)} = X \cup^{\cdot} a^{\{s_1, s_2, \ldots, s_k\}}$ by Proposition 2(b) and $(X \cup^{\cdot} a^{\{s_j\}})(a) = X(a) \cap \{s_j\} = \{s_j\}$ for each $j = 1, \ldots, k$. Then each EOD $X \cup^{\cdot} a^{\{s_j\}} \to Y$ is provable from Γ by the assumption. Therefore $\Gamma \vdash X \to Y$ follows from a successive application of [OD2]:

$$\frac{X \cup^{\cdot} a^{\{s_1\}} \to Y \quad X \cup^{\cdot} a^{\{s_2\}} \to Y \quad \cdots \quad X \cup^{\cdot} a^{\{s_k\}} \to Y}{X \cup^{\cdot} a^{\{s_1, s_2, \ldots, s_k\}} \to Y} {}_*[\mathrm{OD2}].$$

(b) We can set $U = \{a_1, a_2, \ldots, a_n\}$ because U is a finite set by the definition. We now prove the following implication for all $i = 0, \ldots, n-1$:

$$(*) \quad \forall Z_{i+1} \in K_{i+1} : \Gamma \vdash Z_{i+1} \to Y \ \Rightarrow \ \forall Z_i \in K_i : \Gamma \vdash Z_i \to Y,$$

where K_i $(i = 0, \ldots, n)$ denotes the set of all comparators Z satisfying

$$Z \supseteq^{\cdot} X \text{ and } |Z(a_j)| \leq 1 \text{ for each } j = 1, \ldots, i.$$

For each $Z_i \in K_i$ we have an implication by (a):

$$\forall Z_{i+1} \in K'_{i+1} : \Gamma \vdash Z_{i+1} \to Y \ \Rightarrow \ \Gamma \vdash Z_i \to Y,$$

where K'_{i+1} denotes the set of all comparators Z_{i+1} satisfying

$$Z_{i+1} \supseteq^{\cdot} Z_i \text{ and } |Z_{i+1}(a_{i+1})| \leq 1.$$

Noticing $K'_{i+1} \subseteq K_{i+1}$ we have the desired implication $(*)$ and consequently

$$\forall Z \in K_n : \Gamma \vdash Z \to Y \ \Rightarrow \ \forall Z \in K_0 : \Gamma \vdash Z \to Y.$$

Finally remark that K_n consists of all univalent comparators Z such that $Z \supseteq^{\cdot} X$, and K_0 is the set of all comparators Z such that $Z \supseteq^{\cdot} X$. This completes the proof. □

The section is closed with the soundness of a set of inference rules.

Theorem 1. (Soundness Theorem) *The set $J = \{[FD0], [FD1], [FD2], [OD1], [OD2], [OD3]\}$ of inference rules is sound.*

Proof. Assume that $\Gamma \vdash_J X \to Y$ and $\rho \models \Gamma$. Then $\rho \models X \to Y$, since the basic laws FD0,FD1,FD2,OD1,OD2 and OD3 of EODs in Lemma 1 are satisfied. □

7 Completeness

In this section we show the completeness of the set $J = \{[FD0], [FD1], [FD2], [OD1], [OD2], [OD3]\}$ of inference rules. First we see that all EODs are provable from any improper set of EODs.

Lemma 2. *If a set Γ of EODs is improper, then $\Gamma \vdash_J X \to Y$ for each EOD $X \to Y$.*

Proof. As Γ is improper, there exists an EOD $M \to N \in \Gamma$ such that $U^{\cdot} \supseteq^{\cdot} M$ and $U^{\cdot} \not\supseteq^{\cdot} N$ by Proposition 5. Therefore we have the following derivation in J:

$$\cfrac{\cfrac{\{\theta[U^{\cdot}] \sqsubseteq \theta[M]\}}{U^{\cdot} \to M}\text{[FD0]} \quad M \to N \in \Gamma}{\{\, U^{\cdot} \not\supseteq^{\cdot} N \,\}\cfrac{U^{\cdot} \to N}{X \to Y}\text{[OD3]}}\text{[FD2]}.$$

Hence $\Gamma \vdash_J X \to Y$ for each EOD $X \to Y$. □

By the above lemma, to prove the completeness of the set J of inference rules it is enough to show that for each proper set Γ of EODs and EOD $X \to Y$, if $\Gamma \models X \to Y$ then $X \to Y$ is provable from Γ using $J' = \{[FD0], [FD1], [FD2], [OD1], [OD2]\}$.

To show the completeness we need the following definition.

Definition 11. Let Γ be a set of EODs and X a comparator. The set of all EODs $M \to N \in \Gamma$ such that $X \supseteq^{\cdot} M$ or $X \supseteq^{\cdot} M^t$ will be denoted by Γ_X. Also for an EOD $M \to N \in \Gamma_X$ we define

$$\delta(M \to N) = \begin{cases} N & \text{if } X \supseteq^{\cdot} M \text{ and } X \not\supseteq^{\cdot} M^t, \\ N^t & \text{if } X \not\supseteq^{\cdot} M \text{ and } X \supseteq^{\cdot} M^t, \\ N \cup^{\cdot} N^t & \text{if } X \supseteq^{\cdot} M \text{ and } X \supseteq^{\cdot} M^t \end{cases}$$

and for a subset $\Gamma' \subseteq \Gamma_X$

$$\delta(\Gamma') = \cup^{\cdot}_{M \to N \in \Gamma'} \delta(M \to N).$$

□

Proposition 8. *Let Γ be a set of EODs and $X \to Y$ an EOD. Then the following hold:*

(a) *If $M \to N \in \Gamma_X$, then $\{M \to N\} \vdash_{J'} X \to \delta(M \to N)$.*
(b) $\Gamma \vdash_{J'} X \to \delta(\Gamma_X)$.
(c) *If $\Gamma \vdash_{J'} X \cup^{\cdot} \delta(\Gamma_X) \to Y$, then $\Gamma \vdash_{J'} X \to Y$.*

Proof. (a) Let $M \to N \in \Gamma_X$. In the case of $X \supseteq^{\cdot} M$ and $X \not\supseteq^{\cdot} M^t$ we have $\theta[X] \sqsubseteq \theta[M]$ and

$$\frac{\dfrac{\{\theta[X] \sqsubseteq \theta[M]\}}{X \to M}[\mathrm{FD0}] \quad M \to N}{X \to N}[\mathrm{FD2}].$$

In the case of $X \not\supseteq^{\cdot} M$ and $X \supseteq^{\cdot} M^t$ we have $\theta[X] \sqsubseteq \theta[M^t]$ and

$$\frac{\dfrac{\{\theta[X] \sqsubseteq \theta[M^t]\}}{X \to M^t}[\mathrm{FD0}] \quad \dfrac{M \to N}{M^t \to N^t}[\mathrm{OD1}]}{X \to N^t}[\mathrm{FD2}].$$

In the case of $X \supseteq^{\cdot} M$ and $X \supseteq^{\cdot} M^t$ we have $\theta[X] \sqsubseteq \theta[M]$ and $\theta[X] \sqsubseteq \theta[M^t]$. Hence we have $\{M \to N\} \vdash_{J'} X \to N \cup^{\cdot} N^t$ by $[FD3]$, since $\{M \to N\} \vdash_{J'} X \to N$ and $\{M \to N\} \vdash_{J'} X \to N^t$.
(b) Recall that $\delta(\Gamma_X) = \cup^{\cdot}_{M \to N \in \Gamma_X} \delta(M \to N)$ by the definition. By the virtue of (a), we have $\Gamma \vdash_{J'} X \to \delta(M \to N)$ for all EODs $M \to N \in \Gamma_X$, and so $\Gamma \vdash_{J'} X \to \delta(\Gamma_X)$ by successive application of the union rule [FD3].
(c) The claim is direct from the following proof figure:

$$\frac{\dfrac{\{X \supseteq^{\cdot} X\} \quad \dfrac{\Gamma}{X \to \delta(\Gamma_X)}*[J']\ \{\text{(b)}\}}{X \to X \cup^{\cdot} \delta(\Gamma_X)}[FD1] \quad \dfrac{\Gamma}{X \cup^{\cdot} \delta(\Gamma_X) \to Y}*[J']}{X \to Y}[FD2].$$

□

We show key properties of univalent comparators.

Lemma 3. *Let Γ be a set of EODs and X a univalent comparator over L. Then the following hold:*

(a) *If $M \to N \in \Gamma_X$, then $\theta[X \cup^{\cdot} \delta(M \to N)] = \theta[X] \sqcap \theta^*[\{M \to N\}]$.*
(b) $\theta[X \cup^{\cdot} \delta(\Gamma_X)] = \theta[X] \sqcap \theta^*[\Gamma]$.

Proof. (a) Let $M \to N \in \Gamma_X$. In the case of $X \supseteq^{\cdot} M$ and $X \not\supseteq^{\cdot} M^t$ we have

$$\begin{aligned}
&\theta[X] \sqcap \theta^*[\{M \to N\}] \\
&= \theta[X] \sqcap (\theta[M] \Rightarrow \theta[N]) \sqcap (\theta[M] \Rightarrow \theta[N])^\sharp && \{\text{ Definition 7(b) }\} \\
&= \theta[X] \sqcap (\theta[M] \Rightarrow \theta[N]) \sqcap (\theta[M^t] \Rightarrow \theta[N^t]) && \{\text{ Proposition 1(e) }\} \\
&= \theta[X] \sqcap (\theta[M] \Rightarrow \theta[N]) && \{\text{ Proposition 4(d) }\} \\
&= \theta[X \cup^{\cdot} N] && \{\text{ Proposition 4(c) }\} \\
&= \theta[X \cup^{\cdot} \delta(M \to N)].
\end{aligned}$$

In the case of $X \not\supseteq^{\cdot} M$ and $X \supseteq^{\cdot} M^t$:

$$\begin{aligned}
&\theta[X] \sqcap (\theta[M] \Rightarrow \theta[N]) \sqcap (\theta[M^t] \Rightarrow \theta[N^t]) \\
&= \theta[X] \sqcap (\theta[M^t] \Rightarrow \theta[N^t]) && \{\text{ Proposition 4(d) }\} \\
&= \theta[X \cup^{\cdot} N^t]. && \{\text{ Proposition 4(c) }\}
\end{aligned}$$

In the case of $X \sqsupseteq^{\cdot} M$ and $X \sqsupseteq^{\cdot} M^t$:

$$\begin{aligned} & \theta[X] \sqcap (\theta[M] \Rightarrow \theta[N]) \sqcap (\theta[M^t] \Rightarrow \theta[N^t]) \\ &= \theta[X \cup^{\cdot} N] \sqcap \theta[X \cup^{\cdot} N^t] && \{ \text{ Proposition 4(c) } \} \\ &= \theta[X \cup^{\cdot} N \cup^{\cdot} N^t]. && \{ \text{ Definition 5(a) } \} \end{aligned}$$

(b) First we see that $\theta[X] \sqsubseteq \theta^*[\Gamma - \Gamma_X]$. Let $M \to N \in \Gamma - \Gamma_X$. Then $X \not\sqsupseteq^{\cdot} M$ and $X \not\sqsupseteq^{\cdot} M^t$ and so by Proposition 4(d) we have

$$\theta[X] \sqsubseteq (\theta[M] \Rightarrow \theta[N]) \sqcap (\theta[M^t] \Rightarrow \theta[N^t]),$$

which shows $\theta[X] \sqsubseteq \theta^*[\Gamma - \Gamma_X]$. Therefore it follows from (a) that

$$\begin{aligned} \theta[X \cup^{\cdot} \delta(\Gamma_X)] &= \sqcap_{M \to N \in \Gamma_X} \theta[X \cup^{\cdot} \delta(M \to N)] && \{ \text{ Definition 11 } \} \\ &= \sqcap_{M \to N \in \Gamma_X} (\theta[X] \sqcap \theta^*[\{M \to N\}]) && \{ \text{ (a) } \} \\ &= \theta[X] \sqcap (\sqcap_{M \to N \in \Gamma_X} \theta^*[\{M \to N\}]) \\ &= \theta[X] \sqcap \theta^*[\Gamma_X] && \{ \text{ Definition 7(b) } \} \\ &= \theta[X] \sqcap \theta^*[\Gamma - \Gamma_X] \sqcap \theta^*[\Gamma_X] && \{ \theta[X] \sqsubseteq \theta^*[\Gamma - \Gamma_X] \} \\ &= \theta[X] \sqcap \theta^*[\Gamma] && \{ \text{ Definition 7(b) } \}. \end{aligned}$$

□

Finally we give a completeness theorem for the set J of inference rules for EODs.

Theorem 2. (Completeness Theorem) *The set* $J = \{[FD0], [FD1], [FD2], [OD1], [OD2], [OD3]\}$ *of inference rules is complete.*

Proof. By Lemma 2 it is enough to show that for each proper set Γ of EODs and EOD $X \to Y$, if $\Gamma \models X \to Y$ then $\Gamma \vdash_{J'} X \to Y$. Suppose that Γ is proper and $\Gamma \models X \to Y$. We will prove that $\Gamma \vdash_{J'} X \to Y$. But by Proposition 7(b) it suffices to show that $\Gamma \vdash_{J'} Z \to Y$ for all univalent comparators Z such that $Z \sqsupseteq^{\cdot} X$. Now let Z be a univalent comparator such that $Z \sqsupseteq^{\cdot} X$. Recall that $\Gamma \models X \to Y$ iff $\theta^*[\Gamma] \sqsubseteq \theta[X] \Rightarrow \theta[Y]$ by Proposition 6. From $\theta[Z] \sqsubseteq \theta[X]$ it simply follows that $\theta^*[\Gamma] \sqsubseteq \theta[Z] \Rightarrow \theta[Y]$ by Proposition 1(c), which is equivalent to $\theta[Z] \sqcap \theta^*[\Gamma] \sqsubseteq \theta[Y]$. By the virtue of the last Lemma 3(b) we have $\theta[Z \cup^{\cdot} \delta(\Gamma_Z)] = \theta[Z] \sqcap \theta^*[\Gamma] \sqsubseteq \theta[Y]$ and so $\Gamma \vdash_{J'} Z \cup^{\cdot} \delta(\Gamma_Z) \to Y$ by $[FD0]$. Therefore by Proposition 8(c) we can conclude $\Gamma \vdash_{J'} Z \to Y$, which completes the proof. □

8 Conclusion

In this paper, we improved theory of order dependency by Ginsburg and Hull using comparison systems (an abstraction of the marked attributes in [5]) and relational attribute models in Dedekind categories (a generalization of those in [9] or indiscernibility relations in [10]). The EOD defined here is just an extension of functional dependencies in [9], and the basic laws FD0, FD1, FD2, OD1, OD2 and OD3 of EODs were proved by only fundamental properties of

Heyting algebra. Without using the complicated production rules [5] we could give a simple proof of completeness theorem for EOD.

We believe that the formalization introduced in this paper can be applied to discover useful knowledge from large databases with attribute domains with order relations.

References

1. Brink, C., Kahl, W. and Schmidt, G.(eds.): Relational Methods in Computer Science, Advances in Computing Science. Springer-Verlag, 1997
2. Beeri, C., Fagin, R. and Howard, J.H.: A complete axiomatization for functional and multivalued dependencies in database relations. In Proc. ACM SIGMOD Internat. Conf. on Management of Data, Toronto, 1977, 47–61
3. Buszkowski, W. and Orłowska, E.: Indiscernibility-based formalization of dependencies in information systems. In Orłowska, E. (ed.), Incomplete information : Rough set analysis, Physica Verlag, 1998, 293–315.
4. Freyd, P.J. and Scedrov, A.: Categories, Allegories. North-Holland, 1990.
5. Ginsburg, S. and Hull, R.: Order dependency in the relational model. Theor. Comput. Sci. **26**(1983), 149–195.
6. Jaoua, A., Belkhiter, N., Ounalli, H. and Moukam, T.: Databases. In [1], 197–210.
7. MacCaull, W.: A proof system for dependencies for information relations. Fundamenta Informaticae, **42**(1) (2000) 1–27.
8. MacCaull, W.: A tableaux procedure for the implication problem for association rules. In Orłowska, E. and Szalas, A. (eds.), Relational Methods for Computer Science Applications, Springer-Physica Verlag, 2001, 77–95.
9. Okuma, H. and Kawahara, Y.: Relational aspects of relational database dependencies. Bulletin of Informatics and Cybernetics, **32**(2)(2000), 91–104.
10. Orłowska, E.: Algebraic approach to database constraints. Fundamenta Informaticae, **10** (1987), 57–68.
11. Ounalli, H. and Jaoua, A.: Fuzzy difunctional dependencies. In Proc. 3rd RelMiCS, Hammamet, Tunisia, 1997, 51–62.
12. Schmidt, G. and Ströhlein, T.: Relations and Graphs - Discrete Mathematics for Computer Scientists. Springer-Verlag, 1993.

Double Residuated Lattices and Their Applications

Ewa Orłowska[1] and Anna Maria Radzikowska[2]

[1] Institute of Telecommunications
Szachowa 1, 04–894 Warsaw, POLAND
orlowska@itl.waw.pl

[2] Faculty of Mathematics and Information Science
Warsaw University of Technology
Plac Politechniki 1, 00–661 Warsaw, POLAND
annrad@mini.pw.edu.pl

Abstract. In this paper we introduce a new class of *double residuated lattices*. Basic properties of these algebras are given. Taking double residuated lattices as a basis, we propose a fuzzy generalisation of information relations. We also define several fuzzy information operators and show that some classes of information relations can be characterised by means of these operators.

Keywords: Residuated lattices, Information relations, Information operators, Fuzzy sets, Fuzzy logical operators.

1 Introduction

In this paper we introduce the class of double residuated lattices and propose a generalisation of information relations to the relations based on these lattices. Our approach is motivated, on one hand, by the role that residuated lattices ([8], [21]) play in the fuzzy set theory ([17], [20], [37], see also [16]), and on the other hand, by the rough set-style data analysis ([27]). In a residuated lattice the product operator is an abstract counterpart of a triangular norm (see [36]) which is a generalisation of classical conjunction and determines the intersection operator of fuzzy sets. The union of fuzzy sets is usually defined in a De Morgan-style from the product. However, this does not provide a sufficiently general counterpart of a triangular conorm. Our view is that an adequate generalisation of a triangular norm and a triangular conorm is provided by two independent monoid operators of product and sum, and their corresponding residua and dual residua, respectively, adjoined to a lattice. We refer to such lattices as *double residuated lattices*. Various extensions of residuated lattices motivated by multiple-valued logics, in particular fuzzy logics, can be found in the literature (e.g. [5], [12], [13], [17]). In this paper we mention the analogous extensions of double residuated lattices.

We present multiple-valued generalisation of information relations and information operators determined by these relations. In the classical setting an

H. de Swart (Ed.): RelMiCS 2001, LNCS 2561, pp. 171–189, 2002.

information relation is any relation defined on a set of objects of an information system and determined by the attributes of the information system. Examples of some information relations and their theories can be found e.g., in [2], [6], [9], [24], [26], [38]. A comprehensive exposition of logical and algebraic theories of information relations and their applications is presented in [7]. Multiple-valued generalisations of information relations are developed in [28], [32]. In [28] information relations derived from deterministic information systems are generalised to the relations based on residuated lattices. In [32] and [33] information relations derived from non-deterministic information systems are generalised to the relations based on the real [0,1] interval.

Along the lines of Kripke semantics relations determine modal-like operators which, in turn, are the abstract counterparts to the information operators derived from information systems (see [4], [7]). By information operators we mean the operators acting on sets of objects of an information system and determined by information relations. A general theory of the classical abstract information operators is developed in [9], [10] and [11]. A fuzzy generalisation, based on the interval [0,1], of some information operators, corresponding to rough approximation operators, are investigated in [30] and [31].

A generalisation of information operators to the ones determined by relations based on double residuated lattices is an open problem. In this paper we suggest some operators of that kind. They may be a basis of multiple-valued modal logics and algebras. Moreover, they can be viewed as generalised approximation operators and sufficiency operators, respectively, which are the main tool of the rough set data analysis. We show that various properties of information relations can be expressed with these information operators. This paper is an extension of [29].

2 Double Residuated Lattices

Definition 1. *Let $(W, \leqslant)$ be a poset and let $\bigcirc$ be a binary operation on W. We define the following residua $\rightarrow$ and $\Rightarrow$ of $\bigcirc$:*

- *$\rightarrow$ is the **left residuum of** $\bigcirc$ iff*

$$z \bigcirc x \leqslant y \text{ iff } z \leqslant x \rightarrow y \text{ for all } x, y, z \in W. \tag{1}$$

- *$\Rightarrow$ is the **right residuum of** $\bigcirc$ iff*

$$x \bigcirc z \leqslant y \text{ iff } z \leqslant x \Rightarrow y \text{ for all } x, y, z \in W. \tag{2}$$

*Conditions (1)–(2) are called **residuation conditions**.* □

Clearly, $\rightarrow$ and $\Rightarrow$ are uniquely determined by the residuation conditions. Note that the existence of the residua implies that

- $x \rightarrow y$ is the greatest element in the set $\{z \in W : z \bigcirc x \leqslant y\}$
- $x \Rightarrow y$ is the greatest element in the set $\{z \in W : x \bigcirc z \leqslant y\}$.

Definition 2. *A* ***monoid*** *is a structure of the form* $\mathfrak{M}=(W,\bigcirc,e)$*, where* W *is a non-empty set,* $\bigcirc$ *is an associative binary operation on* W *and* e *is the unit element.* □

Definition 3. *A structure* $\mathfrak{L}=(W,+,\bullet,\bigcirc,\rightarrow,\Rightarrow,0,1,e)$ *where*

- $(W,\bigcirc,e)$ *is a monoid*
- $(W,+,\bullet,0,1)$ *is a bounded lattice with the top element* 1 *and the bottom element* 0
- $\rightarrow$ *and* $\Rightarrow$ *is the left and the right residuum of* $\bigcirc$*, respectively,*

is a ***residuated lattice*** *(R–lattice).* □

Example 1. Let W be a non-empty set and let $\backslash$ and $/$ be the binary operations of the *weakest prespecification* and the *weakest postspecification*, respectively, on the family of binary relations on W (see [19]) defined as follows:

$$Q\backslash R=\{(x,y)\in W\times W:\ \text{for every } z\in W,\ \text{if } (y,z)\in Q \text{ then } (x,z)\in R\}$$
$$R/P=\{(y,z)\in W\times W:\ \text{for every } x\in W,\ \text{if } (x,y)\in P \text{ then } (x,z)\in R\}.$$

It is easy to see that $Q\backslash R=Q\rightarrow R$ and $R/P=P\Rightarrow R$, where $\rightarrow$ and $\Rightarrow$ are the residua of the composition (or the relative product) of relations defined by:

$$P\,;Q=\{(x,y)\in W\times W:\ \text{there is } z\in W \text{ such that } (x,z)\in P \text{ and } (z,y)\in Q\}.$$

The intuitive meaning of these residua is as follows. If the relations are interpreted as sequential programs, and if R is a target relation that is supposed to be obtained by performing, first, a program P and next a program Q, then the residuation conditions guarantee that if $P=Q\backslash R$ or $Q=R/P$, then in both cases we have $P\,;Q\subseteq R$. That is the residua enable us to get the greatest relation P or the greatest relation Q such that given R and Q or R and P, respectively, the composition of P and Q is included in R. In other words, $(Q\backslash R)\,;Q\subseteq R$ and $P\,;(R/P)\subseteq R$. □

Definition 4. *Given a poset* $(W,\leqslant)$ *and a binary operation* $\bigcirc$ *on* W*, we define the following* ***dual residua of*** $\bigcirc$*:*

* $\leftarrow$ *is the* ***dual left residuum of*** $\bigcirc$ *iff*

$$y\leqslant z\bigcirc x \quad \textit{iff} \quad x\leftarrow y\leqslant z \quad \textit{for all } x,y,z\in W \tag{3}$$

* $\Leftarrow$ *is the* ***dual right residuum of*** $\bigcirc$ *iff*

$$y\leqslant x\bigcirc z \quad \textit{iff} \quad x\Leftarrow y\leqslant z \quad \textit{for all } x,y,z\in W. \tag{4}$$

Conditions (3)–(4) are called ***dual residuation conditions.*** □

It is easily noted that the existence of dual residua implies that

* $x\leftarrow y$ is the smallest element in the set $\{z\in W: y\leqslant z\bigcirc x\}$
* $x\Leftarrow y$ is the smallest element in the set $\{z\in W: y\leqslant x\bigcirc z\}$.

Example 2. In the theory of algebras of relations the operator of *relative sum* is defined by:

$$R \dagger Q = \{(x,y) \in W \times W : \text{ for every } z \in W,\ (x,z) \in R \text{ or } (z,y) \in Q\}.$$

We suggest that the dual residua of the relative sum can have an interpretation similar to the specification interpretation of residua mentioned in Example 1. Namely, if $R/\!\!/P = P \Leftarrow R$ and $Q\backslash\!\!\backslash R = Q \leftarrow R$, then $R \subseteq P \dagger (R/\!\!/P)$ and $R \subseteq (Q\backslash\!\!\backslash R) \dagger Q$. This means that if a target relation R is supposed to be obtained from relations P and Q by making their relative sum, then $P = Q\backslash\!\!\backslash R$ or $Q = R/\!\!/P$ are the smallest relations such that, given R and Q or R and P, respectively, $R \subseteq P \dagger Q$. □

Lemma 1. *If* $\odot$ *is commutative then* $\rightarrow\, = \,\Rightarrow$ *and* $\leftarrow\, = \,\Leftarrow$. ■

Definition 5. *A* ***double residuated lattice*** *(DR–lattice) is a structure of the form* $\mathfrak{L} = (W, +, \bullet, \odot, \rightarrow, \Rightarrow, \oplus, \leftarrow, \Leftarrow, 1, 0, \top, \bot)$, *where*

- $(W, +, \bullet, \odot, \rightarrow, \Rightarrow, 1, 0, \top)$ *is an R–lattice*
- $(W, \oplus, \bot)$ *is a monoid*
- $\leftarrow$ *and* $\Leftarrow$ *is the dual left and the dual right residuum of* $\oplus$, *respectively.*

□

The operations $\odot$ and $\oplus$ are called a *product* and a *sum*, respectively.
A DR–lattice $\mathfrak{L} = (W, +, \bullet, \odot, \rightarrow, \Rightarrow, \oplus, \leftarrow, \Leftarrow, 1, 0, \top, \bot)$ is called

- *integral* iff $1 = \top$ and $0 = \bot$
- *commutative* iff $\odot$ and $\oplus$ are commutative
- *complete* iff for every family $(x_i)_{i \in I} \subseteq W$, $\sup_i x_i$ and $\inf_i x_i$ exist.

We will use the following four ***complement operations*** definable in terms of the residua:

$$\neg_l x \stackrel{def}{=} x \rightarrow 0, \qquad \neg_r x \stackrel{def}{=} x \Rightarrow 0$$
$$\ulcorner_l x \stackrel{def}{=} x \leftarrow 1, \qquad \ulcorner_r x \stackrel{def}{=} x \Leftarrow 1.$$

They are generalisations of the pseudo–complement and the dual pseudo–complement in a lattice (see [34]). If $\bullet = \odot$ then $\rightarrow$ is the relative pseudo–complement, $\neg = \neg_l = \neg_r$ is the pseudo–complement and $(W, +, \bullet, \rightarrow, \neg, 1, 0)$ is a Heyting algebra. If in addition $+ = \oplus$ then $\ulcorner = \ulcorner_l = \ulcorner_r$ is the dual pseudo–complement.

DR–lattices are the weakest structures that provide an algebraic framework for degrees of membership to fuzzy sets and, in a more general setting, degrees of certainty of assertions in fuzzy theories. The other classes of algebras, analogous to the algebras studied in [12] and [17], are obtained from DR–lattices by postulating some additional axioms.

Definition 6. *Let* $\mathfrak{L} = (W, +, \bullet, \odot, \rightarrow, \oplus, \leftarrow, 1, 0)$ *be a commutative and integral DR–lattice.* $\mathfrak{L}$ *is called*

- *a* ***DMTL algebra*** *iff for all* $x, y \in W$

$$(x \to y) + (y \to x) = 1 \qquad \textit{prelinearity}$$

- *a* ***DMCL algebra*** *iff for all* $x, y \in W$,

$$(x \leftarrow y) \bullet (y \leftarrow x) = 0 \qquad \textit{dual prelinearity}$$

If $\mathfrak{L}$ *is both a DMTL and a DMCL algebra then it is called a* ***DMCTL algebra.***

- *A DMTL algebra is called a* ***DBTL algebra*** *iff for all* $x, y \in W$

$$x \odot (x \to y) = x \bullet y \qquad \textit{divisibility}$$

- *A DMCL algebra is called a* ***DBCL algebra*** *iff for all* $x, y \in W$

$$x \oplus (x \leftarrow y) = x + y \qquad \textit{dual divisibility}$$

If $\mathfrak{L}$ *is both a DBTL and a DBCL algebra then it is called a* ***DBCTL algebra.*** □

The acronyms *DMTL* and *DMCL* mean ***D**ouble **M**onoidal **t**-norm **L**ogic* and ***D**ouble **M**onoidal t-**c**onorm **L**ogic.* Monoidal t–norm logic MTL has been introduced in [13]. Our proposed class DMTL extends the class of algebras for logic MTL to double residuated lattices with an appropriate axiom. Having a double residuated lattice as a basis we define a t–conorm counterpart to DMTL–algebras, namely the class DMCL. Similarly, in analogy to the algebras for the basic fuzzy logic BL introduced in [17] we define the class DBTL of algebras for a ***D**ouble **B**asic **t**-norm **L**ogic* and the class DBCL of algebras for a ***D**ouble **B**asic t-**c**onorm **L**ogic.* These classes are extensions of BL–algebras to double residuated lattices. The logics corresponding to these classes of algebras can be defined in a natural way.

Definition 7. *Let* $\mathfrak{L} = (W, +, \bullet, \odot, \to, \oplus, \leftarrow, 1, 0)$ *be a commutative and integral DR–lattice. Then* $\mathfrak{L}$ *is called a* ***DDR algebra*** *iff for all* $x, y, z \in W$ *the following condition holds:*

$$x \odot (y \oplus z) \leqslant (x \odot y) \oplus (x \odot z). \quad \square$$

Let $\mathcal{A} = (W, +, \bullet, 1, 0)$ be a lattice with the natural ordering $\leqslant$. By $\mathcal{A}^{-1}$ we mean the lattice obtained from $\mathcal{A}$ by reversing the ordering, that is the natural ordering of $\mathcal{A}^{-1}$ is $\geqslant \; = \; \leqslant^{-1}$. It follows that the join of $\mathcal{A}$ is the meet of $\mathcal{A}^{-1}$, the meet of $\mathcal{A}$ is the join of $\mathcal{A}^{-1}$, the l.u.b. of $\mathcal{A}$ (i.e. 1) is the g.l.b. of $\mathcal{A}^{-1}$ and the g.l.b. of $\mathcal{A}$ (i.e. 0) is the l.u.b. of $\mathcal{A}^{-1}$.
Given a lattice $\mathcal{A} = (W, +, \bullet, 1, 0)$, an extension of $\mathcal{A}$ to an R–lattice (resp. DR–lattice) will be written $(\mathcal{A}, \odot, \to, \Rightarrow, \top)$ (resp. $(\mathcal{A}, \odot, \to, \Rightarrow, \oplus, \leftarrow, \Leftarrow, \top, \bot)$).

Theorem 1. *A structure* $(\mathcal{A}, \odot, \to, \Rightarrow, \oplus, \leftarrow, \Leftarrow, \top, \bot)$ *is a DR–lattice iff both* $(\mathcal{A}, \odot, \to, \Rightarrow, \top)$ *and* $(\mathcal{A}^{-1}, \oplus, \leftarrow, \Leftarrow, \bot)$ *are R–lattices.* ■

Let L be the language of DR–lattices and let τ be a term of L. Furthermore, let $\delta(\tau)$ be a term obtained from τ by replacing $\odot$ (resp. $\to$, $\Rightarrow$, $\bullet$, $+$, 1, 0, $\top$, $\bot$) by $\oplus$ (resp. $\leftarrow$, $\Leftarrow$, $+$, $\bullet$, 0, 1, $\bot$, $\top$). Clearly, $\delta(\tau)$ is a term in the language L.

Proposition 1. *Let* $\mathcal{A} = (W, +, \bullet, 1, 0)$ *be a bounded lattice and let* $\mathfrak{L} = (\mathcal{A}, \odot, \rightarrow, \Rightarrow, \oplus, \leftarrow, \Leftarrow, \top, \bot)$ *be a DR–lattice. Then for all terms* τ_1 *and* τ_2 *of* L *such that* $\oplus$, $\leftarrow$, $\Leftarrow$ *and* $\bot$ *do not occur in them the following conditions are satisfied:*

(i) $\tau_1 = \tau_2$ *is true in* $(\mathcal{A}, \odot, \rightarrow, \Rightarrow, \top)$ *iff*
$\delta(\tau_1) = \delta(\tau_2)$ *is true in* $(\mathcal{A}^{-1}, \oplus, \leftarrow, \Leftarrow, \bot)$

(ii) $\tau_1 \leqslant \tau_2$ *is true in* $(\mathcal{A}, \odot, \rightarrow, \Rightarrow, \top)$ *iff*
$\delta(\tau_1) \geqslant \delta(\tau_2)$ *is true in* $(\mathcal{A}^{-1}, \oplus, \leftarrow, \Leftarrow, \bot)$. ∎

It is well-known that R–lattices form a variety. The respective axioms are the lattice axioms for $(W, +, \bullet, 1, 0)$, monoid axioms for $(W, \odot, \top)$ and $(W, \oplus, \bot)$ and the following:

(A1) $x \leqslant y \rightarrow (x \odot y)$
(A2) $(y \rightarrow z) \odot y \leqslant z$
(A3) $y \leqslant x \Rightarrow (x \odot y)$
(A4) $x \odot (x \Rightarrow z) \leqslant z$
(A5) $x \odot (y + z) = (x \odot y) + (x \odot z)$
(A6) $(x + y) \odot z = (x \odot z) + (y \odot z)$
(A7) $z \rightarrow (x \bullet y) = (z \rightarrow x) \bullet (z \rightarrow y)$
(A8) $z \Rightarrow (x \bullet y) = (z \Rightarrow x) \bullet (z \Rightarrow y)$.

Proposition 2. *The equalities* **(A1)**–**(A8)** *are true in every residuated lattice.*

Proof. Observe that for any isotone unary maps f and g on a poset the following conditions are equivalent:

(i) $f(x) \leqslant y$ iff $x \leqslant g(y)$
(ii) $x \leqslant g(f(x))$ and $f(g(y)) \leqslant y$.

We obtain **(i)** implies **(ii)** taking $x = g(y)$, $y = f(x)$ and applying **(i)** to $f(x) \leqslant f(x)$ and $g(y) \leqslant g(y)$. Now assume **(ii)**. If $f(x) \leqslant y$, then $x \leqslant g(f(x)) \leqslant g(y)$, since g is isotone. The reverse implication can be obtained in a similar way.
It is known that

(iii) in any residuated lattice $\odot$ is isotone in both arguments and $\rightarrow$, $\Rightarrow$ are isotone in the second argument.

Now consider a residuated lattice, that is we have:

(iv) $x \odot y \leqslant z$ iff $z \leqslant y \rightarrow x$
(v) $x \odot y \leqslant z$ iff $y \leqslant x \Rightarrow z$.

(iv) says that for a fixed y the maps $f_y(x) = x \odot y$ and $g_y(z) = y \rightarrow z$ satisfy **(i)**. So by **(ii)** we get **(A1)** and **(A2)**. Similarly, **(v)** says that for a fixed x the maps $h_x(y) = x \odot y$ and $k_x(z) = x \Rightarrow z$ satisfy **(i)**. Again, from **(ii)** we have **(A3)** and **(A4)**.
The equalities **(A5)**–**(A8)** follow from the fact that f_y, g_y, h_x and k_x are isotone, which can be easily derived from **(iii)**. Equivalently, f_y and h_x are join preserving, so **(A5)** and **(A6)** hold. Similarly, g_y and k_x are meet preserving, so **(A7)** and **(A8)** hold. □

The axioms of DR–lattices consist of the equations and the inequalities **(A1)**–**(A8)** together with

- the equations of the form $\delta(\tau_1)=\delta(\tau_2)$ for all $\tau_1=\tau_2 \in \{(\mathbf{A5}),\dots,(\mathbf{A8})\}$, and
- the inequalities $\delta(\tau_1) \geqslant \delta(\tau_2)$ for all $\tau_1 \leqslant \tau_2 \in \{(\mathbf{A1}),\dots,(\mathbf{A4})\}$.

Dual residua of a lattice join are studied by Rauszer ([35]) in the context of Heyting-Brouwer logic. Dual pseudo-complement is one of the operations in double Stone algebras. However, it is a primitive operation there, residuation operations are not on the signature of Stone algebras. In [1] dual residua of a monoid operator are considered and a Kripke semantics for an extension of linear logic with dual residua is developed. Meet and product are used in semantic structures of some substructural logics, e.g. relevant logics ([25]), whose languages usually include two conjunctions. Some recent results on residuated lattices can be found in [3], [18] and [23],

2.1 Triangular Norms and Triangular Conorms

A typical example of monoid operators which play an important role in the fuzzy set theory are triangular norms and triangular conorms ([36]).

A *triangular norm* (t–norm, for short) is a mapping $\star : [0,1]^2 \to [0,1]$ satisfying the following conditions:

(1) $\star$ is commutative and associative
(2) $\star$ is isotone in both arguments, that is for $x \leqslant y$ it holds $x \star z \leqslant y \star z$, where $\leqslant$ is the natural ordering on reals
(3) $1 \star x = x$ for every $x \in [0,1]$.

Well–known t–norms are:

* Zadeh's t–norm $x \star_Z y = \min\{x,y\}$
* algebraic product $x \star_P y = x \cdot y$
* the Łukasiewicz t–norm $x \star_L y = \max\{0, x+y-1\}$
* the drastic t–norm $x \star_D y = x$ if $y=1$, $x \star_D y = y$ if $x=1$ and $x \star_D y = 0$ otherwise.

It can be shown that $\star_Z$ is the largest t–norm, whereas $\star_D$ is the smallest t–norm.

A *triangular conorm* (shortly t–conorm) is a mapping $\circ : [0,1]^2 \to [0,1]$ satisfying the following conditions:

(1) $\circ$ is commutative and associative
(2) $\circ$ is isotone in both arguments
(3) $0 \circ x = x$ for every $x \in [0,1]$.

The most popular t–conorms are:

* the Zadeh's t–conorm $x \circ_Z y = \max\{x,y\}$,
* bounded sum $x \circ_P y = x + y - x \cdot y$,
* the Łukasiewicz t–conorm $x \circ_L y = \min\{1, x+y\}$

* the drastic t–conorm $x \circ_D y = x$ if $y = 0$, $x \circ_D y = y$ if $x = 0$ and $x \circ_D y = 1$ otherwise.

It is well-known that $\circ_Z$ is the smallest t–conorm and $\circ_D$ is the largest t–conorm.

Given a t–norm or a t–conorm $\star$, its *partial mappings* are the mappings $\star_x(y) = x \star y$ for any fixed $x \in [0,1]$ and the mappings $\star_y(x) = x \star y$ for any fixed $y \in [0,1]$. The maping $\star$ is called *left–continuous* (resp. *right–continuous*) iff its partial mappings $\star_x$ (for all $x \in [0,1]$) and $\star_y$ (for all $y \in [0,1]$) are left–continuous (resp. right–continuous).

Let $\mathcal{L} = ([0,1], +, \bullet, \star, \rightarrow, \circ, \leftarrow, 1, 0)$ be an algebra such that $\star$ is a left–continuous t–norm, $\circ$ is a right–continuous t–conorm, $\rightarrow$ is the residuum of $\star$, $\leftarrow$ is the dual residuum of $\circ$, $x \bullet y = \min\{x, y\}$ and $x + y = \max\{x, y\}$.

The following proposition results from properties given in [12], [15], [17] and Theorem 1.

Proposition 3.

(i) *A t–norm has the residuum iff it is left–continuous and a t–conorm has the dual residuum iff it is right–continuous.*

(ii) *The algebra $\mathcal{L}$ defined above is a DR–lattice.*

(iii) *The following identities hold in $\mathcal{L}$:*

$$(x \rightarrow y) + (y \rightarrow x) = 1$$
$$(x \leftarrow y) \bullet (y \leftarrow x) = 0$$
$$x + y = ((x \rightarrow y) \rightarrow y) \bullet ((y \rightarrow x) \rightarrow x)$$
$$x \bullet y = ((x \leftarrow y) \leftarrow y) + ((y \leftarrow x) \leftarrow x).$$

(iv) *If $\star$ is continuous, then the identity $x \star (x \rightarrow y) = x \bullet y$ holds in $\mathcal{L}$.*

(v) *If $\circ$ is continuous, then the identity $x \circ (x \leftarrow y) = x + y$ holds in $\mathcal{L}$.* ∎

For a left–continuous t–norm $\star$, its residuum is called a *residual implication.* Three well–known residual implications, being the residua of $\star_Z$, $\star_P$ and $\star_L$, respectively, are:

* the Gödel implication $x \rightarrow_Z y = \begin{cases} 1 & \text{if } x \leqslant y \\ y & \text{otherwise} \end{cases}$
* the Gaines implication $x \rightarrow_P y = \begin{cases} 1 & \text{if } x \leqslant y \\ \frac{y}{x} & \text{otherwise} \end{cases}$
* the Łukasiewicz implication $x \rightarrow_L y = \min\{1, 1 - x + y\}$.

One can easily check that the dual residua of $\circ_Z$, $\circ_P$, and $\circ_L$, respectively, are given by:

* $x \leftarrow_Z y = \begin{cases} 0 & \text{if } y \leqslant x \\ y & \text{otherwise} \end{cases}$
* $x \leftarrow_P y = \begin{cases} 0 & \text{if } y \leqslant x \\ \frac{y-x}{1-x} & \text{otherwise} \end{cases}$
* $x \leftarrow_L y = \max\{0, y - x\}$.

A generalisation of t-norms to non-commutative operations, called pseudo-t-norms, is studied in ([14]).

3 Properties of DR–Lattices

In the rest of this paper we will consider commutative and integral DR–lattices. They will be written $(W, +, \bullet, \odot, \rightarrow, \oplus, \leftarrow, 1, 0)$.

Below we list some basic properties of the operations in DR–lattices.

Lemma 2. *Let* $\mathfrak{L} = (W, +, \bullet, \odot, \rightarrow, \oplus, \leftarrow, 1, 0)$ *be a commutative and integral DR–lattice. Then the following conditions hold for all* $x, y, z \in W$,

(i) $\rightarrow$ *and* $\leftarrow$ *are antitone in the first and isotone in the second argument*

(ii) $\odot$ *and* $\oplus$ *are isotone in both arguments*

(iii) $\neg$ *and* $\ulcorner$ *are antitone*

(iv) $x \rightarrow x = x \rightarrow 1 = 0 \rightarrow x = 1$, $1 \rightarrow x = x$ | **(iv')** $x \leftarrow x = 1 \leftarrow x = x \leftarrow 0 = 0$, $0 \leftarrow x = x$

(v) $x \odot (x \rightarrow y) \leqslant y$ | **(v')** $y \leqslant x \oplus (x \leftarrow y)$

(vi) $z \leqslant x \rightarrow y$ *iff* $x \leqslant z \rightarrow y$ | **(vi')** $x \leftarrow y \leqslant z$ *iff* $z \leftarrow y \leqslant x$

(vii) $y \leqslant x \rightarrow (x \odot y)$ | **(vii')** $x \leftarrow (x \oplus y) \leqslant y$

(viii) $x \leqslant y$ *iff* $x \rightarrow y = 1$ | **(viii')** $x \leqslant y$ *iff* $y \leftarrow x = 0$

(ix) $x \odot y \leqslant x$ | **(ix')** $x \oplus y \geqslant x$

(x) $x \odot x = 1$ *iff* $x = 1$ | **(x')** $x \oplus x = 0$ *iff* $x = 0$

(xi) $x \rightarrow y \geqslant y$ | **(xi')** $x \leftarrow y \leqslant y$

(xii) $x \rightarrow (y \rightarrow z) = (x \odot y) \rightarrow z$ | **(xii')** $x \leftarrow (y \leftarrow z) = (x \oplus y) \leftarrow z$

(xiii) $x \odot y \leqslant x \bullet y$ | **(xiii')** $x \oplus y \geqslant x + y$

(xiv) $x \odot (y + z) = (x \odot y) + (x \odot z)$ | **(xiv')** $x \oplus (y \bullet z) = (x \oplus y) \bullet (x \oplus z)$

(xv) $x \rightarrow (y \bullet z) = (x \rightarrow y) \bullet (x \rightarrow z)$ | **(xv')** $x \leftarrow (y + z) = (x \leftarrow y) + (x \leftarrow z)$

(xvi) $(x + y) \rightarrow z = (x \rightarrow z) \bullet (y \rightarrow z)$ | **(xvi')** $(x \bullet y) \leftarrow z = (x \leftarrow z) + (y \leftarrow z)$

(xvii) $(x \rightarrow y) \odot (y \rightarrow z) \leqslant (x \rightarrow z)$ | **(xvii')** $(x \leftarrow y) \oplus (y \leftarrow z) \geqslant (x \leftarrow z)$.

Proof. Conditions **(i)**–**(xvii)** are shown in [37]. By way of example we prove **(viii')** and **(xii')**.

(viii') Let $x \leqslant y \oplus 0$. By the dual residuation condition (3) this is equivalent to $y \leftarrow x \leqslant 0$. Obviously, $y \leftarrow x \geqslant 0$, so $y \leftarrow x = 0$.

(xii') By **(v')**, $(x \oplus y) \oplus ((x \oplus y) \leftarrow z) \geqslant z$, which is equivalent to

$$y \oplus (x \oplus ((x \oplus y) \leftarrow z))) \geqslant z.$$

By the dual residuation condition we get $x \oplus ((x \oplus y) \leftarrow z)) \geqslant y \leftarrow z$. Applying again the dual residuation condition we obtain

$$(x \oplus y) \leftarrow z \geqslant x \leftarrow (y \leftarrow z). \tag{5}$$

Furthermore, by **(v')**, $x \oplus (x \leftarrow (y \leftarrow z)) \geqslant y \leftarrow z$, which by **(ii)** gives

$$y \oplus (x \oplus (x \leftarrow (y \leftarrow z))) \geqslant y \oplus (y \leftarrow z) \geqslant z,$$

or equivalently, $(x \oplus y) \oplus (x \leftarrow (y \leftarrow z)) \geqslant z$. By the dual residuation condition we get $(x \oplus y) \leftarrow z \leqslant x \leftarrow (y \leftarrow z)$. Hence, in view of (5) we obtain the result. ∎

The following lemma provides the properties of the complement operations.

Lemma 3. *Let* $\mathfrak{L}=(W, +, \bullet, \odot, \rightarrow, \oplus, \leftarrow, 1, 0)$ *be a commutative and integral DR–lattice. Then for all* $x, y \in W$ *the following conditions hold:*

(i) $x \leqslant \neg\neg x$	**(i')** $x \geqslant \ulcorner\ulcorner x$
(ii) $x \odot \neg x = 0$	**(ii')** $x \oplus \ulcorner x = 1$
(iii) $\neg\neg\neg x = \neg x$	**(iii')** $\ulcorner\ulcorner\ulcorner x = \ulcorner x$
(iv) $x \rightarrow \neg y = \neg(x \odot y)$	**(iv')** $x \leftarrow \ulcorner y = \ulcorner(x \oplus y)$
(v) $x \rightarrow y \leqslant \neg y \rightarrow \neg x$	**(v')** $x \leftarrow y \geqslant \ulcorner y \leftarrow \ulcorner x$
(vi) $\neg x + y \leqslant x \rightarrow y$	**(vi')** $x \bullet \ulcorner y \geqslant y \leftarrow x.$

Proof. By way of example we prove **(iii')**. From **(i')** and Lemma 2**(iii)**,

$$\ulcorner x \leqslant \ulcorner\ulcorner\ulcorner x. \tag{6}$$

Next, by **(ii')**, $\ulcorner x \oplus \ulcorner\ulcorner x = 1 \geqslant 1$, which, by the dual residuation condition, means that $\ulcorner\ulcorner x \leftarrow 1 \leqslant \ulcorner x$, or equivalently, by the definition of $\ulcorner$,

$$\ulcorner\ulcorner\ulcorner x \leqslant \ulcorner x. \tag{7}$$

From (6)–(7) we get the result. ∎

Proposition 4. *For all terms* τ_1, τ_2 *of the language* L *of DR–lattices,* $\tau_1 \rightarrow \tau_2 = 1$ *is true in* L *iff* $(\tau_1 \leftarrow \tau_2) \rightarrow 0 = 1$ *is true in* L.

Proof. Easily follows from Lemma 2(viii) and 2(viii'). ∎

Lemma 4. *Let* $\mathfrak{L}=(W, +, \bullet, \odot, \rightarrow, \oplus, \leftarrow, 1, 0)$ *be a commutative and integral DR–lattice. Then for all families* $(x_i)_{i\in I}$ *and* $(y_i)_{i\in I}$ *of elements of* W *and for every* $z \in W$, *if the respective suprema and infima exist then the following conditions hold:*

(i) $z \odot \sup_i x_i = \sup_i(z \odot x_i)$	**(i')** $z \oplus \inf_i x_i = \inf_i(z \oplus x_i)$
(ii) $z \bullet \inf_i x_i = \inf_i(z \bullet x_i)$	**(ii')** $z + \sup_i x_i = \sup_i(z + x_i)$
(iii) $(z \rightarrow \inf_i x_i) = \inf_i(z \rightarrow x_i)$	**(iii')** $(z \leftarrow \sup_i x_i) = \sup_i(z \leftarrow x_i)$
(iv) $(\sup_i x_i \rightarrow z) = \inf_i(x_i \rightarrow z)$	**(iv')** $(\inf_i x_i \leftarrow z) = \sup_i(x_i \leftarrow z)$
(v) $\inf_i(x_i \rightarrow y_i) \leqslant \sup_i x_i \rightarrow \sup_i y_i$	**(v')** $\sup_i(x_i \leftarrow y_i) \geqslant \inf_i x_i \leftarrow \inf_i y_i$
(vi) $\inf_i(x_i \rightarrow y_i) \leqslant \inf_i x_i \rightarrow \inf_i y_i$	**(vi')** $\sup_i x_i \leftarrow \sup_i y_i \leqslant \sup_i(x_i \leftarrow y_i)$
(vii) $\neg \sup_i x_i = \inf_i \neg x_i$	**(vii')** $\ulcorner \inf_i x_i = \sup_i \ulcorner x_i$
(viii) $\sup_i x_i \leqslant \neg \inf_i \neg x_i$	**(viii')** $\inf_i x_i \geqslant \ulcorner \sup_i \ulcorner x_i.$

Proof. By way of example we prove **(i')**. By Lemma 2**(ii)**, $z \oplus \inf_i x_i \leqslant z \oplus x_i$ for every $i \in I$, so

$$z \oplus \inf_i x_i \leqslant \inf_i(z \oplus x_i). \tag{8}$$

Clearly, for every $i \in I$, $\inf_i(z \oplus x_i) \leqslant z \oplus x_i$. Hence, by the dual residuation condition, we get for every $i \in I$, $z \leftarrow \inf_i(z \oplus x_i) \leqslant x_i$. Then $z \leftarrow \inf_i(z \oplus x_i) \leqslant \inf_i x_i$. Applying again the dual residuation condition we get

$$\inf_i(z \oplus x_i) \leqslant z \oplus \inf_i x_i. \tag{9}$$

(8)–(9) imply the result. ∎

Lemma 5. *In every DMCTL algebra* $\mathfrak{L} = (W, +, \bullet, \odot, \rightarrow, \oplus, \leftarrow, 1, 0)$, *for all* $x, y, z \in W$ *and for every family* $(x_i)_{i \in I}$ *of elements of* W, *if the respective suprema and infima exist then the following conditions hold:*

(i) $x \odot y \leqslant (x \odot x) \bullet (y \odot y)$ **(i')** $x \oplus y \geqslant (x \oplus x) + (y \oplus y)$

(ii) $\sup_i x_i \odot \sup_i x_i = \sup_i(x_i \odot x_i)$ **(ii')** $\inf_i x_i \oplus \inf_i x_i = \inf_i(x_i \oplus x_i)$

(iii) $(x \bullet y) \rightarrow z = (x \rightarrow z) + (y \rightarrow z)$ **(iii')** $(x + y) \leftarrow z = (x \leftarrow z) \bullet (y \leftarrow z)$

(iv) $x \rightarrow (y + z) = (x \rightarrow y) + (x \rightarrow z)$ **(iv')** $x \leftarrow (y \bullet z) = (x \leftarrow y) \bullet (x \leftarrow z)$

(v) $x \bullet (y + z) = (x \bullet y) + (x \bullet z)$.

Proof. By way of example we prove **(i')** and **(ii')**.

(i') Since $\mathfrak{L}$ is a DMCTL algebra, we have:

$$x \oplus y = (x \oplus y) \oplus 0 = (x \oplus y) \oplus ((x \leftarrow y) \bullet (y \leftarrow x)).$$

Hence, by Lemma 2(**v'**) and 2(**xiv'**) we get

$$x \oplus y = (x \oplus y \oplus (x \leftarrow y)) \bullet (x \oplus y \oplus (y \leftarrow x)) \geqslant (x \oplus x) \bullet (y \oplus y).$$

(ii') From **(i')** we have for all $i, j \in I$,

$$x_i \oplus x_j \geqslant (x_i \oplus x_i) \bullet (x_j \oplus x_j) \geqslant \inf_i(x_i \oplus x_i) \bullet \inf_j(x_j \oplus x_j) = \inf_i(x_i \oplus x_i),$$

so $\inf_i \inf_j(x_i \oplus x_j) \geqslant \inf_i(x_i \oplus x_i)$. By Lemma 4(**i'**),

$$\inf_i \inf_j(x_i \oplus x_j) = \inf_i(x_i \oplus \inf_j x_j) = \inf_i x_i \oplus \inf_i x_i.$$

Hence we get

$$\inf_i x_i \oplus \inf_j x_j \geqslant \inf_i(x_i \oplus x_i). \tag{10}$$

On the other hand, for every $i \in I$, $\inf_i x_i \oplus \inf_i x_i \leqslant x_i \oplus x_i$, so

$$\inf_i x_i \oplus \inf_i x_i \leqslant \inf_i(x_i \oplus x_i)$$

which by (10) gives the result. ∎

Lemma 6. *In every DBCTL algebra* $\mathfrak{L} = (W, +, \bullet, \odot, \rightarrow, \oplus, \leftarrow, 1, 0)$ *the following conditions hold for all* $x, y \in W$,

(i) $x \leftarrow y \leqslant y \rightarrow \ulcorner x$ (i') $x \rightarrow y \geqslant y \leftarrow \neg x$.

Proof. Since both conditions can be proved in the similar way, we show (i) only. By Lemma 3(vi'), $y \bullet \ulcorner x \geqslant x \leftarrow y$. Since $\mathfrak{L}$ is a DBTL algebra, $y \bullet \ulcorner x = y \odot (y \rightarrow \ulcorner x)$. So we have $y \odot (y \rightarrow \ulcorner x) \geqslant x \leftarrow y$, which by the dual residuation condition is equivalent to

$$x \oplus (y \odot (y \rightarrow \ulcorner x)) \geqslant y.$$

But $y \odot (y \rightarrow \ulcorner x) \leqslant y \rightarrow \ulcorner x$. Hence

$$x \oplus (y \rightarrow \ulcorner x) \geqslant x \oplus (y \odot (y \rightarrow \ulcorner x)) \geqslant y.$$

Applying again the dual residuation condition we obtain the result. ∎

Lemma 7. *In every DDR algebra* $\mathfrak{L} = (W, +, \bullet, \odot, \rightarrow, \oplus, \leftarrow, 1, 0)$ *the following conditions hold for every* $x \in W$

(i) $\neg x \leqslant \ulcorner x$

(ii) $\neg\neg x \leqslant \ulcorner\ulcorner x$

(iii) $\neg\ulcorner x \leqslant x$.

Proof. By way of example we prove (i). Clearly, $\neg x = \neg x \odot 1 = \neg x \odot (x \oplus \ulcorner x)$ by Lemma 3(ii'). Hence, by the definition of DDR–lattice,

$$\neg x = \neg x \odot (x \oplus \ulcorner x) \leqslant (\neg x \odot x) \oplus (\neg x \odot \ulcorner x).$$

By Lemma 3(ii), $\neg x \odot x = 0$. So we get

$$\neg x \leqslant 0 \oplus (\neg x \odot \ulcorner x) = \neg x \odot \ulcorner x \leqslant \ulcorner x.$$

from Lemma 2(ix). ∎

4 Fuzzy Sets and Fuzzy Relations

Definition 8. *Let* $\mathfrak{L} = (W, +, \bullet, \odot, \rightarrow, \oplus, \leftarrow, 1, 0)$ *be a commutative and integral DR–lattice and let* $\mathfrak{U}$ *be a universe of objects. An* ***$\mathfrak{L}$–fuzzy set in*** $\mathfrak{U}$ *is any mapping* $X : \mathfrak{U} \rightarrow W$. □

The universe $\mathfrak{U}$ will be viewed as the following fuzzy set (in $\mathfrak{U}$): $\mathfrak{U}(u) = 1$ for every $u \in \mathfrak{U}$. The family of all $\mathfrak{L}$–fuzzy sets will be denoted by $\mathcal{F}_{\mathfrak{L}}(\mathfrak{U})$. By a ***fuzzy set in*** $\mathfrak{U}$ we mean an $\mathfrak{L}$–fuzzy set for some commutative and integral DR–lattice $\mathfrak{L}$.

For any commutative and integral DR–lattice $\mathfrak{L}$, for every $\mathfrak{L}$–fuzzy set $X \in \mathcal{F}_{\mathfrak{L}}(\mathfrak{U})$ and for every $u \in \mathfrak{U}$, $X(u)$ is the *membership degree* to which u belongs to X.

The operations on $\mathfrak{L}$–fuzzy sets are defined in the following way. Let $X, Y \in \mathcal{F}_{\mathfrak{L}}(\mathfrak{U})$. Then for every $u \in \mathfrak{U}$,

$$(X \cap_{\mathfrak{L}} Y)(u) \stackrel{def}{=} X(u) \odot Y(u)$$

$$(X \cup_{\mathfrak{L}} Y)(u) \stackrel{def}{=} X(u) \oplus Y(u)$$

$$(\neg_{\mathfrak{L}} X)(u) \stackrel{def}{=} \neg X(u)$$

$$(\ulcorner_{\mathfrak{L}} X)(u) \stackrel{def}{=} \ulcorner X(u).$$

If $\mathfrak{L}$ is fixed then we will simply write $X \cap Y$, $X \cup Y$, $\neg X$, $\ulcorner X$, respectively.

For a commutative and integral DR–lattice $\mathfrak{L}$ and any $\mathfrak{L}$–fuzzy sets $X, Y \in \mathcal{F}_{\mathfrak{L}}(\mathcal{U})$, we will write $X \subseteq_{\mathfrak{L}} Y$ iff $X(u) \leqslant Y(u)$ for every $u \in \mathcal{U}$.

Definition 9. *Given a commutative and integral DR–lattice* $\mathfrak{L} = (W, +, \bullet, \odot, \rightarrow, \oplus, \leftarrow, 1, 0)$ *and a non-empty set* $\mathcal{U}$, *a mapping* $R : \mathcal{U}^n \rightarrow W$, $n \geqslant 2$, *is an* ***$\mathfrak{L}$–fuzzy relation on*** $\mathcal{U}$. *If* $n = 2$ *then* R *is a binary* $\mathfrak{L}$*–fuzzy relation.* □

The family of all binary $\mathfrak{L}$–fuzzy relations on $\mathcal{U}$ will be denoted by $\mathcal{R}_{\mathfrak{L}}(\mathcal{U})$. Let $R \in \mathcal{R}_{\mathfrak{L}}(\mathcal{U})$. R is called

- *reflexive* iff $R(x,x) = 1$ for every $x \in \mathcal{U}$
- *irreflexive* iff $R(x,x) = 0$ for every $x \in \mathcal{U}$
- *symmetric* iff $R(x,y) = R(y,x)$ for all $x, y \in \mathcal{U}$
- *transitive* iff $R(x,y) \odot R(y,z) \leqslant R(x,z)$ for all $x, y, z \in \mathcal{U}$
- *cotransitive* iff $R(x,y) \oplus R(y,z) \geqslant R(x,z)$ for all $x, y, z \in \mathcal{U}$
- *crisp* iff $R(x,y) \in \{0,1\}$ for all $x, y \in \mathcal{U}$.

R is called an $\mathfrak{L}$*–equivalence* relation iff it is reflexive, symmetric and transitive.

Remark 1. Let $\mathfrak{L} = (W, +, \bullet, \odot, \rightarrow, \oplus, \leftarrow, 1, 0)$ be a commutative and integral DR–lattice and let R be a crisp transitive relation on $\mathcal{U}$. By definition, $R(x,y) \odot R(y,z) \leqslant R(x,z)$ for all $x, y, z \in \mathcal{U}$. It is easily noted that

$$\neg R(x,y) \oplus \neg R(y,z) \geqslant \neg R(x,z) \quad \text{for all } x, y, z \in \mathcal{U}.$$

Hence $\neg R$ is cotransitive. In other words, for crisp relations cotransitivity is the property of complements of transitive relations. □

Given a complete DR–lattice $\mathfrak{L} = (W, +, \bullet, \odot, \rightarrow, \oplus, \leftarrow, 1, 0)$, let us define the following $\mathfrak{L}$–fuzzy relations on $\mathcal{F}_{\mathfrak{L}}(\mathcal{U})$: for all $X, Y \in \mathcal{F}_{\mathfrak{L}}(\mathcal{U})$,

(i) $\mathfrak{L}$*–fuzzy inclusion*: $inc_{\mathfrak{L}}(X,Y) \stackrel{def}{=} \inf_{u \in \mathcal{U}}(X(u) \rightarrow Y(u))$

(ii) $\mathfrak{L}$*–fuzzy noninclusion*: $ninc_{\mathfrak{L}}(X,Y) \stackrel{def}{=} \sup_{u \in \mathcal{U}}(Y(u) \leftarrow X(u))$

(iii) $\mathfrak{L}$*–fuzzy compatibility*: $com_{\mathfrak{L}}(X,Y) \stackrel{def}{=} \sup_{u \in \mathcal{U}}(X(u) \odot Y(u))$

(iv) $\mathfrak{L}$*–fuzzy exhaustiveness*: $exh_{\mathfrak{L}}(X,Y) \stackrel{def}{=} \inf_{u \in \mathcal{U}}(X(u) \oplus Y(u))$

(v) $\mathfrak{L}$*–fuzzy indiscernibility*: $ind_{\mathfrak{L}}(X,Y) \stackrel{def}{=} inc_{\mathfrak{L}}(X,Y) \odot inc_{\mathfrak{L}}(Y,X)$

(vi) $\mathfrak{L}$*–fuzzy diversity*: $div_{\mathfrak{L}}(X,Y) \stackrel{def}{=} ninc_{\mathfrak{L}}(X,Y) \oplus ninc_{\mathfrak{L}}(Y,X)$.

For any $X, Y \in \mathcal{F}_{\mathfrak{L}}(\mathcal{U})$, $inc_{\mathfrak{L}}(X,Y)$ (resp. $com_{\mathfrak{L}}(X,Y)$) represents the degree to which X is included in Y (resp. non-disjoint with Y). Similarly, $ninc_{\mathfrak{L}}(X,Y)$ is the degree to which X is *not* included in Y, whereas $exh_{\mathfrak{L}}(X,Y)$ is the degree to which $X \cup_{\mathfrak{L}} Y$ covers the whole universe $\mathcal{U}$. Finally, $ind_{\mathfrak{L}}(X,Y)$ (resp. $div_{\mathfrak{L}}(X,Y)$) is the degree to which X is equal to Y (resp. differs from Y).

The following proposition provides the basic properties of $\mathfrak{L}$–fuzzy relations defined above.

Proposition 5. *For every complete commutative and integral DR–lattice $\mathfrak{L}$,*

(i) $inc_{\mathfrak{L}}$ *is reflexive and transitive*
(ii) $ninc_{\mathfrak{L}}$ *is irreflexive and cotransitive*
(iii) $com_{\mathfrak{L}}$ *and* $exh_{\mathfrak{L}}$ *are symmetric.*
(iv) $ind_{\mathfrak{L}}$ *is an* $\mathfrak{L}$*–equivalence relation*
(v) $div_{\mathfrak{L}}$ *is irreflexive and symmetric.*

Proof. By way of example we show **(ii)** and **(v)**.
(ii) By Lemma 2**(iv')**, for every $X \in \mathcal{F}_{\mathfrak{L}}(\mathcal{U})$,

$$ninc_{\mathfrak{L}}(X,X) = \sup_{u \in \mathcal{U}}(X(u) \leftarrow X(u)) = 0.$$

Hence $ninc_{\mathfrak{L}}$ is irreflexive. Furthermore, for all $X, Y, Z \in \mathcal{F}_{\mathfrak{L}}(\mathcal{U})$,

$$\begin{aligned} ninc_{\mathfrak{L}}(X,Y) \oplus ninc_{\mathfrak{L}}(Y,Z) &= \sup_{u \in \mathcal{U}}(X(u) \leftarrow Y(u)) \oplus \sup_{u \in \mathcal{U}}(Y(u) \leftarrow Z(u)) \\ &\geqslant \sup_{u \in \mathcal{U}}((X(u) \leftarrow Y(u)) \oplus (Y(u) \leftarrow Z(u))). \end{aligned}$$

By Lemma 2**(xvii)**, $(X(u) \leftarrow Y(u)) \oplus (Y(u) \leftarrow Z(u)) = (X(u) \leftarrow Z(u))$. Therefore, $ninc_{\mathfrak{L}}(X,Y) \oplus ninc_{\mathfrak{L}}(Y,Z) \geqslant ninc_{\mathfrak{L}}(X,Z)$, so $ninc_{\mathfrak{L}}$ is cotransitive.

(v) Clearly, $0 \oplus 0 = 0$. Hence, by **(ii)**, we have:

$$div_{\mathfrak{L}}(X,X) = ninc_{\mathfrak{L}}(X,X) \oplus ninc_{\mathfrak{L}}(X,X) = 0,$$

so $div_{\mathfrak{L}}$ is irreflexive. Symmetry of $div_{\mathfrak{L}}$ immediately follows from commutativity of $\oplus$. ■

5 Fuzzy Information Operators

Let $\mathfrak{L}$ be a DR–lattice and let $\mathcal{U}$ be a non-empty set. By an *$\mathfrak{L}$–information operator in* $\mathcal{U}$ we mean any mapping $\Omega : \mathcal{R}_{\mathfrak{L}}(\mathcal{U}) \times \mathcal{F}_{\mathfrak{L}}(\mathcal{U}) \to \mathcal{F}_{\mathfrak{L}}(\mathcal{U})$. The mapping Ω is called a *fuzzy information operator* iff it is an $\mathfrak{L}$–information operator for some DR–lattice $\mathfrak{L}$.

Definition 10. *Let* $\mathfrak{L} = (W, +, \cdot, \odot, \to, \oplus, \leftarrow, 1, 0)$ *be a commutative and integral DR–lattice. We define the following* $\mathfrak{L}$*–information operators: for any* $R \in \mathcal{R}_{\mathfrak{L}}(\mathcal{U})$, *for every* $X \in \mathcal{F}_{\mathfrak{L}}(\mathcal{U})$ *and for every* $u \in \mathcal{U}$,

- $[R]_{\odot}X(u) = \inf_{w \in \mathcal{U}}(R(u,w) \to X(w))$
- $[R]_{\oplus}X(u) = \sup_{w \in \mathcal{U}}(R(u,w) \leftarrow X(w))$
- $\langle R \rangle_{\odot}X(u) = \sup_{w \in \mathcal{U}}(R(u,w) \odot X(w))$
- $\langle R \rangle_{\oplus}X(u) = \inf_{w \in \mathcal{U}}(R(u,w) \oplus X(w))$. □

Let R be an $\mathfrak{L}$–fuzzy relation on $\mathfrak{U}$. For any $u \in \mathfrak{U}$ we write uR to denote the $\mathfrak{L}$–fuzzy set in $\mathfrak{U}$ given by: $(uR)(w) = R(u, w)$. Notice that for every $X \in \mathcal{F}_{\mathfrak{L}}(\mathfrak{U})$ and for every $u \in \mathfrak{U}$,

$$[R]_{\odot}X(u) = inc_{\mathfrak{L}}(uR, X) \qquad [R]_{\oplus}X(u) = ninc_{\mathfrak{L}}(uR, X)$$
$$\langle R\rangle_{\odot}X(u) = com_{\mathfrak{L}}(uR, X) \qquad \langle R\rangle_{\oplus}X(u) = exh_{\mathfrak{L}}(uR, X).$$

Let $\mathcal{A} = (W, +, \bullet, -, 1, 0)$ be a Boolean algebra and let $\mathfrak{L} = (\mathcal{A}, \odot, \to, \oplus, \leftarrow)$ be such that $+ = \oplus$ and $\bullet = \odot$. Clearly, $\mathfrak{L}$ is the double residuated lattice, where $- = \neg = \ulcorner$. Then if R is a crisp relation on $\mathfrak{U}$ then the corresponding operators determined by R coincide with the classical modalities: $[R]_{\odot}$ and $\langle R\rangle_{\odot}$ are necessity and possibility operators $[R]$ and $\langle R\rangle$, respectively, and for $X \subseteq W$:[1]

$$[R]_{\oplus}X = \langle -R\rangle X = [\![R]\!]{-X}$$
$$\langle R\rangle_{\oplus}X = [-R]X = \langle\!\langle R\rangle\!\rangle{-X}.$$

Several basic properties of the above operators are given in the following proposition.

Proposition 6. *For every commutative and integral DR–lattice* $\mathfrak{L} = (W, +, \bullet, \odot, \to, \oplus, \leftarrow, 1, 0)$ *and for every* $R \in \mathcal{R}_{\mathfrak{L}}(\mathfrak{U})$,

(i) $[R]_{\odot}\mathfrak{U} = \langle R\rangle_{\oplus}\mathfrak{U} = \mathfrak{U}, \quad [R]_{\oplus}\emptyset = \langle R\rangle_{\odot}\emptyset = \emptyset$

(ii) *for all* $X, Y \in \mathcal{F}_{\mathfrak{L}}(\mathfrak{U})$ *and for every* $\Omega \in \{[\,]_{\odot}, [\,]_{\oplus}, \langle\,\rangle_{\odot}, \langle\,\rangle_{\oplus}\}$

$$X \subseteq_{\mathfrak{L}} Y \textit{ implies } \Omega(X) \subseteq_{\mathfrak{L}} \Omega(Y)$$

(iii) *for every* $X \in \mathcal{F}_{\mathfrak{L}}(\mathfrak{U})$,

$$[R]_{\odot}X \subseteq_{\mathfrak{L}} \neg\langle R\rangle_{\odot}\neg X, \qquad \ulcorner\langle R\rangle_{\oplus}\ulcorner X \subseteq_{\mathfrak{L}} [R]_{\oplus}X$$
$$\langle R\rangle_{\odot}X \subseteq_{\mathfrak{L}} \neg[R]_{\odot}\neg X, \qquad \ulcorner[R]_{\oplus}\ulcorner X \subseteq_{\mathfrak{L}} \langle R\rangle_{\oplus}X.$$

Proof. **(i)** follows from Lemma 2(**iv**) and **(ii)** results from Lemma 2(**i**)–(**ii**).
(iii) For every $X \in \mathcal{F}_{\mathfrak{L}}(\mathfrak{U})$ and for every $u \in \mathfrak{U}$,

$$\neg\langle R\rangle_{\odot}\neg X(w) = \neg \sup_{w\in\mathfrak{U}}(R(u,w) \odot \neg X(w)) = \inf_{w\in\mathfrak{U}} \neg(R(u,w) \odot \neg X(w))$$

by Lemma 4(**vii**). Next, by Lemma 3(**iv**) and 3(**i**),

$$\neg(R(u,w) \odot \neg X(w)) = (R(u,w) \to \neg\neg X(w) \geqslant R(u,w) \to X(w).$$

Then $\neg\langle R\rangle_{\odot}\neg X(u) \geqslant \inf_{w\in\mathfrak{U}}(R(u,w) \to X(w)) = [R]_{\odot}X(u)$ for every $u \in \mathfrak{U}$, so $[R]_{\odot}X \subseteq_{\mathfrak{L}} \neg\langle R\rangle_{\odot}\neg X$.

Also, for every $X \in \mathcal{F}_{\mathfrak{L}}(\mathfrak{U})$ and for every $u \in \mathfrak{U}$,

$$\ulcorner[R]_{\oplus}\ulcorner X(u) = \ulcorner\sup_{w\in\mathfrak{U}}(R(u,w) \leftarrow \ulcorner X(w)) = \ulcorner\sup_{w\in\mathfrak{U}} \ulcorner(R(u,w) \oplus X(w))$$

by Lemma 3(**iv'**). Finally, by Lemma 4(**viii'**),

[1] We recall that for any $R \subseteq W \times W$, $[\![R]\!]$ and $\langle\!\langle R\rangle\!\rangle$ are the sufficiency and the impossibility operator, respectively, defined in [22] by: for any $X \subseteq W$ and any $w \in W$, $w \in [\![R]\!]X$ iff for every $u \in W$, $u \in X$ implies $(w, u) \in R$, whereas $w \in \langle\!\langle R\rangle\!\rangle X$ iff for some $u \in W$, $(w, u) \notin R$ and $u \notin X$.

$$\ulcorner \sup_{w\in\mathcal{U}} \ulcorner(R(u,w)\oplus X(w)) \leqslant \inf_{w\in\mathcal{U}}(R(u,w)\oplus X(w)) = \langle R\rangle_{\oplus}X(u).$$

Hence we get $\ulcorner[R]_{\oplus}\ulcorner X \subseteq_{\mathfrak{L}} \langle R\rangle_{\oplus}X$.

The remaining inclusions can be proved in the analogous way. ∎

It is well–known that (crisp) information operators are useful tools for characterising particular classes of binary relations. This is also the case for fuzzy information operators. The following theorem provides complete characterisations of some basic binary fuzzy relations.

Theorem 2. *For every commutative and integral DR–lattice* $\mathfrak{L}=(W,+,\bullet,\odot,\rightarrow,\oplus,\leftarrow,1,0)$, *for every* $R\in\mathcal{R}_{\mathfrak{L}}(\mathcal{U})$ *and for every* $X\in\mathcal{F}_{\mathfrak{L}}(\mathcal{U})$ *the following conditions hold:*

(i) *R is reflexive* *iff* $[R]_{\odot}X \subseteq_{\mathfrak{L}} X$
iff $X \subseteq_{\mathfrak{L}} \langle R\rangle_{\odot}X$

(ii) *R is irreflexive* *iff* $X \subseteq_{\mathfrak{L}} [R]_{\oplus}X$
iff $\langle R\rangle_{\oplus}X \subseteq_{\mathfrak{L}} X$

(iii) *R is transitive* *iff* $[R]_{\odot}X \subseteq_{\mathfrak{L}} [R]_{\odot}[R]_{\odot}X$
iff $\langle R\rangle_{\odot}\langle R\rangle_{\odot}X \subseteq_{\mathfrak{L}} \langle R\rangle_{\odot}X$

(iv) *R is cotransitive* *iff* $[R]_{\oplus}[R]_{\oplus}X \subseteq_{\mathfrak{L}} [R]_{\oplus}X$
iff $\langle R\rangle_{\oplus}X \subseteq_{\mathfrak{L}} \langle R\rangle_{\oplus}\langle R\rangle_{\oplus}X$

(v) *R is symmetric* *iff* $\langle R\rangle_{\odot}[R]_{\odot}X \subseteq_{\mathfrak{L}} X$
iff $X \subseteq_{\mathfrak{L}} [R]_{\odot}\langle R\rangle_{\odot}X$
iff $[R]_{\oplus}\langle R\rangle_{\oplus}X \subseteq_{\mathfrak{L}} X$
iff $X \subseteq_{\mathfrak{L}} \langle R\rangle_{\oplus}[R]_{\oplus}X$.

Proof. By way of example we prove **(iv)**.

(⇒) For every $X\in\mathcal{F}_{\mathfrak{L}}(\mathcal{U})$ and for every $u\in\mathcal{U}$,

$$\begin{aligned}
[R]_{\oplus}[R]_{\oplus}X(u) &= \sup_{w\in\mathcal{U}}\left(R(u,w)\leftarrow \sup_{v\in\mathcal{U}}(R(w,v)\leftarrow X(v))\right)\\
&= \sup_{v\in\mathcal{U}}\sup_{w\in\mathcal{U}}(R(u,w)\leftarrow(R(w,v)\leftarrow X(v)))\\
&= \sup_{v\in\mathcal{U}}\sup_{w\in\mathcal{U}}((R(u,w)\oplus R(w,v))\leftarrow X(v))
\end{aligned}$$

by Lemmas 4(**iii'**) and 2(**xii'**). Furthermore, by Lemma 4(**iv'**)

$$\sup_{v\in\mathcal{U}}\sup_{w\in\mathcal{U}}((R(u,w)\oplus R(w,v))\leftarrow X(v)) = \sup_{v\in\mathcal{U}}\left(\inf_{w\in\mathcal{U}}(R(u,w)\oplus R(w,v))\leftarrow X(v)\right).$$

By assumption and Lemma 2(**i**),

$$\sup_{v\in\mathcal{U}}\left(\inf_{w\in\mathcal{U}}(R(u,w)\oplus R(w,v))\leftarrow X(v)\right) \leqslant \sup_{v\in\mathcal{U}}(R(u,v)\leftarrow X(v)) = [R]_{\oplus}X(u).$$

Hence $[R]_\oplus [R]_\oplus X \subseteq_{\mathfrak{L}} [R]_\oplus X$.

($\Leftarrow$) Assume that R is not cotransitive. This means that

$$R(u_0, v_0) \not\leqslant R(u_0, w_0) \oplus R(w_0, v_0) \quad \text{for some } u_0, w_0, v_0 \in \mathfrak{U}.$$

By Lemma 2(**viii'**) this means that

$$R(u_0, w_0) \oplus R(w_0, v_0) \leftarrow R(u_0, v_0) > 0. \tag{11}$$

Consider $X \in \mathcal{F}_{\mathfrak{L}}(\mathfrak{U})$ such that $X(w) = R(u_0, w)$ for every $w \in \mathfrak{U}$. Then we have:

$$[R]_\oplus X(u_0) = \sup_{w \in \mathfrak{U}} (R(u_0, w) \leftarrow R(u_0, w)) = 0$$

by Lemma 2(**iv'**), so

$$\begin{aligned} [R]_\oplus X(u_0) \leftarrow [R]_\oplus [R]_\oplus X(u_0) &= 0 \leftarrow [R]_\oplus [R]_\oplus X(u_0) \\ &= [R]_\oplus [R]_\oplus X(u_0) \end{aligned} \tag{12}$$

by Lemma 2(**iv'**). However, by Lemma 2(**i**) and 2(**xii'**),

$$\begin{aligned} [R]_\oplus [R]_\oplus X(u_0) &= \sup_{w \in \mathfrak{U}} \left(R(u_0, w) \leftarrow \sup_{v \in \mathfrak{U}} (R(w, v) \leftarrow R(u_0, v)) \right) \\ &\geqslant R(u_0, w_0) \leftarrow (R(w_0, v_0) \leftarrow R(u_0, v_0)) \\ &= R(u_0, w_0) \oplus R(w_0, v_0) \leftarrow R(u_0, v_0) > 0 \end{aligned}$$

from (11). Hence, by (12), we get

$$[R]_\oplus X(u_0) \leftarrow [R]_\oplus [R]_\oplus X(u_0) > 0$$

which, by Lemma 2(**viii'**), means that

$$[R]_\oplus [R]_\oplus X(u_0) \not\leqslant [R]_\oplus X(u_0).$$

Hence $[R]_\oplus [R]_\oplus X \not\subseteq_{\mathfrak{L}} [R]_\oplus X$.

The second equivalence can be shown in the similar way. ∎

Acknowledgements. The authors are grateful to Petr Hajek, Peter Jipsen and the anonymous referee for helpful suggestions. This paper was partially supported by the KBN Grant No 8T11C01617. The work was carried out in the framework of the COST Action 274 on *Theory and Applications of Relational Structures as Knowledge Instruments.*

References

1. G. Allwein and M. Dunn (1993). "Kripke models for linear logic". Journal of Symbolic Logic **58**, No 2, pp. 514-545.

2. Ph. Balbiani and E. Orłowska (1999). "A hierarchy of modal logics with relative accessibility relations". Journal of Applied Non-Classical Logics **9**, No 2-3, pp. 303–328, special issue in the memory of George Gargov.
3. K. Blount and C. Tsinakis (2001). "The structure of residuated lattices". Preprint.
4. L. Farinas del Cerro and H. Prade (1986) "Rough sets, fuzzy sets and modal logic". Fuzziness in indiscernibility and partial information. In: A. Di Nola and A. G. Ventre (eds), *The Mathematics of Fuzzy Systems*, Verlag TÜV Rheinland.
5. C. C. Chang (1958) "Algebraic analysis of many-valued logics". Transactions of the American Mathematical Society **88**, pp. 467–490.
6. S. Demri, E Orłowska and D. Vakarelov (1999). "Indiscernibility and complementarity relations in information systems". In: J. Gerbrandy, M. Marx, M. de Rijke and Y. Venema (eds) JFAK. Essays Dedicated to Johan van Benthem on the Occasion of his 50th Birthday. Amsterdam University Press.
7. S. Demri and E Orłowska (2002). *Incomplete Information: Structure, Inference, Complexity.* EATCS Monographs in Theoretical Computer Science, Springer, forthcoming.
8. R. P. Dilworth and N. Ward (1939) "Residuated lattices". Transactions of the American Mathematical Society **45**, pp. 335–354.
9. I. Düntsch and E. Orłowska (2000). "Logics of complementarity in information systems". Mathematical Logic Quarterly **46**, pp. 267–288.
10. I. Düntsch and E. Orłowska (2000). "Beyond modalities: sufficiency and mixed algebras". In: E. Orłowska and A. Szałas (eds) *Relational Methods for Computer Science Applications*, Physica Verlag, Heidelberg, pp. 263–285.
11. I. Düntsch and E. Orłowska (2001). "Algebraic structures for qualitative reasoning". Alfred Tarski Centenary Conference, Warsaw, Poland.
12. F. Esteva and L. Godo (2001) "Monoidal t–norm based logic: towards a logic for left–continuous t–norms". Fuzzy Sets and Systems **124**, pp. 271–288.
13. F. Esteva and L. Godo (2001). "On complete residuated many-valued logics with t-norm conjunction". Proceedings of the 31st International Symposium on Multiple-Valued Logic, Warsaw, Poland, pp. 81–86.
14. P. Flondor, G. Georgescu, and A. Iorgulecu (2001). "Psedo-t-norms and pseudo-BL algebras". Soft Computing **5**, No 5, pp. 355-371.
15. S. Gottwald (2001). *A Treatise on Many-Valued Logics*, Studies in Logic and Computation **9**, Research Studies Press: Baldock, Hertfordshire, England.
16. J. A. Gougen (1967). "L–fuzzy sets". Journal of Mathematical Analysis and Applications **18**, pp. 145–174.
17. P. Hajek (1998). *Metamathematics of Fuzzy Logic*, Kluwer, Dordrecht.
18. J. B. Hart, L. Rafter, and C. Tsinakis (2001). "The structure of commutative residuated lattices". Preprint.
19. C. A. R. Hoare and H. Jifeng (1986). "The weakest prespecification". Part I: Fundamenta Informaticae IX, pp. 51–84. Part II: Fundamenta Informaticae IX, pp. 217–252.
20. U. Höhle and U.P. Klement (eds) (1996). *Non–Classical Logics and their Applications to Fuzzy Subsets*, Kluwer, Dordrecht.
21. U. Höhle (1996). "Commutative, residuated l-monoids". In [20], pp. 53–106.
22. I. Humberstone (1983). "Inaccessible words". Notre Dame Journal of Formal Logic **24**, pp. 346–352.
23. P. Jipsen (2001). "A Gentzen system and decidability for residuated lattices". Preprint.
24. B. Konikowska (1997). "A logic for reasoning about relative similarity". Studia Logica **58**, pp. 185–226.

25. R. K. Meyer and R. Routley (1972). "Algebraic analysis of entailment", Logique et analyse **15**, pp. 407–428.
26. E. Orłowska (1988). "Kripke models with relative accessibility and their application to inferences from incomplete information". In: Mirkowska, G. and Rasiowa, H. (eds), *Mathematical Problems in Computation Theory.* Banach Center Publications **21**, pp. 329–339.
27. E. Orłowska (ed) (1998). *Incomplete Information – Rough Set Analysis,* Studies in Fuzziness and Soft Computing, Springer-Verlag.
28. E. Orłowska (1999). "Many–valuedness and uncertainty". Multiple-Valued Logic **4**, pp. 207–227.
29. E. Orłowska and A. M. Radzikowska (2001). "Information relations and operators based on double residuated lattices". In H. de Swart (ed), Proceedings of the 6th Seminar on Relational Methods in Computer Science *RelMiCS'2001*, pp. 185–199.
30. A. M. Radzikowska and E. E. Kerre (2002) "A comparative study of fuzzy rough sets". Fuzzy Sets and Systems **126** No 2, pp. 137–155.
31. A. M. Radzikowska, E. E. Kerre (2002) "A general calculus of fuzzy rough sets". Submitted.
32. A. M. Radzikowska and E. E. Kerre (2001). "On some classes of fuzzy information relations". Proceedings of the 31st International Symposium on Multiple-Valued Logic, Warsaw, Poland, pp. 75–80.
33. A. M. Radzikowska and E. E. Kerre (2001) "Towards studying of fuzzy information relations". To appear in Proceedings of *EUSFLAT-2001*, Leicester, UK.
34. H. Rasiowa and R. Sikorski (1970). *The Mathematics of Metamathematics,* Warszawa.
35. C. Rauszer (1974). "Semi-Boolean algebras and their applications to intuitionistic logic with dual operations". Fundamenta Mathematicae **83**, pp. 219-249.
36. B. Schweizer and A. Sklar (1983). *Probabilistic Metric Spaces.* North Holland, Amsterdam.
37. E. Turunen (1999) *Mathematics Behind Fuzzy Logic,* Springer–Verlag.
38. D. Vakarelov (1998). "Information systems, similarity relations and modal logics". In [27], pp. 492–550.

Interval Bilattices and Some Other Simple Bilattices

Agata Pilitowska*

Warsaw University of Technology, Department of Mathematics
Plac Politechniki 1, 00-661 Warsaw, Poland
irbis@pol.pl

Abstract. In a number of papers M.Ginsberg introduced algebras called bilattices having two separate lattice structure and one additional basic unary operation. They originated as an algebraization of some non-classical logics that arise in artificial intelligence and knowledge-based logic programming. In this paper we introduce some new class of bilattices which originate from interval lattices and show that each of them is simple. A known simple lattices are used to give other examples of simple bilattices. We also describe simple bilattices satisfying some additional identities so called P-bilattices (or interlaced bilattices).

1 Introduction

In a number of papers M.Ginsberg [6], [7] introduced algebras called bilattices having two separate lattice structure and one additional basic unary operation acting on both lattices in a very regular way. Bilattices originated as an algebraization of some non-classical logics that arise in artificial intelligence and knowledge-based logic programming. The two lattice orderings of a bilattice may be viewed as the degrees of truth and knowledge of possible events. In general, both lattice structures of a bilattice are "almost" not connected to each other. However bilattices appearing in applications usually satisfy some additional conditions. Bilattices were also investigated by Fitting, Romanowska, Avron, Mobasher, Pigozzi, Slutzki, Voutsadakis, Pynko and others.

In [5] M.Fitting described a structure based on intervals of a lattice with one lattice and one semilattice orderings. The main purpose of this paper is to introduce a new class of bilattices which also originates from interval lattices and to show that such bilattices are simple.

In Sect. 1 we collect some useful facts about interval lattices, and bilattices. Interval bilattices, a new class of bilattices which originate from interval lattices, are described in Sect. 2. In Sect. 3 we show that all interval bilattices are simple. In Sect. 4 a known simple lattices are used to give other examples of simple bilattices. Finally, we describe simple P-bilattices (bilattices satisfying some additional identities which were introduced and investigated in [13]. P-bilattices were also investigated under the name of "interlaced" bilattices by other authors.).

* The paper was written within the framework of COST Action 274.

H. de Swart (Ed.): RelMiCS 2001, LNCS 2561, pp. 190–196, 2002.

2 Preliminaries

Let $\mathbf{L} = (L, \vee, \wedge)$ be a lattice with the ordering relation $\leq$. The sublattice $[a, b] = \{x \in L | a \leq x \leq b\}$ of $\mathbf{L}$ is called *an interval* with the end-points a and b. The empty subset of L is also considered as an interval $[\ ,\ ]$. The family of all intervals in $\mathbf{L}$ is denoted by $IntL$.
$IntL$ is a lattice $\mathbf{IntL_k} = (IntL, \circ, +)$ under set inclusion $\leq_k$. We have that $[a, b] \leq_k [c, d]$ if and only if $a \geq c$ and $b \leq d$. The lattice operations are defined by the following formulas:

$$[a,b] \circ [c,d] := \begin{cases} [a \vee c, b \wedge d] & \text{if } a \vee c \leq b \wedge d \\ [\ ,\ ] & \text{otherwise} \end{cases} \tag{1}$$

$$[a,b] + [c,d] := [a \wedge c, b \vee d] \ . \tag{2}$$

The lattice $\mathbf{IntL_k}$ will be called *an interval lattice.* Note that the empty interval $[\ ,\ ]$ is the least element in this lattice. The interval lattices were studied by W.Duthie [2] and by V.Igoshin [8].

Definition 1. *A bilattice* $\mathbf{B} = (B, \wedge, \vee, \circ, +, \neg, 0_1, 1_1, 0_2, 1_2)$ *is an algebra such that*

1. $(B, \wedge, \vee, 0_1, 1_1)$ *and* $(B, \circ, +, 0_2, 1_2)$ *are bounded lattices.*
2. *A negation* $\neg : B \to B$ *is an unary operation on* B *satisfying for all* $x, y \in B$ *the identities:*

$$\neg\neg x = x, \tag{3}$$

$$\neg(x \vee y) = \neg x \wedge \neg y \tag{4}$$

$$\neg(x \wedge y) = \neg x \vee \neg y \tag{5}$$

$$\neg(x + y) = \neg x + \neg y \tag{6}$$

$$\neg(x \circ y) = \neg x \circ \neg y \ . \tag{7}$$

Definition 2. *A bilattice* $\mathbf{B} = (B, \wedge, \vee, \circ, +, \neg, 0_1, 1_1, 0_2, 1_2)$ *satisfying the following identities:*

$$((x \wedge y) \circ z) \wedge (y \circ z) = (x \wedge y) \circ z \tag{8}$$

$$((x \circ y) \wedge z) \circ (y \wedge z) = (x \circ y) \wedge z \tag{9}$$

$$((x \wedge y) + z) \wedge (y + z) = (x \wedge y) + z \tag{10}$$

$$((x + y) \wedge z) + (y \wedge z) = (x + y) \wedge z \tag{11}$$

will be called P-bilattice.

The following results were proved in [13]. (See also [1], [4] and [12].)

Theorem 1. *Let* $\mathbf{L} = (L, \wedge, \vee, 0, 1)$ *be a bounded lattice. On a set* $L \times L$ *we define an algebra* $\mathbf{B}(\mathbf{L}) = (L \times L, \wedge, \vee, \circ, +, \neg, 0_1, 1_1, 0_2, 1_2)$ *as follows: for all* $(x, x'), (y, y')$ *in* $L \times L$,

$$(x, x') \wedge (y, y') := (x \wedge y, x' \vee y') \tag{12}$$

$$(x, x') \vee (y, y') := (x \vee y, x' \wedge y') \tag{13}$$

$$(x, x') \circ (y, y') := (x \wedge y, x' \wedge y') \tag{14}$$

$$(x, x') + (y, y') := (x \vee y, x' \vee y') \tag{15}$$

$$\neg(x, y) := (y, x) \tag{16}$$

$$0_1 := (0, 1), \;\; 1_1 := (1, 0), \;\; 0_2 := (0, 0), \;\; 1_2 := (1, 1) \; . \tag{17}$$

The algebra $\mathbf{B}(\mathbf{L})$ *is a P-bilattice.*

The bilattice $\mathbf{B}(\mathbf{L})$ will be called *a product bilattice associated with the lattice* $\mathbf{L}$.

Theorem 2. *An algebra* $\mathbf{B} = (B, \wedge, \vee, \circ, +, \neg, 0_1, 1_1, 0_2, 1_2)$ *is a P-bilattice if and only if there is a bounded lattice* $\mathbf{L} = (L, \wedge, \vee, 0, 1)$, *such that* $\mathbf{B}$ *is isomorphic to the product bilattice* $\mathbf{B}(\mathbf{L})$ *associated with the lattice* $\mathbf{L}$.

3 A Construction of Interval Bilattices

Let $\mathbf{L} = (L, \wedge, \vee, \neg, 0, 1)$ be a bounded lattice with a polarity $\neg : L \to L$ and ordering relation $\leq$. Let $Int_0 L$ be the family of all intervals of $\mathbf{L}$ excluding the empty interval. In $Int_0 L$ define two binary operations as follows:

$$[a, b] \wedge [c, d] := [a \wedge c, b \wedge d] \tag{18}$$

$$[a, b] \vee [c, d] := [a \vee c, b \vee d] \; . \tag{19}$$

Proposition 1. *The algebra* $\mathbf{Int_0 L_t} = (Int_0 L, \wedge, \vee, [0, 0], [1, 1])$ *is a bounded lattice.*

Proof. It is evident that if $a \leq b$ and $c \leq d$ then $a \wedge c \leq b \wedge d$ and $a \vee c \leq d \vee d$. Therefore
$[a, b] \vee [a, b] = [a \vee a, b \vee, b] = [a, b]$,
$[a, b] \vee [c, d] = [a \vee c, b \vee d] = [c \vee a, d \vee b] = [c, d] \vee [a, b]$,
$([a, b] \vee [c, d]) \vee [e, f] = [a \vee c, b \vee d] \vee [e, f] = [a \vee c \vee e, b \vee d \vee f] =$
$= [a, b] \vee [c \vee e, d \vee f] = [a, b] \vee ([c, d] \vee [e, f])$,
$[a, b] \vee ([a, b] \wedge [c, d]) = [a \vee b] \vee [a \wedge c, b \wedge d] = [a \vee (a \wedge c), b \vee (b \wedge d)] = [a, b]$.
Simillarly, we can show that the second operation $\wedge$ is idempotent, commutative and associative.
Moreover, for all $[a, b]$ in $Int_0 L$:
$[a, b] \vee [1, 1] = [a \vee 1, b \vee 1] = [1, 1]$,
$[a, b] \wedge [0, 0] = [a \wedge 0, b \wedge 0] = [0, 0]$.
It follows that $\mathbf{Int_0 L_t}$ is a bounded lattice with the least element $[0, 0]$ and the greatest element $[1, 1]$. □

Corollary 1. *The ordering relation* $\leq_t$ *of a lattice* $\mathbf{Int_0L_t}$ *is defined by*

$$[a,b] \leq_t [c,d] \text{ if and only if } a \leq c \text{ and } b \leq d \ . \tag{20}$$

Now let $\mathbf{IntL_t}$ be the extension of the partially ordered set $(Int_0L, \leq_t)$ by the empty interval $[\ ,\ ]$. By definition $[\ ,\ ]$ covers $[0,0]$, is covered by $[1,1]$ and is not comparable with any other interval of $\mathbf{L}$.
In $IntL$ we define one unary operation $\neg : IntL \to IntL$ by

$$\neg[a,b] := [\neg b, \neg a] \tag{21}$$

$$\neg[\ ,\ ] := [\ ,\ ] \ . \tag{22}$$

Note that if $a \leq b$ then $\neg b \leq \neg a$, whence $[\neg b, \neg a]$ is an interval in $\mathbf{L}$.
Proposition 2. *The algebra* $\mathbf{BIntL} = (IntL, \wedge, \vee, \circ, +, \neg, [0,0], [1,1], [\ ,\ ], [0,1])$ *is a bilattice.*
Proof. Since the lattice $\mathbf{L} = (L, \wedge, \vee, \neg, 0, 1)$ is bounded, the lattice $\mathbf{IntL_k}$ has the greatest element $[0,1]$. Hence $\mathbf{IntL_t} = (IntL, \wedge, \vee, [0,0], [1,1])$ and $\mathbf{IntL_k} = (IntL, \circ, +, [\ ,\ ], [0,1])$ are bounded lattices.

Note that the unary operation $\neg : IntL \to IntL$ is an involution

$$\neg\neg[a,b] = \neg[\neg b, \neg a] = [a,b] \ . \tag{23}$$

It reverses the order $\leq_t$ and preserves the order $\leq_k$.
Indeed, if $[a,b] \leq_t [c,d]$ then $a \leq c$ and $b \leq d$ implying $\neg a \geq \neg c$ and $\neg b \geq \neg d$.
Hence $[\neg b, \neg a] \geq_t [\neg d, \neg c]$ and $\neg[a,b] \geq_t \neg[c,d]$.
If $[a,b] \leq_k [c,d]$ then $[a,b] + [c,d] = [a \wedge c, b \vee d] = [c,d]$ implying
$\neg[a \wedge c, b \vee d] = [\neg(b \vee d), \neg(a \wedge c)] = [\neg b \wedge \neg d, \neg a \vee \neg c] =$
$= [\neg b, \neg a] + [\neg d, \neg c] = \neg[a,b] + \neg[c,d] = \neg[c,d]$.
Hence, $\neg[a,b] \leq_k \neg[c,d]$.
It follows that $\mathbf{BIntL}$ is a bilattice. □

Definition 3. *The bilattice* $\mathbf{BIntL} = (IntL, \wedge, \vee, \circ, +, \neg, [0,0], [1,1], [\ ,\ ], [0,1])$ *will be called an interval bilattice.*

Example 1. Let $(\{0,1\}, \wedge, \vee, \neg, 0, 1)$ be two element chain. In this case $IntL = \{[\ ,\ ], [0,0], [1,1], [0,1]\}$ and the interval bilattice $\mathbf{BIntL}$ is isomorphic to four element distributive bilattice $\mathbf{B_4}$.

Lemma 1. *Let* $\mathbf{L} = (L, \wedge, \vee, \neg, 0, 1)$ *be an arbitrary bounded polarity lattice with more than two elements. Then the interval bilattice* $\mathbf{BIntL} = (IntL, \wedge, \vee, \circ, +, \neg, [0,0], [1,1], [\ ,\ ], [0,1])$ *is not a P-bilattice.*
Proof. Let $a \neq 0$ and $a \neq 1$ be an element in L. We have
$(([\ ,\ ] \wedge [0,a]) \circ [a,1]) \wedge ([0,a] \circ [a,1]) = ([0,0] \circ [a,1]) \wedge [0 \vee a, a \wedge 1] =$
$= [0 \vee a, 0 \wedge 1] \wedge [a,a] = [\ ,\ ] \wedge [a,a] = [0,0]$
and
$([\ ,\ ] \wedge [0,a]) \circ [a,1] = [0,0] \circ [a,1] = [\ ,\ ]$.
It follows that identity

$$((x \wedge y) \circ z) \wedge (y \circ z) = (x \wedge y) \circ z \tag{24}$$

is not satisfied in the interval bilattice $\mathbf{BIntL}$. Hence $\mathbf{BIntL}$ is not a P-bilattice. □

4 Congruence Lattices of Interval Bilattices

In this section we give a characterization of the lattice of congruence relations of an interval bilattice. Let **BIntL** be an interval bilattice obtained from some bounded polarity lattice **L**$= (L, \wedge, \vee, \neg, 0, 1)$. The lattice of congruence relations of **BIntL** will be denoted by Con**BIntL**.

Theorem 3. *For any bounded polarity lattice* **L**$= (L, \wedge, \vee, \neg, 0, 1)$, *the congruence lattice* Con**BIntL** *is isomorphic to the 2-element chain.*

Proof. Let a, b, c, d be in L, $d \neq 0$, $d \neq 1$ and θ be a non-zero congruence relation of **BIntL**.
First note that if $[a, b] \leq_k [c, d]$ then $c \leq a \leq b \leq d$ and
$[a, b] \circ [d, d] = [a \vee d, b \wedge d] = [\ ,\ ]$
and
$[c, d] \circ [d, d] = [c \vee d, d \wedge d] = [d, d]$.
Therefore it is easy to see that if $[a, b] \leq_k [c, d]$ and $[a, b]$ and $[c, d]$ are in θ then $[d, d]$ and $[\ ,\ ]$ are in θ too.
It is well known that in a lattice $(L, \wedge, \vee)$, elements x and y are in some congruence relation φ if and only if $x \wedge y$ and $x \vee y$ are in φ.

Since by definition intervals $[d, d]$ and $[\ ,\ ]$ are not comparable with respect to $\leq_t$, it is clear that if $[d, d]$ and $[\ ,\ ]$ lie in θ, then θ also contains $[0, 0]$ and $[1, 1]$, and hence all other intervals.

Moreover, if $[1,1]$ and $[\ ,\ ]$ (or $[0,0]$ and $[\ ,\ ]$) are in θ, then $\neg[1, 1] = [0, 0]$ and $\neg[\ ,\ ] = [\ ,\ ]$ (or $\neg[0, 0] = [1, 1]$ and $[\ ,\ ]$) are in θ too.

It follows that θ is the largest trivial congruence relation. □

Corollary 2. *For any bounded polarity lattice* **L**, *the interval bilattice* **BIntL** *is simple.*

5 Further Examples of Simple Bilattices

In this section we give further examples of simple bilattices.

Example 2. Let n, i and j be natural numbers and let
$Q^n := \{(0,0), (1,1)\} \cup \{(2^{-i}, 0) | 0 \leq i \leq n-1\} \cup \{(0, 2^{-i}) | 0 \leq i \leq n-1\} \cup$
$\cup \{(2^{-i}, 2^{-j}) | 1 \leq i, j \leq n \text{ and } |i-j| \leq 1\}$.
Let $\mathbf{Q^n_k} = (Q^n, \leq_k)$ be the set Q^n with the order $\leq_k$ defined by

$$(a_1, b_1) \leq_k (a_2, b_2) \text{ iff } a_1 \leq a_2 \text{ and } b_1 \leq b_2 \ . \tag{25}$$

And let $\mathbf{Q^n_t} = (Q^n, \leq_t)$ be the set Q^n with the order $\leq_t$ defined by

$$(a_1, b_1) \leq_t (a_2, b_2) \text{ iff } a_1 \leq a_2 \text{ and } b_1 \geq b_2 \ . \tag{26}$$

Obviously, $\mathbf{Q^n_k}$ and $\mathbf{Q^n_t}$ are lattices.

The lattices $\mathbf{Q^n_k}$ and $\mathbf{Q^n_t}$ with the unary operation $\neg(a, b) := (b, a)$ form the bilattice $\mathbf{Q^n}$. First note that the unary operation $\neg : Q^n \to Q^n$ is an involution:

$\neg\neg(a,b)=(a,b)$. Moreover, it reverses the order $\leq_t$ and preserves the order $\leq_k$, because
$(a_1,b_1)\leq_k(a_2,b_2)\Leftrightarrow a_1\leq a_2$ and $b_1\leq b_2\Leftrightarrow$
$\Leftrightarrow\neg(a_1,b_1)=(b_1,a_1)\leq_k(b_2,a_2)=\neg(a_2,b_2)$
and
$(a_1,b_1)\leq_t(a_2,b_2)\Leftrightarrow a_1\leq a_2$ and $b_1\geq b_2\Leftrightarrow$
$\Leftrightarrow\neg(a_1,b_1)=(b_1,a_1)\geq_t(b_2,a_2)=\neg(a_2,b_2)$.
By R.Wille [14] the lattice $\mathbf{Q^n_k}$ is simple. Because if a lattice reduct of a bilattice is simple then the bilattice in question is always simple, too, the bilattice $\mathbf{Q^n}$ is simple.

Example 3. Let $n\geq 2$ be a natural number and $Q_0^{2n}:=Q^n\cup P^n-\{(0,1),(1,0)\}$, where Q^n is the set defined in Ex. 2 and $P^n:=\{(1-x,1-y)|(x,y)\in Q^n\}$.

In Q_0^{2n} define partial order $\leq_t$ as follows:

$$(a,b)\leq_t(c,d)\text{ iff } a\leq c \text{ and } b\leq d\ . \tag{27}$$

Now let $\mathbf{Q^{2n}_t}$ be the extension of the partially ordered set $(Q_0^{2n},\leq_t)$ by a single element (0,1). By definition (0,1) covers $(2^{-n},2^{-n})$, is covered by $(1-2^{-n},1-2^{-n})$ and is not comparable with any other element of the set Q_0^{2n} different from (1,1) and (0,0). It is clear that Q_t^{2n} is a bounded lattice with (1,1) as the greatest element and (0,0) as the least element.

Moreover in $Q^{2n}:=Q_0^{2n}\cup\{(0,1)\}$ define two binary operations as follows:

$$(a,b)\circ(c,d):=\begin{cases}(max(a,c),max(b,d)) & \text{if } (a,b),(c,d)\in Q^n-\{(0,1),(1,1)\}\\ (min(a,c),min(b,d)) & \text{if } (a,b),(c,d)\in P^n-\{(0,1),(0,0)\}\\ (\frac{1}{2},\frac{1}{2}) & \text{otherwise}\end{cases}$$

$$(a,b)+(c,d):=\begin{cases}(min(a,c),min(b,d)) & \text{if } (a,b),(c,d)\in Q^n-\{(0,1),(1,1)\}\\ (max(a,c),max(b,d)) & \text{if } (a,b),(c,d)\in P^n-\{(0,1),(0,0)\}\\ (0,1) & \text{otherwise}\end{cases}\ .$$

It is easy to see that $\mathbf{Q^{2n}_k}=(Q^{2n},\circ,+,(\frac{1}{2},\frac{1}{2}),(0,1))$ is a bounded lattice.

On the set Q^{2n} we define the unary operation

$$\neg(a,b):=(1-b,1-a)\ . \tag{28}$$

It is easy to note that $\mathbf{Q^{2n}_k}$ and $\mathbf{Q^{2n}_t}$ with the above unary operation is a bilattice $\mathbf{Q^{2n}}$. As was shown by R. Wille [14], the lattice $\mathbf{Q^{2n}_t}$ is simple. Hence the bilattice $\mathbf{Q^{2n}}$ is simple too.

The following result was proved in [13].

Lemma 2. *Let* $\mathbf{B}=(B,\wedge,\vee,\circ,+,\neg,0_1,1_1,0_2,1_2)$ *be a P-bilattice isomorphic to the product bilattice* $\mathbf{B(L)}$ *associated with the lattice* $\mathbf{L}=(L,\wedge,\vee)$. *Then the lattice Con*$\mathbf{B}$ *is isomorphic to the lattice Con*$\mathbf{L}$.

As an easy consequence of this lemma we obtain the following corollary.

Corollary 3. *A P-bilattice* $\mathbf{B(L)}$ *is simple if and only if the lattice* $\mathbf{L}$ *is simple.*

References

1. Avron, A.: The structure of interlaced bilattices. Math. Structures Comput. Sci. **6** (1996) 287-299
2. Duthie, W.: Segments of order sets. Trans. Amer. Math. Soc. **51**(1942) 1-14
3. Fitting, M.: Bilattices and the theory of truth. Journal of Philosophical Logic **18**(1989) 225-256
4. Fitting, M.: Bilattices in logic programming. The Twentieth International Symposium on Multiple-Valued Logic (ed. Epstein, G.) IEEE (1990) 238-246
5. Fitting, M.: Kleene's logic, generalized. J. Logic Computat. **1** (1991) 797-810
6. Ginsberg, M.: Multi-valued logics. Proc. AAAI-86, Fifth National Conference on Artificial Intelligence, Morgan Kaufmann Publishers (1986) 243-247
7. Ginsberg, M.: Multivalued logics: A uniform approach to inference in artificial intelligence. Computational Intelligence **4**(1988) 265-316
8. Igoshin, V.: Algebraic characteristic of interval lattices/Russian/. Uspekhi Mat. Nauk **40**(1985) 205-206
9. McKenzie, R., McNulty, G., Taylor, W.: Algebras, Lattices, Varieties. The Wadsworth and Brooks, Monterey (1987)
10. Mobasher, B., Pigozzi, D., Slutzki, G., Voutsadakis, G.: A Duality theory for bilattices. Algebra Universalis **43**(2000) 109-125
11. Odintsov, V.: Congruences on lattices of intervals /Russian/. Mat. Zap. **14**(1988) 102-111
12. Pynko, A.: Regular bilattices. Journal of Applied Non-Classical Logics **10**(2000) 93-111
13. Romanowska, A., Trakul, A.(Pilitowska A.): On the structure of some bilattices. In: Hakowska, K., Stawski, B.(eds.) Universal and Applied Algebra, World Scientific (1989) 235-253
14. Wille, R.: A note on simple lattices. Col. Math. Soc. Janos Bolyai, 14. Lattice Theory, Szeged (1974) 455-462

Interactive Systems: From Folklore to Mathematics

Gheorghe Ştefănescu*

Department of Computer Science, National University of Singapore
gheorghe@comp.nus.edu.sg

Abstract. This paper gives a brief introduction to *finite interactive systems*, an abstract mathematical model of agents' behavior and their interaction. The paper contains the definition of finite interactive systems, examples, and a few simple results. No examples modeling real interactive systems are included, but there are many pointers suggesting how this can be done.

1 Introduction

This paper gives a brief introduction to *finite interactive systems* (FIS*'s*, for short), an abstract mathematical model of agents' behavior and their interaction. The paper contains the definition of finite interactive systems, examples, a few simple results, and pointers suggesting how FIS's may be use to model real interactive systems.

The agents we are talking about are those specified in standard *concurrent object-oriented languages*. The key point is the observation that *grids may be seen as interaction running patterns* for programs written in these languages much in the same way as paths represent possible executions of classical sequential programs. Then, finite interactive systems may be seen as a kind of two-dimensional finite automata melting together a state transforming automaton with a behavior transforming one (this latter transformation is responsible for modeling the interaction of the threads generated by the first automaton).

The main advances of the present paper are:

- it introduces finite interactive systems as a machine-like device for recognizing grid languages;
- it presents a large collection of toy examples illustrating the power of the mechanism; and
- it gives pointers suggesting how finite interactive systems may be use to model real interactive systems as those described by concurrent object-oriented languages.

Parts of this paper are included in the survey paper [Ste02] published in the volume dedicated to 2001 Marktoberdorf Summer School.

---

* On leave from *Departement of Fundamentals of Computer Science, University of Bucharest*

H. de Swart (Ed.): RelMiCS 2001, LNCS 2561, pp. 197–211, 2002.

2 Grids and Regular Expressions

2.1 Grids

A *grid (or planar word)* is a *rectangular* two-dimensional area filled in with atomic letters in a given alphabet V. Their set is denoted by $V^{\dagger\star}$. Each letter in V is to be seen as a two-dimensional atom having its own *northern, southern, western* and *eastern* border. We do not consider any typing mechanism here, hence all atoms' borders are similar, denoted by 1; this border information is naturally extended to general grids specifying the number of atoms lying on that border (rectangle's dimensions). For a grid p, we let $n(p), s(p), w(p)$ and $e(p)$ denote the dimension of its northern, southern, western, and eastern border, respectively.

In the literature, the term *picture* is often used as a substitute for grid or two-dimensional word, especially in the context of *picture languages.* Useful surveys on two-dimensional languages may be found in [InT91], [GiR97], or [LMN98]. However, our view is more semantical, hence we prefer to use the term 'grid' considered as a two-dimensional version of 'path'. Sometimes we will also use the term 'planar word' (as a two dimensional version of 'word'), but never 'picture'.

2.2 Simple Regular Expressions

We present here a simple extension of classical regular expressions [Kle56,Con71] to cope with grids.

Simple two-dimensional regular expressions. The *signature* of simple two-dimensional regular expressions consists of two collections of usual regular operators, sharing the additive part. Specifically, it is

$$+,\ 0,\ \cdot,\ ^\star,\ \mathsf{I},\ \triangleright,\ ^\dagger,\ -$$

where $(+, 0, \cdot, ^\star, \mathsf{I})$ is a usual Kleene signature to be used for the vertical dimension, while $(+, 0, \triangleright, ^\dagger, -)$ is another Kleene signature to be used for the horizontal dimension. Notice that '0', 'I' and '$-$' are zero-ary operators (constants), '$\star$' and '$\dagger$' are unary, while the remaining ones are binary.

The following names are used for the above symbols: '+' is *sum*, '0' is *zero*, '$\cdot$' is *vertical composition*, '$\star$' is *vertical star* (or iterated vertical composition), 'I' is *vertical identity*, '$\triangleright$' is *horizontal composition*, '$\dagger$' is *horizontal star* (or iterated horizontal composition) and '$-$' is *horizontal identity.*

Finally, *simple two-dimensional regular expressions* over an alphabet V (consisting of two-dimensional letters/atoms) are obtained starting from atoms and iteratively applying the operations in the described signature. Formally,

$$E ::= a \mid 0 \mid E + E \mid E \cdot E \mid E^\star \mid \mathsf{I} \mid E \triangleright E \mid E^\dagger \mid -$$

Their set is denoted by $\mathsf{2RegExp}(V)$.

Axiomatization, one-dimensional case. There are many elegant axiomatizations for (one-dimensional) regular languages, including the one presented by Kozen in [Koz94].

A particular class of Kleene algebras is defined by the following set of identities (where we use matrices to simplify the form of the star axioms), see for instance Ch.8 of [Ste00].

Let $\mathbf{M} = (\mathcal{M}(m,n)_{m,n}, +, \cdot, *, 0_{m,n}, \mathsf{I}_n)$ be a family of matrices with operations: (1) $0_{m,n} \in \mathcal{M}(m,n)$, (2) $\mathsf{I}_n \in \mathcal{M}(n,n)$; (3) '+' : $\mathcal{M}(m,n) \times \mathcal{M}(m,n) \to \mathcal{M}(m,n)$; (4) '$\cdot$' : $\mathcal{M}(m,n) \times \mathcal{M}(n,p) \to \mathcal{M}(m,p)$; (5) '$*$' : $\mathcal{M}(n,n) \to \mathcal{M}(n,n)$. $\mathbf{M}$ is a *Kleene algebra* if

- $\mathbf{M}$ is a *semiring of matrices*, namely: (1) with $+, 0$ it is a commutative monoid $(a+b)+c = a+(b+c)$; $a + 0_{m,n} = 0_{m,n} + a = a$; $a+b = b+a$; (2) with $\cdot, \mathsf{I}_p$ it is a category $(a \cdot b) \cdot c = a \cdot (b \cdot c)$; $a \cdot \mathsf{I}_n = a = \mathsf{I}_m \cdot a$; (3) distributivity: $a \cdot (b+c) = (a \cdot b) + (a \cdot c)$; $(a+b) \cdot c = (a \cdot c) + (b \cdot c)$; and (4) zero-laws: $0_{p,m} \cdot a = 0_{p,n}$; $a \cdot 0_{n,p} = 0_{m,p}$.
- It is *idempotent*, i.e., $a + a = a$.
- The following *axioms for star* holds:
  (I) $(\mathsf{I}_n)^* = \mathsf{I}_n$
  (S) $(a+b)^* = (a^* \cdot b)^* \cdot a^*$
  (P) $(a \cdot b)^* = \mathsf{I}_n + a \cdot (b \cdot a)^* \cdot b$
  (Inv) $a \cdot \rho = \rho \cdot b \Rightarrow a^* \cdot \rho = \rho \cdot b^*$, where ρ is matrix over 0,1.

These axioms for Kleene algebra give a correct and complete axiomatization for regular languages.

It is highly desirable to have a similar set of simple and powerful identities for regular grid languages, too.

From expressions to grids. To each expression in $\mathsf{2RegExp}(V)$ one may associate a language of grids over V. To this end, we first describe how composition operations act on grids and give the meaning of the identity operators.

If v, w are grids, then their horizontal composition is defined only if $e(v) = w(w)$ and the result $v \triangleright w$ is the word obtained putting v on left of w. Their vertical composition is defined only if $s(v) = n(w)$ and $v \cdot w$ is the word obtained putting v on top of w.

For each natural number k one may associate two 'empty' grids: the vertical identity ϵ_k having $w(\epsilon_k) = e(\epsilon_k) = 0$ and $n(\epsilon_k) = s(\epsilon_k) = k$ and the horizontal identity λ_k with $w(\lambda_k) = e(\lambda_k) = k$ and $n(\lambda_k) = s(\lambda_k) = 0$.

Now, the interpretation

$$| \ | : \mathsf{2RegExp}(V) \quad \to \quad \mathcal{P}(V^{\dagger *})$$

from expressions to languages of grids is defined by:

- $|a| = \{a\}$; $|0| = \emptyset$; $|E+F| = |E| \cup |F|$;
- $|E \cdot F| = \{v \cdot w : v \in |E| \ \& \ w \in |F|\}$;
- $|E^*| = \{v_1 \cdot \ldots \cdot v_k : k \in \mathbb{N} \ \& \ v_1, \ldots, v_k \in |E|\}$;
- $|\mathsf{I}| = \{\epsilon_0, \ldots, \epsilon_k, \ldots\}$;

– $|E \triangleright F| = \{v \triangleright w : v \in |E| \,\&\, w \in |F|\}$;
– $|E^\dagger| = \{v_1 \triangleright \ldots \triangleright v_k : k \in \mathbb{N} \,\&\, v_1, \ldots, v_k \in |E|\}$;
– $|-| = \{\lambda_0, \ldots, \lambda_k, \ldots\}$.

Before giving some examples, we define a useful flattening operator mapping languages of grids to languages of usual, one-dimensional words.

The flattening operator. The *flattening operator*

$$\flat \; : \; \mathcal{P}(V^{\dagger\star}) \;\; \to \;\; \mathcal{P}(V^\star)$$

maps sets of grids to sets of strings representing their topological sorting. In more details it is defined as follows:

- Each grid w may be considered as an acyclic directed graph $g(w)$ drawing: (1) horizontal edges from each letter to its right neighbor, if this exists, and (2) vertical edges from each letter to its bottom neighbor, if this exists.
- Being acyclic, $g(w)$ and all of its subgraphs have *minimal vertices*, i.e., vertices without incoming arrows.
- The topological sorting procedure selects a minimal vertex, then deletes it and its outgoing edges and repeats this as long as possible. This way a one-dimensional word containing all the atoms of w is obtained.
- Varying the minimal vertex selection in the topological sorting procedure in all possible ways one gets a set of words that is the value of the flattening operator applied to the grid.
- Finally, to define $\flat$ on a language L, take the union of $\flat(w)$, for $w \in L$.

To have an example, let us start with a grid $\begin{smallmatrix}\mathtt{a}&\mathtt{b}&\mathtt{c}&\mathtt{d}\\ \mathtt{e}&\mathtt{f}&\mathtt{g}&\mathtt{h}\end{smallmatrix}$; there is only one minimal element **a** and after its deletion we get $\begin{smallmatrix}&\mathtt{b}&\mathtt{c}&\mathtt{d}\\ \mathtt{e}&\mathtt{f}&\mathtt{g}&\mathtt{h}\end{smallmatrix}$; now there are two minimal elements **b** and **e** to choose from and suppose we choose **b**; what remains is $\begin{smallmatrix}&&\mathtt{c}&\mathtt{d}\\ \mathtt{e}&\mathtt{f}&\mathtt{g}&\mathtt{h}\end{smallmatrix}$; next, from the minimal elements **c** and **e**, we choose **e** and get $\begin{smallmatrix}&\mathtt{c}&\mathtt{d}\\ \mathtt{f}&\mathtt{g}&\mathtt{h}\end{smallmatrix}$; and so on; finally a usual word, say **abecfgdh**, is obtained. Actually,

$\flat(\begin{smallmatrix}\mathtt{a}&\mathtt{b}&\mathtt{c}&\mathtt{d}\\ \mathtt{e}&\mathtt{f}&\mathtt{g}&\mathtt{h}\end{smallmatrix}) =$ {**abcdefgh**, **abcedfgh**, **abcefdgh**, **abcefgdh**, **abecdfgh**, **abecfdgh**, **abecfgdh**, **abefcdgh**, **abefcgdh**, **aebcdfgh**, **aebcfdgh**, **aebcfgdh**, **aebfcdgh**, **aebfcgdh**}.

It is one of our main claims that this *flattening operator is responsible for the well-known state-explosion problem which occurs in the verification of concurrent object-oriented systems.* And one of our main hopes is that *the lifting of the verification techniques from paths to the grids is possible and may avoid this problem.*

A very nice quantitative analysis of the state-explosion generated by the flattening operator is presented in [BCGS01]: If w is a partial grid (i.e., a part of a grid which remains after some letters are already selected/deleted) and

$k_1, \ldots, k_p$ are the numbers representing for each cell the sum of the cells on top of it and on left of it plus 1, then the number of the flattened words is

$$\frac{p!}{k_1 \ldots k_p}$$

Applied to full, square grids of size n, this show that the state-explosion generated by the flattening operator is of order $\mathcal{O}(n^{n^2})$.

Convention: we use terminal type letters `a, b, ...` in grids; they should be identified with the corresponding italic version, used elsewhere.

Examples. A few simple examples of regular expressions are presented below.

1. The expression $(a \cdot d^\star \cdot g) \triangleright (b \cdot e^\star \cdot h)^\dagger \triangleright (c \cdot f^\star \cdot i)$ represents the language of grids $\begin{array}{cccccc} \mathtt{a} & \mathtt{b} & . & . & \mathtt{b} & \mathtt{c} \\ \mathtt{d} & \mathtt{e} & . & . & \mathtt{e} & \mathtt{f} \\ . & . & . & . & . & . \\ . & . & . & . & . & . \\ \mathtt{d} & \mathtt{e} & . & . & \mathtt{e} & \mathtt{f} \\ \mathtt{g} & \mathtt{h} & . & . & \mathtt{h} & \mathtt{i} \end{array}$. Notice that the same language may be represented by the expression $(a \triangleright b^\dagger \triangleright c) \cdot (d \triangleright e^\dagger \triangleright f)^\star \cdot (g \triangleright h^\dagger \triangleright i)$.

2. The language associated with $a^\dagger \cdot b^\dagger$ consists of grids $\begin{array}{l} \mathtt{aa..a} \\ \mathtt{bb..b} \end{array}$. Its flattened version $\flat(a^\dagger \cdot b^\dagger) = \{w \in \{a,b\}^* : |w|_a = |w|_b \ \ \& \ \ \forall w = w'w'' : |w'|_a \geq |w'|_b\}$ is context-free, but not regular ($|w|_a$ denotes the number of the occurrences of a in w).

The slightly extended expression $a^\dagger \cdot b^\dagger \cdot c^\dagger$ represents the language of grids $\begin{array}{l} \mathtt{aa..a} \\ \mathtt{bb..b} \\ \mathtt{cc..c} \end{array}$. Notice that its flattened version is $\flat(a^\dagger \cdot b^\dagger \cdot c^\dagger) = \{w \in \{a,b,c\}^* : |w|_a = |w|_b = |w|_c \ \ \& \ \ \forall \ \ w = w'w'' : |w'|_a \geq |w'|_b \geq |w'|_c\}$. This latter language is not even context-free.

3. Notice that $(a+b)^{\star\dagger} = (a+b)^{\dagger\star}$. This shows that vertical and horizontal stars may commute, provided simple atoms are involved. (This explain our notation $V^{\dagger\star}$ for the set of grids over V: if $V = \{a_1, \ldots, a_n\}$, then $V^{\dagger\star} = (a_1 + \ldots + a_n)^{\dagger\star}$.)

On the other hand, $\begin{array}{l} \mathtt{aa} \\ \mathtt{bb} \end{array} \in (a + b \triangleright b)^{\dagger\star} \setminus (a + b \triangleright b)^{\star\dagger}$, showing that, in general, the stars do not commute.

4. Finally, notice that $a^\dagger \cdot b^\dagger \neq a^\star \triangleright b^\star$, but $\flat(a^\dagger \cdot b^\dagger) = \flat(a^\star \triangleright b^\star)$. This example shows that the flattening mechanism may lose some information by passing from planar to usual languages.

2.3 Extended Regular Expressions

The regular expressions previously described are simple and natural, but they have a number of shortcomings. For instance, their generative power is quite limited, comparing with finite interactive systems. In this subsection we roughly describe a few possible extensions of simple two-dimensional regular expressions.

The separate vertical and horizontal iteration operators are not strong enough to represent some natural languages, for instance the languages L_{sq} consisting of squares of a's. One way to increase the power of the iterating operators is to allows for more freedom within the iteration process.

For L_{sq} one may use an alternating horizontal/vertical concatenation within the iteration process. Indeed, starting with a and iterating a horizontal concatenation on the right with a vertical string $a^\star$ and then a vertical concatenation on the bottom with a horizontal string $a^\dagger \triangleright a$ one precisely gets the language of squares of a's. In other words, L_{sq} is the least solution of the equation $X = a + (X \triangleright a^\star) \cdot (a^\dagger \triangleright a)$.

Another example considers the language L_{spir} consisting of 'spiral words'

```
x   2aa   2aaaa
    2x1   22aa1
    bb1   22x11
          2bb11
          bbbb1
```

.... A corresponding equation describing this language is $X = x + (2 \triangleright a^\dagger) \cdot (2^\star \triangleright X \triangleright 1^\star) \cdot (b^\dagger \triangleright 1)$. While in the above example the iterative process uses concatenation at right and bottom only, in the present case concatenation on all sides is used within the iteration process.

3 Finite Interactive Systems

3.1 Some Motivation

A finite interactive system may be seen as a kind of two-dimensional automaton mixing together a state-transforming machinery with a machinery used for the interaction of different threads of the first automaton. Syntactically, they may be described by graphs as the ones in Fig.1.

While this way of presentation may be somehow useful, it is also misleading.

The first point is that there are not two different automata to be combined, but these two views are melted together to give this new concept of finite interactive systems.

A next point is that the term 'automaton' is not the right one to consider. While there may be some other interpretations, an automaton is generally considered as a state-transforming device. A state is a temporal section of a running system: it gives the information related to the current values of the involved variables at a given temporal point. This information is used to compose the systems, to get more complex behaviors out of simpler ones. Then, using a term as 'two-dimensional automaton' is more or less as considering automata, but with a more complicate, say two-dimensional, state structure.

The situation with finite interactive systems is by far different: one dimension (by convention, the vertical one) is used to model this state transformation, but the other dimension is considered to be a behavior transforming device. Orthogonal to the previous case of states, a behavior gives a spatial section of the running system: it considers information on all the actions a spatially located agent has made during its evolution. This information is used at the temporal interfaces by which the agents interact, to get more complicate intelligent systems out of simpler ones. An agent is considered as a job-transforming device: given a job-request at its input temporal interface, the agent acts, transforms it, and passes the transformed job-request at its output temporal interface.

One feature which is implicitly present in this view is that the systems are open: to run such a system one has to give an appropriate initial state, but also an initial job-request, actually a place where the starting user interacts with the system. The result of the running is a pair consisting of a final state of the system and a trace of the jobs/messages the system passes at an output user.

Some more motivation along this line may be found in Sec.12.5 of [Ste00] dedicated to space-time duality. That section was written before the discovery of finite interactive systems. However, in some sense, it contains in the nutshell the seeds of this concept.

3.2 Finite Interactive Systems – Formal Definitions

After the above general considerations let us give a precise definition of interactive systems and explain it by a concrete example.

Definition. A *finite interactive system* (FIS) is a finite hyper-graph with two types of vertices and one type of (hyper) edges:
— the first type of vertices is distinguished using a labeling with numbers (or lower case letters); such a vertex denotes a *state* (memory of variables);
— the second type of vertices is distinguished using a labeling with capital letters; such a vertex is considered to be a *class* (of job requests);
— the actions/transitions are labeled by letters denoting atoms of grids and are such that: (1) each transition has two incoming arrows: one from a class vertex, the other from a state vertex, and (2) each transition has two outgoing arrows: one to a class vertex, the other to a state vertex.

Some classes/states may be *initial* (in the graphical representation this situation is specified by using small incoming arrows) or *final* (in this case the double circle representation is used). △

An example is given in Fig.1(a). This FIS has four vertices: two classes A and B and two states 1 or 2; moreover, A is an initial class, B is a final class, 1 is an initial state, and 2 is a final state. It also has three transitions labeled by a, b, and c.

Sometimes it may be more convenient to have a simple non-graphical representation for FIS's. One possibility is to represent a FIS by its transitions and to specify which states/classes are initial or final. (In this 'cross' representation of transitions the incoming vertices are those placed at north and west, while the outgoing ones are those from east and south.) For the given example, the interactive system is represented by:

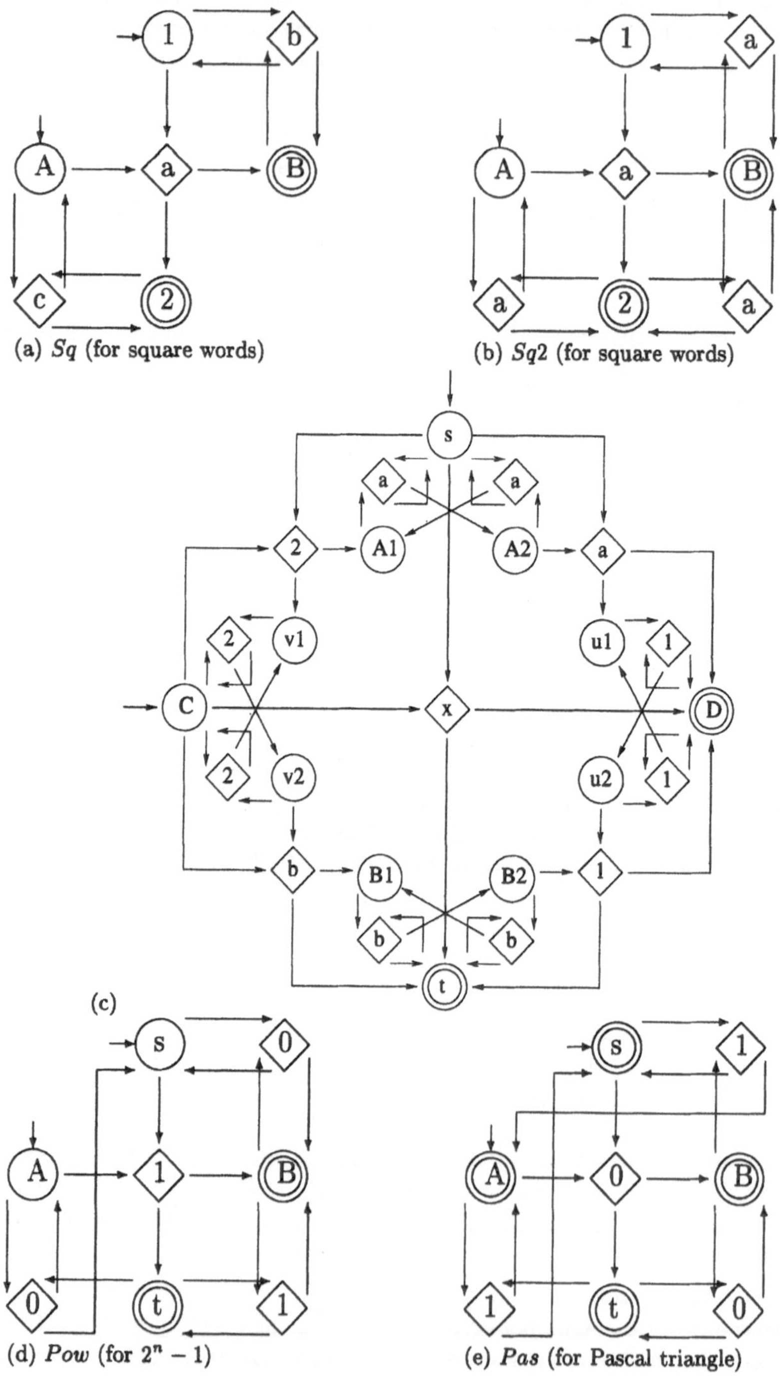

(a) *Sq* (for square words)

(b) *Sq*2 (for square words)

(c)

(d) *Pow* (for $2^n - 1$)

(e) *Pas* (for Pascal triangle)

Fig. 1. Examples of finite interactive systems

— $\begin{array}{ccc} & 1 & \\ A & a & B \\ & 2 & \end{array}$, $\begin{array}{ccc} & 2 & \\ A & c & A \\ & 2 & \end{array}$, and $\begin{array}{ccc} & 1 & \\ B & b & B \\ & 1 & \end{array}$ and the information that

— $A, 1$ are initial states/classes and $B, 2$ are final.

The execution of a transition $\begin{array}{ccc} & 1 & \\ A & a & B \\ & 2 & \end{array}$ has two aspects: By firing action a the state of the corresponding thread is changed from 1 to 2. However, this firing is possible only if a different thread requires the execution of a via class A. After the firing of a, our current thread may requires via class B other transitions to fire.

One may see this as a highly abstract mechanism capturing the atomic parts (methods' invocation) of the running of concurrent object oriented programs. As captured by FIS's, this mechanism of methods' invocation is non-blocking. This means, after the passing of requirements at B the current thread is free to continue its evolution.

Notice that the states are locals, each computation thread having its own set of variables. Similarly, the classes are local, each interactive object having a finite number of methods for its interaction interface.

A simple example dealing with the '8 queens problem' is given in Sec.2.4 of [Ste02].

Running patterns, accepting criteria. Interactive systems may be used to *recognize* grids. The procedure is the following.

Definition. (1) Suppose we are given: a FIS S, a grid w with m lines and n columns, a horizontal string with initial states $b_n = s_1 \dots s_n$, and a vertical string with initial classes $b_w = C_1 \dots C_m$. Then:

- Insert the states $s_1, \dots, s_n$ at the northern border of w (from left to right) and the classes $C_1, \dots, C_m$ at the western border of w (from top to bottom).
- Parse the grid w selecting minimal[1] still unprocessed atoms.
- For each such minimal unprocessed atom a, suppose s is the state inserted at its northern border and C is the class inserted at its western border. Then, choose a transition $\begin{array}{ccc} & s & \\ C & a & C' \\ & s' & \end{array}$ from S (if any), insert the state s' at the southern border of a and the class C' at its eastern border, and consider this atom to be already processed.
- Repeat the above two steps as long as possible.

If finally all the atoms of w were processed, then look at the eastern border b_e and the southern border b_s of the result. If they contain only final classes and final states, respectively, then consider this to be a *successful running for w with respect to the border conditions b_n, b_w, b_e, b_s*.

[1] The notion of minimal atoms we are using here is the one described when the flattening operator has been introduced.

(2) Given a FIS S and four regular languages B_n, B_w, B_e, B_s, a grid w is *recognized by S with respect to the border conditions* B_n, B_w, B_e, B_s if there exists some strings $b_n \in B_n$, $b_w \in B_w$, $b_e \in B_e$, $b_s \in B_s$ and a successful running for w with respect to the border conditions b_n, b_w, b_e, b_s. Let us denote by $L(S; B_n, B_w, B_e, B_s)$ their set.

(3) Finally, a language of grids L is *recognized by S* if there are some regular languages B_n, B_w, B_e, B_s such that $L = L(S; B_n, B_w, B_e, B_s)$.

(4) We simply denote the associated language by $L = L(S)$ when the border conditions are superfluous, i.e., the border languages are the corresponding full languages and actually no border condition is imposed. △

Example. A concrete example may be useful. Consider the FIS Sq in Fig. 1(a) and suppose that no border conditions are imposed (that is, $B_n = \{1\}^*$, $B_w = \{A\}^*$, $B_e = \{B\}^*$, $B_s = \{2\}^*$). A running is:

```
             1 1 1       1 1 1       1 1 1       1 1 1            1 1 1
a b b       Aa b b      AaBb b      AaBbBb      AaBbBb     ...   AaBbBbB
                         2           2 1         2 1              2 1 1
c a b       Ac a b      Ac a b      Ac a b      AcAa b           AcAaBbB
                                                 2                2 2 1
c c a       Ac c a      Ac c a      Ac c a      Ac c a           AcAcAaB
                                                                  2 2 2
```

w $\quad w_0$ $\quad w_1$ $\quad w_2$ $\quad w_3$ $\quad w_9$

The above sequence starts with a grid w. (1) The word obtained bordering w with initial states/classes is w_0. (2) At this stage all atoms are unprocessed, but only one is minimal, i.e., the a in the left-upper corner. In S there is only one transition of the type $\begin{smallmatrix} & 1 & \\ A & a & \ \\ & \ & \end{smallmatrix}$, i.e., $\begin{smallmatrix} & 1 & \\ A & a & B \\ & 2 & \end{smallmatrix}$. After its application one gets w_1. (3) Now there are two minimal unprocessed elements: b (position (1,2) in w) and c (position (2,1) in w). Suppose we choose b. Again, there is only one appropriate transition $\begin{smallmatrix} & 1 & \\ B & b & B \\ & 1 & \end{smallmatrix}$. After its application one gets w_2. (4) Next, the minimal unprocessed elements are b (position (1,3) in w) and c (position (2,1) in w). Suppose we choose c and the corresponding unique transition $\begin{smallmatrix} & 2 & \\ A & c & A \\ & 2 & \end{smallmatrix}$. Then we get w_3. (5) We can continue and finally an accepting running is obtained, leading to w_9.

One may easily see that the recognized language $L(S)$ consists of *square words with a's on the main diagonal, the top-right half filled in with b's and the bottom-left half filled in with c's.*

Finally, one may look at some border conditions. There is only one initial/final state/class, hence the border languages are regular languages over a one letter alphabet. To have an example, if $B_n = (111)^*$, $B_w = (AAAA)^*$, $B_e = B^*$ and $B_s = (22)^*$, then $L(S; B_n, B_w, B_e, B_s)$ is the sub-language of the above $L(S)$ consisting of those squares whose dimension k is a multiple of 12.

State projection and class projection. Some familiar nondeterministic finite automata (NFA's) may be obtained from FIS's neglecting one dimension. The automaton obtained from a FIS S neglecting the class transforming part is call the *state projection* NFA associated with S and it is denoted by $s(S)$. Similarly, the *class projection* NFA, denoted $c(S)$, is the NFA obtained neglecting the state transforming part of S.

The class projection NFA $c(S)$ may be used to define *macro-step transitions* in the original interactive system S. Dually, the state projection NFA $s(S)$ may be used to define *agent behaviors* (seen as job class transformers) in the original interactive system. (See Ch.12.5 of [Ste00] for more comments on this 'job class transformers' view.)

Let us return to the FIS S in Fig.1(a). Its class projection $c(S)$ is given by the transitions $A \xrightarrow{a} B,\quad A \xrightarrow{c} A,\quad B \xrightarrow{b} B$ and the information that A is initial and B is final. The state projection $s(S)$ is defined by transitions $1 \xrightarrow{a} 2,\quad 2 \xrightarrow{c} 2,\quad 1 \xrightarrow{b} 1$ with 1 initial and 2 final. *The* FIS *S in Fig.1(a) has an interesting intersection property $L(S) = L(s(S)) \cap L(c(S))$ showing that, in some sense, the interaction between the state and the class transformations of S is not so strong.*[2]

3.3 Examples

This subsection contains some more toy examples of finite interactive systems. There is no claim that they describe some interesting, real interactive systems. They are merely used to illustrate the mathematical power of the mechanism. While syntactically some of them look quite similar, actually they recognize very unrelated languages.

Squares. Let L_{sq} be the language $\begin{array}{c}\mathtt{a}\end{array}\quad \begin{array}{c}\mathtt{aa}\\ \mathtt{aa}\end{array}\quad \begin{array}{c}\mathtt{aaa}\\ \mathtt{aaa}\\ \mathtt{aaa}\end{array}\quad \begin{array}{c}\mathtt{aaaa}\\ \mathtt{aaaa}\\ \mathtt{aaaa}\\ \mathtt{aaaa}\end{array}\quad \ldots$ consisting of square grids of a's. An example of a FIS recognizing L_{sq} is Sq illustrated in Fig.1(a). Two examples of running patterns for Sq are shown below (the first one is successful; the second one is unsuccessful due to the fact that the southern border has an occurrence of the non-final state 1):

```
 1 1 1             1 1 1 1
AaBaBaB           AaBaBaBaB
 2 1 1             2 1 1 1
AaAaBaB           AaAaBaBaB
 2 2 1             2 2 1 1
AaAaAaB           AaAaAaBaB
 2 2 2             2 2 2 1
```

The same language is recognized by $Sq2$, drawn in the same figure. Making distinction between the various occurrences of a in $Sq2$ one gets another FIS $Sq1$

[2] To avoid some confusions, we emphasize here that the usual languages associated with the projection NFA's are extended to the two-dimensional case. For instance, a grid w is in $L(s(S))$ if every column in w is recognized by $s(S)$ in the usual sense. Similarly for $L(c(S))$, but now the lines are taken into account.

that recognizes the language of square grids with a's on the main diagonal, the top-right triangle filled in with b's, and the bottom-left triangle filled in with c's. ($Sq1$ consists of the initial state 1, the initial class A, the final state 2, the final classes B, and the following transitions $\begin{array}{|c|c|c|} \hline & 1 & \\ \hline A & a & B \\ \hline & 2 & \\ \hline \end{array}$, $\begin{array}{|c|c|c|} \hline & 2 & \\ \hline A & c & A \\ \hline & 2 & \\ \hline \end{array}$, $\begin{array}{|c|c|c|} \hline & 1 & \\ \hline B & b & B \\ \hline & 1 & \\ \hline \end{array}$, $\begin{array}{|c|c|c|} \hline & 2 & \\ \hline B & d & B \\ \hline & 2 & \\ \hline \end{array}$.)

Exponential function. Another interactive system is Pow, drawn in Fig.1(d). It recognizes the language $1 \quad \begin{array}{c} 10 \\ 01 \\ 11 \end{array} \quad \begin{array}{c} 100 \\ 010 \\ 110 \\ 001 \\ 101 \\ 011 \\ 111 \end{array}$ ..., grids whose number of lines is exponential (more precisaly, $2^n - 1$).

Pascal triangle. The FIS Pas in Fig.1(e) recognizes the set of grids over $\{0, 1\}$ whose northern and western borders have alternating sequences of 0 and 1 (starting with 0) and, along the secondary diagonals, it satisfies the recurrence rule of the Pascal triangle, modulo 2 (that is, the value in a cell (i, j) is the sum of the values of the cells $(i - 1, j)$ and $(i, j - 1)$, modulo 2).

In this case, the projection automata (on states and classes, respectively) are reset automata where all states/classes are final, hence they recognize any word. The intersection of their extensions to grids is the full language of grids, very far from the grids with Pascal-triangle property. This may be interpreted as a strong interaction between the state transformation and the class transformation within Pas. It also shows that *the 'intersection' property is not generally valid for the full class of finite interactive systems.*

Spiral grids. The finite interactive system drawn in Fig.1(c) recognizes the language of 'spiral' grids $\mathtt{x} \quad \begin{array}{c} \mathtt{2aa} \\ \mathtt{2x1} \\ \mathtt{bb1} \end{array} \quad \begin{array}{c} \mathtt{2aaaa} \\ \mathtt{22aa1} \\ \mathtt{22x11} \\ \mathtt{2bb11} \\ \mathtt{bbbb1} \end{array}$ ....

3.4 Recognizable Two-Dimensional Languages

It is known that the following statements are equivalent for a grid language L (called *recognizable two-dimensional language*):

1. L is recognized by a *on-line tessellation automaton*;
2. L is defined by a *simple regular expression with intersection and adding homomorphisms*;
3. L is defined by a *local lattice language plus homomorphisms*;
4. L is defined by an *existential monadic second order formula.*

The proof of this and many other informations may be found in the survey papers [GiR97,LMN98]. Alternatively, one may browse for papers starting from the following web page `http://math.uni-heidelberg.de/logic/bb/2dpapers.html`.

Notice that regular expressions and local lattice languags do not cover the class of recognizable languages, but reach this using homomorphisms. This may be seen as an inconvenience of these mechanisms. On the other hand, this problem does not occur for on-line tessellation automata or for our finite interactive systems, as they are closed under homomorphisms.

Our finite interactive systems represent a natural extension of finite automata to two dimensions and, as one may see below, capture recognizability. Find appropriate classes of two-dimensional regular expressions to capture recognizability without using homomorphisms is still an open question. Whether deterministic and (general) finite interactive systems are equivalent is also currently an open question.

Local Lattice Languages (LLL's, for short), also called *tile systems*, are specified using a finite set of finite (rectangular) grids that are allowed as sub-grids of the recognized grids. If one combines this with (letter-to-letter) homomorphisms, then the power of the mechanism does not change by restricting to 2×2 pieces, or even to horizontal and vertical dominoes (i.e., 1×2 and 2×1 pieces).

It is not too difficult to show that the class of grid languages specified by FIS's coincides with the class of recognizable two-dimensional languages:

- One may simulate an LLL by a FIS and the class of languages recognized by FIS's is closed under letter-to-letter homomorphisms. From these simple observations it follows that the class of languages recognized by finite interactive systems contains the class of recognizable two-dimensional languages.
- The converse inclusion is also valid: One can easily see that the running patterns of a FIS S, completed with a special symbols for the blank space, are recognized by a finite tile system. Then, one may use an homomorphism to hide the state/class/blank information around the letters in order to capture $L(S)$.

Hence we get the following theorem: *A set of grids is recognizable by a finite interactive system if and only if it is recognizable by a tiling system.*

This connection gives a way of inheriting many results known for two dimensional languages. Two important ones are: *(1) if one consider the restriction to the top line of the grid languages recognized by* FIS*'s, then one precisely gets context-sensitive string languages; and (2) the emptiness problem for finite interactive systems is undecidable.*

4 Closing Comments

The concept of finite interactive systems presented here has evolved from a study of Mixed Network Algebras, see [Ste00].

Our finite interaction systems may be seen as an extension of finite transition systems [Arn94], Petri nets [GaR92], data-flow networks [BrS01], or asynchronous automata [Zie87], in the sense that a potentially unbound number of processes may interact. In the mentioned models, while the number of processes may be unbounded, their interaction is bounded by the transitions' breadth. This seems to be a key feature for the ability to pass from concurrent to concurrent, object-oriented agents.

Our model has some similarities with the zero-safe nets model studied by Bruni and Montanari (see, e.g., [Bru99], [BrM97]), but the precise relationship still has to be investigated. We also expect that a logic similar to the spatial logic of [CaG00] may be associated with interaction systems and use for the verification of concurrent, object-oriented systems.

References

[Arn94] A. Arnold. *Finite transition systems.* Prentice-Hall, 1994.

[Bor99] Borchert, B. http://math.uni-heidelberg.de/logic/bb/2dpapers.html

[BCGS01] M. Broy, G. Ciobanu, R. Grosu, and G. Stefanescu. *Finite interactive systems: A unified model for agents' behaviour and their interaction.* Draft, December, 2001.

[BrS01] M. Broy and G. Stefanescu. The algebra of stream processing functions. *Theorertical Computer Science*, 258:95–125, 2001.

[Bru99] R. Bruni. *Tile Logic for Synchronized Rewriting of Concurrent Systems.* PhD thesis, Dipartimento di Informatica, Universita di Pisa, 1999. Report TD-1/99.

[BrM97] R. Bruni and U. Montanari. Zero-safe nets, or transition synchronization made simple. *Electronic Notes in Theoretical Computer Science*, vol. 7(20 pages), 1997.

[CaG00] L. Cardeli and A. Gordon. Anytime, anywhere: modal logics for mobile ambients. In: *POPL-2000, Symposium on Principles of Programming Languages*, Boston, 2000. ACM Press, 2000.

[Con71] J.H. Conway. *Regular Algebra and Finite Machines.* Chapman and Hall, 1971.

[GaR92] V. Garg and M.T. Ragunath. Concurrent regular expressions and their relationship to Petri nets. *Theoretical Computer Science*, 96:285–304, 1992.

[GiR97] D. Giammarresi and A. Restivo. Two-dimensional languages. In G. Rozenberg and A. Salomaa, editors, *Handbook of formal languages. Vol. 3: Beyond words*, pages 215–265. Springer-Verlag, 1997.

[GrLS00] R. Grosu, D. Lucanu, and G. Stefanescu. Mixed relations as enriched semiringal categories. *Journal of Universal Computer Science*, 6(1):112–129, 2000.

[GrSB98] R. Grosu, G. Stefanescu, and M. Broy. Visual formalism revised. In *Proceeding of the CSD'98 (International Conference on Application of Concurrency to System Design, March 23-26, 1998, Fukushima, Japan)*, pages 41–51. IEEE Computer Society Press, 1998.

[InT91] K. Inoue and I. Takanami. A survey of two-dimensional automata theory. *Information Sciences*, 55:99–121, 1991.

[Kle56] S.C. Kleene. Representation of events in nerve nets and finite automata. In: C.E. Shannon and J. McCarthy, eds., *Automata Studies, Annals of Mathematical Studies*, vol. 34, 3–41. Princeton University Press, 1956.

[Koz94] D. Kozen. A completeness theorem for Kleene algebras and the algebra of regular events. *Information and Computation* 110:366-390, 1994.

[LMN98] K. Lindgren, C. Moore, and M. Nordahl. Complexity of two-dimensional patterns. *Journal of Statistical Physics* 91:909–951, 1998.

[Ste98] G. Stefanescu. Reaction and control I. Mixing additive and multiplicative network algebras. *Logic Journal of the IGPL*, 6(2):349–368, 1998.

[Ste00] G. Stefanescu. *Network algebra*. Springer-Verlag, 2000.

[Ste01] G. Stefanescu. Kleene algebras of two dimensional words: A model for interactive systems. *Dagstuhl Seminar on "Applications of Kleene algebras"*, Seminar 01081, 19-23 February, 2001.

[Ste02] G. Stefanescu. Algebra of networks: Modeling simple networks, as well as complex interactive systems. In: H. Schwichtenberg and R. Steibrügen, eds., *Proof and System Reliability*, 49-78. Kluwer Academic Publishers, 2002.

[Zie87] W. Zielonka. Notes on finite asynchronous automata. *Theoretical Informatics and Applications*, 21:99–135, 1987.

Relational Constructions in Goguen Categories

Michael Winter

Department of Computer Science
University of the Federal Armed Forces Munich
85577 Neubiberg, Germany
thrash@informatik.unibw-muenchen.de

Abstract. Goguen categories were introduced as a suitable calculus for $\mathcal{L}$-fuzzy relations, i.e., for relations taking values from an arbitrary complete Brouwerian lattice $\mathcal{L}$ instead of the unit interval $[0,1]$ of the real numbers. Such a category may provide some relational constructions as products, sums or subobjects. The aim of this paper is to show that under an assumption on the lattice $\mathcal{L}$ one may require without loss of generality that the related relations are crisp, i.e., all entries are either the least element 0 or the greatest element 1 of $\mathcal{L}$.

1 Introduction

Since the mid-1970's it has become clear that the calculus of relations is a fundamental conceptual and methodological tool in computer science just as much as in mathematics. While computer science applications are evolving rapidly in several areas as in communication, programming, software, data or knowledge engineering, mathematical foundations are needed to provide a basis for existing methods. One important application is the treatment of uncertain or incomplete information. To handle this kind of information, Zadeh [15] introduced the concept of fuzzy sets. Later on, Goguen [4] generalized this concept to $\mathcal{L}$-fuzzy sets and relations for an arbitrary complete Brouwerian lattice $\mathcal{L}$ (or complete Heyting algebra) instead of the unit interval $[0,1]$ of the real numbers. Such structures constitute a Dedekind category introduced in [8].

In [7] the concept of fuzzy relation algebras was introduced as an algebraic formalization of mathematical structures formed by fuzzy relations, e.g., relations with coefficients from the unit interval, on a set with sup-min composition. Those algebras are equipped with a semi-scalar multiplication, i.e., an operation mapping an element from $[0,1]$ and a fuzzy relation to a fuzzy relation. Using this operation, one may characterize when a relation is 0-1 crisp, i.e., all entries are either the least element 0 or the greatest element 1. Unfortunately, such an operation does not exist if the unit interval $[0,1]$ is replaced by an arbitrary complete Brouwerian lattice.

On the other hand, it was shown in [13] that a suitable algebraic formalization for arbitrary $\mathcal{L}$-fuzzy relations demands an extra operator. In particular, it was shown that there is no formula in the theory of Dedekind categories expressing the fact that a given $\mathcal{L}$-fuzzy relation is 0-1 crisp. Therefore, the concept of Goguen categories was introduced. It was shown that the standard model of

H. de Swart (Ed.): RelMiCS 2001, LNCS 2561, pp. 212–227, 2002.

$\mathcal{L}$-fuzzy relations, i.e., the matrix algebra with coefficients from $\mathcal{L}$, is indeed a Goguen category.

A Goguen category may provide some relational constructions as products, sums or subobjects (cf. [11,12]). Such a construction is given by an object together with a set of relations fulfilling some equations. For example, a relational product of two objects A and B is an object $A \times B$ together with two relation $\pi : A \times B \to A$ and $\rho : A \times B \to B$ such that

$$\pi^{\smile};\pi \sqsubseteq \mathbb{I}_A, \quad \rho^{\smile};\rho \sqsubseteq \mathbb{I}_B \quad \pi^{\smile};\rho = \mathbb{T}_{AB}, \quad \pi;\pi^{\smile} \sqcap \rho;\rho^{\smile} = \mathbb{I}_{A\times B}.$$

One may expect that the pairing $Q;\pi^{\smile} \sqcap R;\rho^{\smile}$ of two crisp relations Q and R is crisp again. If the projections are crisp then this is the case since the class of crisp relations is closed under the usual relational operations. Especially in applications, such a property seems to be essential. For example, a fuzzy controller may be described using the following picture.

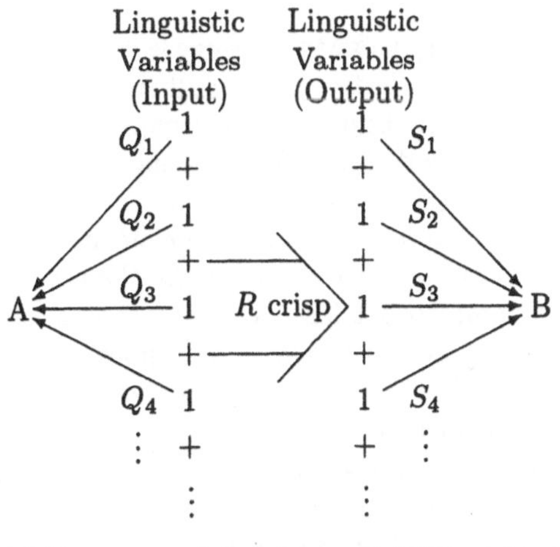

The linguistic variables are modeled by a disjoint union of several copies of a singleton set. The relation R corresponds to the rulebase of the controller, and the relations Q_i and S_i are the fuzzy sets corresponding to every linguistic variable. The controller itself is described by the relational term

$$(\bigsqcup_{i\in I} Q_i^{\smile};\iota_i);R;(\bigsqcup_{j\in J} \iota_j^{\smile};S_j),$$

where ι_i denotes the injection into the disjoint union, followed by a suitable defuzzification. Now, if the injections are non-crisp or A or B is a product with non-crisp projection there would be a fuzzification which is not an integral part of the controller. This fuzzification arises from the specific choice of the product or sum. In this case, reasoning about the controller using the term above seems to be impossible or at least difficult. Unfortunately, there may exists such non-crisp injections or projections. Consider the Boolean lattice $\{0, a, b, 1\}$ and the relations

$$\pi_1 := \begin{pmatrix} 1\,0 \\ a\,b \\ b\,a \\ 0\,1 \end{pmatrix}, \quad \rho_1 := \begin{pmatrix} 1\,0 \\ b\,a \\ a\,b \\ 0\,1 \end{pmatrix}, \quad \pi_2 := \begin{pmatrix} 1\,0 \\ 1\,0 \\ 0\,1 \\ 0\,1 \end{pmatrix}, \quad \rho_2 := \begin{pmatrix} 1\,0 \\ 0\,1 \\ 1\,0 \\ 0\,1 \end{pmatrix}.$$

Both pairs (π_1, ρ_1) and (π_2, ρ_2) constitute a product of two copies of a set with two elements, i.e., they fulfill the equations above. The first pair of relations is not crisp. But, this example also shows that there is a crisp version (π_2, ρ_2) of the product, i.e., there are crisp relations between the same objects fulfilling the same set of equations. In our example, one may require without loss of generality that the projections are crisp. The aim of this paper is to show that under an assumption on the lattice $\mathcal{L}$ such a crisp version always exists.

We assume that the reader is familiar with the basic concepts of allegories [3].

2 Lattices and Antimorphisms

In [14], it was shown that every Goguen category is isomorphic to the category of antimorphisms from a suitable lattice to the full subcategory of crisp relations. We will use this isomorphism throughout this paper. Therefore, we will recall some results from lattice theory (cf. [1]). We denote a lattice and the induced ordering by $(\mathcal{L}, \sqcup, \sqcap)$ resp. $\sqsubseteq$.

An atomic lattice is a lattice such that every element of $x \in \mathcal{L}$ is equal to the union of atoms less than or equal to x. We denote the set of all atoms by $\mathcal{A}$.

An element x of a complete lattice $\mathcal{L}$ is called completely irreducible iff for all subsets of $M \subseteq \mathcal{L}$ the property $\bigsqcup M = x$ implies $x \in M$. Notice, that an atom is completely irreducible which implies that every finite lattice has at least one completely irreducible element.

Suppose $f : \mathcal{L} \to \mathcal{L}$ is a monotone function, i.e., $x \sqsubseteq y$ implies $f(x) \sqsubseteq f(y)$ for all $x, y \in \mathcal{L}$, on a complete lattice $\mathcal{L}$. If a is an element from $\mathcal{L}$ such that $a \sqsubseteq f(a)$ then there exists a least fixedpoint $\mu_f(a)$ of f greater or equal to a, i.e., $f(\mu_f(a)) = \mu_f(a)$, $a \sqsubseteq \mu_f(a)$ and if $f(b) = b$ and $a \sqsubseteq b$ then we have $\mu_f(a) \sqsubseteq b$.

A predicate $\mathfrak{P}$ is called admissible (or continuous) iff for every set $M \subseteq \mathcal{L}$ the property $\mathfrak{P}(x)$ for all $x \in M$ implies $\mathfrak{P}(\bigsqcup M)$. The so-called principle of fixedpoint induction states that for every admissible predicate $\mathfrak{P}$, monotone function $f : \mathcal{L} \to \mathcal{L}$, and $a \in \mathcal{L}$ with $a \sqsubseteq f(a)$ we may conclude from $\mathfrak{P}(a)$ (basis step) and $\mathfrak{P}(b)$ implies $\mathfrak{P}(f(b))$ for every $b \in \mathcal{L}$ (induction step) that $\mathfrak{P}(\mu_f(a))$ holds. For n-ary predicates we get a similar result by taking the n-ary product of the underlying lattices.

The set of all functions between two lattices $(\mathcal{L}_1, \sqcap, \sqcup)$ and $(\mathcal{L}_2, \sqcap, \sqcup)$ may be ordered pointwise. This structure is again a lattice, and will be denoted by $(\mathcal{L}_1 \to \mathcal{L}_2, \sqcup, \sqcap)$. This lattice inherits interesting properties from $\mathcal{L}_2$, i.e., if $\mathcal{L}_2$ is complete Brouwerian then so is $\mathcal{L}_1 \to \mathcal{L}_2$.

A function $f : \mathcal{L}_1 \to \mathcal{L}_2$ is called antitonic iff $x \sqsubseteq y$ implies $f(y) \sqsubseteq f(x)$ for all $x, y \in \mathcal{L}_1$. The set of all antitone functions $\mathcal{L}_1 \overset{\sqsupseteq}{\to} \mathcal{L}_2$ constitutes a sublattice of $\mathcal{L}_1 \to \mathcal{L}_2$. Again, this sublattice inherits interesting properties from $\mathcal{L}_1 \to \mathcal{L}_2$.

A function $f : \mathcal{L}_1 \to \mathcal{L}_2$ is called an antimorphism iff $f(\bigsqcup M) = \bigsqcap f(M)$ for all subsets M of $\mathcal{L}_1$. Every antimorphism is antitonic, and we have $f(0) = f(\bigsqcup \emptyset) = \bigsqcap \emptyset = 1$. Given an element $z \in \mathcal{L}_2$ we denote with $\dot{z}$ the quasi-constant antimorphism induced by z, which is defined by

$$\dot{z}(x) := \begin{cases} 1 \text{ iff } x = 0, \\ z \text{ iff } x \neq 0. \end{cases}$$

The set of all antimorphisms $\mathcal{L}_1 \overset{\text{anti}}{\to} \mathcal{L}_2$ need not be a sublattice of $\mathcal{L}_1 \to \mathcal{L}_2$ or $\mathcal{L}_1 \overset{\sqsupseteq}{\to} \mathcal{L}_2$. But this set forms a closure system, i.e., is closed under arbitrary intersection and contains the greatest function $\dot{1}$, $x \mapsto 1$. Therefore, it induces the following closure operation

$$\tau(f) := \bigsqcap \{h \mid f \sqsubseteq h \text{ and } h \text{ antimorphism}\}.$$

τ is a monotone mapping from $\mathcal{L}_1 \to \mathcal{L}_2$ to $\mathcal{L}_1 \overset{\text{anti}}{\to} \mathcal{L}_2$, and we have $f \sqsubseteq \tau(f)$ and $\tau^2(f) = \tau(f)$ for all f. Now, $\mathcal{L}_1 \overset{\text{anti}}{\to} \mathcal{L}_2$ together with the operations $\dot{\bigsqcup}_{i \in I} f_i := \tau(\bigsqcup_{i \in I} f_i)$ and $\bigsqcap_{i \in I} f_i$ forms a complete lattice.

Unfortunately, the definition of τ gives us no information about the image of $\tau(f)$ for a given value $x \in \mathcal{L}_1$. Therefore, consider the following operation

$$\varphi(f)(x) := \bigsqcup_{\substack{M \subseteq \mathcal{L}_1 \\ \sqcup M = x}} \bigsqcap_{y \in M} f(y).$$

This φ is monotone, expanding and we have the following.

Lemma 1. *Let $\mathcal{L}_1$ be a complete Brouwerian lattice, $\mathcal{L}_2$ a complete lattice and $f : \mathcal{L}_1 \to \mathcal{L}_2$ be antitonic. Then we have $\tau(f) = \mu_\varphi(f)$.*

A proof may be found in [14]. The last lemma gives allows us to prove properties of τ using fixedpoint induction.

3 Dedekind Categories

In the remainder this paper, we use the following notations. To indicate that a morphism R of a category $\mathcal{R}$ has source A and target B we write $R : A \to B$. The collection of all morphisms $R : A \to B$ is denoted by $\mathcal{R}[A, B]$ and the composition of a morphism $R : A \to B$ followed by a morphism $S : B \to C$ by $R; S$. The identity morphism on A is denoted by $\mathbb{I}_A$.

In this section we recall some fundamentals on Dedekind categories [8,9]. Dedekind categories are called locally complete division allegories in [3].

Definition 1. *A Dedekind category $\mathcal{R}$ is a category satisfying the following:*

1. *For all objects A and B the collection of all morphisms (or relations) $\mathcal{R}[A, B]$ is a complete Brouwerian lattice. Meet, join, the induced ordering, the least and the greatest element are denoted by $\sqcap, \sqcup, \sqsubseteq, \perp\!\!\!\perp_{AB}, \top\!\!\!\top_{AB}$, respectively.*
2. *There is a monotone operation $\smile$ (called converse) such that for all relations $Q : A \to B$ and $R : B \to C$ the following holds: $(Q;R)^\smile = R^\smile;Q^\smile$ and $(Q^\smile)^\smile = Q$.*
3. *For all relations $Q : A \to B, R : B \to C$ and $S : A \to C$ the modular law $Q;R \sqcap S \sqsubseteq Q;(R \sqcap Q^\smile;S)$ holds.*
4. *For all relations $R : B \to C$ and $S : A \to C$ there is a relation $S/R : A \to B$ (called the left residual of S and R) such that for all $X : A \to B$ the following holds: $X;R \sqsubseteq S \iff X \sqsubseteq S/R$.*

Corresponding to the left residual, we define the right residual by $Q\backslash R := (R^\smile/Q^\smile)^\smile$. This relation is characterized by $Q;Y \sqsubseteq R \iff Y \sqsubseteq Q\backslash R$.

We will use some basic properties of relations in a Dedekind category throughout the paper without mentioning. These properties and their proofs may be found in [2,3,10,11,12].

Later on, we will investigate whenever a given equation holds in several Dedekind categories. For this purpose, we have to define a language and its meaning. This language is given by the equational theory of a distributive allegory [3], i.e., a relational category which need not be complete nor provide residuals. We require a set of object variables and a set of typed relation variables, i.e., every relation variable is of the form $r : a \to b$ where a and b are object variables.

Definition 2. *The set of terms of type $a \to b$ is defined inductively as follows:*

1. *Every relation variable $r : a \to b$ is a term of type $a \to b$.*
2. *If a is an object variable then $\mathbb{I}_a$ is a term of type $a \to a$.*
3. *If a and b are object variables then $\perp\!\!\!\perp_{ab}$ and $\top\!\!\!\top_{ab}$ are terms of type $a \to b$.*
4. *If t is a term of type $a \to b$ then $t^\smile$ is a term of type $b \to a$.*
5. *If t_1 and t_2 are terms of type $a \to b$ then $t_1 \sqcap t_2$ is a term of type $a \to b$.*
6. *If t_1 and t_2 are terms of type $a \to b$ then $t_1 \sqcup t_2$ is a term of type $a \to b$.*
7. *If t_1 and t_2 are terms of type $a \to b$ resp. $b \to c$ then $t_1;t_2$ is a term of type $a \to c$.*

We usually omit the type of a relation variable, for brevity.

We will compute the value of a term in a distributive allegory. Such an allegory need not be a Dedekind category. For such a structure $\mathcal{R}$, an environment σ over $\mathcal{R}$ is a function mapping every object variable a to an object A of $\mathcal{R}$ and every relation variable $r : a \to b$ to a relation $R : \sigma(a) \to \sigma(b)$. The update $\sigma[R/r]$ of σ at the relation variable $r : a \to b$ with the relation $R : \sigma(a) \to \sigma(b)$ is defined for object variables a and relation variables s by

$$\sigma[R/r](a) := \sigma(a), \quad \sigma[R/r](s) := \begin{cases} \sigma(s) & \text{iff } r \neq s \\ R & \text{iff } r = s \end{cases}$$

As usual we denote a sequence of updates $\sigma[Q/q][R/r][S/s]$ on an environment σ by $\sigma[Q/q, R/r, S/s]$.

Definition 3. *The value $\mathcal{V}_{\mathcal{R}}(t)(\sigma)$ of a term t of type $a \to b$ is defined inductively as follows:*

1. $\mathcal{V}_{\mathcal{R}}(r)(\sigma) := \sigma(r)$,
2. $\mathcal{V}_{\mathcal{R}}(\mathbb{I}_a)(\sigma) := \mathbb{I}_{\sigma(a)}$,
3. $\mathcal{V}_{\mathcal{R}}(\perp\!\!\!\perp_{ab})(\sigma) := \perp\!\!\!\perp_{\sigma(a)\sigma(b)}$,
4. $\mathcal{V}_{\mathcal{R}}(\top\!\!\!\top_{ab})(\sigma) := \top\!\!\!\top_{\sigma(a)\sigma(b)}$,
5. $\mathcal{V}_{\mathcal{R}}(t^{\smile})(\sigma) := (\mathcal{V}_{\mathcal{R}}(t)(\sigma))^{\smile}$,
6. $\mathcal{V}_{\mathcal{R}}(t_1 \sqcap t_2)(\sigma) := \mathcal{V}_{\mathcal{R}}(t_1)(\sigma) \sqcap \mathcal{V}_{\mathcal{R}}(t_2)(\sigma)$,
7. $\mathcal{V}_{\mathcal{R}}(t_1 \sqcup t_2)(\sigma) := \mathcal{V}_{\mathcal{R}}(t_1)(\sigma) \sqcup \mathcal{V}_{\mathcal{R}}(t_2)(\sigma)$,
8. $\mathcal{V}_{\mathcal{R}}(t_1;t_2)(\sigma) := \mathcal{V}_{\mathcal{R}}(t_1)(\sigma);\mathcal{V}_{\mathcal{R}}(t_2)(\sigma)$,

Obviously, the value of a term depends only on the values of variables occurring in it.

Let S be a set of equations, i.e., a set such that every element is of the form $t_1 = t_2$ with t_1 and t_2 terms of equal type. Suppose the variables of S are within $\{r_1, \ldots, r_n\}$. We say that the relations $R_1, \ldots, R_n$ fulfill S, denoted by $R_1, \ldots, R_n \models S$, iff for all equations $t_1 = t_2 \in S$ and all environments σ over $\mathcal{R}$ the following holds

$$\mathcal{V}_{\mathcal{R}}(t_1)(\sigma[R_1/r_1, \ldots, R_n/r_n]) = \mathcal{V}_{\mathcal{R}}(t_2)(\sigma[R_1/r_1, \ldots, R_n/r_n]).$$

In some sense a relation of a Dedekind category may be seen as an $\mathcal{L}$-relation. The lattice $\mathcal{L}$ may equivalently be characterized by the ideal relations, i.e., a relation $J : A \to B$ satisfying $\top\!\!\!\top_{AA};J;\top\!\!\!\top_{BB} = J$, or by the scalar relations.

Definition 4. *A relation $\alpha : A \to A$ is called a scalar on A iff $\alpha \sqsubseteq \mathbb{I}_A$ and $\top\!\!\!\top_{AA};\alpha = \alpha;\top\!\!\!\top_{AA}$. The set of all scalars on A is denoted by $\mathrm{Sc}_{\mathcal{R}}(A)$.*

The notion of ideal elements was introduced by Jónsson and Tarski [5] and the notion of scalars by Furusawa and Kawahara [6].

Notice that the collection of ideal elements on A is isomorphic to the collection of scalars on A via the mappings $\phi(J) := J \sqcap \mathbb{I}_A$ and $\phi^{-1}(\alpha) := \alpha;\top\!\!\!\top_{AA}$.

4 Goguen Categories

In [13], it was shown that we need an additional concept to define a suitable algebraic theory of $\mathcal{L}$-fuzzy relations. Our approach introduces two operations mapping every relation to the greatest crisp relation it contains resp. to the least crisp relation it is included in. We now give the abstract definition.

Definition 5. *A Goguen category $\mathcal{G}$ is a Dedekind category with $\perp\!\!\!\perp_{AB} \neq \top\!\!\!\top_{AB}$ for all objects A and B together with two operations $^{\uparrow}$ and $^{\downarrow}$ satisfying the following:*

1. $R^{\uparrow}, R^{\downarrow} : A \to B$ *for all* $R : A \to B$.
2. $(^{\uparrow}, ^{\downarrow})$ *is a Galois correspondence between* $(\mathcal{G}[A,B], \sqsupseteq)$ *and* $(\mathcal{G}[A,B], \sqsubseteq)$, *i.e.,* $S \sqsupseteq R^{\uparrow} \iff R \sqsubseteq S^{\downarrow}$ *for all* $R, S : A \to B$.
3. $(R^{\smile}; S^{\downarrow})^{\uparrow} = R^{\uparrow\smile}; S^{\downarrow}$ *for all* $R : B \to A$ *and* $S : B \to C$.
4. *If* $\alpha \neq \perp\!\!\!\perp_{AA}$ *is a nonzero scalar then* $\alpha^{\uparrow} = \mathbb{I}_A$.
5. *For all antimorphisms* $f : \mathrm{Sc}_{\mathcal{G}}(A) \overset{\mathrm{anti}}{\to} \mathcal{G}[A,B]$ *such that* $f(\alpha)^{\uparrow} = f(\alpha)$ *for all* $\alpha \in \mathrm{Sc}_{\mathcal{G}}(A)$ *and all* $R : A \to B$ *the following equivalence holds*

$$R \sqsubseteq \bigsqcup_{\substack{\alpha : A \to A \\ \alpha \text{ scalar}}} \alpha; f(\alpha) \iff (\alpha \backslash R)^{\downarrow} \sqsubseteq f(\alpha) \text{ for all } \alpha \in \mathrm{Sc}_{\mathcal{G}}(A).$$

In agreement with our intuition, we define crispness in an arbitrary Goguen category as follows.

Definition 6. *A relation* $R : A \to B$ *of a Goguen category is called crisp iff* $R^{\uparrow} = R$ *(or equivalently* $R^{\downarrow} = R$*). The crisp fragment* $\mathcal{G}^{\uparrow}$ *of* $\mathcal{G}$ *is defined as the collection of all crisp relations of* $\mathcal{G}$.

In [13] it was shown that the sets $\mathrm{Sc}_{\mathcal{G}}(A)$ of scalars on A are isomorphic via the mapping $\alpha \mapsto \top\!\!\!\top_{BA}; \alpha; \top\!\!\!\top_{AB} \sqcap \mathbb{I}_B$. For this reason we identify those sets. We call this set the underlying lattice of $\mathcal{G}$ and denote it by $\mathrm{Sc}[\mathcal{G}]$.

Furthermore, it was shown that $\mathcal{G}^{\uparrow}$ is a subcategory of $\mathcal{G}$ and together with the operations of $\mathcal{G}$ a simple Dedekind category.

Theorem 1. *Let* $\mathcal{L}$ *be an arbitrary complete Brouwerian lattice and* $\mathcal{R}$ *be a simple Dedekind category. Then* $\mathcal{R}^{\mathcal{L}}$ *defined by*

1. *the objects of* $\mathcal{R}^{\mathcal{L}}$ *are the objects of* $\mathcal{R}$,
2. $\mathcal{R}^{\mathcal{L}}[A,B]$ *is the set of all antimorphisms from* $\mathcal{L}$ *to* $\mathcal{R}[A,B]$,
3. *the constants are given by the antimorphisms* $\dot{\mathbb{I}}_A, \dot{\perp\!\!\!\perp}_{AB}$ *and* $\dot{\top\!\!\!\top}_{AB}$,
4. *intersection and transposition are defined componentwise,*
5. $\dot{\bigsqcup_{i \in I}} f_i := \tau(\bigsqcup_{i \in I} f_i)$ *where* $\bigsqcup_{i \in I} f_i$ *is defined componentwise,*
6. $f \,\dot{;}\, g := \tau(f; g)$ *where* $f; g$ *is defined componentwise,*
7. $f \dot{/} h := \dot{\bigsqcup}\{k \mid k \text{ antimorphism and } k \,\dot{;}\, h \sqsubseteq f\}$,
8. $f^{\uparrow} := \dot{R}$ *where* $R := \bigsqcup_{y \neq 0} f(y)$,
9. $f^{\downarrow} := \dot{S}$ *where* $S := f(1)$.

is a Goguen category.

A proof may be found in [14]. Notice, that the crisp relations in $\mathcal{R}^{\mathcal{L}}$ are those functions which are of the form $\dot{R}$ for an R in $\mathcal{R}$. Since the class of crisp relations is closed under all operations union and composition of crisp relations in $\mathcal{R}^{\mathcal{L}}$ can also be computed componentwise.

Furthermore, in [14] the following pseudo-representation theorem was shown.

Theorem 2 (Pseudo-Representation Theorem). *Let* $\mathcal{G}$ *be a Goguen category. Then* ${\mathcal{G}^{\uparrow}}^{\mathrm{Sc}[\mathcal{G}]}$ *is again a Goguen category and* $\mathcal{G}$ *and* ${\mathcal{G}^{\uparrow}}^{\mathrm{Sc}[\mathcal{G}]}$ *are isomorphic.*

For technical reasons, we will use the structure where (2) of Theorem 1 is replaced by the set of all antitone functions, union and composition are defined componentwise. Notice, that this structure is a distributive allegory. A Goguen category of the form $\mathcal{R}^{\mathcal{L}}$ is always a subclass of structure defined above. Therefore, we will denote this structure by $\widetilde{\mathcal{R}^{\mathcal{L}}}$.

Lemma 2. *Let $\mathcal{L}$ be an arbitrary complete Brouwerian lattice and $\mathcal{R}$ be a Dedekind category. For all antitone functions $f, g, f_i : \mathcal{L} \overset{\sqsupseteq}{\to} \mathcal{R}[A, B]$ for all $i \in I$ and $h : \mathcal{L} \overset{\sqsupseteq}{\to} \mathcal{R}[B, C]$ we have*

(1) $\tau(f) \sqcap \tau(g) = \tau(f \sqcap g)$,
(2) $\tau(\bigsqcup_{i\in I} \tau(f_i)) = \tau(\bigsqcup_{i\in I} f_i)$,
(3) $\tau(\tau(f); h) = \tau(f; h)$,
(4) $\tau(f; \tau(h)) = \tau(f; h)$,
(5) $\tau(f; h)^{\smile} = \tau(h^{\smile}; f^{\smile})$.

Proof. (1),(3)-(5) were shown in [14]. The inclusion $\sqsupseteq$ in (2) is trivial. The other inclusion follows from

$$\begin{aligned}
\forall i \in I : \tau(f_i) \sqsubseteq \tau(\bigsqcup_{i\in I} f_i) \quad &\Rightarrow \quad \bigsqcup_{i\in I} \tau(f_i) \sqsubseteq \tau(\bigsqcup_{i\in I} f_i) \\
&\Rightarrow \quad \tau(\bigsqcup_{i\in I} \tau(f_i)) \sqsubseteq \tau^2(\bigsqcup_{i\in I} f_i) = \tau(\bigsqcup_{i\in I} f_i).
\end{aligned}$$

□

For atomic lattices, τ reduces to an operation that is easy to handle.

Lemma 3. *Let $\mathcal{L}_1$ be an atomic complete Brouwerian lattice, $\mathcal{L}_2$ a complete lattice and $f : \mathcal{L}_1 \to \mathcal{L}_2$ be antitonic. Then we have $\tau(f)(x) = \bigsqcap_{\substack{a\in\mathcal{A} \\ a\sqsubseteq x}} f(a)$.*

Proof. $\sqsupseteq$: From $f(a) \sqsubseteq \tau(f)(a)$ we conclude that

$$\bigsqcap_{a\sqsubseteq x} f(a) \sqsubseteq \bigsqcap_{a\sqsubseteq x} \tau(f)(a) = \tau(f)(\bigsqcup_{a\sqsubseteq x} a) = \tau(f)(x)$$

since $\tau(f)$ is an antimorphism.

$\sqsubseteq$: The other inclusion is shown by fixedpoint induction. Therefore, we define the predicate $\mathfrak{P}(h) \;:\Leftrightarrow\; \forall x : h(x) \sqsubseteq \bigsqcap_{a\sqsubseteq x} f(a)$. This predicate is admissible since $\mathfrak{P}(h_i)$ for all $i \in I$ implies

$$\begin{aligned}
&h_i(x) \sqsubseteq \bigsqcap_{a\sqsubseteq x} f(a) \quad \text{for all } i \in I \text{ and } x \in \mathcal{L} \\
&\Rightarrow \bigsqcup_{i\in I} h_i(x) \sqsubseteq \bigsqcap_{a\sqsubseteq x} f(a) \quad \text{for all } x \in \mathcal{L} \\
&\Leftrightarrow \mathfrak{P}(\bigsqcup_{i\in I} h_i).
\end{aligned}$$

The basis step $\mathfrak{P}(f)$ is trivial, and the induction step follows from

$$\begin{aligned}\varphi(h)(x) &= \bigsqcup_{\bigsqcup M=x} \bigsqcap_{y\in M} h(y)\\ &\sqsubseteq \bigsqcup_{\bigsqcup M=x} \bigsqcap_{y\in M} \bigsqcap_{a\sqsubseteq y} f(a)\\ &= \bigsqcup_{\bigsqcup M=x} \bigsqcap_{a\sqsubseteq\bigsqcup M} f(a)\\ &= \bigsqcap_{a\sqsubseteq x} f(a).\end{aligned}$$

From the principle of fixedpoint induction we conclude that $\tau(f)(x) \sqsubseteq \bigsqcap_{a\sqsubseteq x} f(a)$ for all x. □

The last lemma shows that $\tau(f)(a) = f(a)$ for every atom of an atomic lattice. Later on, this property will allow us to construct from given relations fulfilling a set of equations a crisp relation fulfilling the same set of equations.

5 Complete Prime Filters and Proper Lattices

In this section we focus on the problem to find for a given complete Brouwerian lattice $\mathcal{L}$ a suitable atomic lattice $\mathcal{L}'$, i.e., the τ operation for $\mathcal{L}'$ should be a canonical extension of the τ operation for $\mathcal{L}$. Unfortunately, the usual embedding ψ of $\mathcal{L}$ into the powerset of all prime filters does not work. This embedding is not necessarily (upward) continuous such that the least upper bound of a subset $M \subseteq \mathcal{L}$ and union of the set $\psi(y)$ for $y \in M$ need not coincide. As a consequence, φ resp. τ for $\mathcal{L}'$ is not a canonical extension of φ resp. τ for $\mathcal{L}$. We have to switch to a special class of prime filters.

Definition 7. *A subset $F \subseteq \mathcal{L}$ of a complete Brouwerian lattice $\mathcal{L}$ is called a complete prime filter iff*

1. $0 \notin F$,
2. $x \sqcap y \in F$ *iff* $x \in F$ *and* $y \in F$ *for all* $x, y \in \mathcal{L}$,
3. $\bigsqcup M \in F$ *iff* $\exists y \in M : y \in F$ *for all subsets* $M \subseteq \mathcal{L}$.

We will denote the set of all complete prime filters of $\mathcal{L}$ by $\mathcal{F}_\mathcal{L}$. If $\mathcal{F}_\mathcal{L} \neq \emptyset$ we call $\mathcal{L}$ proper.

First, we want to study the class of proper lattices.

Theorem 3. *1. Every linear ordering is proper.*
2. If $\mathcal{L}$ has at least one completely irreducible element then $\mathcal{L}$ is proper.
3. The class of proper lattices is closed under arbitrary products.
4. A complete atomless Boolean algebra is not proper.

Proof. 1. The set $F := \mathcal{L} \setminus \{0\}$ for a linear ordering $\mathcal{L}$ is obviously a complete prime filter.

2. We prove the following equivalence which implies our assertion.

   A principal filter $(x) = \{y \in \mathcal{L} \mid x \sqsubseteq y\}$ is a complete prime filter iff x is completely irreducible.

   $\Rightarrow$: Suppose $\bigsqcup M = x \in (x)$. Then there is a $y \in M \cap (x)$ since (x) is a complete prime filter. We conclude $y \sqsubseteq x$ from $y \in M$ and $x \sqsubseteq y$ from $y \in (x)$. This implies $x = y \in M$.
   $\Leftarrow$: It is sufficient to show that $\bigsqcup M \in (x)$ implies that there is a $y \in M$ with $x \sqsubseteq y$. Define $M_x := \{x \sqcap y \mid y \in M\}$. Then we have $\bigsqcup M_x = \bigsqcup\limits_{y \in M} (x \sqcap y) = x \sqcap \bigsqcup M = x$. This implies $x \in M_x$ since x is completely irreducible. We conclude $x = x \sqcap y \sqsubseteq y$ for a $y \in M$.

3. Suppose F_i is a complete prime filter of the proper lattice $\mathcal{L}_i$ for all $i \in I$. Then $F := \prod\limits_{i \in I} F_i$ is a complete prime filter of $\prod\limits_{i \in I} \mathcal{L}_i$.

4. We just give the sketch of the proof since we did not introduce the notions of prime and ultrafilters. Suppose F is a complete prime filter of a complete atomless Boolean algebra $\mathcal{B}$. Then F is a prime filter and hence an ultrafilter of $\mathcal{B}$. Since ultrafilters are maximal filters it can be shown that $\bigsqcap F$ is an atom or 0. Since $\mathcal{B}$ is atomless, $\bigsqcap F = 0$ follows. Furthermore, F is an ultrafilter implies that $\overline{F} := \{\overline{x} \mid x \in F\}$ is a maximal ideal and $F \cap \overline{F} = \emptyset$. Last but not least, we have $\bigsqcup \overline{F} = \bigsqcup\limits_{x \in F} \overline{x} = \overline{\bigsqcap F} = 1 \in F$ which shows that F is not a complete prime filter. □

Notice, that property (2) of the last theorem implies that every finite lattice is proper. Furthermore, from (1) we conclude that the unit interval $[0, 1]$ of the real numbers is also proper.

Now, we define a function $\psi : \mathcal{L} \to \mathcal{P}(\mathcal{F}_{\mathcal{L}})$ by $\psi(x) := \{F \in \mathcal{F}_{\mathcal{L}} \mid x \in F\}$. This function is not necessarily injective. But we have the following.

Lemma 4. *1.* $\psi(x \sqcap y) = \psi(x) \cap \psi(y)$,
2. $\psi(\bigsqcup M) = \bigcup\limits_{y \in M} \psi(y)$.

Proof. 1. Consider the following computation

$$\begin{aligned} F \in \psi(x \sqcap y) &\iff x \sqcap y \in F \\ &\iff x \in F \text{ and } y \in F \qquad F \text{ is a prime filter} \\ &\iff F \in \psi(x) \text{ and } F \in \psi(y) \\ &\iff F \in \psi(x) \cap \psi(y). \end{aligned}$$

2. The assertion follows immediately from

$$\begin{aligned} F \in \psi(\textstyle\bigsqcup M) &\iff \textstyle\bigsqcup M \in F \\ &\iff \exists y \in M : y \in F \qquad F \text{ is a complete prime filter} \end{aligned}$$

$$\begin{aligned} &\iff \exists y \in M : F \in \psi(y) \\ &\iff F \in \bigcup_{y \in M} \psi(y). \end{aligned}$$ □

For a function $f : \mathcal{L}_1 \to \mathcal{L}_2$ we define $\hat{f} : \mathcal{P}(\mathcal{F}_{\mathcal{L}_1}) \to \mathcal{L}_2$ by

$$\hat{f}(\mathfrak{M}) := \bigsqcup_{\substack{x \in \mathcal{L}_1 \\ \mathfrak{M} \subseteq \psi(x)}} f(x).$$

Let $\mathcal{R}^{\mathcal{L}}$ be a Goguen category. Then the structure $\mathcal{R}^{\mathcal{P}(\mathcal{F}_{\mathcal{L}})}$ is again a Goguen category and $\hat{}$ maps every element of $\mathcal{R}^{\mathcal{L}}$ to an element of $\mathcal{R}^{\mathcal{P}(\mathcal{F}_{\mathcal{L}})}$.

Lemma 5. *Let $\mathcal{L}$ be an arbitrary complete Brouwerian lattice and $\mathcal{R}$ be a Dedekind category. For all antitone functions $f, g, f_i : \mathcal{L} \overset{\exists}{\to} \mathcal{R}[A, B]$ for all $i \in I$ and $h : \mathcal{L} \overset{\exists}{\to} \mathcal{R}[B, C]$ we have*

(1) $\bigsqcup_{i \in I} \hat{f}_i = \widehat{\bigsqcup_{i \in I} f_i}$,
(2) $\widehat{f \sqcap g} = \hat{f} \sqcap \hat{g}$,
(3) $\widehat{f^\smile} = \hat{f}^\smile$,
(4) $\widehat{f ; h} = \hat{f} ; \hat{h}$,
(5) $\hat{f}$ is antitonic,
(6) $\hat{f}(\{F\}) = \bigsqcup_{x \in F} f(x)$.

Proof. 1. Consider the following computation

$$\begin{aligned} (\bigsqcup_{i \in I} \hat{f}_i)(\mathfrak{M}) &= \bigsqcup_{i \in I} \hat{f}_i(\mathfrak{M}) \\ &= \bigsqcup_{i \in I} \bigsqcup_{\mathfrak{M} \subseteq \psi(x)} f_i(x) \\ &= \bigsqcup_{\mathfrak{M} \subseteq \psi(x)} \bigsqcup_{i \in I} f_i(x) \\ &= \bigsqcup_{\mathfrak{M} \subseteq \psi(x)} (\bigsqcup_{i \in I} f_i)(x) \\ &= (\widehat{\bigsqcup_{i \in I} f_i})(\mathfrak{M}). \end{aligned}$$

2. First, we show that $\bigsqcup_{\mathfrak{M} \subseteq \psi(x)} (f(x) \sqcap g(x)) = (\bigsqcup_{\mathfrak{M} \subseteq \psi(x)} f(x)) \sqcap (\bigsqcup_{\mathfrak{M} \subseteq \psi(x)} g(x))$. The inclusion $\sqsubseteq$ is trivial. Suppose $\mathfrak{M} \subseteq \psi(x)$ and $\mathfrak{M} \subseteq \psi(y)$. Then $\mathfrak{M} \subseteq \psi(x) \cap \psi(y) = \psi(x \sqcap y)$. Furthermore, $f(x) \sqcap g(y) \sqsubseteq f(x \sqcap y) \sqcap g(x \sqcap y)$ since f and g are antitonic. We conclude that

$$\begin{aligned} (\bigsqcup_{\mathfrak{M} \subseteq \psi(x)} f(x)) \sqcap (\bigsqcup_{\mathfrak{M} \subseteq \psi(x)} g(x)) &= \bigsqcup_{\mathfrak{M} \subseteq \psi(x)} (f(x) \sqcap (\bigsqcup_{\mathfrak{M} \subseteq \psi(x)} g(x)) \\ &= \bigsqcup_{\mathfrak{M} \subseteq \psi(x)} \bigsqcup_{\mathfrak{M} \subseteq \psi(y)} (f(x) \sqcap g(y)) \\ &\sqsubseteq \bigsqcup_{\mathfrak{M} \subseteq \psi(x \sqcap y)} (f(x \sqcap y) \sqcap g(x \sqcap y)) \end{aligned}$$

$$= \bigsqcup_{\mathfrak{M} \subseteq \psi(x)} (f(x) \sqcap g(x)).$$

Now, consider the following computation

$$\begin{aligned}
(\widehat{f \sqcap g})(\mathfrak{M}) &= \bigsqcup_{\mathfrak{M} \subseteq \psi(x)} (f \sqcap g)(x) \\
&= \bigsqcup_{\mathfrak{M} \subseteq \psi(x)} (f(x) \sqcap g(x)) \\
&= (\bigsqcup_{\mathfrak{M} \subseteq \psi(x)} f(x)) \sqcap (\bigsqcup_{\mathfrak{M} \subseteq \psi(x)} g(x)) \\
&= \hat{f}(\mathfrak{M}) \sqcap \hat{g}(\mathfrak{M}) \\
&= (\hat{f} \sqcap \hat{g})(\mathfrak{M}).
\end{aligned}$$

3. The assertion follows immediately from

$$\begin{aligned}
(\widehat{f^{\smile}})(\mathfrak{M}) &= \bigsqcup_{\mathfrak{M} \subseteq \psi(x)} f^{\smile}(x) \\
&= (\bigsqcup_{\mathfrak{M} \subseteq \psi(x)} f(x))^{\smile} \\
&= \hat{f}(\mathfrak{M})^{\smile} \\
&= (\hat{f}^{\smile})(\mathfrak{M}).
\end{aligned}$$

4. Analogously to (2).
5. If $\mathfrak{M} \subseteq \mathfrak{N}$ we get $\{x \mid \mathfrak{N} \subseteq \psi(x)\} \subseteq \{x \mid \mathfrak{M} \subseteq \psi(x)\}$ and hence

$$\hat{f}(\mathfrak{N}) = \bigsqcup_{\mathfrak{N} \subseteq \psi(x)} f(x) \sqsubseteq \bigsqcup_{\mathfrak{M} \subseteq \psi(x)} f(x) = \hat{f}(\mathfrak{M}).$$

6. We immediately conclude that

$$\hat{f}(\{F\}) = \bigsqcup_{\{F\} \subseteq \psi(x)} f(x) = \bigsqcup_{F \in \psi(x)} f(x) = \bigsqcup_{x \in F} f(x).$$ □

The next lemma will show the key property we will need for our main theorem. Notice, that in the proof of this lemma it is essential that ψ is continuous.

Lemma 6. *If $f : \mathcal{L}_1 \to \mathcal{L}_2$ is antitonic, then we have $\tau(\widehat{\tau(f)}) = \tau(\hat{f})$.*

Proof. $\sqsupseteq$: Since $f \sqsubseteq \tau(f)$ we get $\hat{f} \sqsubseteq \widehat{\tau(f)}$ by Lemma 5(1) and hence $\tau(\hat{f}) \sqsubseteq \tau(\widehat{\tau(f)})$.

$\sqsubseteq$: Consider the property

$$(*) \qquad \widehat{\tau(f)} \sqsubseteq \tau(\hat{f}).$$

From (∗) we immediately conclude that $\tau(\widehat{\tau(\hat{f})}) \sqsubseteq \tau^2(\hat{f}) = \tau(\hat{f})$. We proof (∗) by fixedpoint induction. Therefore, we define the predicate $\mathfrak{P}(h) :\Leftrightarrow \hat{h} \sqsubseteq \tau(\hat{f})$. This predicate is admissible since $\mathfrak{P}(h_i)$ for all $i \in I$ implies

$$\begin{aligned}\forall i \in I : \hat{h}_i \sqsubseteq \tau(\hat{f}) \;&\Rightarrow\; \bigsqcup_{i\in I} \hat{h}_i \sqsubseteq \tau(\hat{f})\\ &\Leftrightarrow\; \widehat{\bigsqcup_{i\in I} h_i} \sqsubseteq \tau(\hat{f}) \quad \text{by Lemma 5(1)}\\ &\Leftrightarrow\; \mathfrak{P}(\bigsqcup_{i\in I} h_i).\end{aligned}$$

The basis step $\mathfrak{P}(f)$ is trivial. Let $\mathfrak{M}$ be a subset of $\mathcal{F}_{\mathcal{L}}$, x such that $\mathfrak{M} \subseteq \psi(x)$ and M a set with $\bigsqcup M = x$. Then we define $P := \{\mathfrak{M} \cap \psi(y) \mid y \in M\}$. Using Lemma 4(2) we conclude that

$$\bigcup P = \bigcup_{y\in M} (\mathfrak{M} \cap \psi(y)) = \mathfrak{M} \cap \bigcup_{y\in M} \psi(y) = \mathfrak{M} \cap \psi(\bigsqcup M) = \mathfrak{M}.$$

Furthermore, we have $h(y) \sqsubseteq \bigsqcup_{\mathfrak{M}\cap\psi(y)\subseteq\psi(z)} h(z)$ for all $y \in M$ which implies $\bigsqcap_{y\in M} h(y) \sqsubseteq \bigsqcap_{\mathfrak{N}\in P} \bigsqcup_{\mathfrak{N}\subseteq\psi(z)} h(z)$. We conclude that

$$(**) \qquad \bigsqcup_{\mathfrak{M}\subseteq\psi(x)} \bigsqcup_{\bigsqcup M = x} \bigsqcap_{y\in M} h(y) \sqsubseteq \bigsqcup_{\bigcup P=\mathfrak{M}} \bigsqcap_{\mathfrak{N}\in P} \bigsqcup_{\mathfrak{N}\subseteq\psi(z)} h(z).$$

Now, the induction step follows from

$$\begin{aligned}\widehat{\varphi(h)}(\mathfrak{M}) &= \bigsqcup_{\mathfrak{M}\subseteq\psi(x)} \varphi(h)(x)\\ &= \bigsqcup_{\mathfrak{M}\subseteq\psi(x)} \bigsqcup_{\bigsqcup M=x} \bigsqcap_{y\in M} h(y)\\ &\sqsubseteq \bigsqcup_{\bigcup P=\mathfrak{M}} \bigsqcap_{\mathfrak{N}\in P} \bigsqcup_{\mathfrak{N}\subseteq\psi(z)} h(z) \qquad \text{by } (**)\\ &= \bigsqcup_{\bigcup P=\mathfrak{M}} \bigsqcap_{\mathfrak{N}\in P} \hat{h}(\mathfrak{N})\\ &\sqsubseteq \bigsqcup_{\bigcup P=\mathfrak{M}} \bigsqcap_{\mathfrak{N}\in P} \tau(\hat{f})(\mathfrak{N}) \qquad \text{induction hypothesis}\\ &= \varphi(\tau(\hat{f}))(\mathfrak{M})\\ &= \tau(\hat{f})(\mathfrak{M}).\end{aligned}$$

From the principle of fixedpoint induction we conclude $\mathfrak{P}(\mu_\varphi(f))$ and hence property (∗). □

6 Equations in Goguen Categories

In this section we will investigate when a given equation holds in several Dedekind/Goguen categories introduced in the previous sections. Furthermore, we will show that if a set of equations is fulfilled by some relations of a Goguen category then there are crisp relations fulfilling the same set of equations.

Lemma 7. *Let $\mathcal{R}^{\mathcal{L}}$ be a Goguen category, t be a term, σ be an environment over $\mathcal{R}^{\mathcal{L}}$ and σ' be an environment over $\widetilde{\mathcal{R}^{\mathcal{L}}}$. Then we have*

1. $\mathcal{V}_{\mathcal{R}^{\mathcal{L}}}(t)(\sigma) = \tau(\mathcal{V}_{\widetilde{\mathcal{R}^{\mathcal{L}}}}(t)(\sigma))$,
2. $\widehat{\mathcal{V}_{\widetilde{\mathcal{R}^{\mathcal{L}}}}(t)(\sigma')} = \mathcal{V}_{\widetilde{\mathcal{R}^{\mathcal{P}(F)}}}(t)(\sigma'')$ *where* $\sigma''(r) := \widehat{\sigma'(r)}$,
3. *if* $x \neq 0$ *then* $(\mathcal{V}_{\widetilde{\mathcal{R}^{\mathcal{L}}}}(t)(\sigma'))(x) = \mathcal{V}_{\mathcal{R}}(t)(\sigma'')$ *where* $\sigma''(r) := \sigma'(r)(x)$.

Proof. 1. We prove the assertion by structural induction. If t is a constant or a variable, the assertion is trivial since the corresponding function are antimorphisms. From Lemma 2 we conclude that
   a) $\tau(f) \sqcap \tau(g) = \tau(f \sqcap g)$,
   b) $\tau(f) \sqcup \tau(g) = \tau(f \sqcup g)$,
   c) $\tau(f) \,\ddot{;}\, \tau(h) = \tau(f; h)$,
   d) $\tau(f)^{\smile} = \tau(f^{\smile})$,
   such that the assertion for all other cases follows immediately.
2. Again, the assertion is proved by structural induction. If t is a constant or a variable, the assertion is trivial. From Lemma 5(1)-(4) the assertion follows for all other cases immediately.
3. The assertion is trivial since all operations are defined componentwise and for $x \neq 0$ we have $\dot{\mathbb{I}}(x) = \mathbb{I}$ and $\dot{\perp\!\!\!\perp}(x) = \perp\!\!\!\perp$. □

Now, we are ready to prove our main theorem.

Theorem 4. *Let $\mathcal{R}^{\mathcal{L}}$ be a Goguen category with a proper lattice $\mathcal{L}$, S be a set of equations with variable within $\{r_1, \ldots, r_n\}$ and $f_1, \ldots, f_n$ be elements of $\mathcal{R}^{\mathcal{L}}$ such that $f_1, \ldots, f_n \models S$. Then there are relations $U_1, \ldots, U_n$ from $\mathcal{R}$ with $\dot{U}_1, \ldots, \dot{U}_n \models S$.*

Proof. Suppose $t_1 = t_2 \in S$ and σ is an environment over $\mathcal{R}^{\mathcal{L}}$. For brevity, let $\tilde{\sigma} := \sigma[f_1/r_1, \ldots, f_n/r_n]$, $\tilde{\sigma}'(r) := \widehat{\tilde{\sigma}(r)}$, $\tilde{\sigma}''(r) := \tilde{\sigma}'(r)(\{F\})$ with $F \in \mathcal{F}_{\mathcal{L}}$ and $h_i := \mathcal{V}_{\widetilde{\mathcal{R}^{\mathcal{L}}}}(t_i)(\tilde{\sigma})$ for $i = 1, 2$. Then we conclude that

$$\begin{array}{ll}
\mathcal{V}_{\mathcal{R}^{\mathcal{L}}}(t_1)(\tilde{\sigma}) = \mathcal{V}_{\mathcal{R}^{\mathcal{L}}}(t_2)(\tilde{\sigma}) & \\
\Leftrightarrow \tau(h_1) = \tau(h_2) & \text{Lemma 7(1)} \\
\Rightarrow \widehat{\tau(h_1)} = \widehat{\tau(h_2)} & \\
\Rightarrow \tau(\widehat{\tau(h_1)}) = \tau(\widehat{\tau(h_2)}) & \\
\Leftrightarrow \tau(\hat{h}_1) = \tau(\hat{h}_2) & \text{Lemma 6} \\
\Leftrightarrow \tau(\mathcal{V}_{\widetilde{\mathcal{R}^{\mathcal{P}(\mathcal{F}_{\mathcal{L}})}}}(t_1)(\tilde{\sigma}')) = \tau(\mathcal{V}_{\widetilde{\mathcal{R}^{\mathcal{P}(\mathcal{F}_{\mathcal{L}})}}}(t_2)(\tilde{\sigma}')) & \text{Lemma 7(2)} \\
\Rightarrow (\mathcal{V}_{\widetilde{\mathcal{R}^{\mathcal{P}(\mathcal{F}_{\mathcal{L}})}}}(t_1)(\tilde{\sigma}'))(\{F\}) = (\mathcal{V}_{\widetilde{\mathcal{R}^{\mathcal{P}(\mathcal{F}_{\mathcal{L}})}}}(t_2)(\tilde{\sigma}'))(\{F\}) & \text{Lemma 3} \\
\Leftrightarrow \mathcal{V}_{\mathcal{R}}(t_1)(\tilde{\sigma}'') = \mathcal{V}_{\mathcal{R}}(t_2)(\tilde{\sigma}'') & \text{Lemma 7(3).}
\end{array}$$

Now, let $U_i := \tilde{\sigma}''(r_i) = \tilde{\sigma}'(r_i)(\{F\}) = (\widehat{\tilde{\sigma}(r_i)})(\{F\}) = \hat{f}_i(\{F\}) = \bigsqcup_{x \in F} f_i(x)$ for $i \in \{1,\ldots,n\}$, δ an environment over $\mathcal{R}^{\mathcal{L}}$, $\tilde{\delta} := \delta[\dot{U}_1/r_1,\ldots,\dot{U}_n/r_n]$ and $\tilde{\delta}'(r) := \tilde{\delta}(r)(x)$. Notice, that $\mathcal{V}_{\widetilde{\mathcal{R}^{\mathcal{L}}}}(t_i)(\tilde{\delta})$ for $i = 1,2$ is crisp, i.e., of the form $\dot{R}$ for a suitable R, since all constants and all values for variable in t_i are crisp. Therefore, this relation is an antimorphism.

For $x = 0$ we immediately conclude that $\mathcal{V}_{\mathcal{R}^{\mathcal{L}}}(t_1)(\tilde{\delta})(x) = \mathbb{T} = \mathcal{V}_{\mathcal{R}^{\mathcal{L}}}(t_2)(\tilde{\delta})(x)$ since a value of a term is an antimorphism. If $x \neq 0$ we get for $i = 1,2$

$$\begin{aligned} \mathcal{V}_{\mathcal{R}^{\mathcal{L}}}(t_i)(\tilde{\delta})(x) &= \tau(\mathcal{V}_{\widetilde{\mathcal{R}^{\mathcal{L}}}}(t_i)(\tilde{\delta}))(x) && \text{Lemma 7(1)} \\ &= \mathcal{V}_{\widetilde{\mathcal{R}^{\mathcal{L}}}}(t_i)(\tilde{\delta})(x) && \mathcal{V}_{\widetilde{\mathcal{R}^{\mathcal{L}}}}(t_i)(\tilde{\delta}) \text{ is an antimorphism} \\ &= \mathcal{V}_{\mathcal{R}}(t_i)(\tilde{\delta}') && \text{Lemma 7(2)} \\ &= \mathcal{V}_{\mathcal{R}}(t_i)(\tilde{\sigma}'') \end{aligned}$$

where the last equality holds since for all variable $r_i \in \{r_1,\ldots,r_n\}$ within t_1 and t_2 we have $\tilde{\delta}'(r_i) = \tilde{\delta}(r_i)(x) = \dot{U}_i(x) = \tilde{\sigma}''(r_i)$. Last but not least, $\mathcal{V}_{\mathcal{R}^{\mathcal{L}}}(t_1)(\delta[\dot{U}_1/r_1,\ldots,\dot{U}_n/r_n]) = \mathcal{V}_{\mathcal{R}^{\mathcal{L}}}(t_2)(\delta[\dot{U}_1/r_1,\ldots,\dot{U}_n/r_n])$ follows. □

Since $\mathcal{G}$ and $\mathcal{G}\!\uparrow^{\mathrm{Sc}[\mathcal{G}]}$ are isomorphic, we get the following corollary.

Corollary 1. *Let $\mathcal{G}$ be a Goguen category with a proper underlying lattice* $\mathrm{Sc}[\mathcal{G}]$, *S be a set of equations with variable within $\{r_1,\ldots,r_n\}$ and $R_1,\ldots,R_n$ relations such that $R_1,\ldots,R_n \models S$. Then there are crisp relations $Q_1,\ldots,Q_n$ with $Q_1,\ldots,Q_n \models S$.*

Since products, sums and subobjects induced by crisp partial identities are defined by equations, we may require without loss of generality that the related relations are crisp.

Unfortunately, we were just able to prove Theorem 4 for Goguen categories $\mathcal{R}^{\mathcal{L}}$ with a proper lattice $\mathcal{L}$. The experiences we made during searching a counterexample to this theorem in general leads us to the following conjecture.

Conjecture 1. Theorem 4 is true for all Goguen categories $\mathcal{R}^{\mathcal{L}}$.

References

1. Birkhoff G.: Lattice Theory. American Mathematical Society Colloquium Publications Vol. XXV (1940).
2. Chin L.H., Tarski A.: Distributive and modular laws in the arithmetic of relation algebras. University of California Press, Berkley and Los Angeles (1951).
3. Freyd P., Scedrov A.: Categories, Allegories. North-Holland (1990).
4. Goguen J.A.: L-fuzzy sets. J. Math. Anal. Appl. 18 (1967), 145-157.
5. Jónsson B., Tarski A.: Boolean algebras with operators, I, II, Amer. J. Math. 73 (1951) 891-939, 74 (1952) 127-162.
6. Kawahara, Y., Furusawa H.: Crispness and Representation Theorems in Dedekind Categories. DOI-TR 143, Kyushu University (1997).
7. Kawahara, Y., Furusawa H.: An Algebraic Formalisation of Fuzzy Relations. Fuzzy Sets and Systems 101 (1999), 125-135.

8. Olivier J.P., Serrato D.: Catégories de Dedekind. Morphismes dans les Catégories de Schröder. C.R. Acad. Sci. Paris 290 (1980) 939-941.
9. Olivier J.P., Serrato D.: Squares and Rectangles in Relational Categories - Three Cases: Semilattice, Distributive lattice and Boolean Non-unitary. Fuzzy sets and systems 72 (1995), 167-178.
10. Schmidt G., Ströhlein T.: Relationen und Graphen. Springer (1989); English version: Relations and Graphs. Discrete Mathematics for Computer Scientists, EATCS Monographs on Theoret. Comput. Sci., Springer (1993).
11. Schmidt G., Hattensperger C., Winter M.: Heterogeneous Relation Algebras. *In:* Brink C., Kahl W., Schmidt G. (eds.), Relational Methods in Computer Science, Advances in Computer Science, Springer Vienna (1997).
12. Winter M.: Strukturtheorie heterogener Relationenalgebren mit Anwendung auf Nichtdetermismus in Programmiersprachen. Dissertationsverlag NG Kopierladen GmbH, München (1998).
13. Winter M.: A new Algebraic Approach to L-Fuzzy Relations Convenient to Study Crispness. INS Information Science 139/3-4 (2001), 233-252
14. Winter M.: Representation Theory of Goguen Categories. Submitted to Fuzzy Sets and Systems.
15. Zadeh L.A.: Fuzzy sets. Information and Control 8 (1965), 338-353.

A Subintuitionistic Logic and Some of Its Methods

Ernst Zimmermann

ernstzimmermann@de.ibm.com

Abstract. We develop a predicate logical extension of a subintuitionistic propositional logic. Therefore a Hilbert type calculus and a Kripke type model are given. The propositional logic is formulated to axiomatize the idea of strategic weakening of Kripke's semantic for intuitionistic logic: dropping the semantical condition of heredity or persistence leads to a nonmonotonic model. On the syntactic side this leads to a certain restriction imposed on the deduction theorem. By means of a Henkin argument strong completeness is proved making use of predicate logical principles, which are only classically acceptable. Semantic tableaux and an embedding into modal logic are defined straightforward.

1 Introduction

In 1994 Greg Restall developed a class of subintuitionistic propositional logics axiomatizing the model theoretical idea, that strategic weakening could be imposed in various ways on Kripke's semantic for intuitionistic logics: the model theoretical conditions of heredity, transitivity, reflexivity and absurdity may be given up in different combinations finally reaching models for a subintuitionistic basic logic. Restall not only described models for these logics, he also gave completeness proofs for 11 subintuitionistic propositional logics.

Closely related to the considerations of Restall are the considerations of Johan van Benthem going back to 1986: with classical modal logic as starting point van Benthem tried to introduce aspects of nonmonotonicity into the models for intuitionistic propositional logic. Nonmonotonicity was his motivation to give up the condition of heredity in Kripke's semantic for intuitionistic propositional logic. Van Benthem was able to give an axiomatization for his nonmonotonic Kripke model too, thus ending in a subintuitionistic logic.

But the considerations of Restall and van Benthem were restricted to the propositional case, and especially Restall has given a summary of difficulties and negative results for the extension of subintuitionistic propositional logic to predicate logic. In this summary Restall has shown at least the direction of a further research: he stated that the usual intuitionistic models in terms of growing domains are not applicable in the case of subintuitionistic logics; instead of growing domains one has to choose for a predicate logical extension of any subintuitionistic logic models with constant domains he pointed out.

Exactly this is the direction I would like to continue the work of Restall and van Benthem: that subintuitionistic propositional logic is regarded as base logic,

H. de Swart (Ed.): RelMiCS 2001, LNCS 2561, pp. 228–240, 2002.

which axiomatizes the transitive, reflexive, but not hereditary or not persistent Kripke model. With regard to the fact, that this logic has two negations, we can claim that we treat two of Restall's subintuitionistic logics. Further, a predicate logical extension of this subintuitionistic propositional logic is formulated. Bearing in mind that only quantifiers over constant domains may give any completeness results, we finally reach at a complete predicate logical extension of the subintuitionistic propositional logic being acceptable only from a classical point of view.

2 Hilbert Type Calculus and Kripke Type Model

Definition 1. The sets V, K, P constitute the *language* of predicate logic: a set V of denumerably many individual variables: $x_1, \ldots, x_k, \ldots \in V$, for $k \in \omega$; a set K of denumerably many individual constants: $a_1, \ldots, a_k, \ldots \in K$, for $k \in \omega$; a set P of predicates, which contains for every $n \in \omega$ a set $\mathbf{P}^n \subset P$ of denumerably many n-ary predicates: $\mathbf{P}^n_1, \ldots, \mathbf{P}^n_k, \ldots$ for $k \in \omega$. The set $\mathbf{P}^0 \subseteq P$ of predicates $\mathbf{P}^0_k$ for $k \in \omega$ is the set of propositional variables; the set of terms is the set T: $t \in T$, if $t \in V$ or $t \in K$; $r, s, t, t_1, t_2, \ldots, t_k, \ldots$ with $k \in \omega$ are the terms, x, y, z are the variables, and a, b, c are the constants, with or without numerical indices. Further the following logical symbols are used: $\rightarrow, \wedge, \vee$ are dyadic connectives; $\forall, \exists$ are monadic quantifiers; (,) are parentheses. The following well formed formulae wff are precisely the well formed formulae: $\mathbf{P}^j_k(t_1, \ldots, t_j)$ is a wff, if $t_1, \ldots, t_j \in T$ and $\mathbf{P}^j_k \in P$; $(A \rightarrow B), (A \wedge B), (A \vee B)$ are wff, if A and B wff; $\forall x A, \exists x A$ are wff, if A is a wff. Parentheses, especially external parentheses are often missing.

Definition 2. $X(s/t)$ has the meaning, that in the expression X – a term or a formula – the term s is substituted in a uniform manner at all places by the term t. A variable x is called *free* in (the formula) A if x is not bound in the formula A by a quantifier. A term t is called *free for* x in (the formula) A, if t is a constant, or if t is a variable and is not bound by a quantifier after substitution.

Definition 3. A *frame* F is a triple $\langle W, R, D\rangle$, where $W \neq \emptyset$, $R \subseteq W \times W$, R is reflexive and transitive and $D \neq \emptyset$. A *model* M as usual relates a frame with a language by some mappings, i.e. a model M is a quadruple $\langle W, R, D, v\rangle$, i.e. a frame F with a function v such that $v(k) \in D$ for arbitrary constants $k \in K$ and $v_\alpha(\mathbf{P}^n_j) \subseteq D^n$ for arbitrary n-ary predicates $\mathbf{P}^n_j \in P$ and arbitrary $\alpha \in W$. Finally a function h is defined mapping the set of individual variables V into D; i.e. $h(x) \in D$ for every $x \in V$. Further on there holds: $h^x_a(y) = h(y)$, if $y \neq x$, and $h^x_a(y) = a$, if $y = x$. Values of terms are defined as follows: $t^{M,h} := v(t)$, if t is a constant, and $t^{M,h} := h(t)$, if t is a variable. Then we extend the functions v and h to arbitrary wff and arbitrary $\alpha \in W$:
$\models^M_\alpha \mathbf{P}^n_j(t_1, \ldots, t_n)[h]$ iff $\langle t_1^{M,h}, \ldots, t_n^{M,h}\rangle \in v_\alpha(\mathbf{P}^n_j)$ for all $\mathbf{P}^n_j \in P$;
$\models^M_\alpha A \rightarrow B[h]$ iff for all $\beta \in W$ with $\alpha R \beta$ holds: $\not\models^M_\beta A[h]$ or $\models^M_\beta B[h]$;
$\models^M_\alpha A \wedge B[h]$ iff $\models^M_\alpha A[h]$ and $\models^M_\alpha B[h]$;
$\models^M_\alpha A \vee B[h]$ iff $\models^M_\alpha A[h]$ or $\models^M_\alpha B[h]$;

$\models^M_\alpha \forall x F[h]$ iff for all $a \in D$ holds: $\models^M_\alpha F[h^x_a]$;
$\models^M_\alpha \exists x F[h]$ iff there is a $a \in D$ such that $\models^M_\alpha F[h^x_a]$.

Definition 4. A wff A is *valid* iff $\models A$; $\models A$ iff for all F, all v, all α and all h $\models^{\langle F,v\rangle}_\alpha A[h]$. A wff A is a *consequence* of a set of wff Γ iff $\Gamma \models A$; $\Gamma \models A$ iff for all F, all v, all α and all h, if $\models^{\langle F,v\rangle}_\alpha \Gamma[h]$, then $\models^{\langle F,v\rangle}_\alpha A[h]$. $\models^M_\alpha \Gamma[h]$ has the meaning, that $\models^M_\alpha C[h]$ for all $C \in \Gamma$.

As a matter of fact the following wffs are not valid: $A \to (B \to A)$, $A \to ((A \to B) \to B)$, $(A \to (B \to C)) \to (B \to (A \to C))$, $A \to (B \to (A \wedge B))$.

Lemma 1 (Coincidence of Substitution and Valuation of Terms).
$\models^M_\alpha A(x/t)[h]$ iff $\models^M_\alpha A[h^x_{t^{M,h}}]$.
Proved by an induction on the degree of A.

Definition 5. The following axiom schemata and rules define the *Hilbert type calculus* of positive subintuitionistic predicate logic.
A1. $(A \to (B \to C)) \to ((A \to B) \to (A \to C))$;
A2. $A \to A$;
A3. $(A \to B) \to (C \to (A \to B))$;
A4./A5. $(A \wedge B) \to A$; $(A \wedge B) \to B$;
A6. $(A \to B) \to ((A \to C) \to (A \to (B \wedge C)))$;
A7./A8. $A \to (A \vee B)$; $B \to (A \vee B)$;
A9. $((A \wedge C) \to B) \to (((A \wedge D) \to B) \to ((A \wedge (C \vee D)) \to B))$;
PL1. $\forall x F \to F(x/t)$, if t is free for x in F;
PL2. $\forall x(F \to G) \to (F \to \forall x G)$, if x is not free in F or not free in G;
PL3. $\forall x(F \vee G) \to (F \vee \forall x G)$, if x is not free in F;
PL4. $F(x/t) \to \exists x F$, if t is free for x in F;
PL5. $\forall x((F \wedge H) \to G) \to ((\exists x F \wedge H) \to G)$, if x is not free in F or not free in H, G;
MP. if A and $A \to B$, then B;
RA. if A and B, then $A \wedge B$;
GR. if F, then $\forall x F$.
Finally we define the relation $\Gamma \vdash A$ for pairs of sets of wffs Γ and wffs A, i.e. we define deducibility from assumptions. $\Gamma \vdash A$ holds iff there is a sequence of wffs $\langle A_1, \ldots, A_k\rangle$ for $k \in \omega$, such that $A = A_k$ and for every $j \leq k$ it holds that either A_j is an instance of an axiom, or $A_j \in \Gamma$, or there are $g, h \leq j$ such that A_j is a result of rule MP. applied on wffs A_g and A_h, or there are $g, h \leq j$ such that A_j is a result of rule RA. applied on wffs A_g and A_h, or there is $h \leq j$ such that A_j is a result of rule GR. applied on wff A_h, and either the quantified variable rule GR. contains is not free in A_h or not free in any A_g with $g \leq h$ and $A_g \in \Gamma$.

Theorem 1 (On Deducibility I).
If $A \in \Gamma$, then $\Gamma \vdash A$;
if $\Gamma \vdash A$ and $\Gamma \vdash A \to B$, then $\Gamma \vdash B$;
if $\Gamma \vdash A$ and $\Gamma \vdash B$, then $\Gamma \vdash A \wedge B$;
if $\Gamma \vdash A$, then $\Gamma \cup \Delta \vdash A$;
if $\Gamma \vdash A$, then there is a finite $\Delta \subseteq \Gamma$ such that $\Delta \vdash A$;

if $\Gamma \vdash A$ and $\Delta \vdash C$ for all $C \in \Gamma$, then $\Delta \vdash A$;
if $\Gamma \vdash A$, then $\Gamma \vdash \forall x A$, if x is not free in A or not free in Γ.
Without Proof.

Definition 6. The usual definition of consistency is assumed, where a set of wff Γ is *consistent* iff there is an A such that $\Gamma \nvdash A$. A special set of formulae $\Gamma_{\forall\to}$ is to be defined: $\Gamma_{\forall\to} = \{\forall x_1,\ldots,\forall x_n(A \to B) \mid \forall x_1,\ldots,\forall x_n(A \to B) \in \Gamma\}$, where quantifier prefixes $\forall x_1,\ldots,\forall x_n$ of implications $A \to B$ may consist of any number n of universal quantifiers or none of them. From time to time $\Gamma \cup \{A_1,\ldots,A_k\} \vdash B$ is abbreviated as $\Gamma, A_1,\ldots,A_k \vdash B$.

Lemma 2. $\Gamma, A, B \vdash C$ iff $\Gamma, A \wedge B \vdash C$.
For the proof use the at once with A4., A5. and RA. provable, usual properties of conjunction: if $\Gamma \vdash A$ and $\Gamma \vdash B$, then $\Gamma \vdash A \wedge B$; if $\Gamma \vdash A \wedge B$, then $\Gamma \vdash A$ and $\Gamma \vdash B$. Then prove $\Gamma, A, B \vdash A \wedge B$ and $\Gamma, A \wedge B \vdash A$ and $\Gamma, A \wedge B \vdash B$. Finally use the structural properties of deducibility.

Theorem 2 (On Deducibility II).
If $\Gamma_{\forall\to} \cup \{A\} \vdash B$, then $\Gamma_{\forall\to} \vdash A \to B$;
if $\Gamma \vdash A \wedge B$, then $\Gamma \vdash A$ and $\Gamma \vdash B$;
if $\Gamma \vdash A$ or if $\Gamma \vdash B$, then $\Gamma \vdash A \vee B$;
if $\Gamma \cup \{C\} \vdash B$ and $\Gamma \cup \{D\} \vdash B$ and $\Gamma \vdash C \vee D$, then $\Gamma \vdash B$;
if $\Gamma \vdash \forall x A$, then $\Gamma \vdash A(x/t)$; if t is free for x in A;
if $\Gamma \vdash A \vee B$, then $\Gamma \vdash A \vee \forall x B$, if x is not free in Γ, A or not free in B;
if $\Gamma \vdash A(x/t)$, then $\Gamma \vdash \exists x A$; if t is free for x in A;
if $\Gamma \cup \{F\} \vdash B$ and $\Gamma \vdash \exists x F$, then $\Gamma \vdash B$, if x is not free in Γ, B or not free in F.

At first the modified deduction theorem is proved: if $\Gamma_{\forall\to} \cup \{A\} \vdash B$, then $\Gamma_{\forall\to} \vdash A \to B$; i.e. it is to be proved: if there is a sequence of wffs $\langle C_1,\ldots,C_k\rangle$ for $k \in \omega$, such that $B = C_k$ and for every $j \leq k$ it holds that either C_j is an instance of an axiom, or $C_j \in \Gamma_{\forall\to} \cup \{A\}$, or there are $g, h \leq j$ such that C_j is a result of rule MP. or RA. or GR. applied on wffs A_g and A_h, with respective variable conditions in case of rule GR.; then there is a sequence of wffs $\langle D_1,\ldots,D_m\rangle$ for $m \in \omega$, such that $A \to B = D_m$ and for every $j \leq m$ it holds that either D_j is an instance of an axiom, or $D_j \in \Gamma_{\forall\to}$, or D_j is a result of rule MP., or of rule RA., or of rule GR. with respective conditions.

It can be shown that for a given sequence of wffs $\langle C_1,\ldots,C_k\rangle$ a sequence of wffs $\langle D_1,\ldots,D_m\rangle$ can be constructed effectively, and that in $\langle D_1,\ldots,D_m\rangle$ at most assumptions from $\langle C_1,\ldots,C_k\rangle$ are occuring.

If B is an instance of an axiom schema, B is an implication, say $B = B_1 \to B_2$, and A3. together with MP. gives the desired result, i.e. $\langle C_1\rangle = \langle B_1 \to B_2\rangle$ and $\langle D_1,\ldots,D_3\rangle = \langle B_1 \to B_2, (B_1 \to B_2) \to (A \to (B_1 \to B_2)), A \to (B_1 \to B_2)\rangle$. If $B \in \Gamma_{\forall\to}$, and B is an implication, then again A3. together with rule MP. leads to the desired result; the respective $\langle D_1,\ldots,D_m\rangle$ can be constructed as in the former case. If $B \in \Gamma_{\forall\to}$ and $B = \forall x_1 \ldots \forall x_n(B_1 \to B_2)$, i.e. B is a multiple universal closure of an implication, we conclude with various instances of PL1. and several applications of MP. $\Gamma_{\forall\to} \vdash (B_1 \to B_2)(x_1/y_1,\ldots,x_n/y_n)$

and $\Gamma_{\forall\rightarrow} \vdash B_1(x_1/y_1, \ldots, x_n/y_n) \rightarrow B_2(x_1/y_1, \ldots, x_n/y_n)$, for $y_1, \ldots, y_n$ not in $\Gamma_{\forall\rightarrow}, B, A$; now we apply once more A3. and MP. to get the desired result with several applications of rule GR. and instances of PL2. If $B = A$ apply A2.; i.e. $\langle C_1 \rangle = \langle A \rangle$ and $\langle D_1 \rangle = \langle A \rightarrow A \rangle$. The last case ends the begin of the induction on the length of the deduction. Observe that assumptions occuring in $\langle D_1, \ldots, D_m \rangle$ are at most assumptions occuring in $\langle C_1, \ldots, C_k \rangle$.

Now consider the induction steps consisting of applications of rules MP., RA. and GR.. At first consider the case B being a result of MP.: apply A1. and make use of the induction base. If B is a result of RA., apply A6. and make use of the induction base. If B is a result of GR., i.e. $B = \forall x C$ for some C, we have $\Gamma_{\forall\rightarrow} \cup \{A\} \vdash \forall x C$; this is a result of $\Gamma_{\forall\rightarrow} \cup \{A\} \vdash C$. We treat this case very carefully. The last claim has with respect to free variables the meaning, that there is a sequence $\langle C_1, \ldots, C_k \rangle$ with $C_k = C$ such that variable x is not free in any C_i with $1 \leq i \leq k$, if $C_i \in \Gamma_{\forall\rightarrow} \cup \{A\}$. Now induction base tells that there exists a sequence $\langle D_1, \ldots, D_m \rangle$ such that $D_m = A \rightarrow C$ and such that assumptions occuring in $\langle D_1, \ldots, D_m \rangle$ are at most assumptions occuring in $\langle C_1, \ldots, C_k \rangle$; i.e. even free variables of assumptions in $\langle D_1, \ldots, D_m \rangle$ are at most free variables of assumptions in $\langle C_1, \ldots, C_k \rangle$. Due to this reason we are allowed to apply GR., to reach $\langle D_1, \ldots, A \rightarrow C, \forall x(A \rightarrow C) \rangle$. If $A = C_i$ for some C_i in $\langle C_1, \ldots, C_k \rangle$ with $1 \leq i \leq k$, x is not free in A, so we reach the desired result with an instance of PL2. and an application of MP.: $\langle D_1, \ldots, A \rightarrow C, \forall x(A \rightarrow C), \forall x(A \rightarrow C) \rightarrow (A \rightarrow \forall x C), (A \rightarrow \forall x C) \rangle$. This last claim has the meaning that $\Gamma_{\forall\rightarrow} \vdash A \rightarrow \forall x C$. If $A \neq C_i$ for all C_i in $\langle C_1, \ldots, C_k \rangle$ with $1 \leq i \leq k$, x may be free in A. But since x is not free in any assumption $C_i \in \Gamma_{\forall\rightarrow}$ for C_i in $\langle C_1, \ldots, C_k \rangle$ with $1 \leq i \leq k$, we choose a variable y not in $\Gamma_{\forall\rightarrow}, A, B$, and substitute y for any free variable x in the sequence to reach $\langle C_1(x/y), \ldots, C_k(x/y) \rangle$, which does not affect the shape of $\Gamma_{\forall\rightarrow}, A$. Induction assumption tells that there is a sequence $\langle D_1, \ldots, D_m \rangle$ such that $D_m = A \rightarrow C_k(x/y)$ and such that the free variables of this last sequence are at most the free variables of $\langle C_1(x/y), \ldots, C_k(x/y) \rangle$. The next step is an application of GR. to reach $\langle D_1, \ldots, A \rightarrow C_k(x/y), \forall y(A \rightarrow C_k(x/y)) \rangle$, where the variable condition for the application of GR. is given. The next wffs in the sequence are an instance of PL2. and a result of MP. $\langle D_1, \ldots, A \rightarrow C_k(x/y), \forall y(A \rightarrow C_k(x/y)), \forall y(A \rightarrow C_k(x/y)) \rightarrow (A \rightarrow \forall y C_k(x/y)), A \rightarrow \forall y C_k(x/y) \rangle$. It remains to change the quantifier in the last formula, and deduction theorem is proved.

Next or elimination is proved: if $\Gamma, C \vdash B$ and $\Gamma, D \vdash B$ and $\Gamma \vdash C \vee D$, then $\Gamma \vdash B$. At first we prove with A9. and the deduction theorem if $A \wedge C \vdash B$ and $A \wedge D \vdash B$, then $A \wedge (C \vee D) \vdash B$; now assume $\Gamma, C \vdash B$ and $\Gamma, D \vdash B$; the finiteness of the deducibility relation guarantees that there is a finite $\Delta \subseteq \Gamma$ such that $\Delta, C \vdash B$ and $\Delta, D \vdash B$; we write this having Lemma 2 in mind as $\bigwedge \Delta \wedge C \vdash B$ and $\bigwedge \Delta \wedge D \vdash B$, where $\bigwedge \Delta$ is the multiple conjunction of all formulae in Δ; then we conclude with the just proved fact $\bigwedge \Delta \wedge (C \vee D) \vdash B$; and then we conclude $\Delta, C \vee D \vdash B$ and $\Gamma, C \vee D \vdash B$; therefore we have that, if $\Gamma, C \vdash B$ and $\Gamma, D \vdash B$, then $\Gamma, C \vee D \vdash B$; this last claim is trivially equivalent to or elimination on the background of the defined deducibility relation.

At last existence elimination is proved: if $\Gamma \cup \{F\} \vdash B$ and $\Gamma \vdash \exists x F$, then $\Gamma \vdash B$, if x is not free in Γ, B or not free in F. Clearly existence elimination is equivalent to: if $\Gamma \cup \{F\} \vdash B$, then $\Gamma \cup \{\exists x F\} \vdash B$, if x is not free in F or not free in Γ, B. To prove this last claim we first deduce the rule if $\vdash (A \wedge B) \rightarrow C$ and x is not free in A or not free in B, C, then $\vdash (\exists x A \wedge B) \rightarrow C$. Therefore assume $\vdash (A \wedge B) \rightarrow C$ and x not to be free in A or not in B, C; with GR. we conclude $\vdash \forall x((A \wedge B) \rightarrow C)$, and with the mentioned variable condition, PL7. and an application of MP. $\vdash (\exists x A \wedge B) \rightarrow C$. Now we assume $\Gamma, A \vdash C$ and x not to be free in A or in Γ, C; then we have a finite $\Delta \subseteq \Gamma$ with $\Delta, A \vdash C$; write this finite Δ with Lemma 2 as $\bigwedge \Delta$, the multiple conjunction of all formulae in Δ, then we have $\bigwedge \Delta, A \vdash C$, $A \wedge \bigwedge \Delta \vdash C$ and with deduction theorem $\vdash (A \wedge \bigwedge \Delta) \rightarrow C$; with the deduced rule and the mentioned variable conditions $\vdash (\exists x A \wedge \bigwedge \Delta) \rightarrow C$; this is equivalent with deduction theorem to $\exists x A, \bigwedge \Delta \vdash C$; at last conclude $\Delta, \exists x A \vdash C$ and $\Gamma, \exists x A \vdash C$.

Theorem 3 (Strong Correctness). If $\Gamma \vdash A$, then $\Gamma \models A$.
The reader is asked to convince himself that all axiom schemata are valid, and all rules of the defined deducibility relation lead to consequences; then the proof of strong correctness goes by induction on the length of a deduction.

Definition 7. Now assume a language L with a set of constants K and a set of predicates P, i.e. $L = \langle K, P \rangle$. Assume the constants from Γ, A, B to be in K and the predicates in P. Γ is *deductively closed* iff if $\Gamma \vdash A$, then $A \in \Gamma$; Γ has the *or property* iff if $A \vee B \in \Gamma$, then $A \in \Gamma$ or $B \in \Gamma$; Γ is *prime* iff Γ is consistent, deductively closed and has the or property; Γ is *existence complete* in the language L iff if $\Gamma \vdash \exists x B$, then there is a $c \in K$ such that $\Gamma \vdash B(x/c)$; Γ is *all complete* in L iff if $\Gamma \vdash B(x/c)$ for all $c \in K$, then $\Gamma \vdash \forall x B$; Γ is *saturated* in L iff Γ is prime, all complete in L and existence complete in L.

Lemma 3 (On saturated sets).
1. If $\Gamma \nvdash B$, then there exists a saturated Δ such that $\Gamma \subseteq \Delta$ and $B \notin \Delta$;
2. if $A \rightarrow B \notin \Gamma$ and Γ is saturated in L, then there exists a saturated Δ in L such that $\Gamma_{\forall \rightarrow} \subseteq \Delta$, $A \in \Delta$ and $B \notin \Delta$.

These two lemmata show the difference to the proof of completeness in the intuitionistic case: there Lemma 3.2 is a corollary to Lemma 3.1, but in the absence of the full deduction theorem it is not; the proof of Lemma 3.2 is the hard case due to the additional fact, that we are not allowed to extend languages.

Proof of Lemma 3.1. Assume $\Gamma \nvdash B$ and assume an enumeration of all disjunctions, an enumeration of all existence formulae, an enumeration of all universal formulae and assume constants, new relative to languages associated with certain given sets of formulae; define a sequence of sets of formulae Δ_k and a sequence of formulae A_k, where $\Delta_0 = \Gamma$ and $A_0 = B$. If $k = 0 \pmod 3$ in Δ_k, take the first disjunction $C \vee D$ in the enumeration, which has not been treated, such that $\Delta_k \vdash C \vee D$: if $\Delta_k \cup \{C\} \vdash A_k$, then $\Delta_{k+1} = \Delta_k \cup \{D\}$ and $A_{k+1} = A_k$; if $\Delta_k \cup \{C\} \nvdash A_k$, then $\Delta_{k+1} = \Delta_k \cup \{C\}$ and $A_{k+1} = A_k$. Or elimination guarantees either $\Delta_k \cup \{D\} \nvdash A_k$ or $\Delta_k \cup \{C\} \nvdash A_k$. If $k = 1 \pmod 3$ in Δ_k, take the first universal formula $\forall x C$ in the enumeration, which has not been treated: if $\Delta_k \vdash A_k \vee \forall x C$, then $\Delta_k \cup \{\forall x C\} \nvdash A_k$ with or elimination, and $\Delta_{k+1} = \Delta_k \cup \{\forall x C\}$ and $A_{k+1} = A_k$; but if $\Delta_k \nvdash A_k \vee \forall x C$, then it holds with Theorem 2 that $\Delta_k \nvdash A_k \vee C(x/y)$ for a $y \notin \Delta_k, A_k, C$, and for a new constant c, which is not in the language of Δ_k, A_k, C we have $\Delta_k \nvdash A_k \vee C(x/c)$; then we define $\Delta_{k+1} = \Delta_k$ and $A_{k+1} = A_k \vee C(x/c)$. If $k = 2 \pmod 3$ in Δ_k, take the first existence formula $\exists x C$ not yet been treated such that $\Delta_k \vdash \exists x C$; with existence elimination it holds that $\Delta_k \cup \{C(x/y)\} \nvdash A_k$ for y not in Δ_k, A_k, C; further it holds for a new constant c, which is not in the language of Δ_k, A_k, C $\Delta_k \cup \{C(x/c)\} \nvdash A_k$, and Δ_{k+1} is defined as $\Delta_k \cup \{C(x/c)\}$ and $A_{k+1} = A_k$. Finally define $\Delta = \bigcup_{0 \leq k} \Delta_k$.

Now prove that Δ has all the desired properties: $\Gamma \subseteq \Delta$, $B \notin \Delta$ because $B \notin \Delta_k$ for all k, and Δ is saturated, i.e. Δ is prime, Δ is all complete and Δ is existence complete. This proof is similar to the proof of a related Henkin lemma in Thomason, 1968, p. 3, with appropriate modifications.

Lemma 4. If Γ is saturated in L, then $\Gamma_{\forall\rightarrow}$ is all complete in L.
Assume $L = \langle K, P\rangle$, $\Gamma_{\forall\rightarrow} \vdash B(x/c)$ for all $c \in K$, and Γ to be saturated in L, show that $\Gamma_{\forall\rightarrow} \vdash \forall x B$; prove this by an induction on the length of the deduction of one $B(x/c_k)$. If $B(x/c_k)$ is an instance of an axiom schema or a generalisation of some, then B as well, and with GR. we conclude $\forall x B$. If $B(x/c_k) \in \Gamma_{\forall\rightarrow}$, then $B(x/c_k)$ is an implication or a (multiple) universal implication; further on $\Gamma \vdash B(x/c)$ for all $c \in K$, and $\forall x B \in \Gamma$ since Γ is all complete in L; further on $\forall x B \in \Gamma_{\forall\rightarrow}$, because of the definition of $\Gamma_{\forall\rightarrow}$ and the fact that $\forall x B$ has to be a (multiple) universal implication. These cases constitute the begin of the induction, now we change to the induction step consisting of applications of rules MP., RA. or GR.

Assume $\Gamma_{\forall\rightarrow} \vdash B(x/c_k)$ to be a result of MP., i.e. there is a C such that $\Gamma_{\forall\rightarrow} \vdash C \rightarrow B(x/c_k)$ and $\Gamma_{\forall\rightarrow} \vdash C$; for all other c_n $\Gamma_{\forall\rightarrow} \vdash B(x/c_n)$. Therefore $\Gamma_{\forall\rightarrow} \cup \{C\} \vdash B(x/c_n)$; with deduction theorem we have $\Gamma_{\forall\rightarrow} \vdash C \rightarrow B(x/c_n)$ for all c_n, i.e. this holds for really all $c \in K$. We treat the most complicated case, where C may contain x as free variable. Deduce for all c $\Gamma_{\forall\rightarrow} \vdash C \rightarrow B(x/y)(y/c)$ for a variable y not free in B, C, and further $\Gamma_{\forall\rightarrow} \vdash (C \rightarrow B(x/y))(y/c)$. Then apply induction assumption on variable y, since induction begin was proved for arbitrary variables and get $\Gamma_{\forall\rightarrow} \vdash \forall y(C \rightarrow B(x/y))$. Predicate logical reasoning gives $\Gamma_{\forall\rightarrow} \vdash C \rightarrow \forall y B(x/y)$; further $\Gamma_{\forall\rightarrow} \vdash \forall y B(x/y)$, and the desired $\Gamma_{\forall\rightarrow} \vdash \forall x B$.

The proof of the lemma is completed by considering $\Gamma_{\forall\rightarrow} \vdash B(x/c_k)$ to be a result of rules RA. and GR.

Proof of Lemma 3.2. Assume $L = \langle K, P\rangle$, $A \rightarrow B \notin \Gamma$ and Γ to be saturated in L; show that there exists a saturated Δ in L such that $\Gamma_{\forall\rightarrow} \subseteq \Delta$, $A \in \Delta$ and $B \notin \Delta$. With the assumption we have $\Gamma \nvdash A \rightarrow B$ and $\Gamma_{\forall\rightarrow} \nvdash A \rightarrow B$ and with the deduction theorem $\Gamma_{\forall\rightarrow} \cup \{A\} \nvdash B$. According to Lemma 4 $\Gamma_{\forall\rightarrow}$ is all complete in L.

Now construct a sequence of sets of formulae Δ_k and a sequence of formulae B_k with $\Delta_k = \Gamma_{\forall\rightarrow} \cup \{A, D_1, \ldots, D_{k-1}\}$, i.e. set $\Delta_0 = \Gamma_{\forall\rightarrow}$ and $\Delta_1 = \Gamma_{\forall\rightarrow} \cup \{A\}$; further $B_0 = B$. Assume an enumeration of all disjunctions, an enumeration of all universal formulae and an enumeration of all existence formulae and continue to define Δ_k as follows. If $k = 0 \pmod 3$ in Δ_k, take the first disjunction $C \vee D$ in the enumeration, which has not been treated, such that $\Delta_k \vdash C \vee D$: if $\Delta_k \cup \{C\} \vdash B_k$, then $\Delta_{k+1} = \Delta_k \cup \{D\}$ and $B_{k+1} = B_k$; if $\Delta_k \cup \{C\} \nvdash B$, then $\Delta_{k+1} = \Delta_k \cup \{C\}$ and again $B_{k+1} = B_k$. Or elimination guarantees either $\Delta_k \cup \{D\} \nvdash B$ or $\Delta_k \cup \{C\} \nvdash B$. If $k = 1 \pmod 3$ in Δ_k, take the first universal formula $\forall x C$ in the enumeration, which has not been treated: if $\Delta_k \vdash B_k \vee \forall x C$, then $\Delta_k \cup \{\forall x C\} \nvdash B_k$ with or elimination, and $\Delta_{k+1} = \Delta_k \cup \{\forall x C\}$ and $B_{k+1} = B_k$; but if $\Delta_k \nvdash B_k \vee \forall x C$, then $\Delta_k \nvdash \forall x(B_k \vee C)$ with PL3. and eventually an appropriate renaming of bound variable x. This can be written as $\Gamma_{\forall\rightarrow} \cup \{A \wedge D_1 \wedge \ldots \wedge D_{k-1}\} \nvdash \forall x(B_k \vee C)$, and $\Gamma_{\forall\rightarrow} \nvdash (A \wedge D_1 \wedge \ldots \wedge D_{k-1}) \rightarrow \forall x(B_k \vee C)$. This is with PL2. $\Gamma_{\forall\rightarrow} \nvdash \forall x((A \wedge D_1 \wedge \ldots \wedge D_{k-1}) \rightarrow (B_k \vee C))$; with all completeness of $\Gamma_{\forall\rightarrow}$ conclude $\Gamma_{\forall\rightarrow} \nvdash ((A \wedge D_1 \wedge \ldots \wedge D_{k-1}) \rightarrow (B_k \vee C))(x/c)$ for some $c \in K$. This amounts finally to $\Delta_k \nvdash B_k \vee C(x/c)$ for a $c \in K$. Now define $\Delta_{k+1} = \Delta_k$ and $B_{k+1} = B_k \vee C(x/c)$. If $k = 2 \pmod 3$ in Δ_k, take the first existence formula $\exists x C$ in the enumeration not yet been treated, such that $\Delta_k \vdash \exists x C$: since $\Delta_k \nvdash B_k$, $\Delta_k \cup \{\exists x C\} \nvdash B_k$. Now each Δ_k can be written as $\Gamma_{\forall\rightarrow} \cup \{A, D_1, \ldots, D_{k-1}\}$, $\Gamma_{\forall\rightarrow}$ being allcomplete; i.e. we have $\Gamma_{\forall\rightarrow} \cup \{A \wedge D_1 \wedge \ldots \wedge D_{k-1} \wedge \exists x C\} \nvdash B_k$; now it holds that $\Gamma_{\forall\rightarrow} \nvdash (A \wedge D_1 \wedge \ldots \wedge D_{k-1} \wedge \exists x C) \rightarrow B_k$; and with an appropriate renamed bound variable x and an application of axiom schema PL5. we reach $\Gamma_{\forall\rightarrow} \nvdash \forall x((A \wedge D_1 \wedge \ldots \wedge D_{k-1} \wedge C) \rightarrow B_k)$; again with all completeness of $\Gamma_{\forall\rightarrow}$ this gives $\Gamma_{\forall\rightarrow} \nvdash ((A \wedge D_1 \wedge \ldots \wedge D_{k-1} \wedge C) \rightarrow B_k)(x/c)$ for some $c \in K$; this amounts after reversing the processes the desired $\Delta_k \cup \{C(x/c)\} \nvdash B$ for one $c \in K$; define here $\Delta_{k+1} = \Delta_k \cup \{C(x/c)\}$ and $B_{k+1} = B_k$.

Finally define $\Delta = \bigcup_{0 \leq k} \Delta_k$, then Δ has all desired properties; deductive closure, or property, nonelementhood of B, existence completeness and all completeness, the proofs are as shown in Lemma 3.1.

Theorem 4 (On Saturated Sets). For all in $L = \langle K, P\rangle$ saturated sets of formulae Γ it holds:
$A \rightarrow B \in \Gamma$ iff for all in L saturated Δ with $\Gamma_{\forall\rightarrow} \subseteq \Delta$: $A \notin \Delta$ or $B \in \Delta$;
$A \wedge B \in \Gamma$ iff $A \in \Gamma$ and $B \in \Gamma$;
$A \vee B \in \Gamma$ iff $A \in \Gamma$ or $B \in \Gamma$;
$\forall x F \in \Gamma$ iff for all $c \in K$ $F(x/c) \in \Gamma$;
$\exists x F \in \Gamma$ iff there is an $c \in K$ such that $F(x/c) \in \Gamma$.
The proofs of these theorems are well done with Lemma 3.2 and the theorems on deducibility 2.

Definition 8. The *canonical model* M for a in $L = \langle K, P\rangle$ saturated set of formulae Γ is a quadruple $\langle W, R, D, v\rangle$ defined as follows:
$W = \{\alpha | \Gamma_{\forall\rightarrow} \subseteq \alpha$ and α is saturated in $L\}$, $\alpha R \beta$ iff $\alpha_{\forall\rightarrow} \subseteq \beta$, $D = K$, $v(c) = c$ for arbitrary $c \in K$, $v_\alpha(\mathbf{P}^n_j) = \{\langle c_1, \ldots, c_n\rangle | c_1, \ldots, c_n \in D, \alpha \in W, \mathbf{P}^n_j(c_1, \ldots, c_n) \in \alpha\}$ for all $\alpha \in W$ and all $\mathbf{P}^n_j \in P$.

Lemma 5 (Canonical Truth). For all α of the canonical model M, all A and all h it holds: $\models^M_\alpha A[h]$ iff $A \in \alpha$.
Proof by induction on the degree of A with the theorems on saturated sets 4.

Theorem 5 (Strong Completeness). If $\Gamma \models C$, then $\Gamma \vdash C$.
For strong completeness assume $\Gamma \nvdash C$, show $\Gamma \not\models C$; Lemma 3.1 gives a saturated $\Delta \supseteq \Gamma$ such that $C \notin \Delta$. With the truth Lemma 5 it holds in the canonical model for Δ for any h: $\not\models^M_\Delta C[h]$ and $\models^M_\Delta \Gamma[h]$.

The developed logic can be extended easily with negation: add $\bot$ as a propositional variable to the underlying language and add the definition $\neg A \leftrightarrow_{df} (A \rightarrow \bot)$. Now, as well as in intuitionistic logic, there are two negations: a strong negation, according to the additional axiom schema $\bot \rightarrow A$, and the additional semantical definition $\not\models^M_\alpha \bot$ for any α; and a weak negation without any additional axiom schema and semantical definition.

In the case of the strong negation $\bot \notin \Gamma$ can be proved for any saturated set Γ, so that the truth lemma may be extended to the new language and the new axiom schema to get further results of correctness and completeness.

The following wff is valid with weak and therefore with strong negation: $(A \rightarrow B) \rightarrow (\neg B \rightarrow \neg A)$. The following two wffs are not valid with strong and therefore not with weak subintuitionistic negation – but they are valid with weak and therefore with strong intuitionistic negation: $A \rightarrow \neg\neg A$, $(A \rightarrow \neg B) \rightarrow (B \rightarrow \neg A)$.

3 Semantic Tableaux

3.1 The Method of Semantic Tableaux

The structure underlying the method of semantic tableaux is an ordered class of objects called tableaux. Every tableau α consists of a tree of signed wff, i.e. wff being true (T) or false (F), the root of the tree at the top. As language for subintuitionistic tableaux presuppose the previously defined language of subintuitionistic logic (may be with $\bot$). The formula trees within the tableaux are at most binary branching, and branches may be infinitely long. A tableau is called to be closed, if every branch of the tree of signed wff is closed; and a branch is closed, if it contains a pair TA and FA for some wff A. Now a wff A is provable with the tableaux method, if it is true in every tableaux under any assignment of constants to variables. The tableaux method proceeds by searching a counterexample: to prove a wff A assume FA under an arbitrary assignment in a tableau, say α, and look for closure of tableau α. Clearly the assignment of truth values (T, F) to wff A in dependence on tableaux α proceeds inductively: basically, as already told, one defines closure of a branch of any tableaux α, if the branch contains a pair TA and FA for some wff A. For the inductive steps one assumes a definition by the following rules, holding for all tableaux α:

$$\alpha \frac{T \quad A \wedge B}{\begin{array}{c} T \quad A \\ T \quad B \end{array}} \qquad \alpha \frac{F \quad A \wedge B}{F \quad A \quad / \quad F \quad B}$$

$$\alpha \frac{T \quad A \vee B}{T \; A \quad / \quad T \; B} \qquad \alpha \frac{F \quad A \vee B}{\begin{matrix} F \; A \\ F \; B \end{matrix}}$$

$(\alpha F \quad \bot)$

$$\alpha \frac{T \quad A \to B}{F \; A \quad / \quad T \; B}$$

$$\frac{\alpha F \quad A \to B}{\beta \quad \begin{matrix} T \; A \\ F \; B \end{matrix}}$$

The rule $FA \to B$ needs some explanation. If a signed wff $FA \to B$ occurs in the formula tree of a tableau α one has to create a new tableau, say β. β has as entries all $TC \to D$ for any C, D, with $TC \to D \in \alpha$ – and additionally TA and FB are entries of β. The following description should be a clear working device: copy the subtree of the signed wff $FA \to B$ from α to β, and delete in this copy all entries not being signed wffs $TC \to D$ for any wff C, D, and at the place of $FA \to B$ put TA and FB.

$\alpha \frac{T \quad \forall x A}{T \quad A^x_a}$ $\alpha \frac{F \quad \forall x A}{F \quad A^x_b}$ where a is any constant, and b is a new constant assigned to variable x.

$\alpha \frac{T \quad \exists x A}{T \quad A^x_b}$ $\alpha \frac{F \quad \exists x A}{F \quad A^x_a}$ where b is a new constant, and a is any constant assigned to variable x.

These rules are nearly self explanatory. All rules except the one and only rule for $FA \to B$ are classically boolean. Given a signed wff TA or FA in a tableau α, with the exception of $FA \to B$, one has to add the respectively listed signed wffs to the tree of wffs in the tableau α, i.e. the respectively listed signed wffs - branching or not branching – have to be added to the end of the related branch of the original wff. To give an example: if $TA \to B$ occurs in the tree of wffs of a tableau α, the signed wffs FA and TB have to be added as branching entries to the end of that branch, in which $TA \to B$ occurs in the formula tree of α. Only the rule $FA \to B$ requires the creation of a new tableau with respect to the described conditions, and hence is not a classical rule.

3.2 Equivalence Theorem

A wff is provable by tableaux iff it is semantically valid.

Prove this from left to right by connecting a closed branch of a signed wff in a tableau with a semantically unsatisfiable set of wff. For the proof from right to left a slight modification of Fitting's idea of a consistency property is sufficient (Fitting, 1983, p. 481 ff.).

3.3 Examples

Two examples of tableaux conclude the section. One disproves $A \to (B \to A)$, the other proves $(A \to B) \to (C \to (A \to B))$.

α		$F \quad (A \to B) \to (C \to (A \to B))$	
β		$T \quad A \to B$	
		$F \quad C \to (A \to B)$	
	$F \quad A$	/	$T \quad B$
γ		$T \quad A \to B$	
		$T \quad C$	
		$F \quad A \to B$	
	$F \quad A$	/	$T \quad B$
δ		$T \quad A \to B$	
		$T \quad A$	
		$F \quad B$	
	$F \quad A$	/	$T \quad B$

α	$F \quad A \to (B \to A)$
β	$T \quad A$
	$F \quad B \to A$
γ	$T \quad B$
	$F \quad A$

4 An Embedding into Modal Logic S4

4.1 A Language of Modal Predicate Logic

Take the language of predicate logic as previously defined, $\perp$ included, and add the logical sign $\Box$, a necessity operator. Add as wffs all expressions $\Box A$, where A is a wff. This is the language of modal predicate logic.

4.2 A Model for Modal Predicate Logic S4 with Barcan's Formulas

In the sequel we are interested in a special, well known system of modal predicate logic, classical modal predicate logic S4 with the formulas of Barcan. Since we are only interested in a semantical description of this logic it suffices to define a respective model. The fact that this model is exactly the model for the described logic can be found in many books on modal logic. For that reason suppose a *model* N as a quadruple $\langle W, R, D, v\rangle$, and define the components of this quadruple

exactly similar to the components of the model M of subintuitionistic logic. Make only one difference: after having defined the functions v and h on terms, extend these definitions of v and h as follows to arbitrary wff and arbitrary $\alpha \in W$:
$\models^N_\alpha \mathbf{P}^n_j(t_1, \ldots, t_n)[h]$ iff $\langle t_1^{N,h}, \ldots, t_n^{N,h} \rangle \in v_\alpha(\mathbf{P}^n_j)$ for all $\mathbf{P}^n_j \in P$;
not $\models^N_\alpha \bot[h]$;
$\models^N_\alpha A \to B[h]$ iff not $\models^N_\alpha A[h]$ or $\models^N_\alpha B[h]$;
$\models^N_\alpha A \wedge B[h]$ iff $\models^N_\alpha A[h]$ and $\models^N_\alpha B[h]$;
$\models^N_\alpha A \vee B[h]$ iff $\models^N_\alpha A[h]$ or $\models^N_\alpha B[h]$;
$\models^N_\alpha \forall x F[h]$ iff for all $a \in D$ holds: $\models^N_\alpha F[h^x_a]$;
$\models^N_\alpha \exists x F[h]$ iff there is a $a \in D$ such that $\models^N_\alpha F[h^x_a]$;
$\models^N_\alpha \Box A[h]$ iff for all $\beta \in W$ with $\alpha R \beta$ holds: $\models^N_\beta A[h]$.

4.3 A Translation T

Define recursively a translation T of the language of predicate logic into the language of modal predicate logic: $T(\mathbf{P}^j_k) = \mathbf{P}^j_k, T(\bot) = \bot, T(A \to B) = \Box(T(A) \to T(B)), T(A \wedge B) = T(A) \wedge T(B), T(A \vee B) = T(A) \vee T(B), T(\forall x A) = \forall x T(A), T(\exists x A) = \exists x T(A)$.

4.4 An Embedding Lemma

Remember M to be the model for subintuitionistic logic, and N for classical modal predicate logic S4 with Barcan's formulas, then it holds – if the valuations of the models M and N are defined to agree on atoms:

$$\models^M_\alpha A[h] \text{ iff } \models^N_\alpha T(A)[h] .$$

Prove this by induction on the degree of A.

From this embedding lemma derive that subintuitionistic validity of a wff A and modal validity of the translation $T(A)$ of this formula agree – for this result presuppose a straightforward definition of modal validity.

References

van Benthem, J.: *Partiality and Nonmonotonicity in Classical Logic*, in: Logique et Analyse, **29** (1986) p. 226-246

Fitting, M.C.: *Intuitionistic Logic, Model Theory and Forcing*, North-Holland, Amsterdam (1969)

Fitting, M.C.: *Proof Methods for Modal and Intuitionistic Logic*, Kluwer, Dordrecht (1983)

Gabbay, D.: *Semantical Investigations in Heyting's Intuitionistic Logic*, Reidel, Dordrecht (1981)

Kripke, S.A.: *Semantical Analysis of Intuitionistic Logic* (1965), in: Crossley, J.N., Dummett, M.A.E. (Eds.): *Formal Systems and Recursive Functions*, North-Holland, Amsterdam (1965), p. 93-130

Restall, G.: *Subintuitionistic Logics*, in: Notre Dame Journal of Formal Logic, **35** (1994) p. 116-129
Thomason, R.H.: *Strong Completeness of Intuitionistic Predicate Logic*, in: Journal of Symbol Logic, **33** (1968) p. 1-7

Remark. The author would like to thank two anonymous referees for critical and competent comments, which certainly helped to improve this paper.

Implementation of Relational Algebra Using Binary Decision Diagrams

Rudolf Berghammer, Barbara Leoniuk, and Ulf Milanese

Institut für Informatik und Praktische Mathematik
Christian-Albrechts-Universität Kiel
Olshausenstraße 40, D-24098 Kiel, Germany

Abstract. We show how relations and their operations can efficiently be implemented by means of Binary Decision Diagrams. This implementation is used in the computer system RelView. To demonstrate the power of the approach, we show how it can be applied to attack computationally hard problems

1 Introduction

In the last decades, relational algebra has widely been used as a very convenient means of dealing with fundamental concepts like graphs, orders, lattices, games, data bases, Petri nets, data types, and semantics in mathematics and computer science. For many examples and references to relevant literature, see [11,6]. One main reason for using relational algebra is that it has a small but surprisingly very powerful set of operations all of which can easily and efficiently be implemented. Thus, a supporting computer system can easily be implemented, too.

RelView (see [1,3] for an overview) is such a tool for the manipulation and visualization of relations and relational programming. It has been developed at Kiel University since 1993. Written in the C programming language, it runs on Sun SPARC workstations and INTEL-based Linux systems[1]. The benefits of RelView are manifold and have been described in [5,1,2,4] for example. But in several applications its use was restricted to small input relations only due to the array-like implementation of relations in which every entry consumes 1 Bit of memory. As an example, the *membership relation* $\in$ between a set X and its powerset 2^X for $|X| = 25$ allocates more than 100 MByte of storage. In this particular case, the memory consumption is exponential in the size of X.

Therefore, in the last three years we have exchanged RelView's implementation of relations and their operations to overcome these problems. We have carried out a number of experiments which showed that a promising alternative is given by Binary Decision Diagrams (shortly: BDDs). The main purpose of this paper is to describe this implementation and to illustrate its advantages and RelView's extended possibilities by means of some examples. Its rest is organized as follows. Sections 2 and 3 present background material on relational algebra and BDDs. The implementation of relational algebra using BDDs

[1] RelView is available free of charge via the World-WideWeb. For more details, see URL http://www.informatik.uni-kiel.de/~progsys/relview.shtml

H. de Swart (Ed.): RelMiCS 2001, LNCS 2561, pp. 241–257, 2002.

is shown in Section 4. In Section 5 we demonstrate how a relation-algebraic problem formulation combined with this implementation technique can be used to solve certain computationally hard problems for medium-sized inputs and present some experimental results. Section 6 contains some concluding remarks.

2 Relations and Relational Description of Sets

We denote the set (also: type) of all relations with domain X and range Y by $[X \leftrightarrow Y]$ and write $R : X \leftrightarrow Y$ instead of $R \in [X \leftrightarrow Y]$. If the sets X and Y are finite and of size m respectively n, then one may consider R as a Boolean matrix with m rows and n columns. Since this interpretation is well suited for many purposes and also used by RELVIEW as one possibility to depict a relation on the screen, in the following we often use matrix terminology and matrix notation. The latter means that we write R_{xy} instead of $\langle x, y\rangle \in R$.

We assume the reader to be familiar with the basic operations on relations, viz. R^T (transposition), $\overline{R}$ (complement), $R \cup S$ (join), $R \cap S$ (meet), $R \cdot S$ (composition; in the sequel abbreviated as RS), $R \subseteq S$ (inclusion), and the special relations O (empty relation), L (universal relation), and I (identity relation). A relation R is called *symmetric* if $R = R^\mathsf{T}$ and *irreflexive* if $R \subseteq \overline{\mathsf{I}}$.

If X is a set, then a subset of it can be described/modeled by a *vector* $v : X \leftrightarrow Y$, a relation satisfying $v = v\mathsf{L}$. Having the Boolean matrix model of relations in mind, $v = v\mathsf{L}$ means that v is row-constant. Consequently, for a vector the range is irrelevant. Therefore, in the following we only consider vectors $v : X \leftrightarrow \mathbf{1}$ with a specific singleton set $\mathbf{1} = \{\perp\}$ as range and omit the second subscript, i.e., write v_x instead of $v_{x\perp}$. In the matrix model then v is a Boolean column vector and describes the subset $\{x \in X : v_x\}$ of its domain.

A relation R is called *univalent* if $R^\mathsf{T} R \subseteq \mathsf{I}$, *injective* if $RR^\mathsf{T} \subseteq \mathsf{I}$, and *total* if $R\mathsf{L} = \mathsf{L}$. As usual, a univalent and total relation is a *mapping*. We can use injective mappings as a second way of describing subsets. Given an injective mapping $R : Y \leftrightarrow X$, we call Y a subset of X given by R. In this case, the vector $R^\mathsf{T}\mathsf{L} : X \leftrightarrow \mathbf{1}$ describes Y in the above sense. Clearly, the construction of an injective mapping $\mathsf{inj}(v) : Y \leftrightarrow X$ from a given vector $v : X \leftrightarrow \mathbf{1}$ describing a subset Y of X is also possible. In combination with the membership relation $\in : X \leftrightarrow 2^X$, such *injective mappings generated by vectors* can be used to enumerate sets of sets. More specifically, if $v : 2^X \leftrightarrow \mathbf{1}$ describes a subset $\mathcal{S}$ of 2^X, then $\mathsf{inj}(v)$ is of type $[\mathcal{S} \leftrightarrow 2^X]$ and – using matrix terminology – we obviously obtain the elements of $\mathcal{S}$ as the columns of the composition $\in \mathsf{inj}(v)^\mathsf{T} : X \leftrightarrow \mathcal{S}$.

3 Binary Decision Diagrams

In recent years, BDDs have emerged as an efficient data structure to manipulate very large Boolean functions. Their introduction is relatively old, but only after the work of Bryant they transformed into a useful tool. For an overview, see [7].

We will present BDDs by means of a small example. Given the ternary Boolean function $f : \mathbb{B}^3 \to \mathbb{B}$ in disjunctive normal form

$$f(x_1, x_2, x_3) = (x_2 \wedge x_3) \vee (x_1 \wedge \overline{x_2} \wedge \overline{x_3}), \qquad (1)$$

its BDD is shown in Figure (a) below. A BDD is a directed acyclic graph with one root and two leaf nodes, the latter labeled with 0 and 1. Each node not being a leaf has two outgoing arcs, labeled 1 and 0 and shown in the drawings as continuous and dotted lines, respectively, oriented from top to bottom. To evaluate f for the assignment of variables $x_1 = 1$, $x_2 = 0$, and $x_3 = 1$, we only have to follow the corresponding directed arcs from the root node. The first node we encounter is labeled with variable x_1, whose value is 1. Given this assignment, the 1-arc must be taken. Next, a node labeled with variable x_2 is found. Since $x_2 = 0$, the 0-arc must be taken now. Finally, the 1-arc for variable x_3 reaches the 0 leaf node, which is the value of f for that assignment, i.e., of $f(1, 0, 1)$.

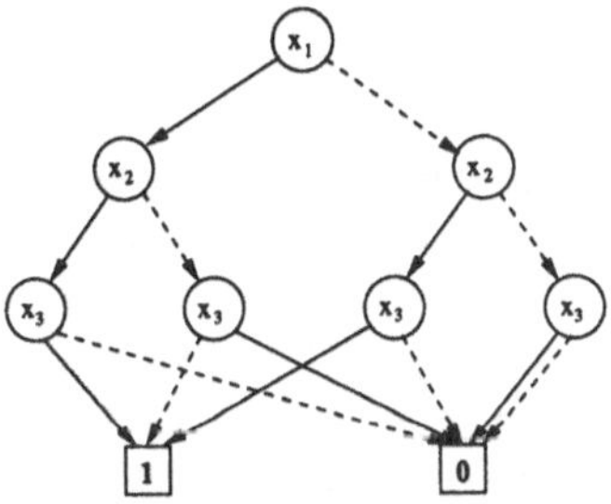

(a) Ordered BDD

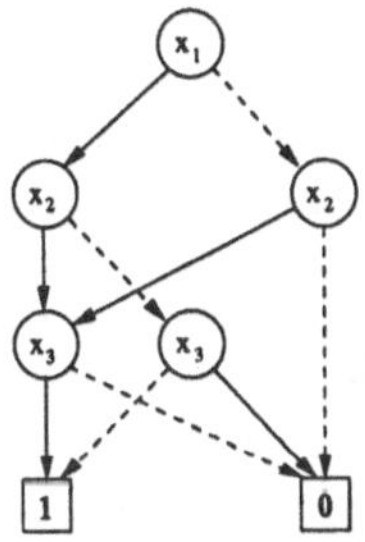

(b) Ordered, reduced BDD

The representation of a Boolean function by means of a BDD is not unique. Figures (a) and (b) depict different BDDs representing the Boolean function f of (1). The BDD in (b) can be obtained from that in (a) by successively applying reduction rules that share isomorphic subgraphs within and eliminate superfluous nodes from the representation. Both BDDs are *ordered.* In an ordered BDD, all variables appear in the same order along all paths from the root to the leaves. If a BDD is ordered and *reduced* (no further reductions can be applied), then we have a Reduced Ordered BDD (shortly: ROBDD). Given a total ordering of variables, a ROBDD is a so-called canonical form, i.e., the representation is unique. See [7]. Figure (b) is a ROBDD with variable ordering $x_1 < x_2 < x_3$.

As shown in [9], the size of a BDD can be exponential in the number of variables; however, ROBDDs are a compact representation of many Boolean functions. Furthermore, Boolean binary operations on them (like disjunction and conjunction) can be calculated in polynomial time in the size of the ROBDDs. Finally, representing Boolean functions as ROBDDs allows to implement many important tests (like implication, equivalence, and satisfiability) in constant time. Henceforth, we will implicitly assume that BDDs are reduced and ordered. Note that each ROBDD node represents, at the same time, a Boolean function whose root is the node itself. This allows to implement BDD packages which manage all BDDs using the same set of variables in only one multi-rooted graph; details can be found in [7]. For the RelView system we used the CUDD package [12] from the University of Colorado, Boulder (USA).

4 Implementation of Relations by ROBDDs

ROBDDs provide a very efficient way to manipulate very large Boolean functions. Therefore, in the following we will present a description of a relation (on, of course, finite sets) as such a function, too. This description is based upon binary encodings of the domain and range of the relation.

We will briefly illustrate our approach by a small example. Assume two sets $X = \{a, b, c\}$ and $Y = \{r, s\}$ and a relation $R : X \leftrightarrow Y$ which is depicted in RelView as the following labeled Boolean 3×2 matrix:

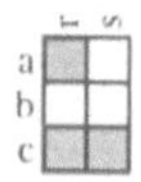

Using binary encodings $c_X : X \to \mathbb{B}^2$ and $c_Y : Y \to \mathbb{B}$ of X and Y, defined by $c_X(a) = 00, c_X(b) = 01, c_X(c) = 10$ respectively $c_Y(r) = 0, c_Y(s) = 1$, the relation R can be interpreted as a ternary partial Boolean function $f : \mathbb{B}^3 \to \mathbb{B}$. The left of the following two tables shows this interpretation of R, where the variables x_1 and x_2 come from the encoding of X and the variable y_1 comes from the encoding of Y. To describe R by a ROBDD, in the next step we use the truth value 0 as a dummy result and pass from the partial function f over to the total function $f_R : \mathbb{B}^3 \to \mathbb{B}$ as indicated in the right of the following tables.

x_1 x_2 y_1	$f(x_1, x_2, y_1)$
0 0 0	1
0 0 1	0
0 1 0	0
0 1 1	0
1 0 0	1
1 0 1	1
1 1 0	–
1 1 1	–

x_1 x_2 y_1	$f_R(x_1, x_2, y_1)$
0 0 0	1
0 0 1	0
0 1 0	0
0 1 1	0
1 0 0	1
1 0 1	1
1 1 0	0
1 1 1	0

Based on a fixed variable ordering, for example $x_1 < x_2 < y_1$, the ROBDD of f_R together with a description of the sets X and Y represents the relation R completely. Omitting this additional information, we get the following ROBDD:

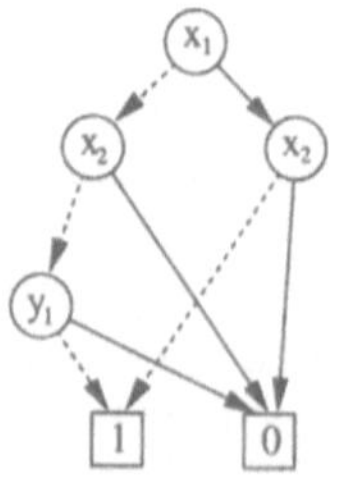

Generally, using binary encodings $c_X : X \to \mathbb{B}^m$ and $c_Y : Y \to \mathbb{B}^n$ (where m equals $\lceil \log |X| \rceil$ and n equals $\lceil \log |Y| \rceil$) a relation $R : X \leftrightarrow Y$ is implemented by the two sizes $|X|$ and $|Y|$ and the ROBDD of the totalized Boolean function $f_R : \mathbb{B}^{m+n} \to \mathbb{B}$, where $f_R(x_1, \ldots, x_m, y_1, \ldots, y_n) = 1$ if and only if the decodings

$c_X^{-1}(x_1, \ldots, x_m)$ and $c_Y^{-1}(y_1, \ldots, y_n)$ are related via R and the variable ordering is $x_1 < \ldots < x_m < y_1 < \ldots < y_n$. To get an impression of possible advantages of this ROBDD representation of relations, one has to consider at least three important aspects.

- First one has to compare the sizes of ROBDDs and of Boolean matrices representing relations typically used in practical applications of the relational framework.
- In a second step, the simplicity of implementing the relational operations based on ROBDDs has to be investigated.
- Finally, the efficiency of these implementations should be analysed by a number of tests and compared with other implementations.

It should be emphasized that the original RELVIEW system with the array-based implementation of relations in combination with some features of the CUDD package provided excellent support for the investigation of all of these aspects.

The first aspect – the size of ROBDDs in comparison with Boolean array implementations – is demonstrated by the following examples. They deal with relations which are helpful if one attacks a computationally hard problem by first generating all candidates and then testing each to see whether it is a solution.

The picture below shows the RELVIEW representation of the membership relation $\in : X \leftrightarrow 2^X$ for the set $X = \{a, b, c, d\}$ as a Boolean 4×16 matrix:

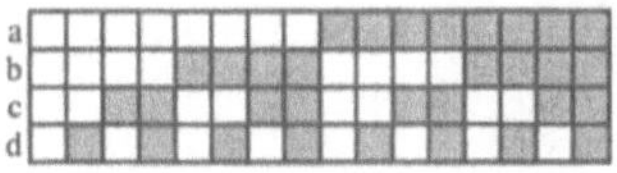

After encoding the elements of X and 2^X we get the following ROBDD; in it the variables x_1 and x_2 encode the 4 elements of the domain X of $\in$ and the variables y_a, y_b, y_c and y_d encode the 16 elements of the range 2^X of $\in$.

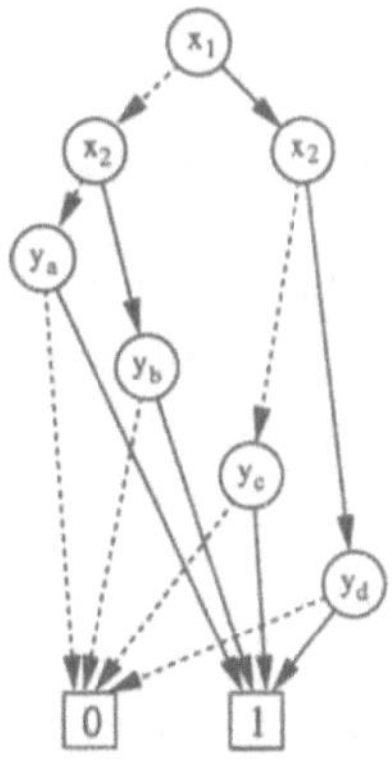

At this point it is essential to note that in the general case each variable used to encode the range of a membership relation corresponds to exactly one row of its Boolean matrix representation as indicated in the above example by their indices. As a consequence, the number of ROBDD nodes of an arbitrary finite membership relation $\in : X \leftrightarrow 2^X$ is linear in the number of elements of its domain X – in contrast to the exponential size of its Boolean matrix representation. In the previously mentioned case of a membership relation on a set with 25 elements we obtain an ROBDD with about 1 KByte only.

As we will see in Section 5, membership relations can help to generate all subsets of a given universum X which are candidates for the solution to a present problem. If this problem is an optimization problem and the task is to compute a subset of X with minimum or maximum size, then a relation $C : 2^X \leftrightarrow 2^X$ may be very helpful which compares the sizes of sets from 2^X, i.e., fulfils C_{UV} if and only if $|U| \leq |V|$ for all U, V from 2^X. The left of the next two pictures shows this *size comparison relation* C for the universum $X = \{a, b, c, d\}$ used already above as a Boolean 16×16 matrix (produced by RelView) and the right-hand picture shows the ROBDD implementation of C. Here the ordering of the sets coincides with the ordering used in the membership example above and is determined by the columns of the Boolean matrix representation of the membership relation $\in : X \leftrightarrow 2^X$. For instance, the first column of the matrix for C corresponds to the empty set, column 6 corresponds to the set $\{b, d\}$, and the last column corresponds to the entire set X.

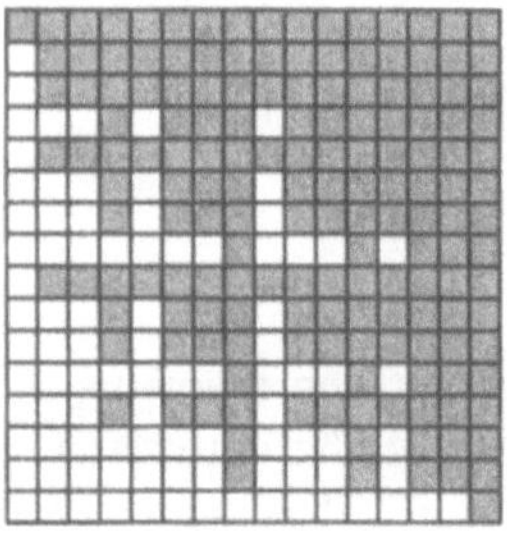

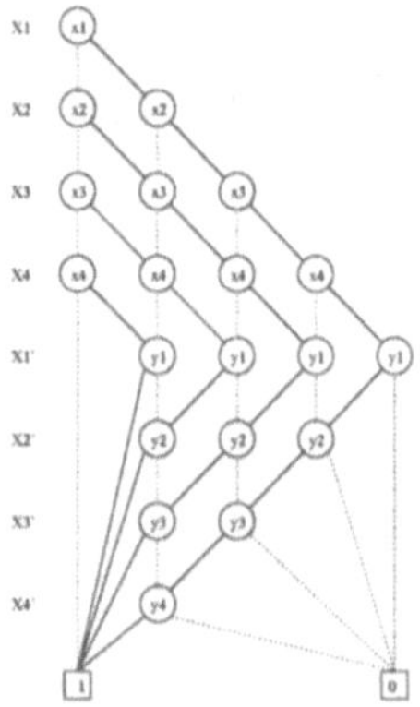

It is noteworthy, that in the general case the size of the Boolean matrix for a size comparison relation $C : 2^X \leftrightarrow 2^X$ is exponential in $|X|$, but the number of ROBDD nodes is polynomial in the size of X.

In the following, we present algorithms which compute the ROBDDs representing the membership relation $\in : X \leftrightarrow 2^X$ and the size comparison relation $C : 2^X \leftrightarrow 2^X$, respectively. The input n of both algorithms is the size of the universum X.

We start with the algorithm MemberBdd for computing the ROBDD of the membership relation. In its subsequent description, the call IthVar(x) returns a 3-node ROBDD representing the Boolean variable x and the call Ite(p,t_1,t_2) implements the conditional expression **if** p **then** t_1 **else** t_2.

```
MemberBdd(n)
  index := ⌈log n⌉;
  for i := 0 to n − 1 do
```

```
3      res[i] := ITHVAR(index + i) od;
4   top_nodes := n;
5   while index > 0 do
6      node := ITHVAR(index − 1);
7      i := 0; j := 0;
8      while i < top_nodes − 1 do
9         res[j] := ITE(node, res[i + 1], res[i]);
10        i := i + 2; j := j + 1 od;
11     if ODD(top_nodes) then res[j] := ITE(node, ZERO, res[i]);
12                            j := j + 1; top_nodes := j fi;
13     index := index − 1 od;
14  return res
```

The algorithm MEMBERBDD creates a ROBDD, which can be subdivided into two parts as seen in the picture above for $X = \{a, b, c, d\}$. By the **for**-loop (l. 2-3) all nodes of the lower part are created. These nodes are used in the **while**-loop for building the upper part in a bottom-up manner. In every run through this loop, *top_nodes* contains the number of nodes which have to be put together, and *index* is the index of the level of nodes which are generated. During the inner **while**-loop (l. 8-10) in each case a new node is produced, the children of which are two nodes from the level below. If *top_nodes* is odd, then the last node from the level below gets combined with the Boolean constant ZERO (l. 11-12).

The ROBDD for the size comparison relation $C : 2^X \leftrightarrow 2^X$ with $n = |X|$ is also built from the bottom to the top; here is the corresponding algorithm:

```
SIZECOMPBDD(n)
1   index := 2 * n − 1;
2   high[0] := ITHVAR(index);
3   for index := 2 * n − 2 downto n do
4      node := ITHVAR(index);
5      low[0] := ITE(node, high[0], ZERO);
6      for i = 1 to 2 * n − index − 2 do
7         low[i] := ITE(node, high[i], high[i − 1]) od;
8      i := 2 * n − index − 1;
9      low[i] := ITE(node, ONE, high[i − 1]);
10     high := low od;
11  for index = n − 1 downto 0 do
12     node := ITHVAR(index);
13     for i = 0 to index do
14        if i ≠ n − 1 then low[i] := ITE(node, high[i], high[i + 1])
15                     else low[i] := ITE(node, high[i], ONE) fi od;
16     high := low od;
17  return low[0]
```

As this program shows, in a first step the single node in the last row of variables is produced (l. 2); in the picture above this node carries the label x'_d. The variable *index* always contains the index of the nodes which are generated in the following steps. In lines 3-10 the lower triangle of the ROBDD is created. Line 5 shows the construction of the node on the very left in each row and in line 9 the node on the row's very right is built. The code between these lines shows the creation of the nodes in between. The upper triangle of the ROBDD is constructed in lines 11-16. Only the node on the very left in the lowest row needs a special treatment (l. 14-15), all other nodes are simply built using the ITE operator.

Having described the construction of two important relational constants represented by ROBDDs, now we deal with the implementation of relational operations. We consider again two examples.

Often it is possible to lead the ROBDD-based implementation of a relational operation back to the operation's component-wise definition. An example of this is the composition of two relations. Let X, Y and Z be finite sets with $|X| = n$, $|Y| = m$ and $|Z| = k$, and let $R : X \leftrightarrow Y$ and $S : Y \leftrightarrow Z$ be relations. Then, the composition $RS : X \leftrightarrow Z$ of R and S is component-wise defined by

$$(RS)_{xz} \iff \exists y \; (R_{xy} \wedge S_{yz})$$

for all x from X and z from Z. In the following, we use the three abbreviations $l_n = \lceil \log n \rceil$, $l_m = \lceil \log m \rceil$, and $l_k = \lceil \log k \rceil$. Using, furthermore, binary encodings $c_X : X \to \mathbb{B}^{l_n}$, $c_Y : Y \to \mathbb{B}^{l_m}$ and $c_Z : Z \to \mathbb{B}^{l_k}$ for X, Y, and Z, we obtain, for the representation of R respectively S, the Boolean functions

$$f_R : \mathbb{B}^{l_n+l_m} \to \mathbb{B} \quad \text{and} \quad f_S : \mathbb{B}^{l_m+l_k} \to \mathbb{B}$$

such that $f_R(r_1, \ldots, r_{l_n}, r_{l_n+1}, \ldots, r_{l_n+l_m}) = 1$ if and only if $c_X^{-1}(r_1, \ldots, r_{l_n})$ and $c_Y^{-1}(r_{l_n+1}, \ldots, r_{l_n+l_m})$ are related via R and $f_S(s_1, \ldots, s_{l_m}, s_{l_m+1}, \ldots, s_{l_m+l_k}) = 1$ if and only if $c_Y^{-1}(s_1, \ldots, s_{l_m})$ and $c_Z^{-1}(s_{l_m+1}, \ldots, s_{l_m+l_k})$ are related via S. If we can identify the variables $r_{l_n+1}, \ldots, r_{l_n+l_m}$ with $s_1, \ldots, s_{l_m}$, hence, the Boolean function for the composition of RS could be computed by

$$f_{RS} = \exists s_1 \ldots \exists s_{l_m} (f_R \wedge f_S)$$

using conjunction and repeated existential quantification of Boolean functions, where the latter operation is defined by $(\exists x_i \, g)(x_1, \ldots, x_n) = 1$ if and only if $g(x_1, \ldots, x_{i-1}, 1, x_{i+1}, \ldots, x_n) = 1$ or $g(x_1, \ldots, x_{i-1}, 0, x_{i+1}, \ldots, x_n) = 1$.

The operations $\wedge$ and $\exists$ are available in all ROBDD packages we know, so we only have to take care of the identity of the variables. Now, one could assume that for the representation of the relations the variables could be chosen in such a way, that the indices of these shared variables are always convenient for composition. Unfortunately, this is not always possible. For example, for a relation R of specific type $[X \leftrightarrow X]$, a so-called *homogeneous* relation, we want to be able to calculate RR. Here we have the simple case that the variables for the representation of the range of R in no way could be the same as those used for its domain, except for the special case that X contains only one element.

Therefore we have decided to encode relations by starting with index 0 and number the following variables consecutively. This leads to a fixed sequence $\mathbf{x}$ of variables, $x_0, x_1, x_2, \ldots$ say, and allows the handling of any arbitrary relation over finite sets using a certain prefix of $\mathbf{x}$. Furthermore, the construction of the relational constants described above concurs with this fixed sequence $\mathbf{x}$ of variables. For the relations R and S mentioned above – respectively the Boolean functions f_R and f_S representing them – this approach means, that variables $r_1, \dots, r_{l_n}$ are equal to $x_0, \dots, x_{l_n-1}$, variables $r_{l_n+1}, \dots, r_{l_n+l_m}$ are identical with $x_{l_n}, \dots, x_{l_n+l_m-1}$, variables $s_1, \dots, s_{l_m}$ are the same as $x_0, \dots, x_{l_m-1}$, and variables $s_{l_m+1}, \dots, s_{l_m+l_k}$ coincide with $x_{l_m}, \dots, x_{l_m+l_k-1}$.

Unfortunately, this fixed encoding causes a problem for the composition RS, because now we have to ensure that the variables representing the range of R are the same as those used for the representation of the domain of S. Let N denote $l_n + l_m + l_k$. Then, this problem can be solved by introducing new Boolean functions f'_R and f'_S, which are defined as follows:

$$f'_R(x_0, \dots, x_{l_n+l_m-1}, \dots, x_{N-1}) = 1 \iff f_R(x_0, \dots, x_{l_n+l_m-1}) = 1$$
$$f'_S(x_{l_m+l_k}, \dots, x_{N-1}, x_0, \dots, x_{l_m+l_k-1}) = 1 \iff f_S(x_0, \dots, x_{l_m+l_k-1}) = 1\,.$$

The computation of these Boolean functions is mainly a renumbering of the indices of the variables. But, because ROBDDs are strongly connected structures with shared nodes, we cannot simply change the variable's indices directly within the ROBDD nodes. Indeed we have to build up a new ROBDD for the Boolean functions f'_R and f'_S.

If we take a look onto the function f'_R, it is obviously, that it is independent of the last l_k parameters. Because such variables do not require any nodes in an ROBDD, it is clear that the ROBDDs for f_R and f'_R are equal.

This is not the case with f'_S: This Boolean function also is independent of some parameters (the first l_n ones), but the needed variables of f'_S have got other indices as the ones of f_S. In fact, every index is incremented by exactly l_n. For the computation of f'_S we have implemented a new ROBDD algorithm, which we called SHIFT. Here is its formulation in the program notation used so far:

```
SHIFT(F, border, n, m)
if F = ONE or F = ZERO
   then res := F
   else index := F → index;
        E := ELSEBDD(F);
        T := THENBDD(F);
        SE := SHIFT(E, border, n, m);
        ST := SHIFT(T, border, n, m);
        if index > border
           then node := ITHVAR(index+n)
           else  node := ITHVAR(index+m) fi;
        res := ITE(node, ST, SE) fi;
return res
```

The first argument F of this algorithm denotes the ROBDD, whose variable indices we want to renumber, the second argument *border* is an index which divides the indices of the variables of F into two sets, the next argument n is the increasement value for the first set of variables, whose indices are less than *border*, and the last argument m is the increasement value for the second set of variables. If F is a constant, then F has got no index, which can be shifted. Otherwise the index of the root of F is stored into *index*. The call ElseBdd(F) returns the ROBDD for $F \wedge \overline{index}$ and the call ThenBdd(F) returns $F \wedge index$. We store these results into new variables T and E and start this algorithm recursively on these ROBDDs. In the next step (l. 8-11) we test whether *index* is situated in the first set of variables or in the second, and construct the result by using the Ite operator.

If one stores additionally already computed results in a table, then it is obviously possible to call algorithm Shift exactly once for every node of F.

After these preparations, we are able to compute the ROBDD F'_S representing f'_S by a call Shift(F_S, l_m, l_n, l_n), where F_S is the ROBDD of f_S. We use l_m as the border, because all variables, whose index is less than l_m, belong to the encoding of the domain of S, and the rest are used for the range of S. This call creates a new ROBDD, which has the same structure as F_S, but all indices are incremented by l_n. As one can see further, f'_R and f'_S have suitable variable indices for the conjunction. Hence, we are allowed to define

$$f'_{RS} = \exists x_{l_n} \dots \exists x_{l_n+l_m-1}(f'_R \wedge f'_S)$$

and, based on this, finally the Boolean function representation f_{RS} for the composition RS by

$$f_{RS}(x_0, \dots, x_{l_n-1}, x_{l_n+l_m}, \dots, x_{N-1}) = 1 \iff f'_{RS}(x_0, \dots, x_{N-1}) = 1\,.$$

It is obvious, that the Boolean function f'_{RS} is independent of the variables in the range from x_{l_n} to $x_{l_n+l_m-1}$. Because of the convention, that all variables are numbered consecutively as $x_0, x_1, x_2, \dots$, we must shift the indices of the variables representing the range of the result into the correct values. This can easily be done by storing the result of the call Shift$(F'_{RS}, l_n, 0, -l_m)$ into F_{RS} with F'_{RS} and F_{RS} representing f'_{RS} and f_{RS}. Altogether, this leads to the following algorithm Compose for the composition of two relations R and S which are implemented as ROBDDs F_R and F_S:

```
Compose(F_R, F_S, l_n, l_m)
  res := Shift(F_S, l_n, l_m, l_m);
  res := And(F_R, res);
  res := Exist(⟨l_n, ..., l_n + l_m − 1⟩, res);
  res := Shift(res, l_n, 0, −l_m);
  return res
```

The above algorithm Shift is also useful for the computation of the transposition of a relation $R : X \leftrightarrow Y$ which is implemented as a ROBDD. Assume $n = |X|$ and $m = |Y|$. We start with the component-wise definition

$$R^{\mathsf{T}}_{xy} \iff R_{yx}$$

for all x from X and y from Y. If we use the abbreviations $l_n = \lceil \log n \rceil$ and $l_m = \lceil \log m \rceil$, then it will immediately lead to the equivalence

$$f_{R^{\mathsf{T}}}(x_{l_n}, \dots, x_{l_n+l_m-1}, x_0, \dots, x_{l_n-1}) = 1 \iff f_R(x_0, \dots, x_{l_n+l_m-1}) = 1\,.$$

For the ROBDD representation of the Boolean function $f_{R^{\mathsf{T}}}$ this means that we only have to exchange the variables used for the encoding of the domain of R with those used for the representation of for its range. This can easily be obtained by a call SHIFT($F_R, l_n, l_m, -l_n$). Within the RELVIEW system, however, we use the following refined version:

```
TRANSPOSE(F_R, l_n, l_m)
  if l_n ≠ 0 and l_m ≠ 0
     then res := SHIFT(F_R, l_n, l_m, -l_n)
     else res := F_R fi;
  return res
```

It takes advantage of the fact that ROBDD representations of relations with domain or range **1**, especially vectors $v : X \leftrightarrow \mathbf{1}$ which frequently occur in relation-algebraic specifications and programs, do not use variables for the singleton set **1**, so that a shifting of variables is not necessary in these cases.

Besides the examples presented in this section, we have developed ROBDD-based algorithms for all the constants and basic operations on relations mentioned in Section 2 and many further constants, operations, and tests typically used in relation-algebraic specifications and programs. Examples of the latter are residuals, symmetric quotients, and constants and operations appearing in relational descriptions of domains (like the two projections $\pi_1 : X \times Y \leftrightarrow X$ and $\pi_2 : X \times Y \leftrightarrow Y$ in the case of direct products), which have already been available in the former, array-based version of RELVIEW (see [1] for details). But we have also extended the system by some new operations like the filtering of subsets with a certain size from a vector $v : 2^X \leftrightarrow \mathbf{1}$, the efficiency of which mainly bases on the new ROBDD implementation. Finally, we have developed an algorithm for the random generation of relations which are implemented as ROBDDs. Among other things, randomly generated relations are very profitable in specification and program testing.

Detailed descriptions of all these algorithms, including the correctness proofs, are presented resp. will be presented in the Ph.D theses [8] and [10].

5 Some Applications

To analyze the efficiency of the ROBDD implementation of relations, we have carried out a lot of experiments for different kinds of problems and with different kinds of randomly generated relations. We have tested relation-algebraic specifications and programs using the new and the former version of the RELVIEW system and have compared both performances. The new ROBDD-implementation proved to be by far superior to the older one, especially if the generate-and-test approach is used to solve computationally hard problems. This is mainly due to the fact that for many such problems the very efficient ROBDD-representation of certain important relations on powersets (membership, inclusion, size comparison, filtering) allows to handle the combinatorial explosion even in the case of medium-sized input relations. In the following, we present three examples from graph-theory and an application for 0/1 matrices. Further examples and the evaluation of the corresponding experiments can be found in [8].

First, we assume a *directed graph* $g = (X, R)$ with the *adjacency relation* $R : X \leftrightarrow X$ on the node set X containing the arcs of g. A set A of nodes is said to be *absorbant* in g if from every node outside of A there is at least one arc leading into A , i.e., if the formula

$$\forall x\ (x \notin A \to \exists y\ (y \in A \land R_{xy}))$$

holds. Furthermore, a set S of nodes is called *stable* in g if no two nodes of S are related via R. This situation is characterized by the formula

$$\forall x\ (x \in S \to \forall y\ (y \in S \to \overline{R}_{xy}).$$

Finally, a *kernel* of g is a set of nodes which is at the same time absorbant and stable. It is an easy exercise to express absorption and stability in terms of relational operations, the universal vector $\mathsf{L} : X \leftrightarrow \mathbf{1}$, and the membership relation $\in : X \leftrightarrow 2^X$; see e.g., [1]. This leads to the relation-algebraic specifications

$$\overline{\mathsf{L}\overline{\in \cup R \in}}^{\mathsf{T}} : 2^X \leftrightarrow \mathbf{1} \quad \text{and} \quad \overline{\mathsf{L}(\in \cap R \in)}^{\mathsf{T}} : 2^X \leftrightarrow \mathbf{1}$$

of the vectors describing the set of absorbant sets respectively the set of stable sets of g. As a consequence, a relation-algebraic specification of the vector $\mathit{kernel}(R) : 2^X \leftrightarrow \mathbf{1}$ describing the set of kernels of g is

$$\mathit{kernel}(R) = \overline{\mathsf{L}\overline{\in \cup R \in}}^{\mathsf{T}} \cap \overline{\mathsf{L}(\in \cap R \in)}^{\mathsf{T}}. \tag{2}$$

Now, assume $g = (X, E)$ to be an *undirected graph*, i.e., the edge set E to consist of sets $\{x, y\}$ of distinct nodes x and y. A set C of nodes is a *clique* of g if each set $\{x, y\}$ of distinct nodes x and y from C forms an edge. If we define the adjacency relation $R : X \leftrightarrow X$ of g by R_{xy} if and only if $\{x, y\}$ is an edge, then C is a clique of g if and only if it is stable in the so-called directed variant $g' = (X, \overline{R} \cap \overline{\mathsf{I}})$ of the complement graph of g. The relation $\overline{R} \cap \overline{\mathsf{I}}$ of g' is symmetric and irreflexive. Hence, the inclusion-maximal stable sets of g' are precisely its kernels; see e.g., [11]. Using (2), we consequently obtain the vector $\mathit{maximalclique}(R) : 2^X \leftrightarrow \mathbf{1}$ describing the inclusion-maximal cliques of g as

$$\mathit{maximalclique}(R) = \mathit{kernel}(\overline{R} \cap \overline{\mathsf{I}}). \tag{3}$$

In our third example we assume $g = (X, E)$ again to be an undirected graph. Then we can associate with g its *incidence relation* $R : E \leftrightarrow X$, defined by R_{ex} if and only if node x is contained in edge e. The relation $RR^{\mathsf{T}} \cap \overline{\mathsf{I}} : E \leftrightarrow E$ is called the *edge adjacency relation* of g and coincides with the adjacency relation of the *line graph* $L(g)$ of g, in which the edges of g are the nodes and two nodes e, f of $L(g)$ form an edge in $L(g)$ if and only if $e \neq f$ and $e \cap f \neq \emptyset$. A *matching* of g is a set M of edges such that no two distinct edges have a node in common. Using the directed variant $g' = (E, RR^{\mathsf{T}} \cap \overline{\mathsf{I}})$ of the line graph $L(g)$, it is easy to enumerate all inclusion-maximal and maximum-size matchings of g. From the definition we immediately obtain that M is a matching of g if and only if it is a stable set in g'. Since $RR^{\mathsf{T}} \cap \bar{\mathsf{I}}$ is again a symmetric and irreflexive relation, we can use specification (2) to compute the vector $maximalmatch(R) : 2^E \leftrightarrow \mathbf{1}$ describing the set of inclusion-maximal matchings of g as

$$maximalmatch(R) = kernel(RR^{\mathsf{T}} \cap \bar{\mathsf{I}})\,. \tag{4}$$

Applying the relation-algebraic specification $greatest(S, v) = v \cap \overline{\overline{S^{\mathsf{T}}}\, v}$ of the greatest elements of the set described by the vector v wrt. the quasi order S (see [11] for example) in combination with specification (4), we get

$$maximummatch(R) = greatest(C, maximalmatch(R)) \tag{5}$$

for the vector $maximummatch(R) : 2^E \leftrightarrow \mathbf{1}$ describing the set of maximum-size matchings of g. In this relation-algebraic specification $C : 2^E \leftrightarrow 2^E$ is the size comparison relation as introduced in Section 4.

As already mentioned, RELVIEW allows to generate random relations, where it is additionally possible to specify the percentage of their entries. Using this feature, a lot of experiments have been carried out with random graphs of various sizes and with various percentages of arcs respectively edges on a Sun Ultra 80 workstation running Solaris 7 at 450 MHz and with 4 Gbyte of main memory. The following table shows some experimental results of the enumeration of cliques for randomly generated symmetric and irreflexive adjacency relations on n elements, which contain approx. 10% respectively 20% and 30% of all entries. These adjacency relations correspond to undirected graphs with n nodes and the same percentages of edges.

nodes	50	100	150	200	250	300	350	400	450
incl-m. 10%	135	751	2253	4833	8966	15701	25196	38858	56898
m.-size 10%	9/4	5/5	36/5	1/6	7/6	14/6	48/6	2/7	124/6
time (sec.)	0.02	1.2	8	31	87	233	628	2437	6145
incl-m. 20%	150	791	2243	4820	8884	15631	25255	38985	–
m.-size 20%	10/4	2/5	26/5	2/6	5/6	7/6	2/7	107/6	–
time (sec.)	0.15	1.4	9	32	90	248	622	2361	–
incl-m. 30%	233	1665	5581	14392	32181	62311	–	–	–
m.-size 30%	10/5	5/6	1/7	246/6	28/7	89/7	182/7	–	–
time (sec.)	0.28	4.1	26	171	784	3056	36282	–	–

Lines 2, 5, 8 show the number of inclusion-maximal cliques, lines 3, 6, 9 the number of maximum-size cliques and their sizes (which can be obtained from the column-wise representation of sets described in Section 2), and lines 4, 7, 10 the time to compute the latter results.

The former array-based version of RELVIEW was able to handle only instances with at most 25 nodes, because for $|X| = 25$ the computation of the membership relation $\in : X \leftrightarrow 2^X$ required more than 100 MByte of memory space. Therefore, it is not possible to compare the results of the above table with the former version. Of course, we have compared both versions also in the case of computations not using membership relations and similar constructions. Again the new version proved to be superior. See [8] fore details.

We close this section with two concrete applications of the RelView system. First, we demonstrate how to compute the permanent

$$\mathsf{per}(A) = \sum_{\pi} \prod_{i=1}^{n} a_{i,\pi(i)}$$

of a $n \times n$ 0/1-matrix, where the sum is over all permutations π on the set $\{1, \ldots, n\}$. The computation of permanents bases on the fact that $\mathsf{per}(A)$ equals the number of perfect matchings of the bipartite undirected graph $g_A = (X, E)$, where $X = \{1, \ldots, 2 * n\}$ and $\{i, j\} \in E$ if and only if $1 \leq i \leq n < j \leq 2 * n$ and $a_{i,j-n} = 1$. (A matching M is said to be *perfect* if every node is incident with at least one edge from M.) Now, we interpret the following 15×15 Boolean RelView-matrix as 0/1-matrix A:

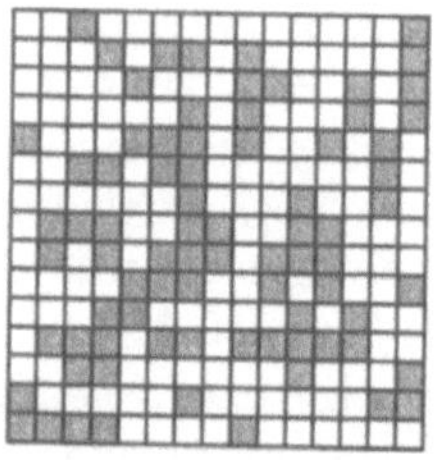

Then the undirected bipartite graph $g_A = (X, E)$ has 30 nodes and 79 edges. On the screen of RelView this graph is depicted as in the following picture:

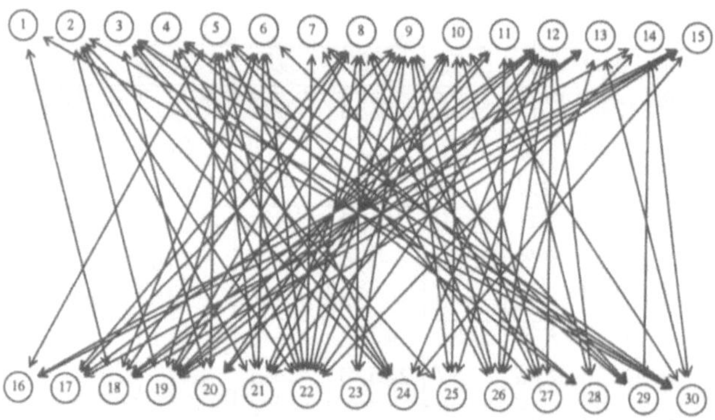

The line graph $L(g_A)$ of g_A has even 79 nodes and 396 edges. This graph is too large to be reproduced here. Using the relation-algebraic specification (5), on the Sun workstation mentioned above RelView needs 541 seconds to compute a vector $v : 2^E \leftrightarrow \mathbf{1}$ which describes the 32844 greatest stable sets of $L(g_A)$, i.e., the 32844 maximum-size matchings of g_A. To test that one of these matchings consists of 15 edges can be done very fast. One only has to select a so-called *point* $p : 2^E \leftrightarrow \mathbf{1}$ contained in v (see e.g., [2] for details), which describes a single maximum-size matching of g_A as an element of the powerset 2^E, and then to transform it into the vector $\in p : E \leftrightarrow \mathbf{1}$ which describes the same matching.

Hence, we get $\mathsf{per}(A) = 32844$. Here it should be mentioned that even 89 seconds suffice to compute this result if one directly works with the relation-algebraic specification of stable sets and the filter operation mentioned at the end of Section 4. More details on the relation-algebraic computation of permanents can be found in [8].

As the second application we consider two queens problems on a chess board. Consider the following RELVIEW picture:

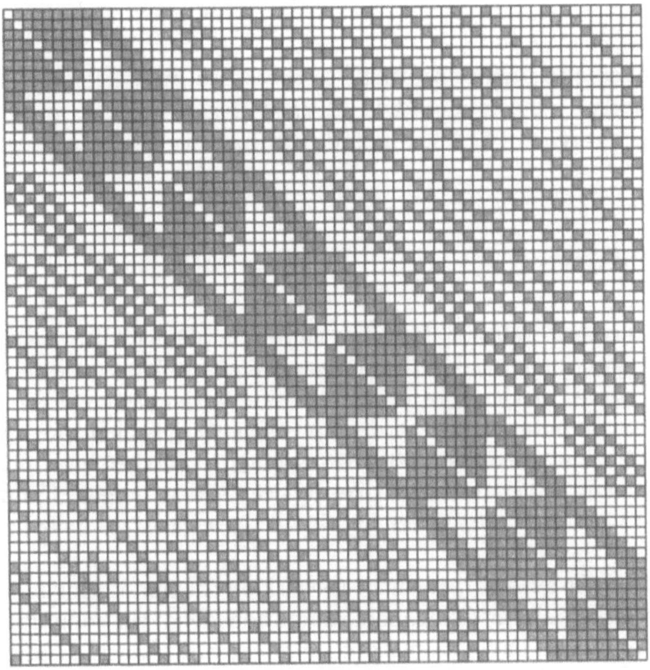

If we describe the possible moves of a queen on a chess board as a relation $R : X \leftrightarrow X$ on a set X representing the 64 squares of the chess board, then this Boolean matrix exactly represents the relation R. Using the RELVIEW system, it is very easy to check that the directed graph $g = (X, R)$ has exactly 10118 kernels, of which 92 have the maximal size 8. The latter kernels correspond exactly to the 92 solutions of the well-known 8-queens problem. I.e., each kernel represents a set of 8 queens which are placed on the chess board in such a way that no two of them are enemies.

Kernels of size 8 are greatest stable sets of the directed graph $g = (X, R)$. We have also considered the dual problem dealing with smallest absorbant sets of g. It corresponds to the following question: How can the entire chess board be checked by as few queens as possible and how many solutions exist? It is well-known that 5 queens suffice; therefore the problem is also called the 5-queens problem. RELVIEW needs some seconds to compute the 4860 solutions to this problem. One of them is shown in the following picture:

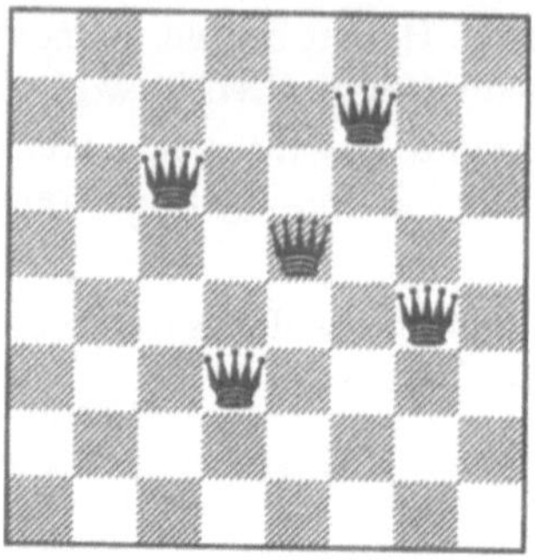

In [8] one finds also further solutions to these two problems on chess boards of other sizes than 8 × 8 computed by RelView. For example, 5 queens also suffice to check the entire 11 × 11 chess board. But for this board size there exist only 2 solutions.

6 Conclusion

We have shown how relational algebra can be implemented by using ROBDDs. This implementation is used in the new version of RelView and we have demonstrated the power of this tool in combination with relational formulations by solving some computationally hard graph-theoretic problems for medium-sized graphs. We have applied the RelView system to further hard problems from graph-theory and other domains (like lattice theory, games, Condition/Event Petri nets, automata theory), too, and in all these cases we have had very positive experience.

Concerning the system, presently we work on the fine-tuning of the ROBDD-implementation of relations since some operations (e.g., the computation of $\mathsf{inj}(v)$ and some operations of relational domains) are not always as efficient as they should be. We also plan to extend the language of the system, for instance, by an assert-command to allow flexible invariance tests. Another future extension of the system concerns the improvement of RelView's debugging and visualization features. But, of course, we are also interested in further applications. Among other things, our current investigations concern the use of RelView in software architecture and re-engineering.

References

1. Behnke R., Berghammer R., Schneider P.: Machine support of relational computations: The Kiel RelView system. Bericht Nr. 9711, Inst. für Inf. und Prak. Math., Univ. Kiel (1997)
2. Behnke R., Berghammer R., Hoffmann T., Leoniuk B., Schneider P.: Applications of the RelView system. In: Berghammer R., Lakhnech Y. (eds.): Tool support for system specification, development and verification. Advances in Computing, Springer, 33-47 (1999)

3. Behnke R., Berghammer R., Meyer E., Schneider P.: RelView – A system for calculating with relations and relational programming. In: Astesiano E. (ed.): Proc. Conf. on Fundamental Approaches to Software Engineering (FASE '98), LNCS 1382, Springer, 318-321 (1998)
4. Berghammer R., Hoffmann T.: Modelling sequences within the RelView system. J. of Univ. Comp. Sci. 7, 107-123 (2001)
5. Berghammer R., Karger B. von, Ulke C.: Relation-algebraic analysis of Petri nets with RelView. In: Margaria T., Steffen B. (eds.): Proc. Workshop on Tools and Applications for the Construction and Analysis of Systems (TACAS '96), LNCS 1055, Springer, 49-69 (1996)
6. Brink C., Kahl W., Schmidt G. (eds.): Relational methods in computer science. Advances in Comp. Sci., Springer (1997)
7. Bryant R.E.: Symbolic Boolean manipulation with ordered binary decision diagrams. ACM Comp. Surv. 24, 293-318 (1992)
8. Leoniuk B.: ROBDD-based implementation of relational algebra with applications (in German). Ph.D. thesis, Inst. für Inf. und Prak. Math., Univ. Kiel (2001)
9. Liaw H.-T., Lin C.-S.: On the OBDD representation of general Boolean functions. IEEE Transactions on Comp. 41, 661-664 (1992)
10. Milanese U.: On the implementation of a ROBDD-based tool for the manipulation and visualization of relations (in German). Ph.D. thesis, Inst. für Inf. und Prak. Math., Univ. Kiel (to appear)
11. Schmidt G., Ströhlein T.: Relations and graphs. Discrete Mathematics for Computer Scientists, EATCS Monographs on Theor. Comp. Sci., Springer (1993)
12. Somenzi, F.: CUDD: CU decision diagram package; Release 2.3.0. `http://vlsi.colorado.edu/~fabio/CUDD/cuddIntro.html` (1998)

Calculating a Relational Program for Transitive Reductions of Strongly Connected Graphs

Rudolf Berghammer and Thorsten Hoffmann

Institut für Informatik und Praktische Mathematik
Christian-Albrechts-Universität Kiel
Olshausenstraße 40, D-24098 Kiel

Abstract. Based on a generic program for computing minimal sets, we present a simple relational program for computing transitive reductions of strongly connected directed graphs. It uses a precomputation phase and can easily be implemented in quadratic running time. The presentation is done as an exercise in generic programming and for combining the Dijkstra-Gries method and relational algebra to derive graph algorithms.

1 Introduction

Since many years it is widely accepted that the traditional ad-hoc approach to program construction does not lead to reliable software and, therefore, a more systematic and formal approach is required. One such approach is program derivation which means that a program is calculated from a formal specification. In the case of imperative programs, specifications consist of preconditions and postconditions and program derivations are mainly based on loop invariants. The latter are usually hypothesized as modifications of postconditions and programs are constructed in such a way that invariants are established and maintained. Based on work of Floyd and Hoare, the use of such techniques started with Dijkstra; see his textbook [6]. Later, especially Gries elaborated Dijkstra's ideas and formulated the fundamental principle that "a program and its correctness proof should be developed hand-in-hand with the proof usually leading the way" (cf. [7], p. 164). Therefore, nowadays one often speaks of the Dijkstra-Gries program development method.

To support formal program derivation, adequate algebraic frameworks are very helpful. For problems on graphs – which constitute a very important domain in algorithmics, see [1,5] just to cite two well-known textbooks on this topic – relational algebra plays a prominent role. Concerning the modeling side, this is due to the fact that relations and graphs are essentially the same and there are many simple and elegant ways to describe fundamental graph-theoretic objects and properties with relation-algebraic means. With regard to proofs, the use of relational algebra has the advantage to reduce the danger of doing wrong proof steps drastically, to clarify the proof structure frequently, and to open the possibility for mechanical support. Finally, the relation-algebraic framework also supports prototyping and validation tasks in a significant manner.

In this paper we want to illustrate the formal derivation of a relational program by combining ideas from generic programming, the Dijkstra-Gries program

H. de Swart (Ed.): RelMiCS 2001, LNCS 2561, pp. 258–275, 2002.

development method, graph searching techniques, and relational algebra. We consider the computation of a transitive reduction of a strongly connected directed graph, a problem which is, for instance, discussed in [11,14]. In the second paper a quite complicated linear algorithm is presented, which, however, in the meantime has shown to be not correct; see [9]. The relational program we will develop is very simple. It can be performed with the RELVIEW tool [2] almost as it stands and its implementation in quadratic running time in a conventional programming language is trivial.

The rest of the paper is organized as follows: Section 2 contains some basic notions of relational algebra and graph theory and shows how to model certain graph-theoretic objects resp. properties with relation-algebraic means. Similar to the approach of Ravelo in [12], then in Section 3 we derive a generic imperative program that computes for a finite set a minimal subset satisfying a certain predicate. We instantiate this program in Section 4 to compute a transitive reduction of a strongly connected directed graph. In doing so, we use a combination of relational depth-first and breadth-first graph searching programs in a precomputation phase to improve the running time of the instantiation as well as the quality of the computed result in view of the number of its arcs. Section 5 contains a short conclusion.

2 Relations and Graphs

Given sets $X, Y \neq \emptyset$, the set of all relations with domain X and range Y is denoted by $[X \leftrightarrow Y]$ and we write $R : X \leftrightarrow Y$ instead of $R \in [X \leftrightarrow Y]$. We assume the reader is familiar with the basic operations on relations, viz. R^{T} (transposition), $\overline{R}$ (negation), $R \cup S$ (union), $R \cap S$ (intersection), RS (composition), $R \subseteq S$ (inclusion, subrelation test), and the special relations O (empty relation), L (universal relation), and I (identity relation). The set-theoretic operations $\overline{}$, $\cup$, $\cap$, $\subseteq$ and the constants O, L form a complete Boolean lattice. Further well-known laws of relations are e.g., $R^{\mathsf{T}^{\mathsf{T}}} = R$, $Q(R \cap S) \subseteq QR \cap QS$, and $(RS)^{\mathsf{T}} = S^{\mathsf{T}} R^{\mathsf{T}}$. The theoretical framework for such laws holding is that of a relational algebra. As constants and operations of this algebraic structure we have those of the set-theoretic relations; its axioms are those of a complete Boolean lattice for $\overline{}$, $\cup$, $\cap$, $\subseteq$, O, L, those of a monoid for composition and I, the equivalence of $Q^{\mathsf{T}} \overline{S} \subseteq \overline{R}$, $QR \subseteq S$ and $\overline{S} R^{\mathsf{T}} \subseteq \overline{Q}$ (in [13] called *Schröder equivalences*), and the equivalence of $R \neq \mathsf{O}$ and $\mathsf{L}R\mathsf{L} = \mathsf{L}$ (in [13] called *Tarski rule*[1]). In the proofs we shall mention only the latter two axioms and their "non-obvious" consequences. Well-known laws like those presented above or in Sections 2.1 until 2.3 of [13] remain unmentioned.

A relation R, for which domain and range coincide, is called *reflexive* if $\mathsf{I} \subseteq R$ and *transitive* if $RR \subseteq R$. The least transitive relation containing R is its *transitive closure* $R^+ = \bigcup_{i \geq 1} R^i$ and the least reflexive and transitive relation

[1] Since we include the Tarski rule we deal with relational algebras which nowadays normally are refered to as "simple".

containing R is its *reflexive-transitive closure* $R^* = \mathsf{I} \cup R^+ = \bigcup_{i \geq 0} R^i$, where $R^0 = \mathsf{I}$ and $R^{i+1} = RR^i$ for all $i \in \mathbb{N}$. Later on, we will apply the implication

$$S\mathsf{L}S \subseteq S \implies (R \cup S)^* = R^* \cup R^*SR^*, \tag{1}$$

which is an immediate consequence of the well-known star-decomposition rule $(Q \cup R)^* = (Q^*R)^*Q^*$, and the equation $R^*R^* = R^*$. A (now arbitrary) relation R is said to be *univalent* (or a partial function) if $R^\mathsf{T}R \subseteq \mathsf{I}$ and *total* if $R\mathsf{L} = \mathsf{L}$, which again is equivalent to $\mathsf{I} \subseteq RR^\mathsf{T}$. R is called *injective* if R^T is univalent and *surjective* if R^T is total.

Since we want to solve a concrete graph-theoretic problem and use relational algebra only as an adequate framework for this task, we work within its standard model of set-theoretic relations. Now let's make the transition to graphs, where we likewise assume the reader is familiar with the basic notations. Directed graphs and relations are essentially the same since the relation $R : V \leftrightarrow V$ of a *directed graph* $g = (V, R)$ with *vertex set* $V \neq \emptyset$ describes its *arc set*. This implies immediately that each set of arcs of g is a subrelation of R. If we identify a singleton set with its only element, we can also model a single arc of g by a specific subrelation of R. An exact characterization is given below. To reason relation-algebraically also about subsets and elements of the vertex set of g, we have to encode these objects as relations, too. Subsets of V can be modeled by vectors. A *vector* is a relation v satisfying $v = v\mathsf{L}$ (one conventionally uses lower-case letters in this context). This definition implies that for describing vectors the range is irrelevant. Therefore, one usually uses a singleton set $\mathbf{1}$ such that $v : V \leftrightarrow \mathbf{1}$ describes the set $\{x \in V : \langle x, \perp \rangle \in v\}$, where $\perp$ is the element of $\mathbf{1}$. To give an example, if $v : V \leftrightarrow \mathbf{1}$ describes the subset X of V, then $R^\mathsf{T}v : V \leftrightarrow \mathbf{1}$ describes the set of all successors of vertices of X and hence $R^\mathsf{T}\mathsf{L} : V \leftrightarrow \mathbf{1}$ describes the set of all endpoints of arcs of g. A vector is said to be a *point* if it is injective and surjective. For $p : V \leftrightarrow \mathbf{1}$ these properties mean that it describes a singleton set, i.e., a vertex of g if we identify $\{x\}$ with x.

Relational algebra is very helpful for describing many graph-theoretic properties. Let again $g = (V, R)$ be a directed graph. In this paper we shall use the following relation-algebraic characterizations: The graph g is *strongly connected* if and only if $R^* = \mathsf{L}$ and *circuit-free* if and only if $R^+ \subseteq \overline{\mathsf{I}}$. A point $r : V \leftrightarrow \mathbf{1}$ is said to be a *root* of g if and only if $r\mathsf{L} \subseteq R^*$. This property means that from the vertex described by r each vertex of g can be reached via a path. Hence, g is a *(directed) tree* (also: *arborescence*) with root r if and only if $R^+ \subseteq \overline{\mathsf{I}}$, $RR^\mathsf{T} \subseteq \mathsf{I}$, and $r\mathsf{L} \subseteq R^*$. Finally, g is a subgraph of a directed graph $h = (W, S)$ if and only if $V = W$ and $R \subseteq S$.

In the standard model of relational algebra the so-called point axiom of [13] holds. It says that for every $R \neq \mathsf{O}$ there exist points p, q such that $pq^\mathsf{T} \subseteq R$. As a consequence, for each vector $v \neq \mathsf{O}$ there exists a point p fulfilling $p \subseteq v$. The choice of an atom $atom(R)$ (i.e., a relation of the form pq^T) contained in $R \neq \mathsf{O}$ and a point p contained in a vector $v \neq \mathsf{O}$ are fundamental for a relational approach to graph algorithms since they correspond to the choice of an arc from a set of arcs resp. a vertex from a set of vertices. Our demands on $atom(R)$ are:

$$atom(R) \subseteq R \qquad atom(R)\mathsf{L} \text{ and } atom(R)^\mathsf{T}\mathsf{L} \text{ are points.} \tag{2}$$

From the axioms (2) we get the inclusion (a proof can be found in [4])

$$atom(R)\mathsf{L}atom(R) \subseteq atom(R)\,. \tag{3}$$

As relation-algebraic axioms for $point(v)$ we demand the following properties:

$$point(v) \subseteq v \qquad\qquad point(v) \text{ is a point.} \tag{4}$$

Note that $atom : [X \leftrightarrow X] \to [X \leftrightarrow X]$ and $point : [X \leftrightarrow \mathbf{1}] \to [X \leftrightarrow \mathbf{1}]$ are (deterministic, partial) functions in the usual mathematical sense. Each call yields the same object so that e.g., $atom(R) = atom(R)$ holds. Of course, the above axiomatizations (2) and (4) allow different realizations. For instance, the specific implementation of *atom* in RelView uses that the system deals only with finite sets which are totally ordered by an internal enumeration. A call $atom(R)$ then yields the relation $\{\langle x, y\rangle\}$ which consists of the lexicographically smallest pair $\langle x, y\rangle$ of R.

3 A Generic Program for Computing Minimal Subsets

For the following we assume a finite set M, a predicate $\mathcal{P}$ on the powerset 2^M of M, and an element R from 2^M. Furthermore we suppose that $\mathcal{P}$ is *upwards-closed*, which means that for all X, Y from 2^M if X is a subset of Y and $\mathcal{P}(X)$ is true, then $\mathcal{P}(Y)$ is also true. Note that this property implies the negation $\neg\mathcal{P}$ of $\mathcal{P}$ to be *downwards-closed*, i.e., if $Y \supseteq X$ and $\mathcal{P}(Y)$ is false, then $\mathcal{P}(X)$ is false.

In this section we shall systematically derive a generic imperative program that computes – for R as its input – a minimal subset of R that satisfies the predicate $\mathcal{P}$. If we use the variable A as the program's output, then a formal specification of the requirements on A is given by the postcondition

$$post(R, A) \;\triangleq\; \mathcal{P}(A) \wedge A \subseteq R \wedge \forall X \in 2^A : \mathcal{P}(X) \to X = A\,.$$

For the derivation of the program we combine the Dijkstra-Gries program development method with set theory and logic. In doing so, for a set X from 2^M and an element x from M we write $X \setminus x$ instead of $X \setminus \{x\}$ and $X \cup x$ instead of $X \cup \{x\}$ to enhance readability. Heading towards a program with an initialization followed by a while-loop, the derivation is carried out in three parts.

First we have to develop a loop invariant. Here we follow the most commonly used technique of generalizing the postcondition (for details see [7]). The corresponding calculation introduces a new variable B and looks as follows:

$$\begin{array}{ll} & post(R, A) \\ \iff & \mathcal{P}(A) \wedge A \subseteq R \wedge \forall X \in 2^A : \mathcal{P}(X) \to X = A \\ \iff & \mathcal{P}(A) \wedge A \subseteq R \wedge \forall X \in 2^A : X \neq A \to \neg\mathcal{P}(X) \\ \iff & \mathcal{P}(A) \wedge A \subseteq R \wedge \forall x \in A : \neg\mathcal{P}(A \setminus x) \\ \Longleftarrow & \mathcal{P}(A) \wedge A \subseteq R \wedge B = \emptyset \wedge B \subseteq A \wedge \forall x \in A \setminus B : \neg\mathcal{P}(A \setminus x)\,. \end{array}$$

Only the direction "$\Longleftarrow$" of the third step is non-trivial and needs an additional explanation: If $X \subset A$, then there exists an element x from A such that $X \subseteq A \setminus x$

and hence $\neg\mathcal{P}(A \setminus x)$ implies $\neg\mathcal{P}(X)$ since $\neg\mathcal{P}$ is downwards-closed. Guided by the above implication, now we choose $B = \emptyset$ as exit condition of the while-loop and the following loop invariant:

$$inv(R, A, B) \triangleq \mathcal{P}(A) \wedge A \subseteq R \wedge B \subseteq A \wedge \forall x \in A \setminus B : \neg\mathcal{P}(A \setminus x).$$

In the second part of the derivation, we now have to consider the initialization of the variables A and B. We do this in combination with the choice of a suitable precondition, i.e., a property on R which is sufficiently general and implies additionally that the initialization establishes the invariant. Since we want to compute a minimal subset of R satisfying $\mathcal{P}$, it seems to be reasonable to require that there exists a subset X of R such that $\mathcal{P}(X)$ is true. Due to the upwards-closedness of $\mathcal{P}$, however, the existence of X is equivalent to the fact that the program's input R satisfies $\mathcal{P}$. Hence, we choose

$$pre(R) \triangleq \mathcal{P}(R)$$

as precondition. This works and yields an initialization which assigns R to both variables A and B. Here is the formal justification:

$$\begin{array}{ll} & pre(R) \\ \iff & \mathcal{P}(R) \wedge R \subseteq R \wedge \forall x \in \emptyset : \neg\mathcal{P}(R \setminus x) \\ \iff & \mathcal{P}(R) \wedge R \subseteq R \wedge R \subseteq R \wedge \forall x \in R \setminus R : \neg\mathcal{P}(R \setminus x) \\ \iff & inv(R, R, R). \end{array}$$

Having achieved an initialization, in the last part of the program derivation we have to elaborate the body of the loop in such a way that its execution maintains the invariant and ensures termination of the loop. To this end, we suppose that B is non-empty and b is an arbitrary element contained in it. We consider two cases. First we assume $\mathcal{P}(A \setminus b)$ to be true. Then we have:

$$\begin{array}{ll} & inv(R, A, B) \\ \iff & \mathcal{P}(A) \wedge A \subseteq R \wedge B \subseteq A \wedge \forall x \in A \setminus B : \neg\mathcal{P}(A \setminus x) \\ \implies & \mathcal{P}(A) \wedge A \subseteq R \wedge B \setminus b \subseteq A \setminus b \wedge \forall x \in A \setminus B : \neg\mathcal{P}(A \setminus x) \\ \implies & \mathcal{P}(A \setminus b) \wedge A \setminus b \subseteq R \wedge B \setminus b \subseteq A \setminus b \wedge \\ & \qquad \forall x \in A \setminus B : \neg\mathcal{P}((A \setminus b) \setminus x) \\ \implies & \mathcal{P}(A \setminus b) \wedge A \setminus b \subseteq R \wedge B \setminus b \subseteq A \setminus b \wedge \\ & \qquad \forall x \in (A \setminus B) \setminus b : \neg\mathcal{P}((A \setminus b) \setminus x) \\ \iff & \mathcal{P}(A \setminus b) \wedge A \setminus b \subseteq R \wedge B \setminus b \subseteq A \setminus b \wedge \\ & \qquad \forall x \in (A \setminus b) \setminus (B \setminus b) : \neg\mathcal{P}((A \setminus b) \setminus x) \\ \iff & inv(R, A \setminus b, B \setminus b). \end{array}$$

This calculation uses in the third step that $\mathcal{P}(A \setminus b)$ is true and the predicate $\neg\mathcal{P}$ is downwards-closed. In the remaining steps only basic laws of set theory and logic and the definition of the invariant are applied. Now we deal with the remaining case of $\mathcal{P}(A \setminus b)$ being false. Here we obtain

$$
\begin{array}{ll}
& inv(R,A,B) \\
\iff & \mathcal{P}(A) \wedge A \subseteq R \wedge B \subseteq A \wedge \forall x \in A \setminus B : \neg\mathcal{P}(A \setminus x) \\
\implies & \mathcal{P}(A) \wedge A \subseteq R \wedge B \setminus b \subseteq A \wedge \forall x \in A \setminus B : \neg\mathcal{P}(A \setminus x) \\
\iff & \mathcal{P}(A) \wedge A \subseteq R \wedge B \setminus b \subseteq A \wedge \forall x \in (A \setminus B) \cup b : \neg\mathcal{P}(A \setminus x) \\
\iff & \mathcal{P}(A) \wedge A \subseteq R \wedge B \setminus b \subseteq A \wedge \forall x \in A \setminus (B \setminus b) : \neg\mathcal{P}(A \setminus x) \\
\iff & inv(R,A,B \setminus b),
\end{array}
$$

where again the assumption of the case, i.e., that $\mathcal{P}(A \setminus b)$ is false, is used in the derivation's third step and basic laws of set theory and logic in combination with the definition of the invariant are applied otherwise.

In view of the above implications we have to change in each run of the while-loop the value of B into $B \setminus b$ and, provided $\mathcal{P}(A \setminus b)$ is true, the value of A into $A \setminus b$. This can be achieved by a conditional and leads – in a syntax which is quite similar to that of the RelView system – to the following program *Minimum*. In this program it is assumed that the call *elem*(B) of the pre-defined operation *elem* yields an element of the non-empty set B. Termination of the program is obvious since M is finite, hence B is initially finite, too, and the cardinality of this set is strictly decreased by every run through the while-loop.

```
Minimum(R)
  DECL A, B, b
  BEG  A, B := R, R;
       WHILE B ≠ ∅ DO
         b := elem(B);
         IF P(A \ b) THEN A, B := A \ b, B \ b
                     ELSE B := B \ b FI OD
       RETURN A
  END.
```

The running time of the program *Minimum* mainly depends on the costs of evaluating the predicate $\mathcal{P}$ and the number of runs of the while-loop. In many cases the evaluation of $\mathcal{P}$ can be improved if $\mathcal{P}$ is decremental in the sense that

$$\mathcal{P}(X \setminus x) \iff \mathcal{P}(X) \wedge \mathcal{Q}(X,x) \tag{5}$$

for all non-empty $X \in 2^M$ and $x \in X$, where $\mathcal{Q}$ is an additional predicate on $2^M \times M$. Since $\mathcal{P}(A)$ is part of the invariant $inv(R,A,B)$, in this case the condition $\mathcal{P}(A \setminus b)$ of the conditional of *Minimum* can be replaced by $\mathcal{Q}(A,b)$ and this reduces the costs if the evaluation of $\mathcal{Q}$ is less expensive than that of $\mathcal{P}$.

The use of (5) is motivated by [12], where the dual property "to be incremental" is applied for computing maximal sets. Exceeding the work of [12], in the following we also consider the number of runs of the while-loop of *Minimum*. It is obvious that this number equals the cardinality of the input set R. Therefore, it seems to be a good idea to refine the initialization by a precomputation phase which yields, of course only if the precondition *pre*(R) holds, a set S such that $\mathcal{P}(S)$ is true, $S \subseteq R$, and the cardinality of S is much smaller than that of R.

Since the conjunction of $\mathcal{P}(S)$ and $S \subseteq R$ is equivalent to $inv(R, S, S)$, this leads – in combination with requirement (5) – to the following correct refinement:

```
Minimum'(R)
   DECL S, A, B, b
   BEG  ≫ precomputation phase ≪;
        { P(S) ∧ S ⊆ R }
        A, B := S, S;
        WHILE B ≠ ∅ DO
          b := elem(B);
          IF Q(A, b) THEN A, B := A \ b, B \ b
                     ELSE B := B \ b FI OD
        RETURN A
   END.
```

In the next section we will use a relation-algebraic instantiation of this refined version to compute transitive reductions of strongly connected directed graphs.

4 Computing a Transitive Reduction

During this section we suppose that $g = (V, R)$ is a fixed directed graph with finite vertex set $V \neq \emptyset$ and relation $R : V \leftrightarrow V$. Then computing a *transitive reduction* of g means to determine a directed graph $r_* = (V, A)$ such that A is a minimal subrelation of R satisfying $A^* = R^*$, i.e., r_* is an arc-minimal subgraph of g with the same information as g concerning reachability. This task is a fundamental optimization problem with many applications (some of them are sketched in [11] for example) since a transitive reduction of a directed graph can be seen as a minimal storage representation of its reachability information.

A transitive reduction of g with a minimum number of arcs is called a *minimum equivalent digraph* of g. In contrast to the problem of computing a transitive reduction this problem is NP-hard. Approximation algorithms for minimum equivalent digraphs have been developed in [10]. Of course, each algorithm for computing a transitive reduction can be seen as an approximation algorithm for minimum equivalent digraphs, too. We will come back to this later.

Now let the supposed graph g in addition be strongly connected, i.e., $R^* = \mathsf{L}$ be true. If we choose the universe M of Section 3 as the set $[V \leftrightarrow V]$ and the predicate $\mathcal{P}(X)$ of the generic program *Minimum'* as

$$\mathcal{P}(X) \quad :\Longleftrightarrow \quad X^* = \mathsf{L}, \tag{6}$$

then this instantiation solves for the input relation R the problem of computing the relation of a transitive reduction of the directed graph g since its precondition exactly describes that g is strongly connected, its postcondition exactly describes that A is the relation of a transitive reduction of g, and $X^* = \mathsf{L}$ is upwards-closed. Of course, we have to replace the empty set $\emptyset$, the choice $elem(B)$, and the difference $A \setminus b$ resp. $B \setminus b$ by their relation-algebraic counter-parts O, $atom(B)$, and $A \cap \overline{b}$ resp. $B \cap \overline{b}$.

The predicate $\mathcal{P}$ of (6) is also decremental in the sense of (5) since for all non-empty relations $X : V \leftrightarrow V$ and all $x = atom(X)$ the equation

$$(X \cap \overline{x})^* = \mathsf{L} \iff X^* = \mathsf{L} \wedge x \subseteq (X \cap \overline{x})^* \tag{7}$$

holds. A proof of "$\Longrightarrow$" is trivial and the remaining direction "$\Longleftarrow$" is shown by the calculation

$$\begin{aligned} \mathsf{L} &= X^* && \text{assumption} \\ &= ((X \cap \overline{x}) \cup x)^* \\ &= (X \cap \overline{x})^* \cup (X \cap \overline{x})^* x (X \cap \overline{x})^* && \text{(1) and (3)} \\ &\subseteq (X \cap \overline{x})^* \cup (X \cap \overline{x})^* (X \cap \overline{x})^* (X \cap \overline{x})^* && \text{assumption} \\ &= (X \cap \overline{x})^* . \end{aligned}$$

Guided by the right-hand side of equivalence (7) we now choose the implication $x \subseteq (X \cap \overline{x})^*$ as $\mathcal{Q}(X, x)$.

To complete the instantiation of the generic program *Minimum'* we still have to find a precomputation phase which establishes $S^* = \mathsf{L}$ and $S \subseteq R$. Here we follow an idea mentioned in [11] and seek for subgraphs $h_1 = (V, T_1)$ of g and $h_2 = (V, T_2)$ of $g^\mathsf{T} = (V, R^\mathsf{T})$ with a common root $r : V \leftrightarrow \mathbf{1}$. Then the precomputation phase can consist of the assignment of $T_1 \cup T_2^\mathsf{T}$ to S since

$$\mathsf{L} = \mathsf{L}r^\mathsf{T}r\mathsf{L} = (r\mathsf{L})^\mathsf{T}r\mathsf{L} \subseteq (T_2^*)^\mathsf{T}T_1^* = (T_2^\mathsf{T})^*T_1^* \subseteq (T_1 \cup T_2^\mathsf{T})^*$$

follows with the help of $r^\mathsf{T}r \neq \mathsf{O}$ and the Tarski-rule, the root properties of r wrt. h_1 and h_2, and two simple laws of reflexive-transitive closures, and

$$T_1 \cup T_2^\mathsf{T} \subseteq R \cup R^{\mathsf{T}\mathsf{T}} = R$$

is trivial. For reasons of efficiency, the result S of the precomputation phase should also contain as few arcs as possible. Therefore, we seek in addition for subgraphs h_1 and h_2 which are trees. If we postpone this task to two relational programs *Tree*$_1$ and *Tree*$_2$ which both have Q and r as inputs, T as output, and are specified by the precondition

$$pre(Q,r) \triangleq Q^* = \mathsf{L} \wedge r = r\mathsf{L} \wedge rr^\mathsf{T} \subseteq \mathsf{I} \wedge r^\mathsf{T}\mathsf{L} = \mathsf{L}$$

(the latter three conjuncts say that r is a point) and the postcondition

$$post(Q,T,r) \triangleq T \subseteq Q \wedge r\mathsf{L} \subseteq T^* \wedge TT^\mathsf{T} \subseteq \mathsf{I} \wedge T^+ \subseteq \overline{\mathsf{I}}$$

(here the latter three conjuncts say that T is the relation of a tree with root r), then in combination with the point property of the choice $point(\mathsf{L})$, i.e., axioms (4), we immediately obtain correctness of the following relational program *TransRed* for computing the relation of a transitive reduction of g:

```
TransRed(R)
  DECL S, A, B, b, r
  BEG  r := point(L);
       S := Tree1(R,r) ∪ Tree2(R^T,r)^T;
       A,B := S,S;
       WHILE B ≠ O DO
         b := atom(B);
         IF b ⊆ (A ∩ b̄)* THEN A,B := A ∩ b̄, B ∩ b̄
                          ELSE B := B ∩ b̄ FI OD
       RETURN A
  END.
```

The union of the two trees in this program contains at most $2 * |V| - 2$ arcs. A transitive reduction of g has at least $|V|$ arcs what in turn implies that the while-loop of *TransRed* is executed not more than $|V| - 2$ times. Hence, we obtain an overall running time $\mathcal{O}(|V|^2)$ if we use two programs $Tree_1$ and $Tree_2$ with running times $\mathcal{O}(|V|^2)$ and if we can test the condition $b \subseteq (A \cap \overline{b})^*$ of *TransRed* in $\mathcal{O}(|V|)$ steps. The latter task, however, is trivial since the inclusion $b \subseteq (A \cap \overline{b})^*$ describes that the sink of b can be reached from its source via arcs from $(A \cap \overline{b})^*$ and this reachability problem can be solved in running time $\mathcal{O}(|A|)$ if A is implemented by successor lists.

One possibility to test $b \subseteq (A \cap \overline{b})^*$ in time $\mathcal{O}(|A|)$ is depth-first-search (usually abbreviated with DFS). But this well-known graph searching technique is also a possibility to solve the first of the just mentioned tasks. From [3] we immediately obtain that the following relational DFS program is correct wrt. the above precondition $pre(Q,r)$ and postcondition $post(Q,T,r)$:

```
Dfs(Q,r)
  DECL T, b, g, p, w
  BEG  T,b,g,p := O,O,r,r;
       WHILE Q^T r ∩ ‾(b ∪ r) ≠ O DO
         w := Q^T p ∩ ‾(b ∪ g);
         IF w = O THEN b,g,p := b ∪ p, g ∩ p̄, Tp
                  ELSE q := point(w);
                       g,T,p := g ∪ q, T ∪ pq^T, q FI OD
       RETURN T
  END.
```

Of course, we are particularly interested in transitive reductions with few arcs, i.e., in approximations of minimum equivalent digraphs. To get an impression how the relational program *TransRed* with *Dfs* as instantiation of both $Tree_1$ and $Tree_2$ behaves wrt. this expectation, we have implemented it in RELVIEW and performed numerous experiments. In each experiment we fixed the number N of vertices and generated random graphs – more exactly: their relations – using a specific basic operation of RELVIEW, where we varied the number of arcs from 10% to 100% of all N^2 possible arcs. The following Figure 1 shows one of the results. Here we have $N = 200$, the density of the input relation of *TransRed* is listed at the x-axis, and the cardinality of the computed transitive reduction is listed at the y-axis:

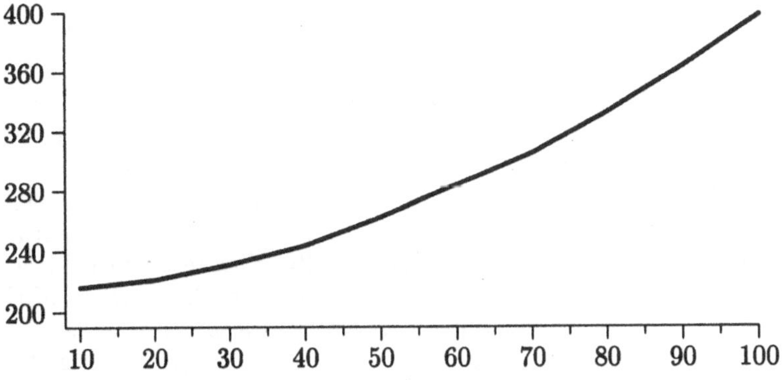

Fig. 1. Density/cardinality-diagram for $N = 200$ and DFS precomputation

The curves of all other experiments are rather similar to that curve. They start at 10% a little bit above N, are nearly linear, and end at 100% a little bit below $2 * N$. Hence, the behaviour of the program does not correspond to our expectation because the number of arcs of a transitive reduction of $g = (V, R)$ obviously should decrease if the density of R increases. Altogether, DFS seems not to be the right choice for the precomputation phase of *TransRed*.

A second well-known approach for constructing a spanning subtree of a strongly connected directed graph from a given root is breadth-first-search, usually abbreviated with BFS. To get a relational formulation of this strategy, we start with the following relational program:

$$
\begin{array}{l}
\mathit{Reach}(Q, r) \\
\quad \texttt{DECL } x, y \\
\quad \texttt{BEG } \; x, y := r, Q^{\mathsf{T}} r \cap \overline{r}; \\
\qquad\quad \texttt{WHILE } y \neq \mathsf{O} \texttt{ DO} \\
\qquad\qquad x := x \cup y; y := Q^{\mathsf{T}} x \cap \overline{x} \texttt{ OD} \\
\qquad\quad \texttt{RETURN } x \\
\quad \texttt{END}.
\end{array}
$$

Given a directed graph and a vertex, this program computes the vertices which are reachable from the given vertex. In terms of relational algebra this means that it is correct wrt. the precondition

$$pre(Q,r) \;\triangleq\; r = r\mathsf{L} \wedge rr^{\mathsf{T}} \subseteq \mathsf{I} \wedge r^{\mathsf{T}}\mathsf{L} = \mathsf{L}$$

and the postcondition

$$post(Q,r,x) \;\triangleq\; x = (Q^*)^{\mathsf{T}} r \,.$$

A formal derivation of the program *Reach* can be found in [4]. It combines relational algebra and the Dijkstra-Gries program development method and uses the following loop invariant:

$$inv(Q,r,x,y) \;\triangleq\; r \subseteq x \wedge x \subseteq (Q^*)^{\mathsf{T}} r \wedge y = \overline{x} \cap Q^{\mathsf{T}} x \,.$$

If one takes a closer look at the above reachabililty program, then one can see that it successively computes the vertices of the different layers of a BFS tree with root r. This is the reason why we use it as starting point of the relational BFS program we are interested in. Our idea is to introduce a new variable T, to compute a BFS tree with its help by adding some lines of code to the initialization and the body of the loop, and to return, finally, T as result instead of x. The initialization of T by xy^{T} is obvious. To enlarge T by the vertices of the next layer by every run through the while-loop, we insert a second while-loop between the assignments $x := x \cup y$ and $y := Q^{\mathsf{T}} x \cap \overline{x}$. This while-loop scans all vertices from the last layer (which is described by the vector y) and inserts arcs to their successors in the next layer (which is described by the vector $Q^{\mathsf{T}}(x \cup y) \cap \overline{x \cup y}$) in such a way that each vertex of the next layer is an endpoint of exactly one inserted arc. Thus, we arrive at the following relational BFS program:

$Bfs(Q,r)$
  DECL a, p, w, x, y, z, T
  BEG $x, y := r, Q^{\mathsf{T}} r \cap \overline{r}$;
    $T := xy^{\mathsf{T}}$;
    WHILE $y \neq \mathsf{O}$ **DO**
      $x, w, z := x \cup y, Q^{\mathsf{T}}(x \cup y) \cap \overline{x \cup y}, \mathsf{O}$;
      WHILE $z \neq w$ **DO**
        $p := point(y)$;
        $a := Q^{\mathsf{T}} p \cap \overline{x \cup z}$;
        $y, T, z := y \cap \overline{p}, T \cup pa^{\mathsf{T}}, z \cup a$ **OD**;
      $y = Q^{\mathsf{T}} x \cap \overline{x}$ **OD**
    RETURN T
  END.

This program is correct wrt. the precondition $pre(Q,r)$ and the postcondition $post(Q,T,r)$; a formal proof is given in the appendix of this paper.

We have also implemented *Bfs* in RELVIEW and used as instantiation of both $Tree_1$ and $Tree_2$ in the precomputation phase of *TransRed*. The experiments we have performed show that this version is even worser than the first version using the DFS program. As an example, in Figure 2 we present again the density/cardinality-diagram for $N = 200$ vertices; all other curves are rather similar to the curve of the figure.

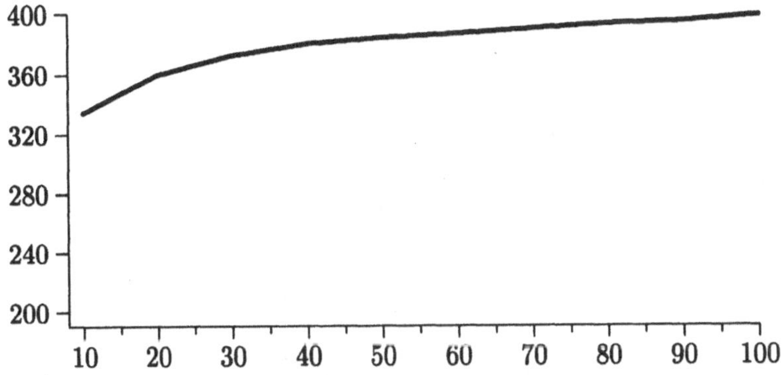

Fig. 2. Density/cardinality-diagram for $N = 200$ and BFS precomputation

Both versions of the relational program *TransRed* we have investigated until now show the same tendency, viz. the denser the input relation gets the more arcs are contained in the result. This is in contrast to our expectation that the number of arcs of a transitive reduction should decrease if the density of the underlying directed graph increases.

The reason for the bad behaviour of our hitherto relational programs gets clear if one compares them with the approximation algorithms for minimum equivalent digraphs presented in [10]. These algorithms compute approximations which are composed of a few large cycles only. In contrast to this, when using the BFS strategy for computing spanning subtrees $h_1 = (V, T_1)$ of g and $h_2 = (V, T_2)$ of g^T with a common root r, then h_1 contains shortest paths leading from r to every other vertex of g and h_2^T contains shortest path leading from every vertex of g except r to r. Thus, the result of precomputation phase of *TransRed*, which consists of the union of the relations T_1 and ${T_2}^\mathsf{T}$, contains a lot of short cycles which are getting shorter if the graph g gets denser. The same phenomenon appears when the DFS strategy is applied twice in the precomputation phase of *TransRed*. The computation of h_1 following the DFS strategy leads to long paths from r to every other vertex of g, and in the same manner h_2^T contains long paths back to r. Therefore, a lot of vertices lie on both paths which in turn implies that the union of T_1 and T_2^T contains many short cycles. As in the case of DFS, these are getting shorter if the graph gets denser.

Now, to obtain long cycles the idea is to use the DFS strategy to compute h_1, which then contains long paths from r to every other vertex of g. For the computation of h_2, however, the use of the BFS strategy seems to be profitable, because then $h_2^\top$ contains shortest paths from every vertex of g except r back to r and, hence, the value $T_1 \cup T_2^\top$ of S in *TransRed* leads to long cycles and better results. Again we have tested this approach with the RELVIEW tool and arrived, for example, in the case $N = 200$ at the diagram of Figure 3; the diagrams of all other experiments with the two relational programs *Dfs* and *Bfs* for computing T_1 resp. T_2 look rather similar.

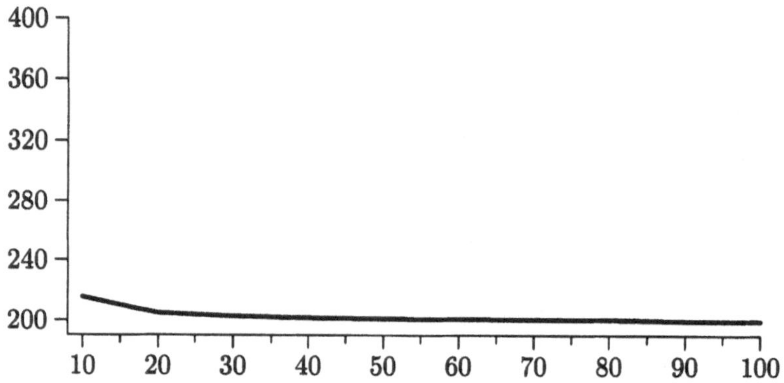

Fig. 3. Density/cardinality-diagram for $N = 200$ and DFS/BFS precomputation

Summing up, the experiments suggest that the combination of DFS and BFS is the right choice for the computation of a transitive reduction with as few arcs as possible using the program *TransRed*. In contrast to the other two approaches here the resultant transitive reduction has much fewer arcs. Furthermore we obtain the expected tendency that the result is getting better when the graph gets denser.

5 Conclusion

In this paper we have developed a relational program for computing the transitive reduction of strongly connected directed graphs as a specific instantiation of a generic program for computing minimal subsets. It uses a precomputation phase which combines DFS and BFS. When regarding it as an approximation algorithm for minimum equivalent digraphs, then the approximation factor is 2. This means that the computed transitive reduction contains at most twice as much arcs as a minimum equivalent digraph[2]. But our numerous practical tests with the RELVIEW system have shown that for randomly generated relations

[2] In the case of minimalization problems an approximation algorithm has an approximation factor f if for all inputs the size c of the computed result is bounded by $c \leq f * opt$, where opt is the size of an optimal result.

all results are very close to a minimum equivalent digraph, i.e., the "average approximation factor" is very close to 1.

In the course of the diploma thesis [9] we have compared our approach with the approximation algorithms of [10]. We have implemented the best algorithm of [10], which computes an approximation of a minimum equivalent digraph in nearly linear time with approximation factor f between $\frac{\pi^2}{6} \approx 1.64$ and $\frac{\pi^2+1}{6} \approx 1.81$. But in practical tests this algorithm leads to results which contain in the average $\frac{3*n}{2}$ arcs, where n is the cardinality of the minimum equivalent digraphs. In the concrete case of $N = 200$ vertices, which we have considered in detail in Section 4, this means that the result contains in the average 300 arcs. This is far away from the optimum and also worse than the results our final program yields. Furthermore these results are no transitive reductions.

To overcome the disadvantages of both approaches, in [9] they have been combined by replacing the DFS/BFS precomputation phase in *TransRed* with a call of the algorithm of [10]. This, finally, leads to an algorithm with the theoretical approximation factor of [10] and our good practial results.

References

1. Aho A.V., Hopcroft J.E., Ullman J.D.: The design and analysis of computer algorithms. Addison-Wesley (1974).
2. Behnke R. et al.: Applications of the RELVIEW system. In: Berghammer R., Lakhnech Y. (eds.): Tool support for system specification, development and verification. Advances in Computing, Springer, 33-47 (1999).
3. Berghammer R., Hoffmann T.: Relational depth-first-search with applications. In: Desharnais J. (ed.): Proc. 5th Int. Seminar on Relational Methods in Computer Science, Laval Univ., Dept. d' Informatique, 11-20 (2000). Extended version in: Information Sciences 139, 167-186 (2001).
4. Berghammer R.: Combining relational calculus and the Dijkstra-Gries method for deriving relational programs. Information Sciences 119, 155-171 (1999).
5. Cormen T.H., Leiserson C.E., Rivest R.L.: Introduction to algorithms. The MIT Press (1990).
6. Dijkstra E.W.: A discipline of programming. Prentice-Hall (1976).
7. Gries D.: The science of computer programming. Springer (1981).
8. Hoffmann T.: Case studies in relational program development by selected strategies for graph searching (in German). Ph.D. thesis, Institut für Informatik und Praktische Mathematik, Universität Kiel (2002).
9. Kasper C.: Investigating algorithms for transitive reductions and minimum equivalent digraphs (in German). Diploma thesis, Institut für Informatik und Praktische Mathematik, Universität Kiel (2001).
10. Khuller S., Raghavachari B., Young N.: Approximating the minimum equivalent digraph. SIAM Journal on Computing 24, 859-972 (1995).
11. Noltemeier H.: Reduktion von Präzedenzstrukturen. Zeitschrift für Operations Res. 20, 151-159 (1976).
12. Ravelo J.: Two graph algorithms derived. Acta Informatica 36, 489-510 (1999).
13. Schmidt G., Ströhlein T.: Relations and graphs. Springer (1993).
14. Simon K.: Finding a minimal transitive reduction in a strongly connected digraph within linear time. In: Nagl M. (ed.): Proc. 15th Workshop on Graph-Theoretic Concepts in Computer Science, LNCS 411, Springer, 245-259 (1990).

Appendix

In this appendix we focus on the formal verification of the relational BFS program from Section 4. First we start with some preparatory steps. A relation T is called a *singleton forest* if the following three formulae hold:

$$TT^{\mathsf{T}} \subseteq \mathsf{I} \qquad T^{+} \subseteq \overline{\mathsf{I}} \qquad (T \cup T^{\mathsf{T}})\mathsf{L}(T \cup T^{\mathsf{T}}) \subseteq (T \cup T^{\mathsf{T}})^{*}.$$

Graph-theoretically this means that $t = (V, T)$ is a forest in which every pair of non-isolated vertices is connected via an undirected path. If additionally r is a point fulfilling $r = T\mathsf{L} \cap \overline{T^{\mathsf{T}}\mathsf{L}}$, we call T a singleton forest with root r. Obviously, this is a generalization of the definition of rooted trees from Section 2.

In this paper we only deal with relations on finite sets. We can describe finiteness of a set X in relational algebra by demanding that the membership relation $\in : X \leftrightarrow 2^X$ is Noetherian. But this characterization is often unwieldy. Therefore, we have decided to use the postulate

$$(T^*)^{\mathsf{T}}(T\mathsf{L} \cap \overline{T^{\mathsf{T}}\mathsf{L}}) = (T \cup T^{\mathsf{T}})\mathsf{L} \tag{8}$$

for all singleton forests T since this is exactly what we need in the subsequent verification. Using graph-theoretic terminology, the equality means: If there exist a root of $t = (V, T)$, then exactly all non-isolated vertices are reachable from it. Otherwise, all vertices are isolated. Note that (8) does not hold for arbitrary relations. For example take the successor-relation on the integers, which obviously is a singleton forest. In this case the left-hand side of (8) is the empty vector, but its right-hand side is the universal vector.

Before we start the formal program verification, we collect two properties of singleton forests. Their proofs can be found in [8]. Let p and r be points, v be a vector, and T be a singleton forest with root r.

(i) If $p \subseteq \overline{v}$, then the relation pv^{T} is a singleton forest with root p.

(ii) If $p \subseteq (T \cup T^{\mathsf{T}})\mathsf{L}$ and $v \subseteq \overline{(T \cup T^{\mathsf{T}})\mathsf{L}}$, then the following properties hold:

(a) $(T \cup pv^{\mathsf{T}})^{*} = T^{*}(pv^{\mathsf{T}})^{*}$ (b) $T \cup pv^{\mathsf{T}}$ is a singleton forest with root r.

After these preparatory steps we begin with the verification of the relational program *Bfs*, where, see Section 4,

$$pre(Q, r) \triangleq Q^{*} = \mathsf{L} \wedge r = r\mathsf{L} \wedge rr^{\mathsf{T}} \subseteq \mathsf{I} \wedge R^{\mathsf{T}}\mathsf{L} = \mathsf{L}$$

is the precondition and

$$post(Q, T, r) \triangleq T \subseteq Q \wedge r\mathsf{L} \subseteq T^{*} \wedge TT^{\mathsf{T}} \subseteq \mathsf{I} \wedge T^{+} \subseteq \overline{\mathsf{I}}$$

is the postcondition. The program uses two nested while-loops. Consequently, we need two invariants, an invariant $inv_{outer}(Q, T, r, x, y)$ for the outer while-loop and a second invariant $inv_{inner}(Q, T, r, w, x, y, z)$ for the inner while-loop.

Here $inv_{outer}(Q,T,r,x,y)$ is an extension of the invariant $inv(Q,r,x,y)$ of the program *Reach* and consists of the conjunction of the upcoming properties:

$$r \subseteq x \tag{9}$$

$$x \subseteq (Q^*)^{\mathsf{T}} r \tag{10}$$

$$y = Q^{\mathsf{T}} x \cap \overline{x} \tag{11}$$

$$T \subseteq Q \tag{12}$$

$$x \cup y = (T^*)^{\mathsf{T}} r \tag{13}$$

$$T \text{ is a singleton forest with root } r. \tag{14}$$

The invariant $inv_{inner}(Q,T,r,w,x,y,z)$ of the inner while-loop consist of the conjunction of the next five properties:

$$T \subseteq Q \tag{15}$$

$$x \cup z = (T^*)^{\mathsf{T}} r \tag{16}$$

$$T \text{ is a singleton forest with root } r \tag{17}$$

$$w = Q^{\mathsf{T}} x \cap \overline{x} \tag{18}$$

$$y \subseteq (T^*)^{\mathsf{T}} r. \tag{19}$$

The verification of the relational BFS program is done in five steps. First we focus on the inner invariant and show its establishment and maintenance. Afterwards we do the same with the invariant of the outer while-loop. Finally, we prove that $y = \mathsf{O}$ and the validity of the outer invariant imply the postcondition. During all these steps we are allowed to use the precondition $pre(Q,r)$.

We start the correctness proof with the establishment of the inner invariant by the assignment to x, w, and z in front of the inner while-loop. Here we are allowed to use the properties of $inv_{outer}(Q,T,r,x,y)$. The establishment of the formulae (15), (17), and (19) follow directly from the validity of (12), (14), and (13). With the aid of equation (13) and the initialization of x by $x \cup y$ resp. z by O, the validity of formula (16) for the latter values is shown by

$$x \cup y \cup \mathsf{O} = x \cup y = (T^*)^{\mathsf{T}} r\,.$$

Finally, the establishment of formula (18) is a direct consequence of the assignment $w := Q^{\mathsf{T}} x \cap \overline{x}\,$.

In the second step we show the maintenance of the inner invariant starting with inclusion (15). Using that p is a point, we get $pa^{\mathsf{T}} \subseteq pp^{\mathsf{T}} Q \subseteq Q$ from the assignment to a which in turn yields $T \cup pa^{\mathsf{T}} \subseteq Q$ due to $T \subseteq Q$. The maintenance of formulae (16) and (17) can be shown with property (ii). Therefore we have to check whether p and a (instead of v) fulfill the assumptions. Obviously, both inclusions $p \subseteq (T \cup T^{\mathsf{T}})\mathsf{L}$ and $a \subseteq \overline{(T \cup T^{\mathsf{T}})\mathsf{L}}$ follow with postulate (8) from $p \subseteq y$, (19), and (17) resp. $a \subseteq \overline{x \cup z}$ and (16). Consequently the assumptions of (ii) are fulfilled and part (a) of (ii) helps us to prove equation (16) for the new values of the variables z and T as follows:

$$
\begin{aligned}
x \cup z \cup a &= (T^*)^{\mathsf{T}} r \cup a && x \cup z = (T^*)^{\mathsf{T}} r\\
&= (T^*)^{\mathsf{T}} r \cup a\mathsf{L} && a \text{ vector}\\
&= (T^*)^{\mathsf{T}} r \cup a p^{\mathsf{T}} (T^*)^{\mathsf{T}} r && (19) \Rightarrow p \subseteq (T^*)^{\mathsf{T}} r \Rightarrow \mathsf{L} \subseteq p^{\mathsf{T}} (T^*)^{\mathsf{T}} r\\
&= (T^* \cup T^* p a^{\mathsf{T}})^{\mathsf{T}} r &&\\
&= (T^* (p a^{\mathsf{T}})^*)^{\mathsf{T}} r && (pa^{\mathsf{T}})^* = \mathsf{I} \cup pa^{\mathsf{T}}\\
&= ((T \cup p a^{\mathsf{T}})^*)^{\mathsf{T}} r && \text{(a) of (ii)}
\end{aligned}
$$

The maintenance of property (17) follows directly from part (b) of (ii). Since the values of x and w remain unchanged by every run through the inner while-loop, maintenance of equation (18) is trivial. The maintenance of the last formula (19) of the inner invariant, finally, follows from

$$y \cap \overline{p} \subseteq y \subseteq (T^*)^{\mathsf{T}} r \subseteq ((T \cup pa^{\mathsf{T}})^*)^{\mathsf{T}} r,$$

where the validity of $y \subseteq (T^*)^{\mathsf{T}} r$ is applied in the second step. This completes the maintenance proof of the invariant of the inner while-loop.

In the next two steps we concentrate on the invariant of the outer while-loop. Here we only have to look at the formulae (12)-(14), because the other properties (9)-(11) are exactly the conjuncts of the invariant of the relational reachability program *Reach*. The corresponding correctness proofs can be found in [4].

We start with the proof that the formulae (12)-(14) are established by the assignments to x, y, and T in front of the outer while-loop. The assignments to T, x, and y imply $T = xy^{\mathsf{T}} \subseteq rr^{\mathsf{T}}Q$, which, in combination with the point property of r, establishes inclusion (12). With aid of the assignments to x and T, equation $(xy^{\mathsf{T}})^* = \mathsf{I} \cup xy^{\mathsf{T}}$, and $x^{\mathsf{T}}x = r^{\mathsf{T}}r = \mathsf{L}$ (which is again a consequence of the point property of r) we get (13) from

$$(T^*)^{\mathsf{T}} r = (T^*)^{\mathsf{T}} x = ((xy^{\mathsf{T}})^*)^{\mathsf{T}} x = (\mathsf{I} \cup yx^{\mathsf{T}})x = x \cup yx^{\mathsf{T}}x = x \cup y\mathsf{L} = x \cup y\,.$$

The establishment of formula (14) is an immediate consequence of the point property of r and property (i).

Step four of the verification is the maintenance proof of the outer invariant. Here we are allowed to use the properties of $inv_{inner}(Q, T, r, w, x, y, z)$ and the exit condition of the inner while-loop. The maintenance of formula (12) follows directly from the validity of (15). In combining the exit condition $z = w$ with equation (18) we get $z = \overline{x} \cap Q^{\mathsf{T}}x$. Now, with the help of this formula, equation $y = \overline{x} \cap Q^{\mathsf{T}}x$, and formula (16), the maintenance of property (13) can be shown with by following calculation:

$$x \cup y = x \cup (\overline{x} \cap Q^{\mathsf{T}}x) = x \cup z = (T^*)^{\mathsf{T}} r\,.$$

The last property (14) for the new value of T ensues from property (17).

As the fifth step of the verification it remains to show that the postcondition $post(Q, T, r)$ follows from the invariant of the outer while-loop and this loop's exit condition $y = \mathsf{O}$. We only have to show the second conjunct $r\mathsf{L} \subseteq T^*$ of $post(Q, T, r)$ because the other formulae are immediate consequences of (12) and

(14). From equation (13) and $y = \mathsf{O}$ we get $x = (T^*)^\mathsf{T} r$ and $x = (Q^*)^\mathsf{T} r$ also holds due to the correctness of the reachability program. Putting these equations together and using the strong connectivity of Q and the point property of r we obtain the desired inclusion by

$$r\mathsf{L} = rr^\mathsf{T}\mathsf{L} = rr^\mathsf{T}Q^* = rr^\mathsf{T}T^* \subseteq T^* .$$

Thereby the correctness proof of the relational BFS program is complete.

Calculating Church-Rosser Proofs in Kleene Algebra

Georg Struth

Institut für Informatik, Universität Augsburg
Universitätsstr. 14, D-86135 Augsburg, Germany
Tel:+49-821-598-3109, Fax:+49-821-598-2274,
struth@informatik.uni-augsburg.de

Abstract. We present simple calculational proofs of Church-Rosser theorems for equational theories, quasiorderings and non-symmetric transitive relations in Kleene algebra. We also calculate the abstract part of two standard proofs of Church-Rosser theorems in the λ-calculus and further central statements of rewriting. Since proofs avoid deduction, in particular induction, and large parts are amenable to automata, the approach is suited for mechanization. Since proofs algebraically reconstruct precisely the usual diagrams, they are also very natural for a human. In all considerations, Kleene algebra is an excellent means of abstraction.

Keywords: Kleene algebra, rewriting, λ-calculus, Church-Rosser theorem, formal mathematics.

1 Introduction

A main concern of formal mathematics and formal methods is the combination of simple readable specifications with powerful proof search. This combination depends on several principles. First, the combination of formal and informal rigor. Specifications and proofs should be precise, concise, natural and revealing both for machines and for humans. Second, the balance of expressive and deductive power. These properties are mutually exclusive. Third, the replacement of deduction by calculation. Routine work, decision procedures, algorithms should be regarded prior to creativity, like for instance induction. Algebraic abstraction is a key to these principles, especially to the second and third one. Abstracting, one may divide and conquer, hide proof complexity in bridge lemmata between different levels, gain structural insight and obtaining simple, natural, revealing specifications and proofs at all levels.

Here, we demonstrate these principles by calculating proofs of Church-Rosser theorems in Kleene algebra. These include transitive relations, quasi-orderings, equational theories and abstract parts of the two standard techniques for the λ-calculus, namely that of Tait-Martin-Löf (c.f. [2]) and Takahashi [26] and Barendregt's indexing technique [2]. We also calculate proofs of the commutation lemma and commutative union lemma of Hindley and Rosen. Church-Rosser

H. de Swart (Ed.): RelMiCS 2001, LNCS 2561, pp. 276–290, 2002.

theorems are central to term rewriting and the λ-calculus with important applications in functional programming, universal algebra, computer algebra and mechanized reasoning. Proofs usually treat properties of (first-order or λ-) terms and those of the associated rewrite relation separately. Bridge lemmata describe their coupling. Reasoning at the relational level is supported by diagrams; their semantics can be given in terms of logic and binary relations. Textbook proofs are usually only informally rigorous. They use diagrams without any safe rules for their transformation or combination. The literature contains several errors even with rather simple statements. Formal Church-Rosser proofs with interactive proof checkers are considered challenging. To obtain at least formal rigor, one typically proves ad hoc structural lemmata about diagrams in a bottom-up way, using induction and essentially higher-order logic. Proof complexity and therefore our second and third principle is usually disregarded.

Unlike previous approaches, we proceed top-down via two abstraction steps. We first translate the logical specifications of Church-Rosser theorems into point-free relational ones. We then extract the relevant algebraic properties of set-theoretic relations. Since only the regular operations (union, composition, reflexive transitive closure) occur in Church-Rosser theorems, Kleene algebra is a more focused target theory than, for instance, relation algebra. Syntactically, a Kleene algebra is a semiring with an additional Kleene star operation modeling iteration or fixed point computation. The Kleene star is defined roughly as the least solution of a fixed point equation, but neither as a limit of an approximation sequence nor as a supremum of an infinite sum of powers. Semantically, Kleene algebra is precisely the algebra of regular events. Since also the regular operations on set-theoretic relations is among its models, properties of Kleene algebra can immediately be transfered to rewrite relations in a top down way.

Kleene algebra yields simple formally and informally rigorous proofs of Church-Rosser theorems. The first claim follows in particular from our implementation of Kleene algebra and proofs with the Isabelle proof checker [25]. The second claim holds, since our proofs formally reconstruct precisely the diagrammatic proofs and thereby give an algebraic semantics to diagrams. Moreover they use only natural and general properties of regular operations. Expressive and deductive power are well-balanced. Deduction is completely replaced by calculation. There is no induction and higher-order reasoning involved, fixed points are computed instead. By the strong connection between Kleene algebra and regular languages, large parts of proofs are even amenable to automata.

Beyond Church-Rosser theorems, the replacement of "regular" induction (the fragment of induction captured by automata) by first-order fixed point computation seems of general interest for formal methods, since first-order proof search methods and even decision procedures can be used then. Relational specifications and proofs are at the heart of methods like Z and B [22,1].

We have tried to include the most important proofs in the paper. A full formal treatment can be found in an extended version [24].

The remainder is organized as follows. Section 2 and section 3 introduce some Church-Rosser and Kleene algebra basics. Section 4 recalls further properties of Kleene algebra for the Church-Rosser proofs for rewriting with quasiorderings and non-symmetric transitive relations in section 5. Section 6 introduces a conversion operation and functions, to specialize the theorems of section 5 to equational rewriting and to prepare for two classical proofs of the Church-Rosser theorem in the λ-calculus in section 7 and section 8, with particular emphasis on mechanized reasoning. Section 9 contains a concluding discussion.

2 Church-Rosser Basics

Let $(A, \to)$ be a *rewrite system* of a set A and a binary set-theoretic relation $\to$ on A. We usually write A for $(A, \to)$ and $a \to b$. The relation $\leftarrow$ denotes the converse of $\to$, the relations $\to^+$, $\to^*$ and $\leftrightarrow^*$ denote the smallest transitive relation, quasiordering and equivalence containing $\to$. Juxtaposition of arrows denotes relational composition. The *diamond property* $\Diamond(x_1, x_2, x_3)$ and the *Church-Rosser property* $\bigtriangledown(x_1, x_2)$ are defined by

$$\Diamond(x_1, x_2, x_3) \quad \Leftrightarrow \quad (x_1 \to^* x_2 \wedge x_1 \to^* x_3 \Rightarrow \exists x_4 . x_2 \to^* x_4 \wedge x_3 \to^* x_4), \qquad (1)$$

$$\bigtriangledown(x_1, x_2) \quad \Leftrightarrow \quad (x_1 \leftrightarrow^* x_2 \Rightarrow \exists x_3 . x_1 \to^* x_3 \wedge x_2 \to^* x_3). \qquad (2)$$

The rewrite system A *has the diamond property,* if $\forall x_1, x_2, x_3 \in A . \Diamond(x_1, x_2, x_3)$ and the *Church-Rosser property,* iff $\forall x_1, x_2 \in A. \bigtriangledown (x_1, x_2)$. The properties are usually visualized by the diagrams

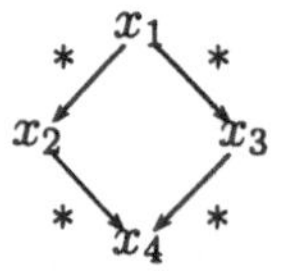

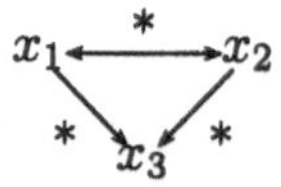

which are often also decorated with universal or existential quantifiers or different kinds of arrows to point out the relation with (1) and (2).

Theorem 1 (Church-Rosser). *A rewrite system has the diamond property, iff it has the Church-Rosser property.*

Proof. We only sketch the standard proof that the diamond property implies the Church-Rosser property. It is by induction on the number of rewrite steps in $\leftrightarrow^*$. In the base case, the Church-Rosser property trivially holds for the zero-step sequence $t \leftrightarrow t$. The induction step is visualized by the diagrams

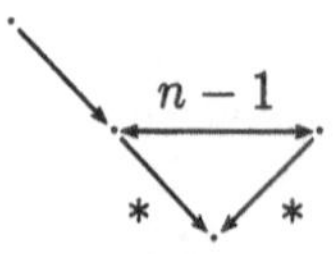

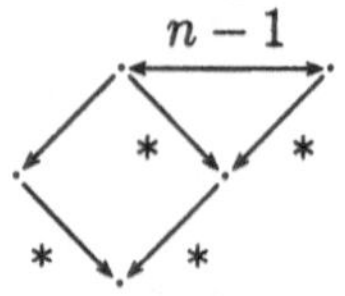

Only the right-hand diagram depends on an assumption, which can be shown to follow from the diamond property. □

The Church-Rosser and the diamond property can be expressed purely relationally, hence in a more abstract point-free way.

Lemma 1. *Let A be a rewrite system.*

(i) $\forall x_1, x_2, x_3 \in A. \Diamond(x_1, x_2, x_3) \Leftrightarrow \leftarrow^*\rightarrow^* \subseteq \rightarrow^*\leftarrow^*$,
(ii) $\forall x_1, x_2 \in A. \bigtriangledown(x_1, x_2) \Leftrightarrow (\rightarrow \cup \leftarrow)^* \subseteq \rightarrow^*\leftarrow^*$.

Lemma 1 is proven using the definitions of set-theoretic union, relational composition and the different closures of $\rightarrow$.

In section 3 we introduce an algebra to reason rigorously about the relevant algebraic properties of these relations.

3 Kleene Algebra

A Kleene algebra is a structure for modeling sequential composition, non-deterministic choice and iteration or fixed point computation. The field has been pioneered by Conway [5] in the context of the algebra of regular events. Besides formal languages and automata, Kleene algebras also arise, for instance, in the context of relation algebra (c.f. [20]) and logics, analysis and construction of programs (c.f. [11]). We follow Kozen's definition [13]. These structures are called *Kozen semirings* in [3]. For concise proofs of all statements in this text see [24]. Some formal proofs with the Isabelle proof checker can be found in [25].

A *semiring* is an algebra $(A, \cdot, +, 0, 1)$ of a set A, two binary operations of multiplication and addition and two constants 0 and 1. Thereby $(A, +, 0)$ is a commutative monoid, $(A, \cdot, 1)$ a monoid, multiplication distributes over addition from the left and right and 0 is both a left and a right annihilator ($0a = a0 = 0$). We often write A for $(A, \cdot, +, 0, 1)$ and ab for $a \cdot b$.

The relation $\leq$ defined on a semiring A by $a \leq b \Leftrightarrow (\exists c \in A).a + c = b$ is a quasiordering. It is a partial ordering, if A is *idempotent*, that is $a + a = a$ holds for every $a \in A$ and $a \leq b \Leftrightarrow a + b = b$. Addition and (left and right) multiplication are monotonic operations with respect to $\leq$ on A; 0 is the minimal element. If A is idempotent, it is a semilattice with respect to addition and least upper bounds. Thus for all $a, b, c \in A$,

$$a + b \leq c \Leftrightarrow a \leq c \wedge b \leq c. \tag{3}$$

A *Kleene algebra* $(A, \cdot, +, ^*, 0, 1)$ is an idempotent semiring $(A, \cdot, +, 0, 1)$ endowed with an additional unary operation * (*Kleene star*) that satisfies

$$1 + aa^* \leq a^*, \quad (4) \qquad ac + b \leq c \Rightarrow a^*b \leq c, \quad (6)$$

$$1 + a^*a \leq a^*, \quad (5) \qquad ca + b \leq c \Rightarrow ba^* \leq c, \quad (7)$$

for all $a, b, c \in A$. Thus a^*b and ba^* are the least pre-fixed points of the monotonic mappings $\lambda x.b + ax$ and $\lambda x.b + xa$. a^* is also a least fixed point of $\lambda x.1{+}ax$, $\lambda x.1{+}xa$; the inequalities (4) and (5) can be strengthened to equations. Moreover, this fixed point is uniquely defined. (6) and (7) are algebraic variants of fixed point induction. Computationally we use them as star elimination rules.

Kleene algebra characterizes the Kleene star algebraically (or combinatorially) and not analytically as a limit of an approximation sequence or a supremum of an infinite sum of powers. Our proofs are therefore simple induction-free calculations (often finite combinatorics) that are well-suited for automation. Since relations satisfy infinite distributivity laws and continuity axioms, proofs in Kleene algebras are more general than relational ones.

We define the *Kleene plus* by $a^+ = aa^*$. Obviously, $a^* = 1 + a^+ = a^+ + a^*$. There are pre-fixed point laws like (4) to (7).

Lemma 2. *let A be a Kleene algebra. Then for all* $a, b, c \in A$

$$a(a^+ + 1) \leq a^+, \quad (8) \qquad a(b + c) \leq c \Rightarrow a^+ b \leq c, \quad (10)$$
$$(a^+ + 1)a \leq a^+, \quad (9) \qquad (b + c)a \leq c \Rightarrow ba^+ \leq c. \quad (11)$$

The Kleene plus and star are monotonic operations on a Kleene algebra.

4 Some Properties of Kleene Algebra

We now review some well-known properties of Kleene algebra that are important for the Church-Rosser proofs. The first statement is a completeness result.

Theorem 2 ([13]). *There is a deductive system for the universal Horn theory of Kleene algebra such that an equation between Kleene algebra terms follows from the axioms of Kleene algebra iff the two terms denote the same regular set.*

Completeness results for related structures have been given, for instance, in [5, 3]. Remind that there is no finite equational axiomatization of the algebra of regular events [19]. By theorem 2, every identity in Kleene algebra can be decided by automata. In contrast, Kleene algebra is incomplete for the Horn theory of regular events [13] and the Horn theory of Kleene algebra is undecidable (by unsolvability of the uniform word problem for semigroups).

The second statement guarantees that our Church-Rosser proofs in Kleene algebra are adequate for set-theoretic relations.

Theorem 3. *The set-theoretic relations on a given set A are a Kleene algebra.*

Thereby, $+$ is interpreted as set-theoretic union, $\cdot$ as relational composition, * as the reflexive transitive closure operation, 0 as the empty, 1 as the identity relation and $\leq$ as set inclusion.

The third and fourth statement collect some standard properties of the Kleene star and plus. Most of them are well-known also for relation algebra, regular languages or automata (c.f. [5,9,13,20]).

Lemma 3. *Let A be a Kleene algebra. For all* $a, b, c \in A$,

$$1 = 1^*, \quad (12) \qquad a^{**} = a^*, \quad (16)$$
$$1 \leq a^*, \quad (13) \qquad ac \leq cb \Rightarrow a^*c \leq cb^*, \quad (17)$$
$$a^*a^* = a^*, \quad (14) \qquad cb \leq ac \Rightarrow cb^* \leq a^*c, \quad (18)$$
$$a \leq a^*, \quad (15) \qquad (a + b)^* = a^*(ba^*)^*. \quad (19)$$

In particular, $(ab)^*a = a(ba)^*$ and $a^*a = aa^*$ are simple consequences of (17).

Lemma 4. *Let A be a Kleene algebra. For all $a, b, c \in A$,*

$$a^+a^+ \leq a^+, \qquad (20)$$
$$a \leq a^+, \qquad (21)$$
$$a^+ \leq a^*, \qquad (22)$$
$$aa^+ = a^+a \leq a^+, \qquad (23)$$
$$a^+a^* = a^*a^+ = a^+, \qquad (24)$$
$$ac \leq cb \Rightarrow a^+c \leq cb^+, \qquad (25)$$
$$cb \leq ac \Rightarrow cb^+ \leq a^+c, \qquad (26)$$
$$(a+b)^+ = (a^*b)^+a^* + a^+. \qquad (27)$$

5 Church-Rosser Theorems

We now present the Church-Rosser theorems. They are further discussed in the remaining sections. In all Kleene algebra computations, inequational reasoning pays. (6) and (7)—as star elimination rules—are the working horses. Also monotonicity and (3) (splitting into subgoals) are abundantly used.

Proposition 1. *Let A be a Kleene algebra. For all $a, b \in A$,*

(i) $ba^* \leq a^*b^* \Rightarrow (a+b)^* \leq a^*b^*$.
(ii) $b^*a^* \leq a^*b^* \Rightarrow (a+b)^* \leq a^*b^*$.
(iii) $ba \leq ab \Rightarrow (a+b)^* \leq a^*b^*$.

Proof. (ad i) By (6) it suffices to show $(a+b)a^*b^* + 1 \leq a^*b^*$; thus, by (3),

$$1 \leq a^*b^*, \qquad (28)$$
$$aa^*b^* \leq a^*b^*, \qquad (29)$$
$$ba^*b^* \leq a^*b^*. \qquad (30)$$

(28) follows immediately from (13): $1 \leq a^* = a^*1 \leq a^*b^*$.

(29) follows immediately from (15), (14) and monotonicity of multiplication: $a \leq a^*$ implies $aa^*b^* \leq a^*a^*b^* = a^*b^*$.

(30) is the only inequality that depends on the assumptions and therefore is beyond automata theory. $ba^*b^* \leq a^*b^*b^* = a^*b^*$, by the assumption, (14) and monotonicity of multiplication.

(ad ii) It suffices to show $b^*a^* \leq a^*b^* \Rightarrow ba^* \leq a^*b^*$[1]. This is straightforward from (15) and monotonicity of multiplication.

(ad iii) It suffices to show $ba \leq ab \Rightarrow ba^* \leq a^*b^*$. By(17), (15) and the assumption, $ba^* \leq a^*b \leq a^*b^*$. □

The proof of proposition 1 (i) is simple and natural. The first steps are obvious. Fixed-point induction (axiom 6) eliminates the Kleene star in $(a+b)^*$ and (3) splits into three subgoals. Comparing with the proof of theorem 1, we see the precise correspondence between (28), (29) and (30) with the base case and the two cases of the induction step. In particular we are almost automatically lead to the right assumption in (30). (28) and (29) are trivial; they can be decided

[1] In fact, this statement is a logical equivalence.

by automata. The properties used for proving (28), (29) and (30) are again simple and natural. They formalize exactly the arguments that are implicitly used in the diagrammatic proof of theorem 1. Besides the semiring arithmetics and monotonicity we use (13), (15),(14), which encode reflexivity, extensivity and transitivity of a^*. Thus our calculations yield precisely formally rigorous algebraic reconstructions of the informally rigorous diagrammatic proofs.

Corollary 1. *In proposition 1, the inequality $(a+b)^* \leq a^*b^*$ can be strengthened to an equality, since $a^*b^* \leq (a+b)^*$.*

Proof. By (19), it suffices to show that $a^*b^* \leq a^*(ba^*)^*$. By (13), $1 \leq a^*$. By monotonicity of multiplication and the Kleene star, therefore $b \leq ba^*$, $b^* \leq (ba^*)^*$ and $a^*b^* \leq a^*(ba^*)^*$. □

Theorem 4 (Church-Rosser). *Let A be a Kleene algebra. For all $a, b \in A$,*

$$b^*a^* \leq a^*b^* \Leftrightarrow (a+b)^* \leq a^*b^*.$$

Proof. By proposition 1 (ii), it remains to prove that $(a+b)^* \leq a^*b^*$ implies $b^*a^* \leq a^*b^*$. By corollary 1 and the hypothesis, $b^*a^* \leq (a+b)^* \leq a^*b^*$. □

All calculations for the Church-Rosser theorems—starting from scratch—are straightforward and induction-free. They often use fixed-point induction (axioms (6), (7)) instead. This simplicity also appears in our formalization with the proof checker Isabelle [25]. We generally reach a high degree of automation.

Proposition 1 and theorem 4 can be generalized to sums $\sum_{i=0}^{n} a_i$ and products $\prod_{i=0}^{n} a_i$ in more than two variables. The proofs (c.f. [24]) use induction on the number of generators and proposition 1 as base cases. The induction is not difficult, but proofs become longer and require some bookkeeping.

Proposition 2. *Let A be a Kleene algebra. For all $a_0, \ldots, a_n \in A$.*

(i) $a_j^*a_i^* \leq a_i^*a_j^*$ *for all* $0 \leq i < j \leq n$ *implies* $(\sum_{i=0}^{n} a_i)^* \leq \prod_{i=1}^{n} a_i^*$.
(ii) $a_ja_i \leq a_ia_j$ *for all* $0 \leq i < j \leq n$ *implies* $(\sum_{i=0}^{n} a_i)^* \leq \prod_{i=1}^{n} a_i^*$.

Proposition 2 (i) is related to Church-Rosser theorems modulo an equivalence relation [12]. Proposition 2 (ii) is an algebraic completeness proof of bubble sort.

We now calculate Church-Rosser proofs for the Kleene plus.

Proposition 3. *Let A be a Kleene algebra. For all $a, b \in A$*

(i) $ba^* \leq a^*b^+ + a^+b^* \Rightarrow (a+b)^+ \leq a^*b^+ + a^+b^*$.
(ii) $b^+a^* \leq a^*b^+ + a^+b^* \Rightarrow (a+b)^+ \leq a^*b^+ + a^+b^*$.
(iii) $ba \leq ab \Rightarrow (a+b)^+ \leq a^*b^+ + a^+b^*$.

Proof. (ad i) By (10) it suffices to show $(a+b)(a^*b^+ + a^+b^*) \leq a^*b^+ + a^+b^*$, thus by distributivity and (3),

$$aa^*b^+ \leq a^*b^+ + a^+b^*, \quad (31) \qquad ba^*b^+ \leq a^*b^+ + a^+b^*, \quad (33)$$

$$aa^+b^* \leq a^*b^+ + a^+b^*, \quad (32) \qquad ba^+b^* \leq a^*b^+ + a^+b^*. \quad (34)$$

(31) holds, since $aa^*b^+ \leq a^*b^+$ by (15) and (14) and $a^*b^+ \leq a^*b^+ + a^+b^*$ by definition of $\leq$.

(32) holds, since $aa^+b^* \leq a^+b^*$ by (23) and monotonicity of multiplication. Moreover, $a^+b^* \leq a^*b^+ + a^+b^*$ by definition of $\leq$.

(33) depends on the assumption and therefore is beyond automata theory.

$$\begin{aligned} ba^*b^+ &\leq a^*b^+b^+ + a^+b^*b^+ && \text{by assumption, distributivity} \\ &\leq a^*b^+ + a^+b^*b^* && \text{by (20), (22), monotonicity of + and } \cdot \\ &= a^*b^+ + a^+b^* && \text{by (14), monotonicity of + and } \cdot \end{aligned}$$

(34) also depends on the assumption.

$$\begin{aligned} ba^+b^* &\leq ba^*b^* && \text{by (22)} \\ &\leq a^*b^+b^* + a^+b^*b^* && \text{by assumption, distributivity} \\ &\leq a^*b^+ + a^+b^* && \text{by (24), (14), monotonicity of + and } \cdot \end{aligned}$$

(ad ii) It suffices to show $b^+a^* \leq a^*b^* \Rightarrow ba^* \leq a^*b^*$. This is straightforward from (21) and monotonicity of multiplication.

(ad iii) It suffices to show $ba \leq ab \Rightarrow ba^+ \leq a^*b^+ + a^+b^*$. By(25), (21), the assumption, monotonicity of addition and multiplication and the definition of $\leq$, we obtain $ba^+ \leq a^+b \leq a^+b^* \leq a^+b^* + a^*b^+$. □

Theorem 5 (Church-Rosser). *Let A be a Kleene algebra. For all $a, b \in A$,*

$$b^+a^* \leq a^*b^+ + a^+b^* \Leftrightarrow (a+b)^+ \leq a^*b^+ + a^+b^*.$$

Proof. By proposition 3 (ii), it remains to show that $(a+b)^+ \leq a^*b^+ + a^+b^*$ implies $b^+a^* \leq a^*b^+ + a^+b^*$. We show $b^+a^* \leq (a+b)^*$. The rest follows from transitivity and the hypothesis. By (27) it suffices to show that $b^+a^* \leq (a^*b)^+a^* + a^+$ and therefore $b^+a^* \leq (a^*b)^+a^*$ by definition of $\leq$. By (15), $1 \leq a^*$. By monotonicity of multiplication and the Kleene plus, therefore $b \leq a^*b$, $b^+ \leq (a^*b)^+$ and $b^+a^* \leq (a^*b)^+a^*$. □

Obviously, the proofs for the Kleene plus are somewhat harder than those with the Kleene star. Like in the proof of proposition 2, proposition 3 yields base cases for inductive proofs of Church-Rosser statements with more than two generators.

Theorem 4 and theorem 5 are the basis for non-symmetric rewriting with quasiorderings and non-symmetric transitive relations [23]. They can be connected with the term structure by bridge lemmata that are usually called *critical pair lemmata*. The inequalities $b^*a^* \leq a^*b^*$ are examples of *semicommutation* properties (as opposed to *commutation* properties like $ab = ba$). As we will see in the next section, their interpretations in the relational model generalize the diamond property. Non-symmetric Church-Rosser theorems generalize their equational counterparts.

6 Conversion and Functions

To obtain precisely an algebraic variant of the Church-Rosser theorem (theorem 1), we add a *converse* operator. A *Kleene algebra with converse* [6] is a tuple $(A, +, \cdot, ^*, ^\circ, 0, 1)$, where $(A, +, \cdot, ^*, 0, 1)$ is a Kleene algebra and the unary operation $^\circ$ is a contravariant involution that distributes with $+$ and *:

$$a^{\circ\circ} = a, \qquad (35)$$
$$(a+b)^\circ = a^\circ + b^\circ, \qquad (36)$$
$$(ab)^\circ = b^\circ a^\circ, \qquad (37)$$
$$a^{*\circ} = a^{\circ *}, \qquad (38)$$
$$a + aa^\circ a = aa^\circ a. \qquad (39)$$

It is easy to show that the converse is monotonic and that $1 = 1^\circ$ and $0 = 0^\circ$.

Kleene algebras with converse allow the definition of partial and total functions, injections or surjections. We will need functions in section 7. An element a of a Kleene algebra A is *total* or *entire*, if $1 \le aa^\circ$, *functional* or *simple*, if $a^\circ a \le 1$ and a *function*, if it is total and functional. From these definitions one can derive many of the usual properties of functions. We need only two of them.

Lemma 5. *The properties simple, entire, map are preserved by composition.*

Lemma 6. *Let A be a Kleene algebra with converse. For all $a, b, f \in A$ and f a function,*

$$af \le b \Leftrightarrow a \le bf^\circ, \qquad (40) \qquad\qquad f^\circ a \le b \Leftrightarrow a \le fb. \qquad (41)$$

The following corollary, interpreted in the relational model according to theorem 3, yields the Church-Rosser statement of theorem 1, using the relational abstraction of lemma 1.

Corollary 2. *Let A be a Kleene algebra with converse. For all $a \in A$,*

$$a^{\circ *}a^* \le a^*a^{\circ *} \Leftrightarrow (a + a^\circ)^* \le a^*a^{\circ *}.$$

Proof. Let $b = a^\circ$ in theorem 4 (i). □

A point-free induction-based proof of this theorem in the relational model has been given in [20]. Similarly, the Church-Rosser theorems for three generators specialize to a variant of a Church-Rosser theorem modulo a congruence.

Corollary 3. *Let A be a Kleene algebra with converse. For all $a, e \in A$,*

$$ea^* \le a^*e^*, a^{\circ *}a^* \le a^*a^{\circ *}, a^{\circ *}e^* \le e^*a^{\circ *} \Rightarrow (a + e + a^\circ)^* \le a^*e^*a^{\circ *}.$$

Proof. Let $n = 2$, $a_0 = a$, $a_1 = e$ and $a_2 = a^*$ in theorem 2 (i). □

The transition from Church-Rosser theorems without to those with converse did not use any algebraic properties of the converse. It worked by mere substitution. Thus corollary 2 and its extension to Church-Rosser modulo are simple instances of proposition 1 and 2. However, the presented properties of converse are used in the following section.

7 The Church-Rosser Theorem of the λ-Calculus

In the Church-Rosser theorem of the λ-calculus, the hypothesis of our Church-Rosser theorem is shown. In this section, we reconstruct the proof in Barendregt [2] (pp.279-283). There some bridge lemmata deal with the term structure. We abstract them such that large parts of the proofs are again Kleene algebra.

Let Λ be the set of λ-terms on a set of variables X and let $\rightarrow_\beta$ be the associated reduction relation. $s[t/x]$ denotes substitution of a term t for all occurrences of the variable x in a term s. The proof in [2] is based on redex-indexing to trace their behavior in reduction sequences. Intuitively, an indexed λ-term is obtained from a λ-term by indexing some of its redices $(\lambda x.s)t$ as $(\lambda_i x.s)t$ in $\mathbb{N}$. Λ' denotes the set of indexed λ-terms. Obviously $\Lambda \subseteq \Lambda'$. We extend $\rightarrow_\beta$ to $\rightarrow_{\beta'} = \rightarrow_{\beta_0} + \rightarrow_{\beta_1}$ on indexed lambda terms such that $\rightarrow_{\beta_0}$ acts on the indexed and $\rightarrow_{\beta_1}$ on the non-indexed part. We (recursively) define the mappings $\pi : \Lambda' \longrightarrow \Lambda$ that forgets all indices and $\sigma : \Lambda' \longrightarrow \Lambda$ that β-reduces all indexed redices in a term from inside out:

$$\begin{aligned}\sigma(x) &= x\\ \sigma(\lambda x.s) &= \lambda x.\sigma(s)\\ \sigma(st) &= \sigma(s)\sigma(t), \qquad \text{if } s \neq \lambda_i x.s',\\ \sigma((\lambda_i x.s)t &= \sigma(s)[\sigma(t)/x].\end{aligned}$$

We need types for modeling $\rightarrow_\beta$, $\rightarrow_{\beta'}$, π and σ in Kleene algebra. These can be borrowed, for instance from allegories [10], see also [24]. Here, we leave the type-reasoning implicit. In [2] there are three relevant properties of λ-terms.

1. There is no difference between performing a β'-reduction on an indexed term and forgetting the indices then or forgetting the indices and then performing a β-reduction (Lemma 11.1.6 (i) in [2]). This is expressed algebraically as
$$\pi \rightarrow_\beta = \rightarrow_{\beta'} \pi. \tag{42}$$
2. Every β'-reduction of an indexed term followed by a σ-application can be simulated by a σ-application followed by a β-reduction. This is expressed as
$$\rightarrow_{\beta'} \sigma \leq \sigma \rightarrow_\beta. \tag{43}$$
By (41) this is equivalent to $\sigma^\circ \rightarrow_{\beta'} \sigma \leq \rightarrow_\beta$ (Lemma 11.1.7 (i), (ii) in [2]).
3. All σ-applications to indexed terms can be simulated by forgetting the indices followed by β-reductions. This is expressed as
$$\sigma \leq \pi \rightarrow_\beta^* \tag{44}$$
By (41) this is equivalent to $\pi^\circ\sigma \leq \rightarrow_\beta^*$ (Lemma 11.1.8 in [2]).

It is easy to check that these bridge lemmata are well-typed. The remainder of the Church-Rosser theorem is then simple Kleene algebra, which again precisely reconstructs the usual diagrammatic reasoning.

Lemma 7. *Let A be a (typed) Kleene algebra with converse and assume that (42), (43) and (44) holds for $\rightarrow_\beta, \rightarrow_{\beta'}, \pi, \sigma \in A$.*

$$\pi \rightarrow_\beta^* = \rightarrow_{\beta'}^* \pi, \quad (45) \qquad \sigma^\circ \rightarrow_{\beta'}^* \sigma \leq \rightarrow_\beta, \quad (47)$$

$$\rightarrow_{\beta'}^* \sigma \leq \sigma \rightarrow_\beta^*, \quad (46) \qquad \pi \leq \sigma \leftarrow_\beta. \quad (48)$$

Proof. (45) and (46) follow immediately from (42) and (43) by (17).

(47) immediately follows from (46) by (41).

For (48), note that $\pi^\circ \leq \rightarrow_\beta^* \sigma^\circ$ is equivalent to (44) by (40). Thus

$$\pi = \pi^{\circ\circ} \leq (\rightarrow_\beta^* \sigma^\circ)^\circ = \sigma \leftarrow_\beta^*$$

by properties of the converse. □

The following is an algebraic version of the strip lemma (lemma 11.1.9 in [2]).

Proposition 4 (Strip Lemma). *Under the assumptions of lemma 7,*

$$\leftarrow_\beta \rightarrow_\beta^* \leq \rightarrow_\beta^* \leftarrow_\beta^*.$$

Proof. Index the redex where the one-step expansion occurs. At this point, therefore, $\rightarrow_\beta = \pi^\circ \sigma$ (which is consistent with (44)).

$$\leftarrow_\beta \rightarrow_\beta^* = (\pi^\circ \sigma)^\circ \rightarrow_\beta^* = \sigma^\circ \pi \rightarrow_\beta^* = \sigma^\circ \rightarrow_{\beta'}^* \pi \leq \sigma^\circ \rightarrow_{\beta'}^* \sigma \leftarrow_\beta^* \leq \rightarrow_\beta^* \leftarrow_\beta^*,$$

by the assumption, properties of the converse, (45), (48) and (47). □

Corollary 4 (Church-Rosser). *Under the assumptions of lemma 7.*

$$(\rightarrow_\beta + \leftarrow_\beta)^* \leq \rightarrow_\beta^* \leftarrow_\beta^*.$$

Proof. By the strip lemma (lemma 4) and the abstract Church-Rosser result of corollary 2, which itself is a corollary of proposition 1. □

8 More Church-Rosser Calculations

In this section we calculate the abstract part of the Church-Rosser theorem of the λ-calculus along the lines of the Tait-Martin-Löf (c.f. [2]) and Takahashi [26] method. This method is nowadays considered standard, in particular for the various approaches to mechanized proof checking, notice only [21,17,18]. Our previous proof uses more diagrammatics, the standard one does more work at the term level. The standard proof is shorter, whereas that of the last section—according to Barendregt—is more perspicuous. A key distinction at the abstract level is that all previous proofs use induction and therefore are essentially higher-order, whereas ours is strictly first-order and often even at the level of finite automata. Nipkow's Isabelle implementation [18] is most concerned with the distinction between the abstract and the term level and therefore closest to ours. He uses a logical definition of a square diagram for proving some simple point-free lemmata for abstract reasoning. The properties of the Kleene star that he needs are probably derived from Isabelle theories about relations, although this is not explicitly mentioned in the text.

We now show that all abstract statements needed for Church-Rosser proofs (c.f. [18]) with the Tait-Martin-Löf and Takahashi method are again simple calculations in Kleene algebra. We fist show two simple auxiliary facts.

Lemma 8. *Let A be a Kleene algebra. For all $a, b \in A$,*

$$b^* = (b+1)^*, \qquad (49) \qquad\qquad (a+b)^* = (a^*+b^*)^*. \qquad (50)$$

Proof. For (49), $(b+1)^* = b^*(1b^*)^* = b^*b^{**} = b^*b^* = b^*$, by (19), (16) and (14).

For (50), $(a^*+b^*)^* = a^{**}(b^*a^{**})^* = a^*(b^*a^*)^* = (a+b^*)^*$ by (19) and (16). Elimination of b^* is similar. □

The next statement is essential for the Church-Rosser proof using the Tait-Martin-Löf and Takahashi method (lemma 6 in [18]).

Lemma 9. *Let A be a Kleene algebra with converse. For all $a, b \in A$.*

$$a^\circ a \leq aa^\circ, b \leq a, a \leq b^* \Rightarrow b^{\circ *}b^* \leq b^*b^{\circ *}.$$

Proof. We first show that $b \leq a$ and $a \leq b^*$ imply $a^* = b^*$. First $b^* \leq a^*$ follows from $b \leq a$ by monotonicity of the Kleene star. Moreover $a^* \leq b^{**} = b^*$ follows from $a \leq b^*$ by monotonicity of the Kleene star and (16).

Applying (17) twice to $a^\circ a \leq aa^\circ$ yields $a^{\circ *}a^* \leq a^*a^{\circ *}$. Moreover $a^* = b^*$ implies $a^{\circ *} = b^{\circ *}$ by monotonicity of the converse and (38). Therefore, replacing equals by equals, $b^{\circ *}b^* \leq b^*b^{\circ *}$. □

The next statement collects some alternatives to the diamond property.

Lemma 10. *Let A be a Kleene algebra. For all $a, b, c \in A$,*

(i) $ab \leq c(a+1), b \leq c \Rightarrow (a+1)b \leq c(a+1)$,
(ii) $ba \leq a^*(b+1) \Rightarrow b^*a^* \leq a^*b^*$,
(iii) $ba \leq (a+1)(b+1) \Rightarrow b^*a^* \leq a^*b^*$.

Proof. (ad i) $(a+1)b = ab+b \leq c(a+1)+b \leq c(a+1)+c = c(a+1+1) = a(a+1)$, by distributivity, the assumption, monotonicity and idempotence.

(ad ii)

$$\begin{aligned} ba \leq a^*(b+1) &\Rightarrow (b+1)a \leq a^*(b+1) && \text{by (i), (15)} \\ &\Rightarrow (b+1)a^* \leq a^{**}(b+1) && \text{by (17)} \\ &\Rightarrow (b+1)^*a^* \leq a^*(b+1)^* && \text{by (16), (17)} \\ &\Rightarrow b^*a^* \leq a^*b^* && \text{by(49)} \end{aligned}$$

(ad iii) Since $(a+1) \leq a^*$ by (15) and (49) this follows immediately from (ii) and monotonicity of multiplication. □

Lemma 10 (ii) (and (iii)) is known as the *Hindely-Rosen commutation lemma* (lemma 5 in [18]). Finally, we also calculate the *Hindley-Rosen commutative union theorem* (part (ii) of lemma 11, lemma 8 in [18]), which serves to show the Church-Rosser property in presence of β- and η-reduction.

Lemma 11. *Let A be a Kleene algebra with converse. For all $a, b, c \in A$,*

(i) $c^*a^* \leq a^*c^*, c^*b^* \leq b^*c^* \Rightarrow c^*(a+b)^* \leq (a+b)^*c^*$,
(ii) $a^{\circ*}a^* \leq a^*a^{\circ*}, b^{\circ*}b^* \leq b^*b^{\circ*}, a^{\circ*}b^* \leq b^*a^{\circ*} \Rightarrow$
$$\Rightarrow (a+b)^{\circ*}(a+b)^* \leq (a+b)^*(a+b)^{\circ*}.$$

Proof. (ad i) For $c^*(a+b)^* \leq (a+b)^*c^*$, it suffices to show $c^*(a^*+b^*)^* \leq (a^*+b^*)^*c^*$ by (50) and $c^*a^* + c^*b^* \leq a^*c^* + b^*c^*$ by (17) and distributivity. This last inequality follows immediately from the hypotheses and monotonicity of addition.

(ad ii) First note that $(a+b)^{\circ*} = (a^{\circ*}+b^{\circ*})^*$ by (36) and (50). Now for $(a+b)^{\circ*}(a+b)^* \leq (a+b)^*(a+b)^{\circ*}$ it suffices to show

$$(a^{\circ*}+b^{\circ*})^*(a+b)^* \leq (a+b)^*(a^{\circ*}+b^{\circ*})^* \tag{51}$$

by (17). Instantiating (ii) with $c = a^\circ$ and $c = b^\circ$, respectively, yields

$$a^{\circ*}a^* \leq a^*a^{\circ*}, a^{\circ*}b^* \leq b^*a^{\circ*} \Rightarrow a^{\circ*}(a+b)^* \leq (a+b)^*a^{\circ*}, \tag{52}$$
$$b^{\circ*}a^* \leq a^*b^{\circ*}, b^{\circ*}b^* \leq b^*b^{\circ*} \Rightarrow b^{\circ*}(a+b)^* \leq (a+b)^*b^{\circ*}. \tag{53}$$

Clearly, the hypotheses of (52) and (53) are hypotheses of (ii). In particular

$$b^{\circ*}a^* = b^{*\circ}a^{*\circ\circ} = (a^{*\circ}b^*)^\circ = (a^{\circ*}b^*)^\circ \leq (b^*a^{\circ*})^\circ = (b^*a^{*\circ})^\circ = a^{*\circ\circ}b^{*\circ} = a^*b^{\circ*}$$

follows from $a^{\circ*}b^* \leq b^*a^{\circ*}$ by properties of the converse. (51) then follows from monotonicity of addition and distributivity. □

Like in previous sections, all proofs are again simple, short and completely calculational. They cover precisely the abstract part of reasoning for the Tait-Martin-Löf and Takahashi proof and precisely reconstruct the usual diagrammatic reasoning. Compared to the proof in section 7, the arguments at the term level are more involved here. The key idea is as follows: Use lemma 9 to define a relation $\rightarrow_\iota$ between $\rightarrow_\beta$ and $\rightarrow_\beta^*$ with the diamond property. Then $\rightarrow_\beta^*$ has the Church-Rosser property. $\rightarrow_\iota$ is a kind of parallel nested β-reduction. For proving its diamond property, the Tait-Martin-Löf method uses induction on $\rightarrow_\iota$ and a case analysis on the relative positions of reductions. The Takahashi method uses a complete development to recursively contract all redices in a term. All this is beyond the applicability of Kleene algebra. We refer the reader to the literature cited above.

For the Church-Rosser theorem for β and η reduction, it remains to show the diamond property of $\rightarrow_\eta^*$ and semicommutation of $\rightarrow_\beta^{\circ*}$ with $\rightarrow_\eta^*$, in order to apply the Hindley-Rosen commutative union lemma. This is again by induction on the term structure, using the Hindely-Rosen commutation lemma.

The calculation of the Hindely-Rosen commutation lemma and commutative union lemma are interesting in their own right. By the results of this section we see that Kleene algebra is not restricted to one particular proof technique, but can handle the abstract part of the classical Church-Rosser proofs. It might even be the case that Kleene algebra suffices to prove all properties of abstract reduction or term rewrite systems that do not involve well-foundedness.

9 Discussion

The Church-Rosser theorems in section 5 are interesting for the foundations of rewriting. Theorem 4 is the Church-Rosser theorem for rewriting with quasiorderings [16], proposition 3 that for rewriting with non-symmetric transitive relations [23]. The commutation and semicommutation relations in the hypotheses are not only relevant to Church-Rosser theorems. They can be used to express independence or precedence in execution sequences or Mazurkiewicz traces, in imperative and concurrent programms [14,15,4,7].

Kleene algebra is related to certain allegories (c.f. [10]). Allegories can be understood either as categories with relations as arrows or as typed or heterogeneous relation algebras. In particular, *semicommuting* relational diagrams play the role of categorical functional diagrams. Also the Church-Rosser diagrams are nothing but semicommuting allegorial diagrams via their Kleene algebra semantics. See [24] for further discussion. This suggests to develop a generic machinery for compiling diagrammatic rewriting proofs to Kleene algebra or similar structures for proof checking or, conversely, visualizing (machine generated) proofs in these structures in terms of diagrams.

Another natural question is how to calculate for instance Newman's lemma or the Church-Rosser theorem modulo in Kleene algebra. In [24] we use allegorial tabulation techniques for a Kleene algebra extended by an $^{\omega}$-operation, to model infinite iteration and well-foundedness. A proof of Newman's lemma in a different extension of Kleene algebra appears in [8]. We believe that Newman's lemma is an excellent test example for extensions of Kleene algebra to infinite behavior.

Finally, Church-Rosser proofs in Kleene algebra are also well-suited for automation. In [25] we have specified Kleene algebra in the Isabelle proof-checker and formalized all proofs up to theorem 4. It might be even more interesting to implement Kleene algebra in an automated first-order prover coupled with a decision procedure for automata.

References

1. J.-R. Abrial. *The B-Book.* Cambridge University Press, 1996.
2. H. P. Barendregt. *The Lambda Calculus.* Studies in Logic and the Foundations of Mathematics. North-Holland, revised edition, 1984.
3. S. L. Bloom and Z. Ésik. *Iteration Theories.* EATCS Monographs on Theoretical Computer Science. Springer-Verlag, 1991.
4. E. Cohen. Separation and reduction. In R. Backhouse and J. N. Oliveira, editors, *Proc. of Mathematics of Program Construction, 5th International Conference, MPC 2000*, volume 1837 of *LNCS*, pages 45–59. Springer-Verlag, 2000.

5. J. H. Conway. *Regular Algebra and Finite State Machines*. Chapman and Hall, 1971.
6. S. Crvenkovič, I. Dolinka, and Z. Ésik. The variety of Kleene algebras with conversion in not finitely based. *Theoretical Computer Science*, 230:235–245, 2000.
7. V. Diekert and G. Rozenberg, editors. *The Book of Traces*. World Scientific, 1995.
8. H. Doornbos, R. C. Backhouse, and J. van der Woude. A calculation approach to mathematical induction. *Theoretical Computer Science*, 179:103–135, 1997.
9. S. Eilenberg. *Automata, Languages and Machines*, volume A. Academic Press, 1974.
10. P. Freyd and A. Scedrov. *Categories, Allegories*. North-Holland, 1990.
11. D. Harel, D. Kozen, and J. Tiuryn. *Dynamic Logic*. MIT Press, 2000.
12. J.-P. Jouannaud and H. Kirchner. Completion of a set of rules modulo a set of equations. *SIAM J. Comput.*, 15:1155–1194, 1986.
13. D. Kozen. A completeness theorem for Kleene algebras and the algebra of regular events. *Information and Computation*, 110(2):366–390, 1994.
14. D. Kozen. Kleene algebra with tests and commutativity conditions. In T. Margaria and B. Steffen, editors, *Proc. of TACAS'96*, volume 1055 of *LNCS*, pages 14–33. Springer-Verlag, 1996.
15. D. Kozen and M.-C. Patron. Certification of compiler optimizations using Kleene algebra with tests. In J. Lloyd, V. Dahl, U. Furbach, M. Kerber, K.-K. Lau, C. Palmadessi, L.-M. Pereira, V. Sagiv, and P.-J. Stuckey, editors, *1st International Conference on Computational Logic (CL2000)*, volume 1861 of *LNAI*, pages 568–582. Springer-Verlag, 2001.
16. J. Levy and J. Agustí. Bi-rewrite systems. *J. Symbolic Comput.*, 22:279–314, 1996.
17. J. McKinna and R. Pollack. Some lambda calculus and type theory formalized. *J. Automated Reasoning*, 23(3–4):373–409, 99.
18. T. Nipkow. More Church-Rosser proofs (in Isabelle/HOL). *J. Automated Reasoning*, 26(1):51–66, 2001.
19. V. N. Redko. On defining relations for the algebra of regular events. *Ukrainian Mathematical Journal*, 16:120–126, 1964.
20. G. W. Schmidt and T. Ströhlein. *Relations and Graphs: Discrete Mathematics for Computer Scientists*. EATCS Monographs on Theoretical Computer Science. Springer-Verlag, 1993.
21. N. Shankar. A mechanical proof of the Church-Rosser theorem. *Journal of the ACM*, 35(3):475–522, 1988.
22. J. M. Spivey. *Understanding Z*. Cambrigde University Press, 1988.
23. G. Struth. *Canonical Transformations in Algebra, Universal Algebra and Logic*. PhD thesis, Institut für Informatik, Universität des Saarlandes, 1998.
24. G. Struth. Church-Rosser proofs in Kleene algebra and allegories. Technical Report 146, Institut für Informatik, Albert-Ludiwigs-Universität Freiburg, 2001. http://www.informatik.uni-freiburg.de/~struth/publications.html.
25. G. Struth. Isabelle-specification and proofs of Church-Rosser theorems. http://www.informatik.uni-freiburg.de/~struth/publications.html, 2001.
26. M. Takahashi. parellel reductions in λ-calculi. *Information and Computation*, 118(1):120–127, 1995.

On the Definition and Representation of a Ranking

Kim Cao-Van and Bernard De Baets

Department of Applied Mathematics, Biometrics and Process Control
Ghent University, Coupure links 653, B-9000 Gent, Belgium
{Kim.CaoVan,Bernard.DeBaets}@rug.ac.be

Abstract. In this paper, we discuss how a proper definition of a ranking can be introduced in the framework of supervised learning. We elaborate on its practical representation, and show how we can deal in a sound way with reversed preferences by transforming them into uncertainties within the representation.

Keywords: Ranking, Reversed preference, Supervised learning.

1 Introduction

Supervised learning has been studied by many research groups, largely coming from statistics, machine learning and information systems science. In these studies, the problems of classification (discrete) and regression (continuous) have received a lot of attention. More recently, the problem of ranking has made its appearance on the scene because of the wide variety of applications it can be used for.

Ranking can be interpreted as monotone classification or monotone regression. The addition of the word "monotone" to the definition is, however, less trivial than it seems. And the problems this addition to the definition entails in the mathematical model used to deal with classification or regression are even more persistent. In this paper, we will focus on discrete models, in other words, we will discuss how we can deal with monotone classification, which is equivalent to monotone ordinal regression.

2 Problems with Earlier Proposals

The aim of supervised learning is to discover a function $f : \Omega \to \mathcal{D}$ based on a finite set of example pairs $(a, f(a))$ with $a \in \Omega$. If $\mathcal{D}$ is finite, then f is referred to as a classification. Generally, the objects $a \in \Omega$ are described by means of a finite set $Q = \{q_1, \ldots, q_n\}$ of attributes $q : \Omega \to \mathcal{X}_q$. Therefore, to each $a \in \Omega$ corresponds a vector $\mathbf{a} = (q_1(a), \ldots, q_n(a)) \in \mathcal{X} = \prod_{q \in Q} \mathcal{X}_q$ (called the **measurement space**), and the problem is then restated as learning the function f based on examples $(\mathbf{a}, f(a))$ with $a \in \Omega$. Although this new definition is not less restrictive if handled with care, it does tend to encourage a

H. de Swart (Ed.): RelMiCS 2001, LNCS 2561, pp. 291–299, 2002.

more narrow view, where $f(a)$ is interpreted as $f(\mathbf{a})$. This may lead to conflicting situations, since it is possible that $a, b \in \Omega$, $\mathbf{a} = \mathbf{b}$ but $f(a) \neq f(b)$. We will use the term *doubt* to refer to such a situation.

The problem of ranking is generally formulated as a classification problem in the narrow view, with the additional restriction that it has to be monotone, i.e. for all $\mathbf{x}, \mathbf{y} \in \mathcal{X}$ we must have that $\mathbf{x} \leq_{\mathcal{X}} \mathbf{y}$ implies $f(\mathbf{x}) \leq_{\mathcal{D}} f(\mathbf{y})$, where $(x_1, \ldots, x_n) \leq_{\mathcal{X}} (y_1, \ldots, y_n)$ if and only if $x_i \leq_{\mathcal{X}_{q_i}} y_i$ for $i = 1, \ldots, n$, and the relations $\leq_{\mathcal{X}_q}$ on $\mathcal{X}_q$ and $\leq_{\mathcal{D}}$ on $\mathcal{D}$ are complete orders. Again, conflicting situations may arise, which we will refer to as *reversed preference* (see Section 6). Some authors [4,5] impose some additional restrictions, such as demanding the training data to fulfill the monotonicity requirement, to make sure these conflicts do not occur anymore. Others [1] propose a form of naive conflict resolution.

However, a fundamental flaw in this definition is that it is formulated as a restriction not on the original definition $f : \Omega \to \mathcal{D}$ of a classification, but on its operationalization $f : \mathcal{X} \to \mathcal{D}$, which was introduced in function of the description of the objects. Yet another problem is that ranking is not merely a restriction, but can also be seen as a generalization of classification, in which the equality relation is replaced by an order relation. For example, in the formulation of a ranking given above, we see that "$\mathbf{x} \leq_{\mathcal{X}} \mathbf{y}$ implies $f(\mathbf{x}) \leq_{\mathcal{D}} f(\mathbf{y})$" is an extension of "$\mathbf{x} = \mathbf{y}$ implies $f(\mathbf{x}) = f(\mathbf{y})$". It is well known that different points of view on a basic definition may lead to completely different extensions, so, if possible, the most intrinsic definition should be chosen. In our case, this means $f : \Omega \to \mathcal{D}$ is preferred.

3 Classification and Ranking

Thus, following the preceding discussion, we define a **classification** in Ω as the assignment of the objects belonging to Ω, to some element, called a **class label**, in a finite universe $\mathcal{D}$, which we will call the **decision space**. The class labels can be identified with their inverse image in the object space Ω, where they constitute a partition. We will call these inverse images **(object) classes**. For any class label $d \in \mathcal{D}$, we denote the corresponding class by $C_d := f^{-1}(d)$, and $\mathrm{Cl} := \{C_d \mid d \in \mathcal{D}\}$. So, the set of all classifications in Ω stands in one-to-one correspondence to the set of all partitions of Ω (which is equivalent to the set of all equivalence relations on Ω).

If we want to define a ranking based on this definition, it becomes clear that there is no room for a concept such as monotonicity since Ω has no inherent structure such as $\mathcal{X}$. Still, we have to plant the seeds for it, such that monotonicity will appear naturally when the measurement space $\mathcal{X}$ is introduced as a representation of Ω. This can be done, following the ideas from [3,6], by returning to the semantics behind ranking, which declares that the higher an object's rank, the more it is preferred.

We can model this preferential information by a complete preorder. A *preorder* R on Ω is a binary relation R on Ω that is reflexive and transitive. It is called *complete* if for all $a, b \in \Omega$ we have aRb or bRa. A preorder can be seen as a special kind of *weak* (also called *large*) *preference relation* [7] S (which is only reflexive), and where the expression aSb stands for "a is at least as good as b" (it is also said that "a *outranks* b"). Given a complete preorder S, we can define a *strict preference relation* P by aPb if and only if aSb and not bSa, and an *indifference relation* I by aIb if and only if aSb and bSa. We clearly have that $S = P \cup I$.

Definition 1. *A* **ranking** *in Ω is a classification $f : \Omega \to \mathcal{D}$, together with an order $\geq_d$ on $\mathcal{D}$. We denote this ranking by $(f, \geq_d)$. Moreover, the order $\geq_d$ defines a weak preference relation S on Ω as follows:*

$$(\forall s \in \mathcal{D})(\forall r \in \mathcal{D})(\forall a \in C_s)(\forall b \in C_r)(aSb \iff s \geq_d r),$$

where $C_d = f^{-1}(d)$ is the class associated with the class label $d \in \mathcal{D}$.

We will only consider **complete** rankings, where $\geq_d$ is a complete order on $\mathcal{D}$, i.e. $(\mathcal{D}, \geq_d)$ is a chain. This is in line with most of the current problems considered in supervised learning. In this way, we have a specific preference structure on Ω linked with the classes. For $a, b \in \Omega$, assume $a \in C_s$ and $b \in C_r$ with $s, r \in \mathcal{D}$. Then we have

$$aPb \iff s >_d r \qquad \text{and} \qquad aIb \iff s = r.$$

So, the set of all rankings in Ω stands in one-to-one correspondence to the set of all weak preference relations S on Ω that are a complete preorder. The classes are formed by the indifference relation I which is an equivalence relation (transitivity holds since S is a complete preorder). Hence, the indifference relation I determines the classification.

4 Representing a Classification

The previous definitions are not really useful in practice since they relate to a universe Ω that is in essence just an enumeration of all the objects. To access some of the interesting properties of the objects, we fall back on a set of attributes Q. In this way, we can represent each object $a \in \Omega$ by a vector $\mathbf{a} = (q_1(a), \ldots, q_n(a)) \in \widehat{\Omega} \subseteq \mathcal{X}$, where $\widehat{\Omega}$ is the set of all measurement vectors corresponding to objects in Ω. This leads also to a representation $\hat{f}$ of the classification $f : \Omega \to \mathcal{D}$ in the following way:

$$\begin{aligned} &\hat{f} : \widehat{\Omega} \to 2^{\mathcal{D}}, \\ &\mathbf{x} \mapsto \hat{f}(\mathbf{x}) = \{f(a) \mid a \in \Omega \wedge \mathbf{a} = \mathbf{x}\}, \end{aligned}$$

where $2^{\mathcal{D}}$ is the power set of $\mathcal{D}$, i.e. the set of all subsets of $\mathcal{D}$. Thus, the representation of a classification is again a classification, but now in the space $\widehat{\Omega} \subseteq \mathcal{X}$, with classes $\mathbf{C}_D := \hat{f}^{-1}(D)$, where $D \subseteq \mathcal{D}$ (we may define a classification $\hat{f}^* : \Omega \to 2^{\mathcal{D}}$ by setting $\hat{f}^*(a) = \hat{f}(\mathbf{a})$)[1]. Moreover, if $\Omega \cong \mathcal{X}$, then $\hat{f} \cong f$.

[1] Since $\hat{f}$ and $\hat{f}^*$ both express the same idea, we will not restrain ourselves of mixing their usage.

This property of isomorphism states that the representation $\hat{f}$ is a very natural one. We certainly want to keep this property in the case of rankings, therefore we will denote the representation of a ranking by $(\hat{f}, Z)$ where Z is some relation on $2^{\mathcal{D}}$, that should be isomorphic to $\geq_d$ if $\Omega \cong \mathcal{X}$.

Moreover, the more general observation that representing a classification results back into a classification is also a very desirable property. We know that the real problem is one of classification, but we will work with a representation, so it would be against our intuition that this representation would become something different from a classification. In the same line of thinking, we would like this property, if possible, to hold for rankings as well.

5 Representing a Ranking

Let $(f, \geq_d)$ be a (complete) ranking. A first remark concerns the range of $\hat{f}$ when we are dealing with rankings. Because of the ordinal nature of the class labels, it is only meaningful to attach intervals to objects. Therefore, we define

$$\begin{aligned} \hat{f} : \widehat{\Omega} &\to \mathcal{D}^{[2]} = \{[r, r'] \mid (r, r') \in \mathcal{D}^2 \wedge r \leq_d r'\}, \\ \mathbf{x} &\mapsto [\hat{f}_\ell(\mathbf{x}), \hat{f}_r(\mathbf{x})], \end{aligned} \tag{1}$$

where

$$\begin{aligned} \hat{f}_\ell(\mathbf{x}) &= \inf\{f(a) \mid a \in \Omega \wedge \mathbf{a} = \mathbf{x}\}, \\ \hat{f}_r(\mathbf{x}) &= \sup\{f(a) \mid a \in \Omega \wedge \mathbf{a} = \mathbf{x}\}. \end{aligned}$$

This entails that the relation Z should be defined on $\mathcal{D}^{[2]}$. There are some other restrictions we should impose. The relation Z should be (i) reflexive, to ensure that equal intervals are also treated equally, (ii) an extension of $\geq_d$, as discussed previously, and (iii) meaningful, just as $\geq_d$ has a meaning in terms of a preference relation. So, we would like to have an interpretation such as

$$\hat{f}(\mathbf{a}) Z \hat{f}(\mathbf{b}) \iff a\widehat{S}b, \tag{2}$$

where $a\widehat{S}b$ means: *based on the information derived from Q and $\hat{f}$, we conclude that a is at least as good as b.* Lastly, closely related to the previous, we would like that (iv) Z does not depend on f, but only on $\geq_d$. Remark that condition (i) has an additional advantage since it enables us to interpret Z as a weak preference relation.

Starting from (ii) and (iii), we will try to extend the semantics behind $\geq_d$. The two following equivalent expressions are the most straightforward way (that enables generalization) to capture these semantics:

$$s \geq_d r \iff (\forall (a,b) \in \Omega^2)((a,b) \in C_s \times C_r \Rightarrow aSb) \tag{3}$$

$$\iff (\exists (a,b) \in \Omega^2)((a,b) \in C_s \times C_r \Rightarrow aSb). \tag{4}$$

These expressions can be easily generalized by replacing $s \geq_d r$ with $D\,Z\,D'$, where $D, D' \in \mathcal{D}^{[2]}$, and C_s (resp. C_r) with $\mathbf{C}_D \neq \emptyset$ (resp. with $\mathbf{C}_{D'} \neq \emptyset$). If we impose reflexivity, and write $D = [s_1, s_2]$, $D' = [r_1, r_2]$, this finally results in

$$[s_1, s_2]\, Z_1\, [r_1, r_2] \iff s_1 \geq_d r_2\,,$$

as a generalization of (3), and

$$[s_1, s_2]\, Z_2\, [r_1, r_2] \iff s_2 \geq_d r_1\,,$$

as a generalization of (4). Relation Z_1 is an order (if we write $\geq_{D_1}$ instead of Z_1, we simply have $[r_1, r_2] \leq_{D_1} [s_1, s_2] \iff r_2 \leq_d s_1$), and relation Z_2 is an interval order[2]. It is clear that these two relations fulfill conditions (i), (ii) and (iv). However, in both cases there are some problems with condition (iii). We have that $[1, 3]$ and $[1, 2]$ are uncomparable w.r.t. Z_1. However, we would prefer an object with label $[1, 3]$ over another with label $[1, 2]$ if that is all we know about these objects, so, following expression (2), we want $[1, 3] Z [1, 2]$. Whereas Z_1 is too restrictive, the relation Z_2 is too loose, being indifferent between $[1, 2]$ and $[1, 3]$.

This means we need to find something in between the generalizations of expressions (3) and (4). Yet another approach, guided by the previous observations, is to first state the desired semantics and to translate them afterwards into expressions of the right form. In that way we define $D\,Z\,D'$ if and only if D is an improvement over D' or D' is a deterioration compared to D. We will now translate this into mathematical expressions. Assuming that $\mathbf{C}_D$ and $\mathbf{C}_{D'}$ are non-empty, we say that D is an **improvement** of D' if and only if

$$\begin{cases} (\exists a \in \Omega)(\forall b \in \Omega)((\mathbf{a}, \mathbf{b}) \in \mathbf{C}_D \times \mathbf{C}_{D'} \Rightarrow aSb) \\ (\forall a \in \Omega)(\exists b \in \Omega)((\mathbf{a}, \mathbf{b}) \in \mathbf{C}_D \times \mathbf{C}_{D'} \Rightarrow aSb)\,. \end{cases} \tag{5}$$

Likewise, D' is a **deterioration** of D if and only if

$$\begin{cases} (\exists b \in \Omega)(\forall a \in \Omega)((\mathbf{a}, \mathbf{b}) \in \mathbf{C}_D \times \mathbf{C}_{D'} \Rightarrow aSb) \\ (\forall b \in \Omega)(\exists a \in \Omega)((\mathbf{a}, \mathbf{b}) \in \mathbf{C}_D \times \mathbf{C}_{D'} \Rightarrow aSb)\,. \end{cases} \tag{6}$$

It immediately strikes that all these expressions can be seen as intermediate to the generalizations of expressions (3) and (4). If we write $D = [s_1, s_2]$ and $D' = [r_1, r_2]$, it can be shown quite easily that

$$(5) \iff (6) \iff ((r_1 \leq_d s_1) \wedge (r_2 \leq_d s_2))\,.$$

Hence, we find that Z is an order, which we will denote by $\geq_D$, defined as follows:

Definition 2. *Let $D, D' \in \mathcal{D}^{[2]}$. If we write $D = [r_1, r_2]$ and $D' = [s_1, s_2]$, we put*

$$[r_1, r_2] \leq_D [s_1, s_2] \iff ((r_1 \leq_d s_1) \wedge (r_2 \leq_d s_2))\,.$$

[2] An **interval order** is a reflexive, complete and Ferrers relation. A relation R on X is called **Ferrers** if $(aRb \wedge cRd) \Rightarrow (aRd \vee cRb)$, for any $a, b, c, d \in X$. The Ferrers property can be seen as a relaxation of transitivity.

It is clear that now all four conditions are met. It should be noted that the order $\leq_D$ was derived from the premise that we only have access to intervals of values to reach a decision. If other information would be available, other orderings might prevail. For example, distributional information might lead to a stochastic ordering, or if risk aversion underlies the decision, then we would end up with a complete order based on leximin. Finally, remark that the order $\leq_D$ is well known and turns $(\mathcal{D}^{[2]}, \leq_D)$ into a complete lattice[3].

6 The Monotonicity Constraint

Up to now, we have given a definition of a (complete) ranking $(f, \geq_d)$ and have shown how it is represented by the (not necessarily complete) ranking $(\hat{f}, \geq_D)$. Note that we did not need any form of monotonicity for the definition or the representation. Monotonicity will arise in a natural way when taking into account the attributes used to describe the properties of objects.

Let us first turn back to classifications. Even if we assume a classification f to be deterministic in the sense that any object $a \in \Omega$ is assigned to exactly one class with label in $\mathcal{D}$, we still cannot guarantee that we have $\hat{f}^*(a) \in \mathcal{D}$ for all $a \in \Omega$. This is a consequence of the possible occurrence of *doubt*.

Definition 3.

(i) *We say there is* **doubt** *between the classification f and the set of attributes Q if*

$$(\exists(a,b) \in \Omega^2)(\mathbf{a} = \mathbf{b} \wedge f(a) \neq f(b)).$$

(ii) *We say there is* **doubt** *inside the representation $\hat{f}$ if*

$$(\exists \mathbf{x} \in \widehat{\Omega})(|\hat{f}(\mathbf{x})| > 1).$$

It is clear that these two notions of doubt coincide. In the case of doubt, the function $\hat{f}^*$ can assign a set of labels to an object, denoting it is not possible to label the object with one specific class label based on the measurement vector. Remark that this does not result in a conflict, as the one discussed in Section 2, when using the representation $\hat{f}$.

In the context of ranking, the attributes have a specific interpretation, and are usually referred to as criteria. A **criterion** is defined as a mapping $c : \Omega \to (\mathcal{X}_c, \geq_c)$, where $(\mathcal{X}_c, \geq_c)$ is a chain, such that it appears meaningful to compare two objects a and b, according to a particular point of view, on the sole basis of their evaluations $c(a)$ and $c(b)$. In this paper, we will only consider **true criteria**, where the induced weak preference relation is a complete preorder defined by $aS_cb \Leftrightarrow c(a) \geq_c c(b)$. We assume to have a finite set $C = \{c_1, \ldots, c_n\}$ at our disposal.

[3] Because this order is the best known order on $\mathcal{D}^{[2]}$, it is the most evident one to consider. In that spirit, Potharst [5] also introduced this order in the setting of rankings, however, without being aware of its semantics.

The **dominance relation**[4] $\rhd$ on Ω w.r.t. C is defined by

$$a \rhd b \iff \begin{cases} (\forall c \in C)(aS_c b) \\ (\exists c \in C)(aP_c b) \end{cases}$$

for any $a, b \in \Omega$. It is said that a **dominates** b. We may also write $b \lhd a$, saying that b **is dominated by** a. We say that a **weakly** dominates b, $a \trianglerighteq b$, if only the first condition holds. Since we are working with true criteria we have that $a \trianglerighteq b$ is equivalent with $\mathbf{a} \geq_{\mathcal{X}} \mathbf{b}$. A basic principle coming from Multicriteria Decision Aid (MCDA) [6] is that $a \trianglerighteq b \Rightarrow aSb$. On the other hand we have that $aSb \iff f(a) \geq_d f(b)$. Merging all these expressions we find in a natural way the monotonicity constraint

$$\mathbf{a} \geq_{\mathcal{X}} \mathbf{b} \Rightarrow f(a) \geq_d f(b)\,.$$

Remark that this constraint does not tolerate the presence of doubt since it advocates $\mathbf{a} = \mathbf{b} \Rightarrow f(a) = f(b)$. Thus, it is too restrictive for applications in supervised learning. The reason for this lies in the fact that we have adopted a principle from MCDA without considering its context: build a ranking based on the set C of criteria. This is a different setting than for supervised learning where we try to reconstruct a ranking based on the set C. In the former, the set C is a framework, in the latter, this same set C is a restriction. We can solve this by applying the same principle but with the additional demand that we restrict our knowledge to the information we can retrieve from C. In that case, we define the dominance relation on $\widehat{\Omega} \subseteq \mathcal{X}$, resulting in $\mathbf{x} \trianglerighteq \mathbf{y}$ if and only if $\mathbf{x} \geq_{\mathcal{X}} \mathbf{y}$, and the principle becomes $\mathbf{x} \trianglerighteq \mathbf{y} \Rightarrow \mathbf{x}\hat{S}\mathbf{y}$. Together with (2), this finally leads to the monotonicity constraint

$$\mathbf{x} \geq_{\mathcal{X}} \mathbf{y} \Rightarrow \hat{f}(\mathbf{x}) \geq_D \hat{f}(\mathbf{y})\,.$$

In this case, doubt is tolerated since $\mathbf{x} = \mathbf{y} \Rightarrow \hat{f}(\mathbf{x}) = \hat{f}(\mathbf{y})$ is a trivial demand, regardless of the cardinality of $\hat{f}(\mathbf{x})$. This also means that the monotonicity constraint reduces to

$$\mathbf{x} >_{\mathcal{X}} \mathbf{y} \Rightarrow \hat{f}(\mathbf{x}) \geq_D \hat{f}(\mathbf{y})\,. \tag{7}$$

Definition 4. *We say there is* **reversed preference** *inside the representation* $(\hat{f}, \geq_D)$ *if*

$$(\exists(\mathbf{x},\mathbf{y}) \in \widehat{\Omega}^2)(\mathbf{x} >_{\mathcal{X}} \mathbf{y} \wedge \hat{f}(\mathbf{x}) <_D \hat{f}(\mathbf{y})).$$

It has to be noted that people can accept doubt in a classification, but they will not accept reversed preference in a ranking.

---

[4] In the literature, the dominance relation is usually denoted by Δ. Because of the symmetrical nature of the symbol Δ, we feel it does not clearly denote its meaning and prefer to use the notation $\rhd$.

Table 1. A simple ranking $(f, \leq)$.

	a_1	a_2	a_3	a_4	a_5	a_6
c	2	1	4	3	6	5
f	1	2	3	4	5	6

Table 2. The consistent representation $(\tilde{f}, \leq_D)$ of $(f, \leq)$.

	$\mathbf{a}_2$	$\leq_{\mathcal{X}}$	$\mathbf{a}_1$	$\leq_{\mathcal{X}}$	$\mathbf{a}_4$	$\leq_{\mathcal{X}}$	$\mathbf{a}_3$	$\leq_{\mathcal{X}}$	$\mathbf{a}_6$	$\leq_{\mathcal{X}}$	$\mathbf{a}_5$
$\hat{f}$	2		1		4		3		6		5
$\tilde{f}$	$[1,2]$	$\leq_D$	$[1,2]$	$\leq_D$	$[3,4]$	$\leq_D$	$[3,4]$	$\leq_D$	$[5,6]$	$\leq_D$	$[5,6]$

7 Transforming Reversed Preference into Doubt: The Consistent Representation

As just mentioned, the occurrence of reversed preference in $\hat{f}$ is not satisfactory. This can be solved by redefining f (or in terms of supervised learning, alter the training data) in such a way that the reversed preferences disappear. A very drastic solution could be to demand that $\mathbf{a} <_{\mathcal{X}} \mathbf{b} \Rightarrow f(a) \leq_d f(b)$ for the training data. Another possibility is to redefine C until the resulting $\hat{f}$ behaves monotonically according to (7). All these proposals have an invasive character, and might even be unfeasible in certain circumstances. We therefore propose another, non-invasive method, which uses all available information, and results in the closest possible consistent representation by defining a mapping $\tilde{f}$ such that $(\tilde{f}, \leq_D)$[5] does not contain reversed preferences anymore. Essentially, we enlarge the uncertainty intervals in a minimal way such that there are no more violations against the monotonicity requirement (7). In other words, we transform the unacceptable reversed preferences into acceptable doubt.

Definition 5. *Let $(\hat{f}, \leq_D)$ be a representation of a ranking in Ω. We now define the mapping $\tilde{f} : \widehat{\Omega} \to \mathcal{D}^{[2]}$ as follows: let $\mathbf{x} \in \widehat{\Omega}$, we set*

$$\tilde{f}(\mathbf{x}) := [\min_{\mathbf{y} \in [\mathbf{x})} \hat{f}_\ell(\mathbf{y}), \max_{\mathbf{y} \in (\mathbf{x}]} \hat{f}_r(\mathbf{y})].$$

where $[\mathbf{x}) = \{\mathbf{x}' \in \widehat{\Omega} \mid \mathbf{x} \leq_{\mathcal{X}} \mathbf{x}'\}$ and $(\mathbf{x}] = \{\mathbf{x}' \in \widehat{\Omega} \mid \mathbf{x}' \leq_{\mathcal{X}} \mathbf{x}\}$. We call $(\tilde{f}, \leq_D)$ the **consistent representation** *of the ranking $(f, \leq_d)$.*

If there is no reversed preference inside $(\hat{f}, \leq_D)$, then $\tilde{f}$ coincides with $\hat{f}$.

Let us demonstrate this on a small example taken from [2]. Assume we have $\Omega = \{a_1, \ldots, a_6\}$, a single criterion $c : \Omega \to (\{1, \ldots, 6\}, \leq)$ and a ranking $(f, \leq)$ with $f : \Omega \to \{1, \ldots 6\}$, as shown in Table 1. There is no doubt, so $\Omega \cong \widehat{\Omega}$. Table 2 lists the consistent representation of this ranking.

[5] We prefer to write a ranking as $(f, \leq_d)$ instead of $(f, \geq_d)$, which was only used in this paper to facilitate the connection with notions from preference modelling.

8 Conclusion

We have paved the path to deal in a mathematically and semantically sound way with rankings in the context of supervised learning. We have given a definition, a representation when dealing with attributes (in which case we are on the territory of ordinal regression), and a (consistent) representation when dealing with criteria.

References

1. A. Ben-David, *Automatic generation of symbolic multiattribute ordinal knowledge-based DSSs: methodology and applications.* Decision Sciences **23** (1992), 1357–1372.
2. G. Gediga and I. Düntsch, *Approximation quality for sorting rules*, submitted.
3. S. Greco, B. Mattarazo and S. Słowiński, *Rough set theory for multicriteria decision analysis.* European J. Oper. Res. **129** (2000), 1–47.
4. K. Makino, T. Suda, H. Ono, T. Ibaraki, *Data analysis by positive decision trees.* IEICE Trans. Inf. and Syst. **E82-D** (1999), 76–88.
5. R. Potharst, J.C. Bioch, *Decision trees for ordinal classification.* Intelligent Data Analysis **4** (2000), 97–112.
6. B. Roy, *Multicriteria Methodology for Decision Aiding.* Kluwer Academic Publishers, Dordrecht, 1996.
7. Ph. Vincke, *Basic concepts of preference modelling.* In: Readings in Multiple Criteria Decision Aid (C. Bana e Costa, ed.), Springer Verlag, Berlin-Heidelberg, 1990, pp. 101–118.

Tangent Circle Algebras*

Ivo Düntsch[1] and Marc Roubens[2]

[1] Department of Computer Science, Brock University,
St. Catherines, Ontario, L2S 3A1, Canada, duentsch@cosc.brocku.ca
[2] Institute of Mathematics, Université de Liège, B-4000 Liège, Belgium,
M.Roubens@ulg.ac.be

Abstract. In relational reasoning, one is concerned with the algebras generated by a given set of relations, when one allows only basic relational operations such as the Boolean operations, relational composition, and converse. According to a result by A. Tarski, the relations obtained in this way are exactly the relations which are definable in the three–variable fragment of first order logic. Thus, a relation algebra is a first indicator of the expressive power of a given set of relations.
In this paper, we investigate relation algebras which arise in the context of preference relations. In particular, we study the tangent circle orders introduced in [1].

1 Introduction

Nontransitive preferences have been investigated in decision theory and preference modelling [2, 3]. Preference structures like interval orders, semiorders [4, 5, 6, 7] and tangent circle orders [1, 8, 9] lack transitivity in their symmetric or/and asymmetric parts.

Preferences are binary relations, i.e. sets of ordered pairs of elements from a given set, and thus they can be manipulated by algebraic operations such as the set theoretic $\cap, \cup, -$. Other natural operations on relations are composition and converse which are defined below. A set of binary relations which is closed under these operations and contains the empty relation, the universal relation, and the identity, is called an *algebra of binary relations* (BRA). A. TARSKI has shown that the relations obtained by applying these operations to a given set of operations are exactly those which are definable in the three variable fragment of first order logic. At a first glance, this fragment may seem to be too weak to express reasonable properties of binary relations, but, indeed, it can be quite powerful. For example, symmetry, antisymmetry, transitivity, reflexivity can all be expressed by algebraic equations such as

$$R \text{ is transitive if and only if } R \circ R \subseteq R,$$

* Co-operation for this work was supported by EU COST Action 274 "Theory and Applications of Relational Structures as Knowledge Instruments" (TARSKI), www.tarski.org

H. de Swart (Ed.): RelMiCS 2001, LNCS 2561, pp. 300–313, 2002.

where $\circ$ is relational composition. In some domains, all relations, which are first order definable from a given set of relations can be obtained in this way [10, 11].

Tangent circle orders were introduced in [1] as a generalisation of the concept of interval orders [see also 8]: If $\mathfrak{C} = \{C_a : a \in A\}$ is a collection of closed disks in the Euclidean plane tangent to the x–axis from above, they define the following two binary relations on A:

$$aPrb \iff C_a \text{ lies totally to the left of } C_b, \quad (1)$$

$$a\mathsf{I}b \iff C_a \cap C_b \neq \emptyset. \quad (2)$$

If $aPrb$, then b is said to be *strictly preferred over* a, and if $a\mathsf{I}b$, then a is called *indifferent to* b. The pair $\langle Pr, \mathsf{I}\rangle$ is called a *tangent circle order on* A. It is very easy to build an example where $aPrc, cPrb$ and $a\mathsf{I}b$ [see also 9, Figure 6]. The study of tangent circle orders can be motivated by the following problem. n balls are given in a three dimensional space XYZ and they are tangent to the same horizontal plane XOY. We can construct two tangent circle orders : one of them represents the projection of the balls on the plance XOZ and the second represents their projections on the plane YOZ. The question is to elaborate a ranking of the n balls which are picked up by a robot without altering their integrity and without collisions. The study of the properties of tangent circle orders and their numerical representation is helpful for solving this problem [see 1].

Another type of circle order represented in some plane has been studied by Scheinerman and Wierman [12], in particular the inclusion relation.

In this paper, we investigate the algebras of binary relations generated by these relations and some of its variants. In particular, we consider the algebras generated by I as above on the collections of all open disks and all closed disks in the Euclidean plane. In the sequel, we interpret the term "circle" to mean "disk".

2 Binary Relations and Their Algebras

A binary relation on a set U is just a subset of $U \times U$. If $R, S \subseteq U \times U$, and $x, y, z \in U$, we will usually write xRy for $\langle x, y\rangle \in R$, and $xRySz$ for xRy and yRz. The *range of x in R* is the set

$$Rx \stackrel{\text{def}}{=} \{y \in U : xRy\}.$$

We denote the set of all binary relations on U by $Rel(U)$; clearly, $Rel(U)$ is a Boolean algebra under the usual set operations $\cap, \cup, -$ with smallest element $\emptyset$ and largest element $V \stackrel{\text{def}}{=} U \times U$. We also consider the following operations on $Rel(U)$:

$$R \circ S = \{\langle x, y\rangle : (\exists z \in U)[xRzSy]\}, \quad \text{Composition} \quad (3)$$

$$R^{\smile} = \{\langle x, y\rangle : yRx\}. \quad \text{Converse} \quad (4)$$

The relational powers of R are defined by

$$R^1 = R, \; R^{n+1} = R^n \circ R. \tag{5}$$

The *transitive closure of* R is the union of all powers of R, and we denote it by R^*.

An additional distinguished constant is the identity relation $1'$. The structure

$$\langle Rel(U), \cap, \cup, -, \emptyset, V, \circ, \breve{}, 1' \rangle$$

is called the *full algebra of binary relations on* U. Any subalgebra of $Rel(U)$ is called an *algebra of binary relations* (BRA). We usually identify algebras with their base set and write $A \leq Rel(U)$ if A is a subalgebra of $Rel(U)$. If $\{R_i : i \in I\} \subseteq Rel(U)$, then $\langle R_i \rangle_{i \in I}$ is the subalgebra of $Rel(U)$ generated by $\{R_i : i \in I\}$. Each $S \in \langle R_i \rangle_{i \in I}$ is called *RA – definable from* $\{R_i : i \in I\}$.

The expressiveness of BRAs corresponds to a fragment of first order logic, and the following fundamental result is due to A. Tarski [see 13]:

Theorem 1. *If* $R_0, \dots, R_k \in Rel(U)$, *then* $\langle R_0, \dots, R_k \rangle$ *is the set of all binary relations on* U *which are definable in the (language of the) relational structure* $\langle U, R_0, \dots, R_k \rangle$ *by first order formulas using at most three variables, two of which are free.*

$A \leq Rel(U)$ is called *integral*, if $1'$ is an atom of A. Each finite BRA is complete and atomic, and it is completely described by its *composition table* such as Table 1 in the following sense [14]: If $R_0, \dots, R_n$ are the atoms of A, then

1. $R_i\breve{}$ is an atom for each $i \leq n$,
2. If $R_i \cap (R_j \circ R_k) \neq \emptyset$, then $R_i \subseteq (R_j \circ R_k)$
3. Each R_i is either contained in $1'$ or disjoint from it.

The result of the composition of two elements of A is a set of atoms, and we write this in matrix form. If A is integral, we omit column and row $1'$. For example,

Table 1. A composition table

$\circ$	PP	$PP\breve{}$	PO	DC
PP	PP	V	PP, PO, DC	DC
$PP\breve{}$	$-DC$	$PP\breve{}$	$PP\breve{}, PO$	$PP\breve{}, PO, DC$
PO	PP, PO	$PP\breve{}, PO, DC$	V	$PP\breve{}, PO, DC$
DC	PP, PO, DC	DC	PP, PO, DC	V

the algebra determined by Table 1 has the atoms $PP, PP\breve{}, PO, DC, 1'$; we see that PO and DC are symmetric relations, and that $1'$ is an atom. The entry at position $\langle PP\breve{}, PP \rangle$ for example, means that

$$PP\breve{} \circ PP = -DC = PP \cup PP\breve{} \cup PO \cup 1'.$$

Table 2. Closed circle algebra $\mathcal{C}_c$

$\circ$	TPP	$TPP^{\smile}$	$NTPP$	$NTPP^{\smile}$	PO	EC	DC
TPP	PP	$-(NTPP \cup NTPP^{\smile})$	$NTPP$	$-P$	$-P^{\smile}$	EC, DC	DC
$TPP^{\smile}$	$1', TPP, TPP^{\smile}, PO$	$PP^{\smile}$	$PP^{\smile}, PO$	$NTPP^{\smile}$	$PP^{\smile}, PO$	$PP^{\smile}, PO, EC$	$-P$
$NTPP$	$NTPP$	$-P^{\smile}$	$NTPP$	V	$-P^{\smile}$	DC	DC
$NTPP^{\smile}$	$PP^{\smile}, PO$	$NTPP^{\smile}$	$-(EC \cup DC)$	$NTPP^{\smile}$	$PP^{\smile}, PO$	$PP^{\smile}, PO$	$-P$
PO	PP, PO	$-P$	PP, PO	$-P$	V	$-P$	$-P$
EC	PP, PO, EC	$EC \cup DC$	PP, PO	DC	$-P^{\smile}$	$-(NTPP \cup NTPP^{\smile})$	$-P$
DC	$-P^{\smile}$	DC	$-P^{\smile}$	DC	$-P^{\smile}$	$-P^{\smile}$	V

If a cell contains e.g. Q, R this is is to be interpreted as $Q \cup R$ and $P = PP \cup 1'$.

For the arithmetic and other properties of BRAs we invite the reader to consult [11].

Suppose that $C \in Rel(U)$; C is called a *contact relation*, if

(6) $\qquad C$ is reflexive and symmetric,

(7) $\qquad Cx = Cy$ implies $x = y$.

A contact relation C generates a partial ordering P on U by

(8) $$xPy \overset{\text{def}}{\Longleftrightarrow} Cx \subseteq Cy.$$

P is called the *part of relation of* C. It was noted in [15] that $P = -(C \circ (-C))$, and thus, P is term definable from C by the relation algebraic operations. In the sequel, we will write PP for the asymmetric part of P, i.e. $PP = P \cap -1'$. Two elements x, y are called *comparable*, if xPy or $xP^{\smile}y$, otherwise, *incomparable*. The relation of comparability is denoted by CP, i.e.

$$CP = P \cup P^{\smile}.$$

A BRA $A \leq Rel(U)$ generated by a contact relation will be called a *contact relation algebra* (CRA). The motivation for the name comes from spatial reasoning, and contact binary relations were first defined in the framework of qualitative geometry [16, 17]. Primary examples for contact relations are the following:

1. Suppose that U is the set of all open circles in the Euclidean plane, and that for $x, y \in U$,

   (9) $$xCy \overset{\text{def}}{\Longleftrightarrow} x \cap y \neq \emptyset.$$

   If xCy, we say that x and y *are in contact with each other*. The relation algebra $\mathcal{C}$ generated by C has the structure given in Table 1, with $C = PP \cup PP^{\smile} \cup PO \cup 1'$. It is not hard to see that PP is the $\subsetneq$ relation on U.
2. If we let U be the collection of all non–degenerate closed circles in the Euclidean plane, and define C as in (9), we obtain the algebra whose composition is given in Table 2, with $C = EC \cup PO \cup NTPP \cup NTPP^{\smile} \cup TPP \cup TPP^{\smile} \cup 1'$. Some of the atoms are pictured in Figure 1. If $a, b \in U$ are tangential to each other, then they are connected, but there is no $c \in U$ which is below both a and b. Thus, tangentiality is RA definable in this domain.

   Many examples of contact relation algebras can be found in [18].

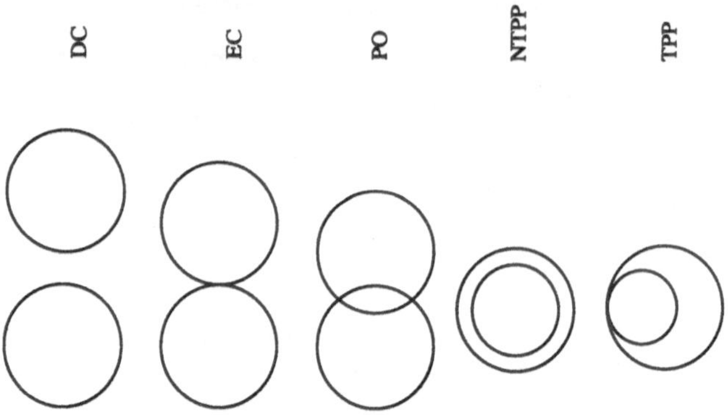

Fig. 1. Closed circle relations

3 Undirected Tangent Circle Algebras

In this section we investigate the BRA generated by the relation C as defined in (9) on the sets of all open or closed circles in the Euclidean plane which are tangent to the x–axis from above. This relation is the relation I defined in (2), and thus, it corresponds to the undirected (symmetric) part of a tangent circle order.

For an (open or closed) circle a, we let x_a be its x – coordinate, and define y_a analogously. If $a \in \mathfrak{C}$, then the radius of a is equal to y_a, and the points on the perimeter of a satisfy the equation

$$(x - x_a)^2 + y^2 - 2y \cdot y_a = 0. \tag{10}$$

With some abuse of notation we also denote the coordinates of a point p in the plane by $\langle x_p, y_p \rangle$. We let $m(a) = \langle x_a, y_a \rangle$ be the centre of a. The Euclidean distance between the centres of two circles a, b is denoted by $d(a, b)$, i.e.

$$d(a, b) = \sqrt{(x_a - x_b)^2 + (y_a - y_b)^2}. \tag{11}$$

It is easy to see that a is tangent to b iff $d(a, b) = y_a + y_b$.

The next result will be useful in the sequel [for a similar result, see 1]:

Theorem 2. *Let a, b be (open or closed) circles tangent to the x–axis from above. The centres of all circles tangent to a and the x–axis from above lie on a parabola with vertex $\langle x_a, 0 \rangle$ and focus $\langle x_a, y_a \rangle = m(a)$. Furthermore,*

$$|\mathrm{cl}(a) \cap \mathrm{cl}(b)| = \begin{cases} 0, & \textit{iff } 4 \cdot y_a \cdot y_b \lneq (x_b - x_a)^2, \\ 2^{\aleph_0}, & \textit{iff } 4 \cdot y_a \cdot y_b \gneq (x_b - x_a)^2. \end{cases}$$

Here, $\mathrm{cl}(a)$ is the closure of a in the usual topology on $\mathbb{R}^2$.

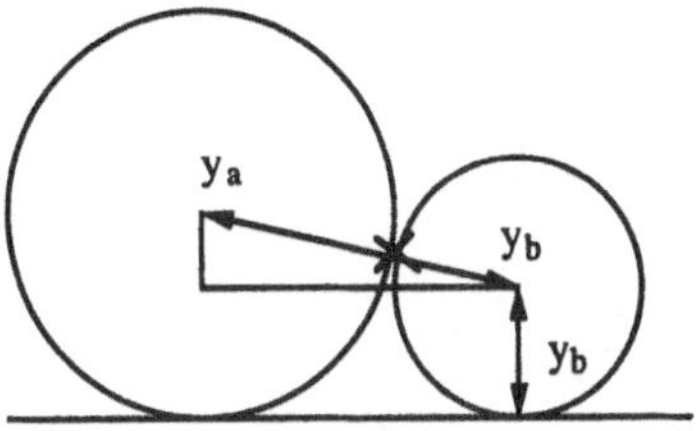

Fig. 2. Tangent circles

Proof. We only prove the first part, whence the second part follows immediately. Let b be tangent to a and the x–axis with $y_b \gneq 0$; then, $m(b)$ has the same distance to a as to the x – axis; in other words,

$$(x_b - x_a)^2 + (y_b - y_a)^2 = (y_a + y_b)^2, \tag{12}$$

see Figure 2. Equation (12) is equivalent to

$$(x_b - x_a)^2 = 4 \cdot y_a \cdot y_b, \tag{13}$$

which proves our claim.

Corollary 1. *Let a, b (open or closed) circles tangent to the x–axis from above.*

1. *If $x_a \lneq x_b$ there is some $c \in \mathfrak{C}$ with $x_a \lneq x_c \lneq x_b$ which is tangent to a and b.*
2. *If $x_a \lneq x_b$ there is some $c \in \mathfrak{C}$ with $x_a \lneq x_c \lneq x_b$ which is disjoint to a and b.*
3. *If $x_a \lneq x_b$ there is some $c \in \mathfrak{C}$ with $x_a \lneq x_c \lneq x_b$ which is intersects both a and b.*
4. *If $a \cap b = \emptyset$, there exists some $c \in \mathfrak{C}$ such that $a \subseteq c$ and c is tangent to b.*

Proof. We only prove 1. By Theorem 2, we have

$$(x_c - x_b)^2 = 4 \cdot y_b \cdot y_c,$$
$$(x_c - x_a)^2 = 4 \cdot y_a \cdot y_c.$$

Solving both equations for $4 \cdot y_c$, equating, and solving for 0, we obtain

$$(x_c - x_b)^2 \cdot y_a - (x_c - x_a)^2 \cdot y_b = 0 \tag{14}$$

$$(y_a - y_b) \cdot x_c^2 + 2(x_a \cdot y_b - x_b \cdot y_a) \cdot x_c + x_b^2 \cdot y_a - x_a^2 \cdot y_b = 0 \tag{15}$$

If $y_a = y_b$, a straightforward computation shows that $x_c = \frac{x_a + x_b}{2}$. Otherwise,

$$x_c^2 + \frac{2(x_a \cdot y_b - x_b \cdot y_a) \cdot x_c + x_b^2 \cdot y_a - x_a^2 \cdot y_b}{y_a - y_b} = 0, \tag{16}$$

and therefore,

(17)

$$x_c = \frac{x_b \cdot y_a - x_a \cdot y_b}{y_a - y_b} \pm \frac{\sqrt{(x_b \cdot y_a - x_a \cdot y_b)^2 - (x_b^2 \cdot y_a - x_a^2 \cdot y_b) \cdot (y_a - y_b)}}{y_a - y_b}$$

(18)

$$= \frac{x_b \cdot y_a - x_a \cdot y_b \pm \sqrt{y_a \cdot y_b} \cdot (x_b - x_a)}{y_a - y_b}.$$

If w.l.o.g. $y_a \lneq y_b$, then the centre of one of the two resulting points will satisfy $x_c \lneq x_a$, and c will touch a and b from "above". For the other point, we have $x_a \lneq x_c \lneq x_b$, and it will touch a and b from "below".

Let $\mathfrak{C}$ be the set of closed circles in the Euclidean plane which are tangent to the x–axis from above, and C be the relation C on $\mathfrak{C}$ defined by (9), i.e.

$$aCb \overset{\text{def}}{\iff} a \cap b \neq \emptyset.$$

We let P be the $\subseteq$ relation, and set $DC \overset{\text{def}}{=} -C$. Now,

Theorem 3. *C is a contact relation on $\mathfrak{C}$.*

Proof. Clearly, C is reflexive and symmetric; it remains to show that $a \subseteq b \iff Ca \subseteq Cb$.

"$\Rightarrow$": Suppose that $a \subseteq b$. If $c \cap a \neq \emptyset$, then, clearly, $c \cap b \neq \emptyset$.

"$\Leftarrow$": Let $a \not\subseteq b$. If $b \subsetneq a$, then $x_b = x_a$ and $y_b \lneq y_a$. Choose c in such a way that $y_c = y_b$, and $x_c = x_b + 2 \cdot y_b$. If $aDCb$ and w.l.o.g. $x_a \lneq x_b$, then choose c such that $x_c = x_a - y_a$ and $y_c = y_a$. In both cases, aCc and $b(-C)c$.

Finally, let $a \cap b \neq \emptyset$, and a, b incomparable; suppose w.l.o.g. that $x_a \lneq x_b$. By Cor. 1, there is some $c \in \mathfrak{C}$ which is tangent to both a and b, and for which $x_a \lneq x_c \lneq x_b$. The circle d with $x_d = x_c - \frac{1}{2} \cdot y_a$ and $y_d = y_c$ intersects a and is disjoint from b.

Besides P and PP, we define the following additional relations:

$O = P^{\smile} \circ P$	common part
$PO = C \cap -(P \cup P^{\smile})$	partial overlap

All these relations are term–definable from C by the relational operators. Furthermore,

Lemma 1. *Let $a, b \in \mathfrak{C}$.*

1. *aOb iff a and b are comparable, i.e. $O = P \cup P^{\smile}$.*
2. *$P^{\smile} \circ P = P \circ P^{\smile}$*

Proof. 1. Let $c \in \mathfrak{C}$. Then,

$$aP^{\smile}cPb \iff x_a = x_c = x_b \text{ and } y_c \leq y_a, y_b$$
$$\iff a \text{ and } b \text{ are comparable.}$$

2. Just as 1.

Table 3. The tangent closed circle algebra $\mathcal{T}_c$

$\circ$	PP	$PP^{\smile}$	PO	DC
PP	PP	O	$-O$	DC
$PP^{\smile}$	O	$PP^{\smile}$	PO	$-O$
PO	PO	$-O$	V	$-P$
DC	$-O$	DC	$-P^{\smile}$	V

The RA generated by C in $\mathfrak{C}$ has the composition given in Table 3; the computations are straightforward. We observe that, unlike the closed circle algebra $\mathcal{C}_c$, the algebra $\mathfrak{C}$ loses the ability to express that two circles are tangential to each other.

It is therefore somewhat surprising, that in the domain $\mathcal{D}$ of open circles tangent to the x–axis, tangentiality is RA expressible: Suppose that $aCb \iff a \cap b \neq \emptyset$ in $\mathcal{D}$. It is not hard to show that C is a contact relation on $\mathcal{D}$ with P being set inclusion. Let $aNTDb$ iff $cl(a) \cap cl(b) = \emptyset$. Observe that $NTD \subsetneq DC$, and set $TD \stackrel{\text{def}}{=} DC \cap -NTD$. Then, $aTDb$ iff a and b are tangential to each other. NTD, and hence, TD are definable from C:

Lemma 2. $PP \circ DC = NTD$.

Proof. "$\subseteq$": Let $aPPcDCb$; then,

$$x_a = x_c,\ y_a \lneq y_c,$$
$$4 \cdot y_c \cdot y_b \leq (x_b - x_c)^2,$$

and thus,

$$4 \cdot y_b \cdot y_a \lneq 4 \cdot y_b \cdot y_c \leq (x_b - x_c)^2 = (x_b - x_a)^2.$$

Using Theorem 2 and Cor. 1, this shows that $aDCb$, and a and b are not tangent to each other.

"$\supseteq$": Suppose that $aNTDb$, i.e. that $4 \cdot y_a \cdot y_b \lneq (x_b - x_a)^2$. Since the reals are dense, we can choose some y such that

$$4 \cdot y_a \cdot y_b \lneq 4 \cdot y \cdot y_b \lneq (x_b - x_a)^2.$$

Then, c with $x_c = x_a$ and $y_c = y$ satisfies $aPPcDCb$.

The complete composition of the algebra generated by C on $\mathcal{D}$ is given in Table 4. Recall that $CP = P \cup P^{\smile}$ is the relation of comparability.

Note that $\mathcal{T}_c$ is isomorphic to the subalgebra of $\mathcal{T}_o$ generated by $C' = C \cup TD$. Obviously,

$$aC'b \iff cl(a) \cap cl(b) \neq \emptyset.$$

Table 4. The tangent open circle algebra $\mathcal{T}_o$

$\circ$	PP	$PP^{\smile}$	PO	TD	NTD
PP	PP	CP	$-CP$	NTD	NTD
$PP^{\smile}$	CP	$PP^{\smile}$	PO	PO	PO, DC
PO	PO	$-CP$	V	$PP^{\smile}, PO, DC$	$PP^{\smile}, PO, DC$
TD	PO	NTD	PP, PO, DC	$-(PP \cup PP^{\smile})$	$-P$
NTD	PO, DC	NTD	PP, PO, DC	$-P^{\smile}$	V

Tangent circle orders are but one of a family of parameterised preference structures as investigated by [9] with Pr and I as defined in (1) and (2). In the rest of the Section we shall briefly digress to other algebras in this family. First, we look at the algebra generated by the set $\mathcal{I}$ of closed intervals of the real line. Observe that we can regard an interval $[a, b]$ as the projection of that $c \in \mathfrak{C}$ with centre $x_c = \frac{a+b}{2}$ and y_c $\frac{|a-b|}{2}$.

A temporal interpretation of the possible positions which such intervals can have with respect to each other has been given in [19]; these are shown in Table 5.

Table 5. Interval relations

before:	$\{\langle [q,r],[q',r'] \rangle : q < r < q' < r', q,r,q',r' \in \mathbb{R}\}$
meets:	$\{\langle [q,r'],[q',r] \rangle : q < r = q' < r', q,r,q',r' \in \mathbb{R}\}$
overlaps:	$\{\langle [q,r],[q',r'] \rangle : q < q' < r < r', q,r,q',r' \in \mathbb{R}\}$
starts:	$\{\langle [q,r],[q',r'] \rangle : q = q' < r < r', q,r,q',r' \in \mathbb{R}\}$
ends:	$\{\langle [q,r],[q',r'] \rangle : q' < q < r = r', q,r,q',r' \in \mathbb{R}\}$
contains:	$\{\langle [q,r],[q',r'] \rangle : q < q' < r' < r, q,r,q',r' \in \mathbb{R}\}$

If we forget the direction and let $aCb \iff a \cap b \neq \emptyset$ then we obtain an algebra which has the same composition as the open circle algebra of Table 1, and where

$$
\begin{aligned}
PP &= \text{contains} \cup \text{starts} \cup \text{ends}, \\
PO &= \text{meets} \cup \text{meets}^{\smile} \cup \text{overlaps} \cup \text{overlaps}^{\smile}, \\
DC &= \text{before} \cup \text{before}^{\smile},
\end{aligned}
$$

If we consider the set $\mathfrak{D}$ of squares tangent to the x–axis from above, then it is easy to see that the relation algebras generated by these squares instead of intervals are just the algebras of intervals given above.

Yet another geometrical interpretation of a preference structure are the diamond orders; here, the objects are parallelograms touching the x–axis from above, and having the same angle at this point.

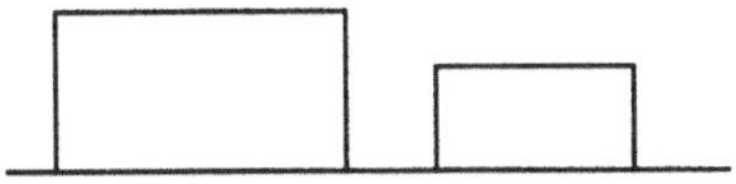

Fig. 3. Tangent squares

4 Directed Tangent Circle Algebras

In this Section we look at the relation which can be obtained by the full preference structure $\langle Pr, \mathsf{I}\rangle$ on $\mathfrak{C}$; recall that $\mathsf{I} = C$. The table of $\mathcal{T}_c$ shows that Pr is not RA – definable from C; however, we can obtain Pr by splitting DC into two disjoint asymmetric parts in the following way: Define $Pr \subseteq Rel(\mathfrak{C})$ by

$$aPrb \overset{\text{def}}{\iff} a \cap b = \emptyset \text{ and } x_a \lneq x_b.$$

Then, clearly, Pr is the preference relation of (1). Moreover,

Theorem 4. *1. $DC = Pr \cup Pr^{\smile}$.*
2. Pr is asymmetric.

Our task will be to find which relational properties of tangent circles are RA expressible from Pr on $\mathfrak{C}$. Since DC, and thus C, are RA definable from Pr, all relations of $\mathcal{T}_c$ can be expressed. PO splits as well: Set $POL \overset{\text{def}}{=} (PP^{\smile} \circ Pr) \cap PO$.

Lemma 3. *$aPOLb \iff aPOb$ and $x_a \lneq x_b$.*

Proof. "$\Rightarrow$": Let $aPP^{\smile}cPrb$ and $aPOb$; then, $x_a = x_c$, and $x_c \lneq x_b$.

"$\Leftarrow$": Let $aPOb$ and $x_a \lneq x_b$. Choose $c \in \mathfrak{C}$ such that

$$x_c = x_a,\ y_c = \frac{(x_b - x_a)^2}{5 \cdot y_b}.$$

Then, $x_c \lneq x_b$, and

$$4 \cdot y_c \cdot y_b = 4 \cdot \frac{(x_b - x_a)^2}{5 \cdot y_b} \cdot y_b \lneq (x_b - x_a)^2 = (x_b - x_c)^2;$$

this proves $cPrb$. Finally, $aPOb$ implies that $4 \cdot y_a \cdot y_b \gneq (x_b - x_a)^2$, and therefore, $4 \cdot y_c \cdot y_b \lneq (x_b - x_a)^2$ shows that $y_c \lneq y_a$.

Note that $PO = POL \cup POL^{\smile}$. Furthermore,

Lemma 4. *1. $POL \cup Pr$ is the transitive closure of Pr.*
2. $POL \cup Pr$ is the transitive closure of POL.

Proof. 1. We show that $Pr \circ Pr = POL \cup Pr$:

"$\subseteq$": Let $aPrbPrc$; then, $x_a \lneq x_c$. If $a \cap c \neq \emptyset$, then $aPOLc$, otherwise, $aPrc$.

"$\supseteq$": If $a(POL \cup Pr)b$, then $x_a \lneq x_b$. Thus, there exists some $c \in \mathfrak{C}$ with $x_a \lneq x_c \lneq x_b$ disjoint from both a and b by Cor. 1.(2)

2. It is enough to show that $POL \circ POL = POL \cup Pr$:

"$\subseteq$": Let $aPOLbPOLc$; then, $x_a \lneq x_c$. If $a \cap c \neq \emptyset$, then $aPOLc$, otherwise, $aPrc$.

"$\supseteq$": Let $a(POL \cup Pr)b$. Then, $x_a \lneq x_b$, and thus there exists some $c \in C$ with $x_a \lneq x_c \lneq x_b$ which partially overlaps both a and b by Corollary 1.

Note that

$$x_a \lneq x_b \text{ iff } a(POL \cup \mathrm{Pr})b. \tag{19}$$

Somewhat surprisingly, Pr distinguishes the horizontal lengths for incomparable circles as well:

Lemma 5. *Let $x_a \lneq x_b$. Then,*

1. $aPOL \circ Pr^{\smile}b \iff y_b \lneq y_a$.
2. $aPr^{\smile} \circ POLb \iff y_a \lneq y_b$.
3. $a - [(POL \circ Pr^{\smile}) \cup (Pr^{\smile} \circ POL)]b \iff y_a = y_b$.

Proof. We only prove 1., since 2. is proved analogously.

"$\Rightarrow$": Let $aPOLcPr^{\smile}b$. Then, $x_a \lneq x_b \lneq x_c$, which implies that $(x_c - x_b)^2 \lneq (x_c - x_a)^2$. Furthermore,

$$4 \cdot y_a \cdot y_c \geq (x_c - x_a)^2, \tag{20}$$

$$4 \cdot y_b \cdot y_c \lneq (x_c - x_b)^2. \tag{21}$$

If $y_a \leq y_b$, then

$$4 \cdot y_a \cdot y_c \leq 4 \cdot y_b \cdot y_c \lneq (x_c - x_b)^2 \lneq (x_c - x_a)^2, \tag{22}$$

which contradicts (20).

"$\Leftarrow$": Suppose that $y_b \lneq y_a$. We are looking for some $c \in \mathfrak{C}$ such that $x_b \lneq x_c$ and

$$4 \cdot y_b \cdot y_c \lneq (x_c - x_b)^2,$$
$$(x_c - x_a)^2 \leq 4 \cdot y_a \cdot y_c,$$

i.e.

$$\frac{(x_c - x_a)^2}{4 \cdot y_a} \leq y_c \lneq \frac{(x_c - x_b)^2}{4 \cdot y_b}.$$

Some straightforward arithmetic shows that we can choose x_c such that

$$\frac{2\cdot(y_a\cdot x_b - y_b\cdot x_a)}{y_a - y_b} \lneq x_c,$$

and the density of $\mathbb{R}$ will give us a suitable y_c.

Using the properties of the converse operation, we immediately obtain

Corollary 2. *Let $x_a \gneq x_b$. Then,*

1. $aPOL^{\smile} \circ Prb \iff y_b \lneq y_a$.
2. $aPr \circ POL^{\smile}b \iff y_a \lneq y_b$.
3. $a - [(POL^{\smile} \circ Pr) \cup (Pr \circ POL^{\smile})]b \iff y_a = y_b$.

This leads to six new RA definable relations and to their converses:

(23) $$aPOLEb \overset{\text{def}}{\iff} aPOLb \text{ and } y_a = y_b,$$

(24) $$aPOLLb \overset{\text{def}}{\iff} aPOLb \text{ and } y_a \lneq y_b,$$

(25) $$aPOLGb \overset{\text{def}}{\iff} aPOLb \text{ and } y_a \gneq y_b,$$

(26) $$aPrEb \overset{\text{def}}{\iff} aPrb \text{ and } y_a = y_b,$$

(27) $$aPrLb \overset{\text{def}}{\iff} aPrb \text{ and } y_a \lneq y_b,$$

(28) $$aPrGb \overset{\text{def}}{\iff} aPrb \text{ and } y_a \gneq y_b.$$

It follows that Pr is able to classify the circles according to the relative positions of their centres.

For $n = 1, 2, 3, \ldots$ let

(29) $$aPrE_nb \overset{\text{def}}{\iff} x_a \lneq x_b \text{ and } y_a = y_b \text{ and } 2n\cdot y_a \lneq x_b - x_a \leq 2(n+1)\cdot y_a.$$

Theorem 5. *$aPrE_nb$ if and only if $aPrE^nb$ and not $aPrE^{n+1}b$.*

Proof. This will follow immediately from

$$aPrE^nb \iff x_a \lneq x_b,\ y_a = y_b,\ 2n\cdot y_a \lneq x_b - x_a.$$

Let $n = 1$; then, $aPrEb$, and this is the case iff $x_a \lneq x_b$, $y_a = y_b$ and $a \cap b = \emptyset$. The condition now follows from Theorem 2.

Suppose the claim is true for $n \geq 1$, and that $aPrE^{n+1}b$. Then, there is some c such that $aPrE^ncPrEb$. Now,

$$\begin{aligned} aPrE^{n+1}b &\iff aPrE^ncPrEb \\ &\iff x_a \lneq x_c \lneq x_b,\ y_a = y_c = y_b,\ 2n\cdot y_a \lneq x_c - x_a,\ 2\cdot y_a \lneq x_b - x_c, \\ &\iff x_a \lneq x_c \lneq x_b,\ y_a = y_c = y_b,\ 2n\cdot y_a + 2\cdot y_a \lneq x_b - x_a, \\ &\iff x_a \lneq x_c \lneq x_b,\ y_a = y_c = y_b,\ 2(n+1)\cdot y_a \lneq x_b - x_a, \end{aligned}$$

which proves our claim.

This shows that the BRA generated by PrE is infinite.

Table 6. Centre discerning relations

$x_a = x_b : \begin{cases} y_a \lneq y_b & \text{iff } aPPb, \\ y_a = y_b & \text{iff } aI'b, \\ y_a \gneq y_b & \text{iff } aPP^{\smile}b \end{cases}$	$x_a \lneq x_b : \begin{cases} y_a \lneq y_b & \text{iff } a(POLL \cup PrL)b, \\ y_a = y_b & \text{iff } a(POLE \cup PrE)b, \\ y_a \gneq y_b & \text{iff } a(POLR \cup PrR)b \end{cases}$

5 Conclusion

In this paper, we have investigated the algebras of binary relations generated by the relations of preference models. It turns out, that the RA formalism, which corresponds to a weak fragment of first order logic, is surprisingly expressive. We have shown that in this formalism, the indifference relation is generated from the preference Pr. Moreover, Pr can differentiate circles by the relative position of their centres, as depicted in Table 6, and also by their distance relative to their diameter.

This contribution clarifies the links and differences that exist between the interval orders and semiorders representations on one hand, and tangent circle orders representations on the other hand. As shown in [8], on finite sets the latter satisfy the property

(30) $$aPrbPrcPrd \Rightarrow aPrc \text{ or } bPrd,$$

which the authors call *quasi-transitivity*. It may be interesting to know under which conditions a binary relation P which satisfies (30), and for which $-(P \cup P^{\smile})$ is a contact relation, can be represented as a tangent circle order.

References

[1] Abbas, M., Vincke, P.: Tangent circle orders : Numerical representation and properties. Working paper 94/06, SMG, Free University of Brussels (1994) `pvincke@smg.ulb.ac.be`.

[2] Fishburn, P.C.: Intransitive indifference with unequal in difference intervals. J. Math. Psychology **7** (1970) 144–149

[3] Fishburn, P.C.: Interval orders and interval graphs. John Wiley, New York (1985)

[4] Bridges, D.S., Mehta, G.B.: Representation of preference orderings. In: Lecture Notes in Economics and Mathematical Systems. Volume 422. Springer-Verlag, Berlin (1995)

[5] Fishburn, P.C.: Nontransitive preferences in decision theory. J. of Risk and Uncertainty **7** (1991) 113–134

[6] Roberts, F.: Measurement Theory. Volume 7 of Encyclopaedia of Mathematics. Addison-Wesley, Reading, Mass. (1979)

[7] Roubens, M., Vincke, P.: Preference Modelling. Volume 250 of Lecture Notes in Economics and Mathematical Sciences. Springer-Verlag, Berlin (1985)

[8] Abbas, M., Pirlot, M., Vincke, P.: Tangent circle orders : a simple non transitive order structure. Manuscript, MATHRO, Faculté polytechnique de Mons (2001) `Marc.Pirlot@fpms.ac.be`.

[9] Fodor, J., Roubens, M.: Parametrized preference structures and some geometrical interpretation. J. Multi-Crit. Decis. Anal. **6** (1997) 253–258
[10] Andréka, H., Düntsch, I., Németi, I.: Binary relations and permutation groups. Mathematical Logic Quarterly **41** (1995) 197–216
[11] Jónsson, B.: The theory of binary relations. In Andréka, H., Monk, J.D., Németi, I., eds.: Algebraic Logic. Volume 54 of Colloquia Mathematica Societatis János Bolyai. North Holland, Amsterdam (1991) 245–292
[12] Scheinerman, E., Wierman, J.: On circle containment orders. Order **4** (1988) 315–318
[13] Tarski, A., Givant, S.: A formalization of set theory without variables. Volume 41 of Colloquium Publications. Amer. Math. Soc., Providence (1987)
[14] Jónsson, B.: Maximal algebras of binary relations. Contemporary Mathematics **33** (1984) 299–307
[15] Düntsch, I., Wang, H., McCloskey, S.: A relation algebraic approach to the Region Connection Calculus. Theoretical Computer Science **255** (2001) 63–83
[16] de Laguna, T.: Point, line and surface as sets of solids. The Journal of Philosophy **19** (1922) 449–461
[17] Nicod, J.: Geometry in a sensible world. Doctoral thesis, Sorbonne, Paris (1924) English translation in *Geometry and Induction*, Routledge and Kegan Paul, 1969.
[18] Düntsch, I., Wang, H., McCloskey, S.: Relation algebras in qualitative spatial reasoning. Fundamenta Informaticae (2000) 229–248
[19] Allen, J.F.: Maintaining knowledge about temporal intervals. Communications of the ACM **26** (1983) 832–843

Author Index

Lecture Notes in Computer Science

For information about Vols. 1–2465

please contact your bookseller or Springer-Verlag

Vol. 2466: M. Beetz, J. Hertzberg, M. Ghallab, M.E. Pollack (Eds.), Advances in Plan-Based Control of Robotic Agents. Proceedings, 2001. VIII, 291 pages. 2002. (Subseries LNAI).

Vol. 2467: B. Christianson, B. Crispo, J.A. Malcolm, M. Roe (Eds.), Security Protocols. Proceedings, 2001. IX, 241 pages. 2002.

Vol. 2468: J. Plaice, P.G. Kropf, P. Schulthess, J. Slonim (Eds.) (Eds.), Distributed Communities on the Web. Proceedings, 2002. XI, 305 pages. 2002.

Vol. 2469: W. Damm, E.-R. Olderog (Eds.), Formal Techniques in Real-Time and Fault-Tolerant Systems. Proceedings, 2002. X, 455 pages. 2002.

Vol. 2470: P. Van Hentenryck (Ed.), Principles and Practice of Constraint Programming – CP 2002. Proceedings, 2002. XVI, 794 pages. 2002.

Vol. 2471: J. Bradfield (Ed.), Computer Science Logic. Proceedings, 2002. XII, 613 pages. 2002.

Vol. 2473: A. Gomez-Perez, V.R. Benjamins, Knowledge Engineering and Knowledge Management. Proceedings, 2002. XI, 402 pages. 2002. (Subseries LNAI).

Vol. 2474: D. Kranzlmüller, P. Kacsuk, J. Dongarra, J. Volkert (Eds.), Recent Advances in Parallel Virtual Machine and Message Passing Interface. Proceedings, 2002. XVI, 462 pages. 2002.

Vol. 2475: J.J. Alpigini, J.F. Peters, A. Skowron, N. Zhong (Eds.), Rough Sets and Current Trends in Computing. Proceedings, 2002. XV, 640 pages. 2002. (Subseries LNAI).

Vol. 2476: A.H.F. Laender, A.L. Oliveira (Eds.), String Processing and Information Retrieval. Proceedings, 2002. XI, 337 pages. 2002.

Vol. 2477: M.V. Hermenegildo, G. Puebla (Eds.), Static Analysis. Proceedings, 2002. XI, 527 pages. 2002.

Vol. 2478: M.J. Egenhofer, D.M. Mark (Eds.), Geographic Information Science. Proceedings, 2002. X, 363 pages. 2002.

Vol. 2479: M. Jarke, J. Koehler, G. Lakemeyer (Eds.), KI 2002: Advances in Artificial Intelligence. Proceedings, 2002. XIII, 327 pages. (Subseries LNAI).

Vol. 2480: Y. Han, S. Tai, D. Wikarski (Eds.), Engineering and Deployment of Cooperative Information Systems. Proceedings, 2002. XIII, 564 pages. 2002.

Vol. 2483: J.D.P. Rolim, S. Vadhan (Eds.), Randomization and Approximation Techniques in Computer Science. Proceedings, 2002. VIII, 275 pages. 2002.

Vol. 2484: P. Adriaans, H. Fernau, M. van Zaanen (Eds.), Grammatical Inference: Algorithms and Applications. Proceedings, 2002. IX, 315 pages. 2002. (Subseries LNAI).

Vol. 2485: A. Bondavalli, P. Thevenod-Fosse (Eds.), Dependable Computing EDCC-4. Proceedings, 2002. XIII, 283 pages. 2002.

Vol. 2486: M. Marinaro, R. Tagliaferri (Eds.), Neural Nets. Proceedings, 2002. IX, 253 pages. 2002.

Vol. 2487: D. Batory, C. Consel, W. Taha (Eds.), Generative Programming and Component Engineering. Proceedings, 2002. VIII, 335 pages. 2002.

Vol. 2488: T. Dohi, R. Kikinis (Eds), Medical Image Computing and Computer-Assisted Intervention – MICCAI 2002. Proceedings, Part I. XXIX, 807 pages. 2002.

Vol. 2489: T. Dohi, R. Kikinis (Eds), Medical Image Computing and Computer-Assisted Intervention – MICCAI 2002. Proceedings, Part II. XXIX, 693 pages. 2002.

Vol. 2490: A.B. Chaudhri, R. Unland, C. Djeraba, W. Lindner (Eds.), XML-Based Data Management and Multimedia Engineering – EDBT 2002. Proceedings, 2002. XII, 652 pages. 2002.

Vol. 2491: A. Sangiovanni-Vincentelli, J. Sifakis (Eds.), Embedded Software. Proceedings, 2002. IX, 423 pages. 2002.

Vol. 2492: F.J. Perales, E.R. Hancock (Eds.), Articulated Motion and Deformable Objects. Proceedings, 2002. X, 257 pages. 2002.

Vol. 2493: S. Bandini, B. Chopard, M. Tomassini (Eds.), Cellular Automata. Proceedings, 2002. XI, 369 pages. 2002.

Vol. 2495: C. George, H. Miao (Eds.), Formal Methods and Software Engineering. Proceedings, 2002. XI, 626 pages. 2002.

Vol. 2496: K.C. Almeroth, M. Hasan (Eds.), Management of Multimedia in the Internet. Proceedings, 2002. XI, 355 pages. 2002.

Vol. 2497: E. Gregori, G. Anastasi, S. Basagni (Eds.), Advanced Lectures on Networking. XI, 195 pages. 2002.

Vol. 2498: G. Borriello, L.E. Holmquist (Eds.), UbiComp 2002: Ubiquitous Computing. Proceedings, 2002. XV, 380 pages. 2002.

Vol. 2499: S.D. Richardson (Ed.), Machine Translation: From Research to Real Users. Proceedings, 2002. XXI, 254 pages. 2002. (Subseries LNAI).

Vol. 2501: D. Zheng (Ed.), Advances in Cryptology – ASIACRYPT 2002. Proceedings, 2002. XIII, 578 pages. 2002.

Vol. 2502: D. Gollmann, G. Karjoth, M. Waidner (Eds.), Computer Security – ESORICS 2002. Proceedings, 2002. X, 281 pages. 2002.

Vol. 2503: S. Spaccapietra, S.T. March, Y. Kambayashi (Eds.), Conceptual Modeling – ER 2002. Proceedings, 2002. XX, 480 pages. 2002.

Vol. 2504: M.T. Escrig, F. Toledo, E. Golobardes (Eds.), Topics in Artificial Intelligence. Proceedings, 2002. XI, 432 pages. 2002. (Subseries LNAI).

Vol. 2506: M. Feridun, P. Kropf, G. Babin (Eds.), Management Technologies for E-Commerce and E-Business Applications. Proceedings, 2002. IX, 209 pages. 2002.

Vol. 2507: G. Bittencourt, G.L. Ramalho (Eds.), Advances in Artificial Intelligence. Proceedings, 2002. XIII, 418 pages. 2002. (Subseries LNAI).

Vol. 2508: D. Malkhi (Ed.), Distributed Computing. Proceedings, 2002. X, 371 pages. 2002.

Vol. 2509: C.S. Calude, M.J. Dinneen, F. Peper (Eds.), Unconventional Models in Computation. Proceedings, 2002. VIII, 331 pages. 2002.

Vol. 2510: H. Shafazand, A Min Tjoa (Eds.), EurAsia-ICT 2002: Information and Communication Technology. Proceedings, 2002. XXIII, 1020 pages. 2002.

Vol. 2511: B. Stiller, M. Smirnow, M. Karsten, P. Reichl (Eds.), From QoS Provisioning to QoS Charging. Proceedings, 2002. XIV, 348 pages. 2002.

Vol. 2512: C. Bussler, R. Hull, S. McIlraith, M.E. Orlowska, B. Pernici, J. Yang (Eds.), Web Services, E-Business, and the Semantic Web. Proceedings, 2002. XI, 277 pages. 2002.

Vol. 2513: R. Deng, S. Qing, F. Bao, J. Zhou (Eds.), Information and Communications Security. Proceedings, 2002. XII, 496 pages. 2002.

Vol. 2514: M. Baaz, A. Voronkov (Eds.), Logic for Programming, Artificial Intelligence, and Reasoning. Proceedings, 2002. XIII, 465 pages. 2002. (Subseries LNAI).

Vol. 2515: F. Boavida, E. Monteiro, J. Orvalho (Eds.), Protocols and Systems for Interactive Distributed Multimedia. Proceedings, 2002. XIV, 372 pages. 2002.

Vol. 2516: A. Wespi, G. Vigna, L. Deri (Eds.), Recent Advances in Intrusion Detection. Proceedings, 2002. X, 327 pages. 2002.

Vol. 2517: M.D. Aagaard, J.W. O'Leary (Eds.), Formal Methods in Computer-Aided Design. Proceedings, 2002. XI, 399 pages. 2002.

Vol. 2518: P. Bose, P. Morin (Eds.), Algorithms and Computation. Proceedings, 2002. XIII, 656 pages. 2002.

Vol. 2519: R. Meersman, Z. Tari, et al. (Eds.), On the Move to Meaningful Internet Systems 2002: CoopIS, DOA, and ODBASE. Proceedings, 2002. XXIII, 1367 pages. 2002.

Vol. 2521: A. Karmouch, T. Magedanz, J. Delgado (Eds.), Mobile Agents for Telecommunication Applications. Proceedings, 2002. XII, 317 pages. 2002.

Vol. 2522: T. Andreasen, A. Motro, H. Christiansen, H. Legind Larsen (Eds.), Flexible Query Answering. Proceedings, 2002. XI, 386 pages. 2002. (Subseries LNAI).

Vol. 2525: H.H. Bülthoff, S.-Whan Lee, T.A. Poggio, C. Wallraven (Eds.), Biologically Motivated Computer Vision. Proceedings, 2002. XIV, 662 pages. 2002.

Vol. 2526: A. Colosimo, A. Giuliani, P. Sirabella (Eds.), Medical Data Analysis. Proceedings, 2002. IX, 222 pages. 2002.

Vol. 2527: F.J. Garijo, J.C. Riquelme, M. Toro (Eds.), Advances in Artificial Intelligence – IBERAMIA 2002. Proceedings, 2002. XVIII, 955 pages. 2002. (Subseries LNAI).

Vol. 2528: M.T. Goodrich, S.G. Kobourov (Eds.), Graph Drawing. Proceedings, 2002. XIII, 384 pages. 2002.

Vol. 2529: D.A. Peled, M.Y. Vardi (Eds.), Formal Techniques for Networked and Distributed Sytems – FORTE 2002. Proceedings, 2002. XI, 371 pages. 2002.

Vol. 2532: Y.-C. Chen, L.-W. Chang, C.-T. Hsu (Eds.), Advances in Multimedia Information Processing – PCM 2002. Proceedings, 2002. XXI, 1255 pages. 2002.

Vol. 2533: N. Cesa-Bianchi, M. Numao, R. Reischuk (Eds.), Algorithmic Learning Theory. Proceedings, 2002. XI, 415 pages. 2002. (Subseries LNAI).

Vol. 2534: S. Lange, K. Satoh, C.H. Smith (Ed.), Discovery Science. Proceedings, 2002. XIII, 464 pages. 2002.

Vol. 2535: N. Suri (Ed.), Mobile Agents. Proceedings, 2002. X, 203 pages. 2002.

Vol. 2536: M. Parashar (Ed.), Grid Computing – GRID 2002. Proceedings, 2002. XI, 318 pages. 2002.

Vol. 2537: D.G. Feitelson, L. Rudolph, U. Schwiegelshohn (Eds.), Job Scheduling Strategies for Parallel Processing. Proceedings, 2002. VII, 237 pages. 2002.

Vol. 2538: B. König-Ries, K. Makki, S.A.M. Makki, N. Pissinou, P. Scheuermann (Eds.), Developing an Infrastructure for Mobile and Wireless Systems. Proceedings 2001. X, 183 pages. 2002.

Vol. 2540: W.I. Grosky, F. Plášil (Eds.), SOFSEM 2002: Theory and Practice of Informatics. Proceedings, 2002. X, 289 pages. 2002.

Vol. 2541: T. Barkowsky, Mental Representation and Processing of Geographic Knowledge. X, 174 pages. 2002. (Subseries LNAI).

Vol. 2544: S. Bhalla (Ed.), Databases in Networked Information Systems. Proceedings 2002. X, 285 pages. 2002.

Vol. 2545: P. Forbrig, Q, Limbourg, B. Urban, J. Vanderdonckt (Eds.), Interactive Systems. Proceedings 2002. X, 269 pages. 2002.

Vol. 2546: J. Sterbenz, O. Takada, C. Tschudin, B. Plattner (Eds.), Active Networks. Proceedings, 2002. XIV, 267 pages. 2002.

Vol. 2548: J. Hernández, Ana Moreira (Eds.), Object-Oriented Technology. Proceedings, 2002. VIII, 223 pages. 2002.

Vol. 2549: J. Cortadella, A. Yakovlev, G. Rozenberg (Eds.), Concurrency and Hardware Design. XI, 345 pages. 2002.

Vol. 2550: A. Jean-Marie (Ed.), Advances in Computing Science – ASIAN 2002. Proceedings, 2002. X, 233 pages. 2002.

Vol. 2551: A. Menezes, P. Sarkar (Eds.), Progress in Cryptology – INDOCRYPT 2002. Proceedings, 2002. XI, 437 pages. 2002.

Vol. 2555: E.-P. Lim, S. Foo, C. Khoo, H. Chen, E. Fox, S. Urs, T. Costantino (Eds.), Digital Libraries: People, Knowledge, and Technology. Proceedings, 2002. XVII, 535 pages. 2002.

Vol. 2556: M. Agrawal, A. Seth (Eds.), FST TCS 2002: Foundations of Software Technology and Theoretical Computer Science. Proceedings, 2002. XI, 361 pages. 2002.

Vol. 2557: B. McKay, J. Slaney (Eds.), AI 2002: Advances in Artificial Intelligence. Proceedings, 2002. XV, 730 pages. 2002. (Subseries LNAI).

Vol. 2559: M. Oivo, S. Komi-Sirviö (Eds.), Product Focused Software Process Improvement. Proceedings, 2002. XV, 646 pages. 2002.

Vol. 2561: H.C.M. de Swart (Ed.), Relational Methods in Computer Science. Proceedings, 2001. X, 315 pages. 2002.

Vol. 2569: D. Gollmann, G. Karjoth, M. Waidner (Eds.), Computer Security – ESORICS 2002. Proceedings, 2002. XIII, 648 pages. 2002. (Subseries LNAI).